NICKEL AND HUMAN HEALTH

Volume

25

in the Wiley Series in

Advances in Environmental
Science and Technology

JEROME O. NRIAGU, Series Editor

NICKEL AND HUMAN HEALTH

Current Perspectives

Edited by
Evert Nieboer
Department of Biochemistry & Occupational Health Program,
McMaster University, Hamilton, Ontario, Canada

and
Jerome O. Nriagu
National Water Research Institute, Burlington, Ontario, Canada

A WILEY-INTERSCIENCE PUBLICATION
JOHN WILEY & SONS, INC.
New York . **Chichester** . **Brisbane** . **Toronto** . **Singapore**

Copyright © 1992 by John Wiley & Sons, Inc.

Library of Congress Cataloging in Publication Data:
Nickel and human health : current perspectives / edited by Evert
    Nieboer and Jerome O. Nriagu.
        p.   cm.—(Advances in environmental science and technology :
    v. 25)
    Most papers were originally presented at the Fourth International Conference
on Nickel Metabolism and Toxicology, held in Espoo, Finland, in September
1988.
    "A Wiley-Interscience publication."
    Includes bibliographical references.
    ISBN 0-471-50076-3
    1. Nickel—Toxicology—Congresses.   I. Nieboer, Evert.
II. Nriagu, Jerome O.   III. International Conference on Nickel
Metabolism and Toxicology (4th : September 1988 : Espoo, Finland)   IV. Series.
TD 180.A38   vol. 25
[RA1231.N44]
628 s—dc20
[615.9′25625]   91-31093
                                                                    CIP

Printed in the United States of America

10 9 8 7 6 5 4 3 2 1

CONTRIBUTORS

SURESH AGGARWAL, University of Virginia, Medical Center, Department of Pathology, Box 168, Charlottesville, Virginia

ANTERO AITIO, Institute of Occupational Health, Topeliuksenkatu 41 a A, SF-00250 Helsinki, Finland

AAGE ANDERSEN, The Cancer Registry of Norway, Montebello, 0310 Oslo 3, Norway

SISKO ANTILA, Institute of Occupational Health, Topeliuksenkatu 41 a A, SF-00250 Helsinki, Finland

C. BAILLY, INSERM Unite 16, 1 place de Verdun, 59045 Lille Cedex, France

B. BELLMANN, Fraunhofer Institute for Toxicology and Aerosol Research, Nikolai Fuchs Strasse 1, D-3000 Hannover 61, Germany

JANET M. BENSON, Inhalation Toxicology Research Institute, Lovelace Biomedical and Environmental Research Institute, P.O. Box 5890, Albuquerque, New Mexico

AHMAD BORAN, Institute of Occupational Health, University of Birmingham, P.O. Box 363, Birmingham B15 2TT, United Kingdom

A. L. BROOKS, Inhalation Toxicology Research Institute, Lovelace Biomedical and Environmental Research Institute, P.O. Box 5890, Albuquerque, New Mexico

DAVID G. BURT, Inhalation Toxicology Research Institute, Lovelace Biomedical and Environmental Research Institute, P.O. Box 5890, Albuquerque, New Mexico

J. F. CHASTANG, INSERM U 88, 91 bd de l'Hopital, F-75634 Paris Cedex 13, France

YUNG SUNG CHENG, Inhalation Toxicology Research Institute, Lovelace Biomedical and Environmental Research Institute, P.O. Box 5890, Albuquerque, New Mexico

DAVID M. CHICO, Inhalation Toxicology Research Institute, Lovelace Biomedical and Environmental Research Institute, P.O. Box 5890, Albuquerque, New Mexico

SHUJI CHO, Faculty of Home Life Science, Fukuoka Women's University, 1-1 Kasumigaoka, Higashi ku, Fukouka 83, Japan

NELWYN T. CHRISTIE, Institute of Environmental Medicine, New York University Medical Center, 550 First Avenue, New York, New York

KATHLEEN CONWAY, Institute of Environmental Medicine, New York University Medical Center, 550 First Avenue, New York, New York

MAX COSTA, Institute of Environmental Medicine, New York University Medical Center, 550 First Avenue, New York, New York

M. CSICSAKY, Medical Institute for Environmental Hygiene, The University of Düsseldorf, P.O. Box 5634, D-4000 Düsseldorf, Germany

A. M. DECAESTECKER, Institut de Medecine du Travail, 1 Place de Verdun, 59045 Lille Cedex, France

RICHARD DOLL, ICRF Cancer Studies Unit, 65 Banbury Road, Oxford OX2 6PE, United Kingdom

D. DUBOURDIEU, Institut Pasteur, Noumea, New Caledonia

JUNE K. DUNNICK, National Toxicology Program, National Institute of Environmental Health Sciences, Research Triangle Park, North Carolina

DOUG F. EASTON, Institute for Cancer Research, Royal Cancer Hospital, 15 Cotswold Road, Sutton, Surrey SM2 5NG, United Kingdom

Y. EDOUTE, Department of Physiology and Biophysics, Mayo Clinic and Mayo Foundation, Rochester, Minnesota

A. F. EIDISON, Lovelace Inhalation Toxicology Research Institute, U.S. Department of Energy, P.O. Box 5890, Albuquerque, New Mexico

M. R. ELWELL, National Toxicology Program, National Institute of Environmental Health Sciences, P.O. Box 12233, Research Triangle Park, North Carolina

LENNART EMTESTAM, Department of Dermatology, Karolinska Institute Huddinge Hospital, S-141 86 Huddinge, Sweden

PETER H. EVANS, MRC Dunn Nutrition Unit, Milton Road, Cambridge CB4 1XJ, United Kingdom

GEORGE FARRANTS, Laboratory for Experimental Pathology and Image Analysis, Institute for Cancer Research, The Norwegian Radium Hospital, Montebello, N-0310 Oslo 3, Norway

KARI FEREN, Laboratory for Experimental Pathology and Image Analysis, Institute for Cancer Research, The Norwegian Radium Hospital, Montebello, N-0310 Oslo 3, Norway

R. FUHST, Fraunhofer Institute for Toxicology and Aerosol Research, Nikolai Fuchs Strasse 1, D-3000 Hannover 61, Germany

ANN FULLERTON, Department of Pharmaceutics, Royal Danish School of Pharmacy, DK-2100 Copenhagen, Denmark

D. FURON, Institut de Medecine du Travail, 1 place de Verdun, 59045 Lille Cedex, France

BENTE GAMMELGAARD, Royal Danish School of Pharmacy, DK-2100 Copenhagen, Denmark

BAOQI GANG, School of Public Health, Harbin Medical University, 41 Dazhi Avenue, Harbin 150001, People's Republic of China

MARCEL GOLDBERG, INSERM U 88, 91 bd de l'Hopital, F-75634 Paris Cedex 13, France

PAQUERETTE GOLDBERG, INSERM U 88, 91 bd de l'Hopital, F-75634 Paris Cedex 13, France

J. GUEZIEC, Hopital Gaston Bourret, Noumea, New Caledonia

FLETCHER F. HAHN, Inhalation Toxicology Research Institute, Lovelace Biomedical and Environmental Research Institute, P.O. Box 5890, Albuquerque, New Mexico

PATRICK J. HALEY, Inhalation Toxicology Research Institute, Lovelace Biomedical and Environmental Research Institute, P.O. Box 5890, Albuquerque, New Mexico

JOJI HARATAKE, School of Medicine, University of Occupational and Environmental Health, 1-1 Iseigaoka, Yahatanishi ku, Kitakyushu 807, Japan

ROBERT P. HAUSINGER, Departments of Microbiology and Biochemistry, Michigan State University, East Lansing, Michigan

ROGENE F. HENDERSON, Inhalation Toxicology Research Institute, Lovelace Biomedical and Environmental Research Institute, P.O. Box 5890, Albuquerque, New Mexico

J. P. HENICHART, INSERM Unite 16, 1 place de Verdun, 59045 Lille Cedex, France

DAVID A. HEROLD, Department of Pathology, University of Virginia, Health Sciences Center, Charlottesville, Virginia

H. FREDERICK HILDEBRAND, Institut de Medecine du Travail, 1 place de Verdun, 59045 Lille Cedex, France

CHARLES H. HOBBS, Inhalation Toxicology Research Institute, Lovelace Biomedical and Environmental Research Institute, P.O. Box 5890, Albuquerque, New Mexico

A. HOELGAARD, Department of Pharmaceutics, Royal Danish School of Pharmacy, DK-2100 Copenhagen, Denmark

SIDNEY M. HOPFER, Departments of Laboratory Medicine and Pharmacology, University of Connecticut School of Medicine, 263 Farmington Avenue, Farmington, Connecticut

AKIO HORIE, Deceased

M. HUERRE, Institut Pasteur, Noumea, New Caledonia

F. HUTH, Institute of Pathology, Municipal Hospital, Hildesheim, Germany

RICHARD IMBRA, Institute of Environmental Medicine, New York University Medical Center, 550 First Avenue, New York, New York

SIGEKO ISHIMATSU, Department of Environmental Health, University of

Occupational and Environmental Health, 1-1 Iseigaoka, Yahatanishi ku, Kitakyushu 807, Japan

NIRODH JADON, McMaster University Medical Centre, 1200 Main Street West/Room 3H50, Hamilton, Ontario, Canada

JAMES A. JULIAN, Occupational Health Program, McMaster University, 1200 Main Street West, Hamilton, Ontario, Canada

KAZIMIERZ S. KASPRZAK, Laboratory of Comparative Carcinogenesis, National Cancer Institute, FCRDC, Frederick, Maryland

M. KIILUNEN, Institute of Occupational Health, Topeliuksenkatu 41 a A, SF-00250 Helsinki, Finland

MICHAEL KINTER, Department of Pathology, University of Virginia, Health Sciences Center, Charlottesville, Virginia

CATHERINE B. KLEIN, Institute of Environmental Medicine, New York University Medical Center, 550 First Avenue, New York, New York

YASUSHI KODAMA, School of Medicine, University of Occupational and Environmental Health, 1-1 Iseigaoka, Yahatanishi ku, Kitakyushu 807, Japan

F. LAVIGNE, Hospital Gaston Bourret, Noumea, New Caledonia

A. LECLERC, INSERM U 88, 91 bd de l'Hopital, F-75634 Paris Cedex 13, France

GEORGE LUMB, Department of Pathology, Hahnemann University College of Medicine, Philadelphia, Pennsylvania

M. J. MARNE, INSERM U 88, 91 bd de l'Hopital, F-75634 Paris Cedex 13, France

R. MARTINEZ, CNRS UA 234, Université des sciences et Techniques de Lille, F 59655 Villeneuve d'A scq Cedex, France

A. MAS, Department of Biochemistry, University of Toronto, Toronto, Canada

MICHELE A. MEDINSKY, Inhalation Toxicology Research Institute, Lovelace Biomedical and Environmental Research Institute, P.O. Box 5890, Albuquerque, New Mexico

TORKIL MENNÉ, Department of Dermatology, Gentofte Hospital, Niels Andersens vej 65, DK-2900 Hellerup, Denmark

L. P. METCALFE, INCO Europe Ltd., Clydach, Swansea SA6 5QR, United Kingdom

U. MOHR, Fraunhofer Institute for Toxicology and Aerosol Research, Nikolai Fuchs Strasse 1, D-3000 Hannover 61, Germany

LINDSAY G. MORGAN, INCO Europe Ltd., Medical Department, Clydach, Swansea SA6 5QR, United Kingdom

HARTWIG MUHLE, Fraunhofer Institute for Toxicology and Aerosol Research, Nikolai Fuchs Strasse 1, D-3000 Hannover 61, Germany

DAVID C. F. MUIR, Occupational Health Program, McMaster University, 1200 Main Street West, Hamilton, Ontario, Canada

STEPHEN NICKLIN, Department of Immunotoxicology, British Industrial Biological Research Association, Carshalton, Surrey, United Kingdom

EVERT NIEBOER, Department of Biochemistry and Occupational Health Program, McMaster University, 1200 Main Street West, Hamilton, Ontario, Canada

GITTE D. NIELSEN, Department of Environmental Medicine, Odense University Hospital, J.B. Winsløwvej 17, DK-5000 Odense C, Denmark

GUNTER OBERDÖRSTER, University of Rochester, School of Medicine, P.O. Box HSC, Rochester, New York

A. BEN OTMANE, Institut de Medecine du Travail, 1 place de Verdun, 59045 Lille Cedex, France

P. PÄÄKKÖ, University of Oulu, Oulu, Finland

KURT PETERS, Gentofte Hospital, Niels Andersens vej 65, DK-2900 Hellerup, Denmark

JULIAN PETO, Institute of Cancer Research, 15 Cotswold Road, Sutton, Surrey SM2 5NG, United Kingdom

JOHN A. PICKRELL, Inhalation Toxicology Research Institute, Lovelace Biomedical and Environmental Research Institute, P.O. Box 5890, Albuquerque, New Mexico

FRIEDRICH POTT, Medical Institute for Environmental Hygiene, University of Düsseldorf, P.O. Box 5634, D-4000 Düsseldorf, Germany

P. PREDKI, Department of Biochemistry, The University of Toronto, Toronto, Ontario, Canada

ALBRECHT REITH, Laboratory for Experimental Pathology and Image Analysis, Institute for Cancer Research, The Norweigian Radium Hospital, Montebello, N-0310 Oslo 3, Norway

VESA RIIHIMAKI, Institute of Occupational Health, Arinatie 3, SF-00370 Helsinki, Finland

RITA MARIA RIPPE, Medical Institute for Environmental Hygiene, University of Düsseldorf, P.O. Box 5634, D-4000 Düsseldorf, Germany

ROBIN S. ROBERTS, Hamilton Civic Research Centre, Henderson General Division, 711 Concession Street, Hamilton, Ontario, Canada

RICARDO RODRIGUEZ, Laboratory of Comparative Carcinogenesis, National Cancer Institute, FCRDC, Frederick, Maryland

M. ROLLER, Medical Institute for Environmental Hygiene, University of Düsseldorf, P.O. Box 5634, D-4000 Düsseldorf, Germany

M. ROSENBRUCH, Medical Institute of Hygiene, University of Düsseldorf, P.O. Box 5634, D-4000 Düsseldorf, Germany

TOBY G. ROSSMAN, New York University Medical Center, 550 First Avenue, New York, New York

GABOR RUBANYI, Schering AG, Research Center, 1000 Berlin 65, Germany

WILLIAM E. SANFORD, Department of Physical Sciences, Mohawk College of Applied Arts and Technology, Hamilton, Ontario, Canada

BIBUDHENDRA SARKAR, Department of Biochemistry, University of Toronto, Toronto, Ontario, Canada

JOHN SAVORY, University of Virginia, Health Sciences Center, Charlottesville, Virginia

ZHI-CHENG SHI, Department of Occupational Medicine and Occupational Diseases Research Centre, Third Hospital, Beijing Medical University, 49 North Garden Road, Beijing 100083, People's Republic of China

P. SHIRALI, Institut de Medecine du Travail, 1 place de Verdun, F 59045 Lille Cedex, France

BRIAN C. STACE, Department of Biochemistry, University of Surrey, Guildford, Surrey, England

SHANZHOUNG SUN, School of Public Health, Harbin Medical University, 41 Dazhi Avenue, Harbin, People's Republic of China

F. WILLIAM SUNDERMAN, JR., Departments of Laboratory Medicine and Pharmacology, University of Connecticut School of Medicine, 263 Farmington Avenue, Farmington, Connecticut

F. WILLIAM SUNDERMAN, SR., Institute for Clinical Science, Pennsylvania Hospital, Philadelphia, Pennsylvania

S. TAKENAKA, Fraunhofer Institute for Toxicology and Aerosol Research, Nikolai Fuchs Strasse 1, D-3000 Hannover 61, Germany

ISAMU TANAKA, School of Medicine, University of Occupational and Environmental Health, 1-1 Iseigaoka, Yahatanishi ku, Kitakyushu 807, Japan

DOUGLAS M. TEMPLETON, Department of Clinical Biochemistry, The University of Toronto, Banting Institute, 100 College Street, Toronto, Ontario, Canada

YOSHIMA TERAKI, Department of Histology, The Nippon Dental University, School of Dentistry at Niigata, 1-8 Hamaura-Cho, Niigata 951, Japan

KENZABURO TSUCHIYA, School of Medicine, University of Occupational and Environmental Health, 1-1 Iseigaoka, Yahatanishi ku, Kitakyushu 807, Japan

DONNA M. TUMMOLO, Institute of Environmental Medicine, New York University Medical Center, 550 First Avenue, New York, New York

A. UCHIUMI, National Chemical Laboratory for Industry, Tsukuba Research Center, Tsukuba, Ibaragi 305, Japan

NAOKA URANO, Department of Public Health, Teikyo University School of Medicine, Itabashi Ku, Tokyo 173, Japan

V. USHER, INCO Europe Ltd., Clydach, Swansea SA6 5NG, United Kingdom

P. M. VANHOUTTE, Department of Physiology and Biophysics, Mayo Clinic and Mayo Foundation, Rochester, Minnesota

XIN WEI WANG, Institute of Environmental Medicine, New York University Medical Center, 550 First Avenue, New York, New York

MARK WHITE, Institute of Naval Medicine, Alverstoke, Gosport PO12 2DL, United Kingdom

MICHAEL R. WILLS, Department of Pathology, University of Virginia, Health Sciences Center, Charlottesville, Virginia

ZHIJIE YANG, School of Public Health, Harbin Medical University, 41 Dazhi Avenue, Harbin 150001, People's Republic of China

EIJI YANO, Department of Public Health, Teikyo University School of Medicine, Itabashi Ku, Tokyo 173, Japan

H. YEGER, Department of Pathology, University of Toronto, Toronto, Ontario, Canada

INTRODUCTION
TO THE SERIES

The deterioration of environmental quality, which began when mankind first congregated into villages, has existed as a serious problem since the industrial revolution. In the second half of the twentieth century, under the ever-increasing impacts of exponentially growing population and of industrializing society, environmental contamination of the air, water, soil, and food has become a threat to the continued existence of many plant and animal communities of various ecosystems and may ultimately threaten the very survival of the human race. Understandably, many scientific, industrial, and governmental communities have recently committed large resources of money and human power to the problems of environmental pollution and pollution abatement by effective control measures.

Advances in Environmental Sciences and Technology deals with creative reviews and critical assessments of all studies pertaining to the quality of the environment and to the technology of its conservation. The volumes published in the series are expected to service several objectives: (1) stimulate interdisciplinary cooperation and understanding among the environmental scientists; (2) provide the scientists with a periodic overview of environmental developments that are of general concern or that are of relevance to their own work or interests; (3) provide the graduate student with a critical assessment of past accomplishment, which may help stimulate him or her toward the career opportunities in this vital area; and (4) provide the research manager and the legislative or administrative official with an assured awareness of newly developing research work on the critical pollutants and with the background information important to their responsibility.

As the skills and techniques of many scientific disciplines are brought to bear on the fundamental and applied aspects of the environmental issues, there is a heightened need to draw together the numerous threads and to present a coherent picture of the various research endeavors. This need and the recent tremendous growth in the field of environmental studies have clearly made some editorial adjustments necessary. Apart from the changes in style and format, each future volume in the series will focus on one particular theme or timely topic, starting with Volume 12. The author(s) of each pertinent section will be expected to critically review the literature and the most important recent developments in the particular field; to critically evaluate new concepts, methods, and data; and to focus attention on important unresolved or controversial questions and on probable future trends. Monographs embodying the results of unusually extensive and well-rounded investigations will also be published in the series. The net result of the new editorial policy should be more integrative and comprehensive volumes on key environmental issues and pollutants. Indeed, the development of realistic standards of environmental quality for many pollutants often entails such a holistic treatment.

JEROME O. NRIAGU, Series Editor

PREFACE

The environmental and occupational health impacts of nickel and its compounds have been extensively researched, reviewed, and debated. Through the auspices of the Commission on Toxicology, International Union of Pure and Applied Chemistry (Division of Clinical Chemistry), the International Conferences on Nickel Metabolism and Toxicology have helped to make this area of research one of the more advanced, compared to similar activities associated with other metals and metalloids. The present volume contains revised and updated papers from the Fourth International Conference held at Helsinki, Finland, September 5–9, 1988. The conference attracted 115 participants representing 22 countries.

The balance between reviews of specific subjects and reports of original research favors the latter. The coverage is comprehensive but not exhaustive. In a real sense, this volume might be considered a compaion to *Nickel in the Environment* (J. O. Nriagu, editor, Wiley-Interscience, New York), which appeared under the Environmental Science and Technology banner in 1980. The earlier book covers in considerable detail topics pertaining to nickel in the natural environment, including production and uses, global cycling, and presence in the aquatic and terrestrial ecosystems. The treatments of these subjects have not been rendered obsolete. The present volume addresses the following major themes: historical perspective of nickel toxicology, biological utilization of nickel, toxicokinetics in humans, occupational exposure and biological monitoring of workers, renal toxicity, hypersensitivity and immunotoxicology, clinical aspects of nickel carbonyl poisoning, and experimental nickel carcinogenesis. It also summarizes the worldwide epidemiological evidence for the occurrence of nickel-induced nasal and lung cancers among nickel workers. One special feature of this volume is that it not only is interdisciplinary in scope but also reflects situations and research activities related to nickel exposure in many jurisdictions. Consequently, the book should have wide appeal to graduate students, researchers, physicians, and allied health professionals as well as regulatory specialists in universities, research institutes, industry, and government who are involved in nickel toxicology and related health effects.

We thank the staff at John Wiley & Sons for invaluable editorial assistance. Special gratitude is accorded to our distinguished group of contributors as well as to the many individuals and organizations that helped to make the Helsinki Nickel Conference a memorable scientific and social event.

EVERT NIEBOER
JEROME O. NRIAGU

Hamilton, Ontario, Canada
Burlington, Ontario, Canada
July, 1992

CONTENTS

NICKEL AND HUMAN HEALTH

1

HAZARDS FROM EXPOSURE TO NICKEL: A HISTORICAL ACCOUNT

F. William Sunderman, Sr.

Institute for Clinical Science, Pennsylvania Hospital, Philadelphia, Pennsylvania 19107

Plenary Address given at the Fourth International Conference on Nickel Metabolism and Toxicology, Hanasaari Cultural Centre, Espoo, Finland, September 5, 1988.

Nickel and Human Health: Current Perspectives, Edited by Evert Nieboer and Jerome O. Nriagu.
ISBN 0-471-50076-3 © 1992 John Wiley & Sons, Inc.

1. INTRODUCTION

It is my great pleasure to address this assembly of learned scientists on the occasion of the Fourth International Conference on Nickel Metabolism and Toxicology.

When I seriously sat down to prepare remarks for this evening's program, I became aghast as I pondered upon the title that I had submitted. How could I presume to cover such a colossal subject in the course of a brief address? It is true that the hazards of exposure to nickel and nickel compounds have been recognized only within recent decades; however, the number of research contributions and publications on nickel toxicology that have appeared during the past three decades has been substantial.

In my ponderings, I harbored the wish that I could call for help from my esteemed friends and scientific colleagues of World War II who instigated many of our early interests in the toxicology of nickel. These eminent scientists are now deceased. At this time, may I pay tribute to them and to the early pioneering contributions of the International Nickel Company, especially through their medical director, J. Gwynne Morgan; to the U.S. Atomic Energy Commission and the Rohm and Haas Company, the two organizations that supported our research studies; and to the contributors to this and the three previous symposia, with special commendation to my beloved son, Bill Jr.

My memory with respect to nickel toxicology goes back to 1943 when I attended my first nickel meeting at Los Alamos, New Mexico. This initial meeting was arranged out of necessity. It was held for the purpose of devising methods to protect workers in nuclear energy from the hazards of acute and chronic exposure to nickel tetracarbonyl, $Ni(CO)_4$. It was recognized then that nickel carbonyl was one of the most toxic of all gases. At our first meeting, those in attendance set three goals: (1) to obtain accurate toxicity data on nickel carbonyl; (2) to establish programs for the protection and treatment of workers who might inadvertently be exposed; and (3) to undertake studies for the acquisition of knowledge on the metabolism of nickel. It is heartening to note that these initial goals have, in large measure, been realized. And now, 45 years later, the extent to which studies on nickel metabolism have expanded is noteworthy and evokes a modicum of satisfaction.

Although metals have been known and used by man since antiquity, it is noteworthy that it has been only within the past two centuries that nickel was acknowledged to be a primary element. Furthermore, nickel has been refined for large-scale commercial distribution only within the past 100 years.

Nickel is ubiquitous and is estimated to be present in 0.008% of the earth's crust. The core of the earth is believed to be similar in composition to meteorites, which consist of iron–nickel alloys averaging about 8.5% nickel. The presence of nickel may be detected spectrographically in the sun and celestial bodies in which the elements are present as incandescent gases. Nickel in the sun is estimated to be present as 0.0002% by volume.

In an interesting review on the medicinal use of nickel, Koplinski (1911) stated, "Nickel and nickel salts, excepting for the very poisonous nickel carbonyl, have no place or record in toxicology." Koplinski elaborated on the exceptional medicinal value of nickel salts for the treatment of epilepsy, chorea, migraine, and neuralgia. He also noted that "nickel acts as a sedative and tonic of peculiar and elective power in controlling the damaging effects of sexual vice on the nervous system." And now, since World War II, the uses of nickel for medicinal purposes have been completely abandoned, and it has become recognized that, in addition to nickel carbonyl, exposure to nickel and other nickel compounds may be exceptionally deleterious to health.

On the other hand, nickel is probably an essential trace element required for mammalian life (Chapter 2). In 1936, it was suggested that nickel may play a normal physiological role in metabolism (Bertrand and Nakamura, 1936). In 1974, Schwartz indicated that nickel is probably an essential element for the life and health of animals. Within the past decade, nickel has been shown to be a component of four plant enzymes of physiological importance, one of those enzymes being urease (Walsh and Orme-Johnson, 1987).

2. NICKEL TOXICOLOGY

2.1. Historical Items Pertaining to Nickel

A synopsis of historical items pertaining to nickel and its toxicology has been assembled (Table 1.1). A few of the pertinent starred items will be discussed.

The hazards from exposure to nickel appear to have been first recognized during the period of the Renaissance. Agricola (1490–1555) referred to the toxic effects of *Kupfer–nickel* ores on the lungs of workers in the Schneeberg area in Germany (Howard-White, 1963). Nickel does not occur in nature by itself but is associated with cobalt or as an alloy with copper, zinc, iron, or arsenic. In the seventeenth century, the German Erzgeburg miners found a red-colored ore they mistakenly thought was copper and which they named *Kupfer–nickel* (copper–nickel). While possessing the appearance of copper ore, it yielded no copper when treated with the process then used to extract copper. Later, the red-colored ore was identified as nickel arsenide (NiAs). A silver-appearing metal emanating from China during this period was given the trade name German Silver. This is essentially a mixture of nickel, copper, and zinc (Howard-White, 1963).

The word *Kobalt* in German denotes a goblin and *Nickel* a scamp. These evil spirits (*Kobalt* and *Nickel*) were believed to haunt the mines and do harm to the miners. In the mining districts in Germany church services during the seventeenth century included prayers for the protection of miners from *Kobalt* and *Kupfer–Nickel*. Howard-White (1963) notes that the fume-emitting

Table 1.1   Synopsis of Historical Items Pertaining to Nickel and Its Toxicity

---

945   The first references to nickel as an alloy were probably those pertaining to so-called white copper that was wrought by Chinese craftsmen into objects of art and utility. It is presumed that this alloy may date as far back as 300 AD (Liu Hsu, 1963).

*1490–1555   Agricola referred to the toxic effects on lungs of workers by *Kupfer–nickel* bearing ores in the Schneeberg area in Germany (Howard-White, 1963).

*1637   An inexpensive silvery appearing metal was supplied to Europe from the Province of Yunnan in China. This metal was given the trade name German Silver, a mixture of copper, nickel, and other metals. The standard formula was 50% copper, 25% nickel, and 25% zinc (Thien Kung Khai Wu, 1963).

1754   Nickel was recognized as an element by Axel Frederik Cronstedt (1758), who named the metal *nickel*.

1804   Richter (1804) published a paper entitled "Our Absolutely Pure Nickel, Proof That It Is a Noble Metal."

*1820–1890   Nickel was isolated as a metal by Berthier (1820). This isolation stimulated a worldwide search for nickel-containing ores.

1865   Malleable nickel was first produced by Joseph Wharton in Philadelphia, Pennsylvania. Nickel was first used in U.S. coinage in 1865. The U.S. Mint produced a 3-cent coin that contained 75% nickel and 25% copper.

*1884   T. P. Anderson Stuart undertook the first studies in nickel toxicology while working in Oswald Schmeideberg's laboratory in Strasbourg, Germany (Sunderman and Brown, 1985).

*1890   Nickel carbonyl was discovered by Mond, Langer, and Imiche (1890). Its ease of formation and high volatility were found to be properties that are especially useful in separating nickel from its ores.

1891   First case of nickel carbonyl poisoning was reported by McKendrick and Snodgrass (1891).

1892   The Mond (1895) carbonyl process for the extraction of nickel was inaugurated.

1900   Mond Nickel Company was founded at Clydach, near Swansea, Wales.

*1903   The deaths of two workmen were reported to have occurred at the Mond Nickel Works at Clydach, Wales. The report to the coroner's jury by Dr. Tunnicliffe stated: "Death was caused by nickel carbonyl, a substance inhaled from the dust given off during employment" (1903).

*1907–1908   Armit (1908) studied the toxicity of nickel carbonyl in experimental animals and found that nickel was the toxic component and not carbon monoxide.

1929–1933   Skin sensitivity to nickel became recognized (Groa, 1929; Stewart, 1933).

Table 1.1 (*Continued*)

1934 Brandes (1934) reported a case of nickel carbonyl poisoning from "cracking gasoline" with the use of nickel as a catalyst.

1937 Baader (1937) was the first investigator to report the high incidence of pulmonary cancer in nickel workers.

1939 Bayer (1939) reviewed the toxicology of nickel carbonyl and the pathological lesions resulting from exposure.

1943 A nickel conference was held at Los Alamos, New Mexico, for the purpose of devising methods to protect workers in nuclear energy from the hazards of acute and chronic exposure to nickel carbonyl.

1953 Gulf Oil Company accidental exposure of over 100 workmen to nickel carbonyl. The seriously injured were treated with BAL. Two of the workmen died; 20 became critically ill and were hospitalized. With the exception of 3 men, the remainder were able to return to work two months after exposure.

*1957 Sodium diethyldithiocarbamate was discovered to be an effective antidote in the treatment of animals exposed to nickel carbonyl (West and Sunderman, 1957).

1958 Doll (1958) collected data from death records to estimate the incidence of cancer of the lung and nose in nickel workers. He estimated that from 1948 to 1956 the risk of nickel workers dying from cancer of the lung was approximately five times normal. During the same period, the risk of nickel workers dying from cancer of the nose was approximately 150 times normal.

1958 Dithiocarb proved to be an effective antidote in the treatment of human exposure to nickel carbonyl (Sunderman and Sunderman, 1958).

1958 Williams (1958) studied the pathology of the lungs in five nickel workers, four of whom had developed pulmonary carcinoma.

1962 Passey (1962) tabulated 144 deaths from cancer of the lung in nickel workers and computed the average age at death in these workers to be 57.6 years.

1965–1988 Extensive experimental studies were undertaken on the induction of nickel carcinogenesis (Sunderman and Donnelly, 1965; Lumb and Sunderman, 1988).

1969 At Toa Gosei Chemical Industries, Nagoya, Japan, 156 workmen were exposed, of which 137 showed symptoms of intoxication and 96 were hospitalized in the 13 hospitals of Nagoya. The critically ill were treated with Antabuse until supplies of Dithiocarb became available. No deaths occurred, although convalescence in 7 men was protracted.

1984 Dithiocarb was found to have antitumorigenic properties in rats subjected to muscular implantations of nickel subsulfide (Ni_3S_2) (Sunderman et al., 1984).

* Items discussed in detail in the text.

reddish ore had a harmful effect on the health of the miners, so they naturally thought that evil spirits or "Old Nick" himself had been at work.

Nickel was isolated as the metal early in the nineteenth century by Berthier (1820). Following this isolation, a worldwide search for nickel-containing ores was made by a number of commercial companies. In retrospect, some of the difficulties and hardships that were encountered in their efforts to mine the ores are amusing. For example, in trying to recruit laborers in New Caledonia, J. Garland wrote, "The native Kanakas . . . hold strong views as to the folly of work and the stupidity of the white man who is addicted to it" (Howard-White, 1963).

T. P. Anderson Stuart (1884) is credited with undertaking the first systematic investigations in nickel toxicology while working as a graduate student in Schmeideberg's laboratory in Strasburg, Germany. He investigated the physiological properties of the salts of nickel and cobalt metabolism and presented the results of his studies in a doctoral thesis, for which he was awarded a gold medal (Stuart, 1884).

Nickel carbonyl was discovered by Mond and co-workers in 1890 (Mond et al., 1890). The physical and chemical properties of nickel carbonyl were found to be especially useful in separating nickel from its ores, and this finding led to the economical commercial production of nickel (Mond, 1895). The first deaths from nickel carbonyl were reported in the *Lancet* in 1903 (Correspondent Reports). Two workmen died from exposure to nickel carbonyl at the Mond Nickel Works at Clydach. The report to the coroner's jury by Dr. Tunnicliffe stated, "Death was caused by nickel carbonyl, a substance inhaled from the dust given off during employment" (Mott, 1907). It might be noted that in the previous decade a case of nickel carbonyl poisoning had been reported in the same plant (McKendrick and Snodgrass, 1891).

An important study on the toxicity of nickel carbonyl was reported in 1907 (Armit, 1907, 1908). Armit studied the effects of exposure of this compound on rabbits, cats, and dogs and found that *nickel* was the toxic component and *not* carbon monoxide.

A number of toxicological investigations on nickel were reported during the beginning and middle of this century (Table 1.1), and these have been reviewed in some of our earlier papers (Kincaid et al., 1953, 1956; Sunderman, 1954, 1981). A noteworthy discovery was made in our laboratory in 1957 when sodium diethyldithiocarbamate (Dithiocarb) was found to be an effective antidote for nickel carbonyl poisoning (Sunderman, 1959; Sunderman and Sunderman, 1958; Sunderman and West, 1958).

2.2. Toxicity of Nickel in Experimental Animals

Scientific data of the hazards of exposure to nickel and nickel compounds have been obtained in experimental animals only within recent decades. A summary of toxicity values of nickel and its compounds, many of them from our laboratory, is given in Table 1.2.

Table 1.2 Toxicity of Nickel and Its Compounds

Nickel (Colloidal and Powdered)		
Intravenous	Dogs	LD, 10–20 mg/kg
Acute oral	Dogs	Tolerated, 1–3 g/kg
Nickel Salts (Cl, NO_3, SO_4, O)		
Intravenous	Dogs	LD, 10–20 mg/kg
Subcutaneous	Rabbits	LD, 1.3 g/kg
Acute oral	Rats	LD_{50}, 2.0 g/kg
Chronic oral	Cats	Tolerated, 25 mg/kg day for 200 days
Skin application	Human	1:10,000 evokes sensitivity reaction
Nickel Carbonyl, Ni(CO)_4		
Inhalation	Mice	LC_{50}, 0.067 mg/L for 30 min
Inhalation	Rats	LC_{50}, 0.24 mg/L for 30 min
Inhalation	Cats	LC_{50}, 1.9 mg/L for 30 min
Intravenous	Rats	LD_{50}, 22 ± 1.1 mg/kg
Subcutaneous	Rats	LD_{50}, 21 ± 4.2 mg/kg
Intraperitoneal	Rats	LD_{50}, 13 ± 1.4 mg/kg

It will be seen that solutions of colloidal nickel or nickel salts have a high degree of toxicity when given either intravenously or subcutaneously. On the other hand, the ingestion of nickel or nickel salts has a relatively low degree of toxicity. When given orally, it will be seen that dogs are able to tolerate doses of metallic nickel and nickel compounds as high as 3 g/kg of body weight.

The most toxic of all of the compounds of nickel that are encountered in industrial operations is nickel carbonyl. The median lethal dose (LD_{50}) values for 30 min exposure for mice and rats are 0.07 and 0.23 mg/L, corresponding to 10 and 33 ppm, respectively (Sunderman, 1981). It may be noted that there are now more than 180 organonickel compounds commercially available and that toxicity studies have been reported for only a few of them.

2.3. Routes of Exposure to Nickel

Any considerations of nickel toxicology require a knowledge of the physical and chemical characteristics of the nickel compound to which the subject

Table 1.3   Types of Nickel Poisoning

Inhalation [Ni(CO)$_4$, Ni, Ni$_3$S$_2$, NiO, Ni$_2$O$_3$]

Acute: pneumonitis with adrenal cortical insufficiency; hyaline membrane formation; pulmonary edema and hemorrhage; hepatic degeneration; brain and renal congestion

Chronic: cancer of respiratory tract; pulmonary eosinophilia (Loeffler's syndrome); asthma

Skin Contact

Primary irritant dermatitis; allergic dermatitis; eczema

Parenteral (Prosthetic Implantations)

Allergenic reactions; osteomyelitis; osteonecrosis; malignant tumors

Oral

Food and beverage; drugs

has been exposed, the concentration, length and type of exposure, as well as the sensitivity of the host and the presence of disease in the individual. The physiological responses also depend in large measure upon the route by which nickel compounds enter the body and are distributed in the tissues. The routes to which nickel may enter the body and the types of nickel poisoning are given in Table 1.3.

3.   TOXICITY OF NICKEL TETRACARBONYL

Our investigations on the toxicity of Ni(CO)$_4$ were undertaken by means of an especially designed exposure chamber (Sunderman et al., 1956, 1957; Figs. 1.1 and 1.2). The chamber provides constant flow of gas, controllable over a wide range of flow rates. The construction permits the use of a variety of experimental animals that may range in size from mice to 25-kg dogs. A special feature of the chamber is the insertion of baffles in order to provide a uniform distribution of the toxic agent (Stevenson and Sunderman, 1965). For our toxicological investigations, Ni(CO)$_4$ was dissolved at concentration from 0.5 to 20% (w/v) in a 50:50 (v/v) mixture of USP-grade ether and anhydrous ethanol. The solution is introduced into the chamber through a motor-driven Luer-type syringe fitted with a 22-gauge needle. Syringes with capacities from 10 to 100 mL are convenient for producing various concentrations of nickel carbonyl in the chamber, ranging by volume from 1 to approximately 100 ppm in air. The material fed through the syringe drops onto a plug of glass wool supported on a wire screen. No trouble with

Figure 1.1. Exposure chamber.

immediate evaporation is encountered with the 50:50 mixture of ether and alcohol. When absolute alcohol alone was used as the solvent, evaporation was sluggish. When ether alone was used, bubbles formed in the syringe leading to an erratic rate of feeding.

Concentrations of $Ni(CO)_4$ in the chamber, as determined by analysis (Kincaid et al., 1953), normally were approximately 80% of those computed from the known rates of airflow and injection of nickel carbonyl. The analysis was accepted as yielding the correct value.

3.1. Gulf Oil Company Refinery Accident

Within the past 35 years, an opportunity has been afforded to examine personally or to review the medical records of several hundred persons acutely exposed to nickel carbonyl. My first large-scale medical experience in treating such patients occurred in April 1953 when I was called as a consultant to supervise the treatment of workmen accidentally exposed at the Gulf Oil Company refinery at Port Arthur, Texas.

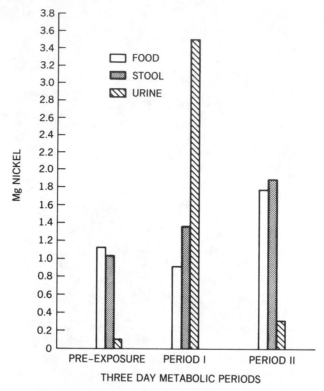

Figure 1.2. Nickel balance studies in dogs before and after exposure to nickel carbonyl.

In this tragic accident, more than 100 men were exposed to nickel carbonyl; 31 of the men required hospitalization, and 2 died. The workmen were exposed during the repair of a reactor in one of the company's chemical plants. The reactor was surrounded by a stall that hindered the dispersal of fumes. Only workmen who entered this stall became ill. The 31 hospitalized employees were treated with BAL* since Dithiocarb had not been discovered to be an effective antidote until 1957.

The opportunity to examine these exposed workmen over a protracted period of time proved to be a valuable experience. Without personal contact with the critically ill patients, it would have been difficult to appreciate the insidious manner in which the symptoms develop in nickel carbonyl poisoning and the severity of the illness that occurs without febrile response.

In Table 1.4 is given a tabulation of symptoms that developed in the exposed workmen, which could be divided into two categories: (1) those with immediate reactions after exposure and (2) those with delayed reactions (Sunderman, 1970). A latent period of 1 to 5 days may occur in which the exposed workmen may be essentially asymptomatic.

*British antilewisite, 2,3-dimercaptopropanol, usually called dimercaprol, or BAL.

Table 1.4   Symptoms of Acute Nickel Carbonyl Poisoning

Immediate	Latent	Delayed
Mild, nonspecific	1–5 days	Constriction in chest
Symptoms disappear when subject removed to uncontaminated air	May be essentially asymptomatic	Chills, sweating
		Shortness of breath
		Unproductive cough
Headache; dizziness		Muscle pains
Sternal and epigastric pains		Weakness, fatigue
Nausea and vomiting occasionally		Gastrointestinal symptoms occasionally
		Convulsions and delirium, sometimes terminally

The most common initial symptoms were dizziness accompanied by frontal throbbing headache and sternal and epigastric pains. More than one-half of the men became nauseated following exposure and many of them vomited. Soon after the exposure, most of the men experienced a sense of constriction in the chest with concomitant shortness of breath and a dry, hacking, unproductive cough. In a number of the patients, the initial symptoms emerged gradually into the more severe, delayed reactions; in others, the transition from one stage to the other was more abrupt. Usually, the delayed stage was ushered in by a paroxysm of coughing. The onset of the severe symptoms varied from 10 hr to six days following exposure.

An almost universal complaint of the critically ill men was that of extreme weakness and ready fatigue. Some stated that they felt too exhausted to breathe, and indeed, such statements were borne out by the fact that in about half of the patients, respirations could not be sustained without pressure oxygen therapy. Breathing in the exposed subjects was rapid and shallow. The pulse rates were increased, but not usually in proportion to the increased respiratory rates.

Seven of the critically ill patients from the Gulf refinery accident developed neurological symptoms; two had convulsions and others became irrational. Diarrhea and abdominal distention occurred in several of the patients two or three days following exposure. This might suggest that nickel was being excreted by the bowel similar to the diarrhea that follows arsenic poisoning; however, no analytical data were obtained to support this concept (Kent and McCance, 1941).

Fever was not a predominant symptom in the critically ill patients. With few exceptions, the temperatures of the patients were not elevated above 101°F. Almost all of the exposed men developed a moderate leucocytosis of 10,000–15,000 leucocytes/cm^3. The two fatalities from the exposure occurred on the fourth and eleventh day after the accident.

Alterations in the serum electrolytes followed the pattern encountered during the precritical period of pneumonia. These included diminished concentrations of sodium and chloride accompanied by increased concentrations of potassium (Sunderman et al., 1961). Similar serum electrolyte changes are observed in acute adrenal cortical insufficiency, and as a consequence, supplemental corticosteroid therapy was administered as an adjunct to antidotal BAL therapy.

In practically all of the cases, convalescence was protracted. On examination two months after exposure, several of the men appeared cyanotic and a few of them had residual tenderness over the hepatic and splenic areas. In most of the exposed workmen, the pulse rates remained elevated above the normal range.

An attending physician unfamiliar with the symptoms following exposure to nickel carbonyl might tend to minimize the seriousness of the illness. Practitioners could readily be led to consider convalescing patients as being unduly apprehensive, or even malingerers, if they were unacquainted with the course of the intoxication.

In retrospect, it is our considered opinion that many of the exposed Gulf Company employees owe their lives to the administration of BAL. This therapy was based upon the investigations undertaken in our laboratory at the time of the accident in which we had shown that BAL increased the tolerance of rats to nickel carbonyl by a factor of approximately 2 (Kincaid et al., 1953).

At the time of the Gulf Oil Company accident, the Rohm and Haas Company was in the midst of constructing a pilot plant in Pasadena, Texas, in which nickel carbonyl was to be used as an intermediary in the synthesis of acrylic esters for the production of plastics.*

Under a patent granted to the company, the $Ni(CO)_4$ in the process was utilized under high pressure, thus increasing the risk of accidental exposure. In reflecting on the accident at the Gulf Oil Company, Otto Haas, one of the two founders of the Rohm and Haas Company, became deeply concerned about the industrial use of nickel carbonyl in his company's operations and told me that he would abandon the project completely if the operation would jeopardize the lives of his workmen. I replied that we had learned to handle plutonium safely during World War II and if adequate safeguards were provided, I felt that the lives of workmen in the nickel carbonyl operation could be adequately protected. During our conversation, I extracted a promise from Mr. Haas that he would permit me to set up the medical safeguards at the plant without interference or rejection by plant management.

At this point, I digress to mention briefly the results of our nickel metabolism studies on dogs since they apply to the safeguards we were about to propose.

---

* Walter Reppe (1949), the German chemist, had made a major contribution in the field of organic chemistry when he discovered the oxo and related reactions with nickel carbonyl as a catalyst.

Three-day metabolic balance studies were undertaken in dogs before and after exposure to $Ni(CO)_4$. It is seen in Figure 1.2 that under normal conditions (preexposure period), approximately 90% of ingested nickel is excreted in the stools and 10% in the urine (Tedeschi and Sunderman, 1957). However, when nickel vapors are inhaled, the converse is observed, that is, most of the inhaled nickel is excreted in the urine and a relatively small percentage of inhaled nickel is excreted in the stools. This observation that there is a sharp increase in the nickel excretion in the urine immediately after exposure and before the onset of any obvious symptoms proved to be of enormous practical importance and led to the development of procedures for detecting accidental exposure to minimal amounts of nickel carbonyl in concentrations too low to produce immediate symptoms.

3.2. Monitoring Workmen Exposed to Nickel Carbonyl

When the time came to set up medical safeguards at the nickel carbonyl operation at the Rohm and Haas plant, I proposed that in addition to monitoring the nickel carbonyl areas with automatic recording detectors and the use of hygienic safeguards (masks, working garments, etc.), each worker would be required, before leaving the plant, to submit a specimen of urine for nickel analysis. As anticipated, this proposal brought an outcry from plant management that this procedure would require the construction of a separate building and the hiring of additional technicians and would substantially increase the costs of maintenance. However, I had had a previous commitment from Mr. Haas, and the plant administrators had to accept the safeguards proposed.

Nickel analysis of urine specimens from the workmen continued throughout the 25 years in which nickel carbonyl was utilized in the Rohm and Haas plant. It was heartening to have administrative officers who objected initially to our program admit publicly in later years that the innovation of submitting urine specimens at the end of the working day had been lifesaving and had "paid off handsomely for the Company." During the 25 years in which nickel carbonyl was used, scores of workmen were found to have elevated concentrations of nickel in their urines. These workmen were given Dithiocarb as a preventive measure as soon as any increased concentration was detected, and practically all of them returned to work the next day symptomless and without complaints. Without Dithiocarb, doubtless a number would have died and/or developed severe undiagnosed pneumonitis within a week or so after exposure.

In the 25 years in which $Ni(CO)_4$ was used in the plant, only one workman lost his life from accidental exposure, and this death could not be attributed to company negligence. The incident is worth noting. Four men working as a crew in the nickel carbonyl area were found to have elevated concentrations of nickel in their urine specimens. Three of the men were contacted immediately after the increased urinary concentrations of nickel were discovered. These three men were given Dithiocarb, and they returned to work symptomless within 48 hr. The fourth man became ill soon after leaving the plant and consulted his private physician who gave him an injection of penicillin and had him admitted to the community hospital before he could be reached by company personnel and given Dithiocarb. The attending physician was unwilling to administer Dithiocarb to his patient, stating that Dithiocarb was listed as an investigational drug and that he was confident in his ability to treat pneumonia. On the fifth day after exposure, the patient became moribund and died without the benefit of Dithiocarb. An autopsy was performed and analyses were made of the concentrations of nickel in the lung and liver tissues from this patient. It is seen in Table 1.5 that the concentrations of nickel in the lung and liver tissues were 11 and 6 times greater than the average concentrations of nickel in lung and liver tissues obtained at autopsy from persons who had died suddenly from causes unrelated to nickel exposure.

In addition to the history, physical examination, autopsy, and chemical measurements in the deceased workman who had not received Dithiocarb, these nickel measurements left no doubt that this workman died of acute nickel carbonyl poisoning.

3.3. Toa Gosei Chemical Company Accident

The second large-scale nickel carbonyl accident in which I became remotely involved occurred in January 1969 at the Toa Gosei Chemical plant in Nagoya, Japan. At the accident, 156 men were exposed to $Ni(CO)_4$, of whom 137 showed symptoms of intoxication. Ninety-six men from the group were hospitalized in 13 hospitals in the Nagoya area. (Nagoya at that time had a population of two million and was well supplied with hospitals.) The Japanese authorities appealed to us for supplies of Dithiocarb, which were immediately dispatched by plane. In the meantime, the attending physicians were advised to treat the critically ill patients with Antabuse (which was available in Japan)

Table 1.5 Analyses of Nickel in Tissues (Sudden Deaths in Previously Healthy Subjects)

Subject	Wet Weight, μg Ni/100 g		Dry Weight, μg Ni/100 g	
	Lung	Liver	Lung	Liver
Stab wound	2.40	0.52	14.6	1.1
Drug intoxication	2.20	0.86	12.1	3.2
Hanging	0.81	0.76	3.3	2.6
Carbon monoxide	0.96	1.32	4.3	4.8
Mean	1.59	0.87	8.6	2.9
Acute nickel carbonyl poisoning	17.3	5.3	115.0	20.7

Table 1.6   Summary of Major Medical Findings in Nagoya Workmen Acutely Exposed to Nickel Carbonyl

1. X-ray examinations of 35 out of 72 exposed workmen (48.6%) revealed abnormal densities that disappeared within six months after the accident.
2. Sixty-two of 96 exposed workmen (64.6%) were found to have abnormal liver function tests within the first two months after exposure. Thereafter, the level decreased.
3. Twenty-six of 92 exposed workmen (27.1%) had more than traces of protein in the urine during the first two months after exposure.
4. The phenolsulfonephthalein and Fishberg renal function tests were normal in all exposed workmen tested.
5. Thirty-seven of the 96 inpatient workmen (38.5%) developed skin lesions that lasted in some cases as long as six moths after exposure.
6. Four of the exposed workmen developed symptoms of encephalopathy that disappeared within three to five weeks without sequela.

until the supplies of Dithiocarb arrived. This decision was based on our published studies showing that the metabolic pathways of Antabuse were similar to Dithiocarb and that Antabuse provided limited protection against the acute toxic effects of nickel carbonyl (Sunderman and West, 1958).

When our supplies of Dithiocarb arrived in Nagoya 72 hr after the accident, seven of the exposed men had become delirious, irrational, and unconscious. In some of the subjects, Dithiocarb had to be administered by stomach tube. However, all of the men treated with Dithiocarb recovered, although for three subjects convalescence was protracted. It is noteworthy that no deaths were reported from the Nagoya accident.

At the invitation of the Toa Gosei Chemical Company, I had an opportunity to review the medical records with the attending physicians and scientists in Nagoya. The Toa Gosei Company had undertaken a detailed clinical investigation of the exposed workmen through several committees composed of specialists in various medical disciplines. The information from the reports I gleaned is summarized in Table 1.6.

To date, more than 375 workmen exposed to nickel carbonyl have received Dithiocarb under our supervision. To our knowledge, no exposed workmen died who received Dithiocarb within the first five days after exposure. The efficacy of Dithiocarb as a specific antidote for acute nickel carbonyl poisoning is now well recognized.

4.   CARCINOGENICITY OF NICKEL

In recent years, the carcinogenicity of nickel and nickel compounds has become of increasing concern. The high incidence of pulmonary cancer occurring in nickel workers was first recognized in 1937 by Baader. Since that time, the relation of nickel to pulmonary carcinogenesis has been the subject of many investigations (Drinker et al., 1924; Loken, 1930; Barnett, 1949; Sunderman et al., 1957; Campbell, 1958; Doll, 1958; Goldblatt, 1958; Hueper, 1958; Morgan, 1958; Williams, 1958; Hatem-Champy, 1961, 1962; Sunderman and Sunderman, 1961, 1963; Gilman, 1962; Gilman and Ruckerbauer, 1962; Hueper et al., 1962; Passey, 1962; Sunderman, 1963; Sunderman and Donnelly, 1965; Sunderman, 1981). The carcinogenicity of nickel has probably been more thoroughly investigated than that of any other element.

In 1958, Doll reported that 35.5% of nickel workers in Wales died of cancer of the lung or upper respiratory tract, whereas the incidence among colliery workers was only 1.5% (Doll, 1958). In 1962, Passey tabulated 144 deaths from cancer of the lung in nickel workers and computed the average age at death in these workers to be 57.6 years. The average length of time that the affected workers were employed in nickel refineries was 27 years, while the average time between the first exposure and death from lung cancer was 30.5 years.

The carcinogenicity of nickel compounds has been documented in experimental animals, including several strains of rats as well as mice, guinea pigs, rabbits, and cats (Payne, 1964). Carcinogenic potency appears to be inversely related to the solubilities of the nickel compounds in water. The carcinogenic compounds that are sparingly soluble in water at 37°C include nickel dust, nickel sulfide, nickel carbonate, nickel oxide, nickel carbonyl, and Ni (II) bisdimethylglyoxime. Soluble nickel salts, such as nickel chloride, nickel sulfate, and nickel ammonium sulfate, have not been shown to be carcinogenic. The lesions that develop after the injection of insoluble nickel compounds into various body tissues are, in the main, malignant fibrous histiocytomas (Lumb and Sunderman, 1988).

The relationship between the inhalation of nickel as $Ni(CO)_4$ and pulmonary cancer has been the subject of a number of investigations from our laboratory (Sunderman et al., 1957; Sunderman and Sunderman, 1961; Sunderman, 1963; Sunderman and Donnelly, 1965). These studies established that pulmonary cancer may be induced in rats exposed to a single heavy concentration of nickel carbonyl as well as in rats exposed to repeated inhalations of sublethal concentrations for a period of a year. It is noteworthy that cancers were not observed in these animals until two or more years after the initial exposure (Sunderman and Donnelly, 1965). The lesions that develop in rats after the inhalations of nickel carbonyl include the common types of pulmonary cancer: squamous cell carcinoma, adenocarcinoma, and anaplastic carcinoma. These tumors closely resemble the cancers that occur in nickel workers.

5. CONCLUDING REMARKS

Nickel is a noble metal, and its uses in furthering the welfare of mankind are legion. However, it must be recognized that under certain conditions, human contacts with the metal and some of its complexes are fraught with

Figure 1.3. Dinosaur ad for oil company.

danger. When considering the uses of nickel, it is an obligation for each of us to anticipate untoward effects that might possibly ensue. I will cite an example. A number of years ago, one of the major oil companies launched a nationwide advertising campaign in the United States through leading newspapers, news magazines, and billboard displays announcing that a nickel compound, which the company was adding to gasoline, was the most important gasoline improvement since World War II (Fig. 1.3). On seeing this advertisement, I wrote to the president of the company stating that undoubtedly the company officials did not realize that the so-called improved gasoline had the potential of spewing deadly nickel carbonyl over the face of America. The company promptly discontinued the addition of nickel to gasoline and quietly recalled the advertising.

Non est vivere, sed valere, vita. Life is not mere living but the enjoyment of health.

REFERENCES

Armit, H. W. (1907, 1908). The toxicology of nickel carbonyl. *J. Hyg.* **7,** 525–551; **8,** 565–600.

Baader, E. W. (1937). Berufskrebs. In C. Adam and D. S. Auler (Eds.), *Neure Ergebnisse auf dem Gebiete der Krebskrankheiten.* Hirzel Verlag, Leipzig, pp. 116–117.

Barnett, G. P. (1949). *Annual Report of the Chief Inspector of Factories for the Year 1948*. Vol. 4. Her Majesty's Stationery Office, London.

Bayer, O. (1939). Contributions to the toxicology, clinical and pathological anatomy of nickel carbonyl poisoning. *Arch. Gewerbepathol. Gewerbehyg.* **9,** 592–606.

Berthier, P. (1820). *Ann. Chim. Phys.* **2,** 14, 52.

Bertrand, G., and Nakamura, H. (1936). Recherches sur l'importance physiologique du nickel et du cobalt. *Bull. Soc. Sci. Hyg. Aliment.* **24,** 238–343.

Brandes, W. W. (1934). Nickel carbonyl poisoning. *J. Am. Med. Assoc.* **102,** 1204–1206.

Campbell, J. A. (1958). Lung tumours in mice and man. *Br. J. Ind. Med.* **15,** 217–223.

Correspondent Reports (1903). Nickel carbonyl poisoning. *Lancet* **1,** 268–269.

Cronstedt, A. F. (1758). *Mineralogie*. Stockholm, p. 218.

Doll, R. (1985a). Cancer of the lung and nose in nickel workers. *Br. J. Ind. Med.* **15,** 217–223.

Doll, R. (1958b). Specific industrial causes. In J. R. Bignall (Ed.), *Carcinoma of the Lung*. Williams & Wilkins, Baltimore, pp. 45–59.

Drinker, K. R., Fairhall, L. T., and Drinker, C. K. (1924). Hygienic significance of nickel. *J. Ind. Hyg. Toxicol.* **6,** 307–356.

Gilman, M. P. W. (1962). Metal carcinogenesis. II. A study of the carcinogenic activity of cobalt, copper, iron and nickel compounds. *Cancer Res.* **22,** 158–162.

Gilman, M.P.W., and Ruckerbauer, G. M. (1962). Metal carcinogenesis. I. Observations on the carcinogenicity of a refinery dust, cobalt, oxide, and colloidal thorium dioxide. *Cancer Res.* **22,** 152–157.

Goldblatt, M. W. (1958). Occupational carcinogenesis. *Br. Med. Bull.* **14,** 136–140.

Groa, K. (1929). Nickelplater's rash. *Urol. Cutan. Rev.* **23,** 606.

Hatem-Champy, S. (1961). Cancers caused by nickel and by a nickel–imidazole complex. *C. R. Acad. Sci.* **253,** 2791–2792.

Hatem-Champy, S. (1962). Affinity of folic acid for nickel and nickel-induced cancer. *C. R. Acad. Sci.* **254,** 1177–1179.

Howard-White, F. B. (1963). *Nickel—An Historical Review*. Methuen, London, p. 24.

Hueper, W. C. (1958). Experimental studies in metal cancerigenesis. IX. Pulmonary lesions in guinea pigs and rats exposed to prolonged inhalation of powdered nickel. *Arch. Pathol.* **65,** 600–607.

Hueper, W. C., Kotin, P., Tabor, E. C., Payne, W. W., Falk, H., and Sawicki, E. (1962). Carcinogenic bioassays on air pollutants. *Arch. Pathol.* **74,** 89–116.

Kent, N. L., and McCance, R. A. (1941). Absorption and excretion of minor elements by man. *Biochem. J.* **35,** 877–883.

Kincaid, J., Strong, J. S., and Sunderman, F. W. (1953). Nickel poisoning. I. Experimental study of the effects of acute and subacute exposure to nickel carbonyl. *AMA Arch. Ind. Hyg. Occup. Med.* **8,** 48–60.

Kincaid, J. F., Stanley, E. L., Beckworth, C. H., and Sunderman, F. W. (1956). Nickel poisoning. III. Procedure for detection, prevention and treatment of nickel carbonyl exposure including a method for the determination of nickel in biological materials. *Am. J. Clin. Pathol.* **26,** 107–119.

Koplinski, L. (1911). On the uses of nickel sulphate in medicine. *Monthly Cyclop. Med. Bull.* **4,** 348–355.

Liu Hsu (1963). "Old History of the Thang Dynasty," quoted by F. B. Howard-White. *Nickel—An Historical Review*. Methuen, London, p. 13.

Loken, A. C. (1950). Carcinoma of the lung in nickel workers. *Tidsskr. Nor. Laegeforen.* **70,** 376–378.

Lumb, G., and Sunderman, F. W. (1988). The mechanism of malignant tumor induction by nickel subsulfide. *Ann. Clin. Lab. Sci.* **18**, 353–366.

McKendrick, J. G., and Snodgrass, S. W. (1891). On the physiological action of carbon monoxide on nickel. *Br. Med. J.* **1**, 1215–1217.

Mond, L. (1895). The history of my process of nickel extraction. *J. Soc. Chem. Inc.* **14**, 945–946.

Mond, L., Langer, C., and Imiche, F. (1890). The action of carbon monoxide on nickel. *J. Chem. Soc. Lond.* **57**, 749–753.

Morgan, J. G. (1958). Some observations on the incidence of respiratory cancer in nickel workers. *Br. J. Ind. Med.* **15**, 224–234.

Mott, F. W. (1907). Carbon monoxide and nickel carbonyl poisoning. *Path. Lab. Lond. Cty. Asylums, Claybury, Essex* **3**, 246.

Passey, R. D. (1962). Some problems of lung cancer. *Lancet* **2**, 107–112.

Payne, W. W. (1964). Carcinogenicity of nickel compounds in experimental animals. *Proc. Am. Soc. Cancer Res.* **5**, 50.

Reppe, J. W. (1949). *Acetylene Chemistry*. Charles A. Meyer, New York.

Richter (1804). Our absolutely pure nickel, proof that it is a noble metal. *Neues. Allgem. Int. Chem.* **2**, 61–72.

Schwartz, K. (1974). Elements newly identified as essential for animals. *Ann. Clin. Lab. Sci.* **4**, 130.

Stevenson, K., and Sunderman, F. W. (1965). A method for testing the uniformity of mixing in flow chambers. *Am. J. Clin. Pathol.* **44**, 303–306.

Stewart, S. G. (1933). Inherent sensitivity of the skin to nickel and cobalt. *Arch. Int. Med.* **51**, 427.

Stuart, T.P.A. (1884). *Arch. Exp. Path. Pharmakol.* **18**, 151–173. Quoted in S. E. Brown and F. W. Sunderman Jr., Eds. (1985). *Progress in Nickel Toxicology*. Proceedings of the Third International Conference. Cambridge University Press, Blackwell Scientific Publications, Cambridge, England, p. 3.

Sunderman, F. W. (1954). Nickel poisoning. II. Studies on patients suffering from acute exposure to vapors of nickel carbonyl. *J. Am. Med. Assoc.* **155**, 889–894.

Sunderman, F. W. (1959). Dithiocarbamates for Treatment of Nickel Poisoning. U. S. Patent 2,876,159.

Sunderman, F. W. (1970). Nickel poisoning. In F. W. Sunderman and F. W. Sunderman Jr. (Eds.), *The Laboratory Diagnosis of Disease Caused by Toxic Agents*. Warren H. Green, Saint Louis, pp. 387–396.

Sunderman, F. W. (1981). Chelation therapy in nickel poisoning. *Ann. Clin. Lab. Sci.* **11**, 1–8.

Sunderman, F. W., and Donnelly, A. J. (1965). Studies of nickel carcinogenesis: Metastasizing pulmonary tumors in rats induced by the inhalation of nickel carbonyl. *Am. J. Pathol.* **46**, 1027–1041.

Sunderman, F. W., and Sunderman, F. W., Jr. (1958). Nickel poisoning. VIII. Dithiocarb: a new therapeutic agent for persons exposed to nickel carbonyl. *Am J. Med. Sci.* **236**, 26–31.

Sunderman, F. W., and Sunderman, F. W., Jr. (1961). Nickel poisoning. XI. Implication of nickel as a pulmonary carcinogen in tobacco smoke. *Am. J. Clin. Pathol.* **35**, 203–209.

Sunderman, F. W., and Sunderman, F. W., Jr. (1963). Studies of nickel carcinogenesis. The subcellular partition of nickel in lung and liver following inhalation of nickel carbonyl. *Am. J. Clin. Pathol.* **40**, 563–575.

Sunderman, F. W., and West, B. (1958). Nickel poisoning. VII. The therapeutic effectiveness of alkyl dithiocarbamates in experimental animals exposed to nickel carbonyl. *Am. J. Med. Sci.* **236**, 15–25.

Sunderman, F. W., Kincaid, J. F., Kooch, W., and Birmelin, E. A. (1956). A constant flow chamber for exposure of experimental animals to gases and volatile liquids. *Am. J. Clin. Pathol.* **26,** 1211–1218.

Sunderman, F. W., Kincaid, J. F., Donnelly, A. J., and West, B. (1957). Nickel poisoning. IV. Chronic exposure of rats to nickel carbonyl: a report after one year of observation. *Arch. Ind. Hyg.* **16,** 480–485.

Sunderman, F. W., Donnelly, A. J., West, B., and Kincaid, J. F. (1959). Nickel poisoning. IX. Carcinogenesis in rats exposed to nickel carbonyl. *Arch. Ind. Hyg.* **20,** 36–41.

Sunderman, F. W., Range, C. L., Sunderman, F. W., Jr., Donnelly, A. J., and Lucyszyn, G. W. (1961). Nickel poisoning. XIII. Metabolic and pathologic changes in acute pneumonitis from nickel carbonyl. *Am. J. Clin. Pathol.* **36,** 477–491.

Sunderman, F. W., Schneider, H. P., and Lumb, G. (1984). Sodium diethyldithiocarbamate administration in nickel-induced malignant tumors. *Ann. Clin. Lab. Sci.* **14,** 1–19.

Sunderman, F. W., Jr. (1963). Studies of nickel carcinogenesis: alterations of ribonucleic acid following inhalation of nickel carbonyl. *Am. J. Clin. Pathol.* **39,** 549–561.

Sunderman, F. W., Jr., and Brown, S. S. (1985). Introduction: 100 years of nickel toxicology. In S. S. Brown and F. W. Sunderman Jr. (Eds.), *Progress in Nickel Toxicology*. Blackwell Scientific, Oxford, pp. 1–6.

Tedeschi, R. E., and Sunderman, F. W. (1957). Nickel poisoning. V. The metabolism of nickel under normal conditions after exposure to nickel carbonyl. *AMA Arch. Ind. Health* **16,** 484–488.

Thien Kung Khai Wu. 1963. Quoted in Howard-White, F. B. *Nickel—An Historical Review.* Methuen, London, p. 14.

Walsh, C. T., and Orme-Johnson, W. H. (1987). Nickel enzymes. *Biochemistry* **26,** 4901–4906.

West, B., and Sunderman, F. W. (1957). Sodium diethyldithiocarbamate in the treatment of nickel poisoning. *Fed. Proc.* **16,** 128.

Williams, W. J. (1958). The pathology of the lungs in five nickel workers. *J. Ind. Med.* **15,** 235–242.

2

BIOLOGICAL UTILIZATION OF NICKEL

Robert P. Hausinger

Departments of Microbiology and Biochemistry,
Michigan State University, East Lansing, Michigan 48824

1. INTRODUCTION

Although nickel is well known to be toxic at elevated concentrations, trace levels of this metal ion are required for several biological processes. For example, four nickel-containing enzymes have been characterized: urease,

Nickel and Human Health: Current Perspectives, Edited by Evert Nieboer and Jerome O. Nriagu.
ISBN 0-471-50076-3   © 1992 John Wiley & Sons, Inc.

methyl coenzyme M reductase, hydrogenase, and carbon monoxide dehy-
rogenase (Thauer et al., 1980; Hausinger, 1987; Walsh and Orme-Johnson,
1987). Other essential, but less well characterized, biological roles for nickel
are known in plants (Dalton et al., 1988), animals (Nieboer et al., 1988;
Nielsen and Ollerich, 1974), and microorganisms (Hausinger, 1987). To supply
their micronutrient requirements for this metal ion, nickel-specific transport
systems have been developed by some organisms (Hausinger, 1987). Each
of these topics involving biological utilization of nickel will be briefly summarized below. Biological aspects of nickel inorganic chemistry will not be
emphasized here but have recently been reviewed by Lancaster (1988).

2. NICKEL-CONTAINING ENZYMES

2.1. Urease

The enzyme urease (urea amidohydrolase, EC 3.5.1.5) hydrolyzes urea to
form ammonia and carbamate, which spontaneously degrades to form a
second molecule of ammonia and carbon dioxide (Andrews et al., 1984;
Blakeley and Zerner, 1984; Mobley and Hausinger, 1989):

$$\underset{H_2N-\overset{O}{\overset{\|}{C}}-NH_2}{} + H_2O \xrightarrow{\text{urease}} NH_3 + HO-\overset{O}{\overset{\|}{C}}-NH_2 \tag{1}$$

$$HO-\overset{O}{\overset{\|}{C}}-NH_2 \rightarrow NH_3 + CO_2 \tag{2}$$

This protein is historically significant because urease from jack bean (*Canavalia
ensiformis*) was the first enzyme crystallized (Sumner, 1926). Nearly 50 years
after its crystallization, the plant enzyme was demonstrated to contain nickel
(Dixon et al., 1975).

Jack bean urease is a large enzyme (M_r = 545,000) comprised of six
identical subunits (M_r = 90,770) of known sequence (Takishima et al., 1988).
Each subunit possesses two tightly bound nickel atoms that can only be
released by subjecting the enzyme to low pH (<5) or to protein denaturants
in the presence of ethylenediaminetetraacetic acid (EDTA) (Dixon et al.,
1980b). Ultraviolet–visible spectroscopic studies of urease in the presence
of the competitive inhibitor 2-mercaptoethanol gave evidence of direct
nickel–sulfur coordination, indicating that substrate urea also binds to the
urease nickel metallocenter (Blakeley et al., 1983; Dixon et al., 1980a).
Zerner and colleagues (Andrews et al., 1984; Blakeley and Zerner, 1984;
Dixon et al., 1980c) have proposed an elegant model for urease catalysis in
which one nickel polarizes the urea carbonyl to facilitate nucleophilic attack
by an activated hydroxide anion bound to the second nickel (Fig. 2.1).

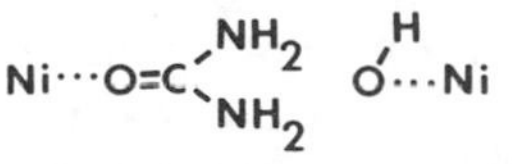

$$Ni\cdots O=C \overset{NH_2}{\underset{NH_2}{\diagdown}} \quad \overset{H}{\underset{}{O}}\cdots Ni$$

Figure 2.1. Speculative model of the urease binickel active site. It is clear that there are two nickel ions per urease catalytic unit, but it has not been demonstrated whether both participate in catalysis at the active site.

Current efforts are directed at testing this model by using biophysical methods to characterize the structure and properties of the jack bean urease nickel active site.

Many other plants, some invertebrate animals, and numerous microorganisms also exhibit urease activity (Andrews et al., 1984). On the basis of the nickel dependence for tobacco, rice, and soybean urease activities, Polacco (1977) proposed that nickel was a universal requirement for plant ureases. In agreement with this hypothesis, nickel-dependent urease activity has been observed in duckweed (Gordon et al., 1978), cowpeas (Walker et al., 1985), and alnus and several other plants (Dalton et al., 1988). Soybean urease has been purified (Polacco and Havir, 1979) and shown to possess two nickel atoms per subunit, as is the case for the jack bean enzyme. The only invertebrate urease that has been purified is that of the land snail *Otala lactea*, which also was demonstrated to contain nickel (McDonald et al., 1980). Several nickel-containing microbial ureases have been purified and characterized (reviewed in Hausinger, 1987; Mobley and Hausinger, 1989). In contrast to the homopolymeric eucaryotic enzymes, bacterial ureases appear to contain three different types of subunits designated α ($M_r = 65,000-73,000$), β ($M_r = 10,000-12,000$), and γ ($M_r = 8,000-10,000$) in an $\alpha_2\beta_4\gamma_4$ stoichiometry (Mobley and Hausinger, 1989). As in the jack bean enzyme, there are two nickel atoms per active site, equivalent to four nickel atoms per bacterial enzyme.

The mechanisms of nickel insertion into urease is unclear but may involve accessory enzymes. By using nickel-free growth media, urease apoenzyme was generated in soybean (Winkler et al., 1983), two algae (Rees and Bekheet, 1982), a cyanobacterium (Mackerras and Smith, 1986), a purple sulfur bacterium (Bast, 1988), and an enteric bacterium (Lee et al., 1990). Urease activity increased in each case when nickel was added, even in the presence of protein synthesis inhibitors. However, when the cells were disrupted to provide an in vitro system, no urease activity developed upon nickel addition. These results suggest that nickel is not simply incorporated into urease apoenzyme, but rather that insertion may involve other factors and may be energy dependent. Genetic evidence is consistent with this view. Three gene products encoded by the bacterial urease operon have been proposed to be involved in some aspect of nickel transport or processing (Molrooney and Hausinger, 1990). In the fungus *Aspergillus nidulans*, four loci are required for urease expression (Mackay and Pateman, 1982): *ure*A is the structural gene for a urea permease, *ure*B encodes the single-subunit enzyme, *ure*C encodes a required product with an unidentified function, and *ure*D is somehow involved with the nickel center. A *ure*D mutation could be overcome by growth in the presence of 0.1 mM NiSO$_4$ (Mackay and Pateman, 1982). Furthermore, two loci are known to be involved in soybean urease maturation (Meyer-Bothling et al., 1987). The gene products encoded at these loci may encode a nickel "insertase" or other type of nickel-processing enzyme.

2.2. Methyl Coenzyme M Reductase

Methyl coenzyme M (CH$_3$–S–CoM) reductase participates in an essential, energy-yielding, methane-producing pathway found in all methanogenic bacteria (Rouviere and Wolfe, 1988). The reductase catalyzes the reaction between CH$_3$–S–CoM and 7-mercapto-N-heptanoyl-O-phospho-L-threonine (HS–HTP) to form methane and an HS–HTP/HS–CoM (2-mercaptoethanesulfonate) mixed disulfide.

$$CH_3\text{–}S\text{–}CoM + HS\text{–}HTP \rightarrow CH_4 + HTP\text{–}S\text{–}S\text{–}CoM \qquad (3)$$

A complex series of other enzymes are involved in disulfide reduction and formation of CH$_3$–S–CoM but will not be discussed here. Rather, we will focus on the CH$_3$–S–CoM reductase that is a nickel-containing enzyme (Ellefson et al., 1982).

Methyl coenzyme M reductase is comprised of three subunit types (M_r of 68,000 for α, 45,000 for β, and 38,500 for γ) in a $\alpha_2\beta_2\gamma_2$ stoichiometry, with two nickel atoms per native molecule (Ellefson et al., 1982). Fully dissociated, purified protein has been successfully reconstituted to yield active enzyme, demonstrating that no accessory enzymes are needed for nickel insertion (Hartzell and Wolfe, 1986). In the same study, the nickel was shown to be bound to the largest subunit.

The methyl coenzyme M reductase nickel center is part of an organometallic coenzyme called F$_{430}$ (Ellefson et al., 1982). So named because it possesses an intense absorbance maximum at 430 nm (ϵ = 23,000 M^{-1} cm^{-1}), F$_{430}$ was originally observed in 1978, well before its role was known (Gunsalus and Wolfe, 1978). The structure of isolated F$_{430}$ was deduced to be a tetrapyrrole by labeling studies (Diekert et al., 1980a,b), and a complete structure of the protein-free chromophore was assigned by using nuclear magnetic resonance (NMR) spectroscopy and other techniques (Pfaltz et al., 1982). The highly reduced tetrapyrrole (illustrated in Fig. 2.2) combines elements of both the corrin and porphyrin ring systems and has been termed a tetrahydrocorphin. In addition to the four pyrrole nitrogen ligands, resonance Raman (Shiemke et al., 1988a) and X-ray absorption (Shiemke et al., 1988b) spectroscopic results are consistent with the presence of two axial ligands in both free and protein-bound F$_{430}$. The coenzyme is thermal labile, resulting in formation of several epimers and an oxidation product labeled F$_{560}$ (Pfaltz et al., 1985). Some of the damaged forms of F$_{430}$ exhibit square-planar geometry around the nickel rather than the native octahedral geometry.

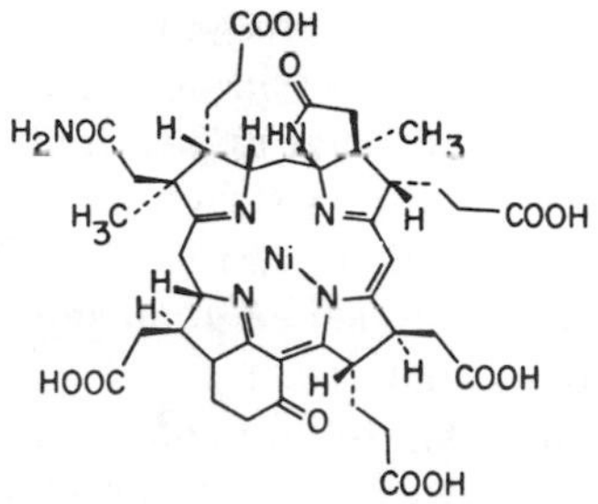

Figure 2.2. Structure of F_{430}. This yellow, nickel tetrahydrocorphin is a cofactor of methyl coenzyme M reductase, an essential enzyme of methanogenic bacteria. In addition to the four pyrrole nitrogens, the coenzyme possesses two unidentified axial ligands.

The biosynthesis of F_{430} has been partially elucidated. As in the case of all tetrapyrroles, 5-aminolevulinic acid is converted in a sequence of reactions to uroporphyrinogen III (Gilles and Thauer, 1983). Dimethylation of this compound occurs by using S-adenosylmethionine (Jaenchen et al., 1981), leading to a sirohydrochlorin derivative (Mucha et al., 1985). Nickel insertion into the ring is followed by two reductive steps, formation of the lactam ring, and closure of the six-membered carbocyclic ring (Pfaltz et al., 1987). Cell extracts are capable of converting the thermally denatured forms of F_{430} into the native coenzyme (Keltjens et al., 1988). Methanogen cells contain significant levels of protein-free F_{430} when grown under nickel-sufficient conditions (Hausinger et al., 1984). During nickel-limited conditions, this free F_{430} was converted to the protein-bound form as expected in a precursor–product relationship (Ankel-Fuchs et al., 1984).

The detailed function of F_{430} in the methyl coenzyme M reductase system is still not clear; however, significant progress in understanding its role has recently been reported by using spectroscopic methods (Albracht et al., 1986, 1988). The F_{430} nickel clearly plays an important role in the oxidation–reduction chemistry of the reaction and probably forms a covalent intermediate with the substrate during catalysis.

2.3. Hydrogenase

Hydrogenases catalyze the reversible oxidation of the simplest molecule, hydrogen gas (Adams et al., 1981):

$$H_2 + A_{ox} \rightleftarrows 2H^+ + A_{red} \qquad (4)$$

In this reaction, A_{ox} and A_{red} represent the oxidized and reduced forms of an electron carrier that may be a cytochrome, quinone, ferredoxin, nicotinamide, or other compound. These enzymes are classified according to their in vivo roles as either hydrogen-consuming or hydrogen-evolving hydrogenases. Many of the hydrogen-consuming enzymes have been shown to contain nickel (Hausinger, 1987).

Evidence for the presence of nickel in hydrogenases from 39 bacteria was summarized by Hausinger (1987) and will not be repeated here. Numerous other microbes are also likely to have nickel-containing hydrogenases, as evidenced by several recent examples (Chen and Yoch, 1987; Muth et al., 1987; Sprott et al., 1987; Teixeira et al., 1987). These microorganisms include methanogenic, hydrogen-oxidizing, sulfate-reducing, phototrophic, nitrogen-fixing, and other bacteria. In many cases, a single microbe was shown to possess multiple nickel-containing hydrogenases; for example, *Escherichia coli* and *Desulfovibrio vulgaris* each contain three distinct nickel-containing hydrogenases (Ballantine and Boxer, 1985; Lissolo et al., 1986).

Although the reaction catalyzed by these enzymes appears to be simple, nickel-containing hydrogenases are very complex proteins. Membrane-bound hydrogenases isolated from several species generally possess two types of subunits, whereas soluble hydrogenases may possess up to four distinct subunit types. In addition to nickel, hydrogenases always include iron–sulfur centers and some may also include selenium, copper, zinc, and flavins as cofactors (Hausinger, 1987). The nickel site in hydrogenases, illustrated in Figure 2.3, is in a distorted octahedral geometry with primarily sulfur ligation (Lindahl et al., 1984). Nickel appears to play a direct role in binding hydrogen and may form a nickel hydride intermediate (Cammack et al., 1987; van der Zwaan et al., 1987). The detailed mechanism of the nickel-dependent reaction remains under active investigation.

Interestingly, nickel appears to be required for the transcription (or perhaps translation) of the hydrogenase genes in *Bradyrhizobium japonicum* (Stults et al., 1986) and *Alcaligenes latus* (Doyle and Arp, 1988). Hydrogenase apoenzyme did not accumulate in these cells when grown under nickel-depleted conditions. In genetic studies with *E. coli*, several genes were shown to be required for hydrogenase activity in addition to the hydrogenase structural genes (Chaudhuri and Krasna, 1987; Waugh and Boxer, 1986; Wu and Mandrand-Berthelot, 1986). Some of the effects of mutations in these genes can be overcome by providing elevated concentrations of nickel in the media. The evidence indicates that multiple accessory factors are required for nickel processing to achieve active hydrogenase.

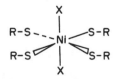

Figure 2.3. Model of the hydrogenase nickel center. Spectroscopic studies have demonstrated that the hydrogenase nickel center possesses octahedral geometry with primarily sulfurous ligands. It is unclear whether the sulfur atoms are derived from inorganic sulfur or they are part of amino acid residues. During catalysis, one of the unidentified axial ligands may be substituted by a hydride.

2.4. Carbon Monoxide Dehydrogenase

A wide range of microorganisms can oxidize carbon monoxide (CO) by the reaction shown in Equation (5), where A_{ox} and A_{red} are the oxidized and reduced forms of an electron carrier:

$$CO + H_2O + A_{ox} \rightleftarrows CO_2 + 2H^+ + A_{red} \qquad (5)$$

Aerobic and anaerobic species use very different enzymes to carry out this reaction. The enzyme derived from aerobes, carbon monoxide–acceptor oxidoreductase, is a protein containing molybdopterin, iron, sulfur and flavin (Meyer and Schlegel, 1983). In contrast, anaerobic bacteria utilize a nickel-containing carbon monoxide dehydrogenase (Hausinger, 1987; Walsh and Orme-Johnson, 1987).

The best characterized CO dehydrogenase is that from *Clostridium thermo-aceticum*, an acetogenic bacterium. This enzyme has an $\alpha_3\beta_3$ composition (M_r of 78,000 and 71,000, respectively) and contains 6 nickel, 3 zinc, and 33 iron atoms in the form of iron–sulfur centers (Ragsdale et al., 1983). The nickel center in this enzyme has been partially characterized by using spectroscopic methods and was clearly shown to be distinct from that of other nickel enzymes (Ragsdale et al., 1985; Cramer et al., 1987; Bastian et al., 1988). The substrate was shown to bind directly to the nickel site. The in vivo role of this enzyme involves acetyl coenzyme A biosynthesis; the oxidation of carbon monoxide is simply an easily measured side reaction (Ljungdahl, 1986; Wood et al., 1986). A complex series of reactions has been demonstrated for this enzyme, all apparently involving the nickel center. A simplified version of the reaction pathway is provided below, where the CO dehydrogenase–bound nickel is shown as Ni, A is an electron carrier, B_{12} is a protein containing vitamin B_{12}, and CoA–SH is coenzyme A:

$$Ni + CO_2 + A_{red} \rightleftarrows Ni{-}CO + A_{ox} + H_2O \qquad (6)$$

$$Ni{-}CO + CH_3{-}B_{12} \rightleftarrows CH_3{-}Ni{-}CO + B_{12} \qquad (7)$$

$$CH_3{-}Ni{-}CO \rightleftarrows CH_3{-}\overset{\displaystyle O}{\overset{\|}{C}}{-}Ni \qquad (8)$$

$$CH_3{-}\overset{\displaystyle O}{\overset{\|}{C}}{-}Ni + CoA{-}SH \rightleftarrows CH_3{-}\overset{\displaystyle O}{\overset{\|}{C}}{-}S{-}CoA + Ni \qquad (9)$$

In brief, CO_2 binds to the enzyme nickel center and is reduced to form bound CO. Methyl group transfer to the nickel takes place, and carbon–carbon bond formation occurs in a carbonylation reaction. Coenzyme A is then acetylated to complete the cyle. The methyl group attached to the vitamin

B_{12} protein is ultimately derived from CO_2; thus, the acetyl coenzyme A pathway represents a novel mode of autotrophic growth (Ljungdahl, 1986; Wood et al., 1986).

Analogous nickel-dependent acetyl coenzyme A synthase enzymes are thought to be present in other acetogens, in autotrophic methanogens, and in sulfate-reducing bacteria (reviewed by Hausinger, 1987). In contrast, a nickel-dependent CO dehydrogenase may carry out nearly the reverse reaction in acetate-utilizing methanogens. The acetoclastic pathway requires activation of acetate to form acetyl coenzyme A or a similar molecule. The acetyl group of this compound is thought to bind to the enzyme nickel center with release of the thiol [Equation (9)]. Carbon–carbon bond cleavage is suggested to occur [Equation (8)], forming bound methyl and CO groups. The methyl group is eventually reduced to methane, whereas CO is oxidized to form CO_2.

A nickel-dependent CO dehydrogenase has been purified from the photosynthetic microorganism *Rhodospirillum rubrum* (Bonam and Ludden, 1987). This enzyme has no apparent role in acetate synthesis or utilization, and the function of CO dehydrogenase in phototrophs is unclear. By growing *R. rubrum* in the absence of nickel, Bonam et al. (1988) were able to obtain an inactive, nickel-deficient form of the enzyme. Purified, inactive protein was successfully reconstituted to full activity by incubation with nickel, demonstrating that no accessory proteins are required for nickel insertion into this enzyme (Bonam et al., 1988).

3.   OTHER BIOLOGICAL ROLES FOR NICKEL

3.1.   Plants

Dalton et al. (1988) have recently reviewed the evidence for a nickel requirement in plants. In some cases, this micronutrient is needed for urease or hydrogenase activity. For example, several studies have shown that nickel is essential for the formation of active plant ureases (Gordon et al., 1978; Klucas et al., 1983; Polacco, 1977); the absence of urease activity resulted in accumulation of toxic concentrations of urea, which led to plant necrosis (Dalton et al., 1988; Eskew et al., 1983; Walker et al., 1985). In soybean–*B. japonicum* symbionts, hydrogenase activity was depressed during nickel-depleted growth (Eskew et al., 1983; Klucas et al., 1983); low levels of the bacterial hydrogenase activity decrease the efficiency of symbiotic nitrogen fixation, resulting in slower plant growth and depressed crop yields. Hydrogenases are also present in some nonsymbiotic plants under certain conditions (Torres et al., 1986), but it is unknown whether this enzyme activity is affected by nickel levels.

Numerous less well defined effects of nickel depletion have been reported to include necrosis of leaves or stems, grain inviability, depressed seedling vigor, and early maturation (Brown et al., 1987; Dalton et al., 1988). In one intriguing example cell culture studies of *Alyssum*, which normally accumulates large amounts of nickel, have demonstrated a growth requirement for this metal ion (Thauer et al., 1980). Because of the broad distribution of plants that exhibit a nickel requirement, it has been suggested that this metal is an essential micronutrient for all higher plants (Brown et al., 1987; Eskew et al., 1983).

3.2. Animals

In 1974, Nielsen and Ollerich (1974) proposed that nickel was an essential trace metal in animals. They summarized studies that demonstrated that nickel-deficient chickens and rats exhibited impaired liver metabolism and morphology. Since that time, a requirement for nickel has been substantiated for many organisms; however, the precise roles of nickel are still unclear for most animals, including humans (Nieboer et al., 1988). One example that is fairly well characterized is described below.

In tunicates, the role of nickel is beginning to be defined at the molecular level. Three tunicates were shown to accumulate high levels of nickel and cobalt in a 7 : 1 ratio (Rayner-Canham et al., 1985). In *Trididemnum solidum*, internalized nickel was found to be associated with a blue-green pigment that was named tunichlorin (Bible et al., 1988). Structural assignment of tunichlorin was based on chemical and spectroscopic methods, and the resulting Ni(II) 2-devinyl-2-hydroxymethylpyropheophorbide structure is shown in Figure 2.4. This nickel chlorin is similar to the chlorophyll-derived tetrapyrroles, with the important distinction that tunichlorin contains nickel. It is not known whether the tunicate itself or an algal symbiont synthesizes tunichlorin nor is the function of the nickel tetrapyrrole established.

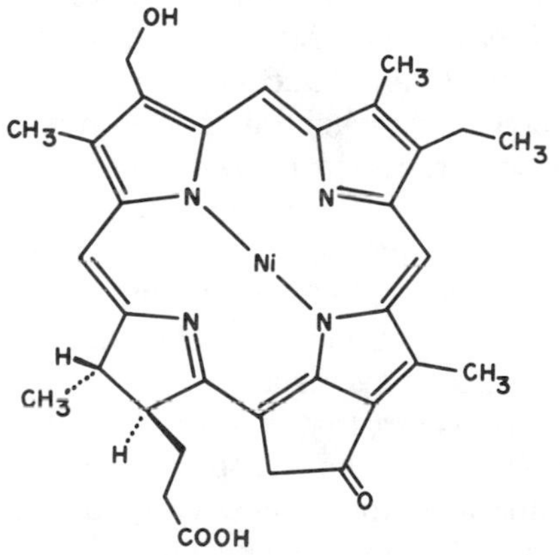

Figure 2.4. Structure of tunichlorin. This blue-green pigment was purified from the tunicate *T. solidum*. No function has yet been assigned to this nickel chlorin.

3.3. Microorganisms

A functionally undefined nickel growth requirement has been observed in three algae, *Chlorella vulgaris* (Bertrand and DeWolfe, 1967), *Chlorella emersonii* (Soeder and Engelmann, 1984), and an *Oscillatoria* species (van Baalen and O'Donnell, 1978). These algae do not possess urease, and it is unclear whether the nickel participants in hydrogen metabolism in these microorganisms.

Two methanogen species, *Methanobacterium bryantii* and *Methanospirillum hungatei*, may have nickel requirements in addition to those previously discussed. In the former case, nickel was required in the medium to prevent cell lysis under ammonia-free conditions (Jarrell et al., 1982). In the latter microorganism, significant nickel levels were observed in the cell membrane and in an external sheath (Sprott et al., 1983).

4. NICKEL TRANSPORT SYSTEMS

The widespread requirement for nickel ion among plants, animals, and micro-organisms means that transport mechanisms must exist to take up this essential ion. Kaltwasser and Frings (1980) have pointed out the important distinction between true nickel transport, which is energy dependent, and passive surface binding of this metal ion. They suggested that extracellular binding may be a prerequisite to internalization of nickel. Many authors have used the term *uptake* when describing surface accumulation of nickel (e.g., Richardson et al., 1980), whereas other authors have used this term to describe true transport, which is discussed below.

For many microorganisms, nickel transport appears to occur via the magnesium transport system as shown by decreased nickel transport in the presence of magnesium. Examples of magnesium-inhibitable nickel transport are known in *Clostridium pasteurianum* (Bryson and Drake, 1988), *E. coli* (Webb, 1970), *Neurospora crassa* (Mohan et al., 1984), *Rhodopseudomonas capsulata* (Takakuwa, 1987), and *Saccharomyces cerevisiae* (Fuhrmann and Rothstein, 1968). In contrast, nickel transport systems that are not inhibited by magnesium have been reported in *Anabaena cylindrica* (Campbell and Smith, 1986), *Bradyrhizobium japonicum* (Stults et al., 1987), *M. bryantii* (Jarrell and Sprott, 1982), *Methanothrix concilii* (Baudet et al., 1988), and soybean seedlings (Cataldo et al., 1978). Furthermore, in the case of *Alcaligenes eutrophus* two mechanisms to transport nickel may exist, only one of which is magnesium inhibited (Lohmeyer and Friedrich, 1987; Tabillion and Kalt-wasser, 1977). Nickel transport that is not inhibited by magnesium may, in some cases, be inhibited by cobalt, copper, or zinc. Independent of the uptake system, the internalization process was found to be energy dependent in all cases except those of *B. japonicum* (Stults et al., 1987) and *M. concilii* (Baudet et al., 1988).

5. CONCLUDING REMARKS

Nickel has clearly been shown to serve several essential functions in plants, animals, and microorganisms. The long delay in recognizing the biological requirement for nickel was due, in part, to the minimal requirements for the metal ion and to the difficulty in removing trace nickel contamination. In addition to the clear roles in ureolysis, methanogenesis, hydrogen metabolism, and acetate metabolism, future studies are likely to discover new biological functions of this metal and to eludicate new nickel-containing enzymes.

ACKNOWLEDGMENTS

Research in the author's laboratory was funded by the Michigan State University Agricultural Experiment Station (article no. 12746) and by Public Health Service Grant AI 22387 from the National Institutes of Health.

REFERENCES

Adams, M. W., Mortenson, L. E., and Chen, J.-S. (1981). Hydrogenase. *Biochim. Biophys. Acta* **594,** 105–176.

Albracht, S.P.J., Ankel-Fuchs, D., van der Zwaan, J. W., Fontijn, R. D., and Thauer, R. K. (1986). A new EPR signal of nickel in *Methanobacterium thermoautotrophicum. Biochim. Biophys. Acta* **870,** 50–57.

Albracht, S.P.J., Ankel-Fuchs, D., Bocher, R., Ellerman, J., Moll, J., van der Zwaan, J. W., and Thauer, R. K. (1988). Five new EPR signals assigned to nickel in methyl-coenzyme M reductase from *Methanobacterium thermoautotrophicum*, strain Marburg. *Biochim. Biophys. Acta* **955,** 86–102.

Andrews, R. K., Blakeley, R. L., and Zerner, B. (1984). Urea and urease. In G. L. Eichhorn and L. G. Marzilli (Eds.), *Advances in Inorganic Chemistry*, Vol. 6. Elsevier Science, New York, pp. 245–283.

Ankel-Fuchs, D., Jaenchen, R., Gebhardt, N. A., and Thauer, R. K. (1984). Functional relationship between protein-bound and free factor F430 in *Methanobacterium. Arch. Microbiol.* **139,** 332–337.

Ballantine, S. P., and Boxer, D. H. (1985). Nickel-containing hydrogenase isoenzymes from anaerobically grown *Escherichia coli. J. Bacteriol.* **163,** 454–459.

Bast, E. (1988). Nickel requirement for the formation of active urease in purple sulfur bacteria (*Chromatiaceae*). *Arch. Microbiol.* **150,** 6–10.

Bastian, N. R., Diekert, G., Niederhoffer, E. C., Teo, B-K., Walsh, C. T., and Orme-Johnson, W. H. (1988). Nickel and iron EXAFS of carbon monoxide dehydrogenase from *Clostridium thermoaceticum* strain DSM. *J. Am. Chem. Soc.* **110,** 5581–5582.

Baudet, C., Sprott, G. D., and Patel, G. B. (1988). Adsorption and uptake of nickel in *Methanothrix concilii. Arch. Microbiol.* **150,** 338–342.

Bertrand, D., and DeWolfe, A. (1967). Le nickel oligoelement dynamique pour les vegetaux superieurs. *C. R. Acad. Sci.* **265,** 1053–1055.

Bible, K. C., Buytendorp, M., Zierath, P. D., and Rinehart, K. L. (1988). Tunichlorin: a nickel chlorin isolated from the Carribean tunicate *Trididemnum solidum*. *Proc. Natl. Acad. Sci. USA* **85**, 4582–4586.

Blakeley, R. L., and Zerner, B. (1984). Jack bean urease: the first nickel enzyme. *J. Mol. Catal.* **23**, 263–292.

Blakeley, R. L., Dixon, N. E., and Zerner, B. (1983). Jack bean urease. VII. Light scattering and nickel(II) spectruym. *Biochim. Biophys. Acta* **744**, 219–229.

Bonam, D., and Ludden, P. W. (1987). Purification and characterization of carbon monoxide dehydrogenase, a nickel, zinc, iron-sulfur protein, from *Rhodospirillum rubrum*. *J. Biol. Chem.* **262**, 2980–2987.

Bonan, D., McKenna, M. C., Stephens, P. J., and Ludden, P. W. (1988). Nickel-deficient carbon monoxide dehydrogenase from *Rhodospirillum rubrum*: *in vivo* and *in vitro* activation by exogenous nickel. *Proc. Natl. Acad. Sci. USA* **85**, 31–35.

Brown, P. H., Welch, R. S., and Cary, E. E. (1987). Nickel: a micronutrient essential for higher plants. *Plant Physiol.* **85**, 801–803.

Bryson, M. F., and Drake, H. L. (1988). Energy-dependent transport of nickel by *Clostridium pasteurianum*. *J. Bacteriol.* **170**, 234–238.

Cammack, R., Patil, D. S., Hatchikian, E. C., and Fernandez, V. M. (1987). Nickel and iron-sulphur centres in *Desulfovibrio gigas* hydrogenase: ESR spectra, redox properties and interactions. *Biochim. Biophys. Acta* **912**, 98–109.

Campbell, P. M., and Smith, G. D. (1986). Transport and accumulation of nickel ions in the cyanobacterium *Anabaena cylindrica*. *Arch. Biochem. Biophys.* **244**, 470–477.

Cataldo, D. A., Garland, T. R., and Wildung, R. E. (1978). Nickel in plants. I. Uptake kinetics using intact soybean seedlings. *Plant Physiol.* **62**, 563–565.

Chaudhuri, A., and Krasna, A. I. (1987). Isolation of genes required for hydrogenase synthesis in *Escherichia coli*. *J. Gen. Microbiol.* **133**, 3289–3298.

Chen, Y.-P., and Yoch, D. C. (1987). Regulation of two nickel-requiring (inducible and constitutive) hydrogenases and their coupling to nitrogenase in *Methylosinus trichosporium* OB3b. *J. Bacteriol.* **169**, 4778–4783.

Cramer, S. P., Eidsness, M. K., Pan, W.-H., Morton, T. A., Ragsdale, S. W., DerVartanian, D. V., Ljungdahl, L. G., and Scott, R. A. (1987). X-ray absorption spectroscopic evidence for a unique site in *Clostridium thermoaceticum* carbon monoxide dehydrogenase. *Inorg. Chem.* **26**, 2477–2479.

Dalton, D. A., Russell, S. A., and Evans, H. J. (1988). Nickel as a micronutrient element for plants. *Biofactors* **1**, 11–16.

Diekert, G., Gilles, H.-H., Jaenchen, R., and Thauer, R. K. (1980a). Incorporation of 8 succinate per mol nickel into factors F_{430} by *Methanobacterium thermoautotrophicum*. *Arch. Microbiol.* **128**, 256–262.

Diekert, G., Jaenchen, R., and Thauer, R. K. (1980b). Biosynthetic evidence for a nickel tetrapyrrolec structure of factor F_{430} from *Methanobacterium thermoautotrophicum*. *FEBS Lett.* **119**, 118–120.

Dixon, N. E., Gazzola, C., Blakeley, R. L., and Zerner, B. (1975). Jack bean urease. A metalloenzyme. A simple biological role. *J. Am. Chem. Soc.* **97**, 4131–4133.

Dixon, N. E., Blakeley, R. L., and Zerner, B. (1980a). Jack bean urease. III. The involvement of active-site nickel ion in inhibition by 2-mercaptoethanol, phosphoramidate, and fluoride. *Can. J. Biochem.* **58**, 481–488.

Dixon, N. E., Gazzola, C., Asher, C. J., Lee, D.S.W., Blakeley, R. L., and Zerner, B. (1980b). Jack bean urease. II. The relationship between nickel, enzymatic activity, and the "abnormal" ultraviolet spectrum. The nickel content of jack beans. *Can. J. Biochem.* **58**, 474–480.

Dixon, N. E., Riddles, P. W., Gazzola, C., Blakeley, R. L., and Zerner, B. (1980c). Jack bean urease. V. On the mechanism of action of urease on urea, formamide, acetamide, N-methylurea, and related compounds. *Can. J. Biochem.* **58**, 1335–1344.

Doyle, C. M., and Arp, D. J. (1988). Nickel affects expression of the nickel-containing hydrogenase of *Alcaligenes latus*. *J. Bacteriol.* **170**, 3891–3896.

Ellefson, W. L., Whitman, W. B., and Wolfe, R. S. (1982). Nickel-containing factor F_{430}: chromophore of the methylreductase of *Methanobacterium*. *Proc. Natl. Acad. Sci. USA* **79**, 3707–3710.

Eskew, D. L., Welch, R. M., and Cary, E. E. (1983). Nickel: an essential micronutrient for legumes and possibly all higher plants. *Science* **222**, 621–623.

Fuhrmann, G. F., and Rothstein, A. (1968). The transport of Zn^{2+}, Co^{2+}, and Ni^{2+} into yeast cells. *Biochim. Biophys. Acta* **163**, 325–330.

Gilles, H., and Thauer, R. K. (1983). Uroporphyrinogen III, an intermediate in the biosynthesis of the nickel-containing factor F_{430} in *Methanobacterium thermoautotrophicum*. *Eur. J. Biochem.* **135**, 109–112.

Gordon, W. R., Schwemmer, S. S., and Hillman, W. S. (1978). Nickel and the metabolism of urea by *Lemna paucicostata* Helgelm. 6746. *Planta* **140**, 265–268.

Gunsalus, R. P., and Wolfe, R. S. (1978). Chromophoric factors F_{342} and F_{430} of *Methanobacterium thermoautotrophicum*. *FEMS Microbiol. Lett.* **3**, 191–193.

Hartzell, P. L., and Wolfe, R. S. (1986). Requirement of the nickel tetrapyrrole F_{430} for *in vivo* methanogenesis: reconstitution of methyl reductase component C from its dissociated subunits. *Proc. Natl. Acad. Sci. USA* **83**, 6726–6730.

Hausinger, R. P. (1987). Nickel utilization by microorganisms. *Microbiol. Rev.* **51**, 22–42.

Hausinger, R. P., Orme-Johnson, W. H., and Walsh, C. (1984). Nickel tetrapyrrole cofactor F_{430}: comparison of the forms bound to methyl coenzyme M reductase and protein free in cells of *Methanobacterium thermoautotrophicum*. *Biochemistry* **23**, 801–804.

Jaenchen, R., Diekert, G., and Thauer, R. K. (1981). Incorporation of methionine-derived methyl groups into factor F_{430} by *Methanobacterium thermoautotrophicum*. *FEBS Lett.* **130**, 133–136.

Jarrell, K. F., and Sprott, G. D. (1982). Nickel transport in *Methanobacterium bryantii*. *J. Bacteriol.* **151**, 1195–1203.

Jarrell, K. F., Colvin, J. R., and Sprott, G. D. (1982). Spontaneous protoplast formation in *Methanobacterium bryantii*. *J. Bacteriol.* **149**, 346–353.

Kaltwasser, H., and Frings, W. (1980). Transport and metabolism of nickel in microorganisms. In J. R. Nriagu (Ed.), *Nickel in the Environment*. Wiley, New York, pp. 463–491.

Keltjens, J. T., Hermans, J.M.H., Rijsdijk, G.J.F.A., van der Drift, C., and Vogels, G. D. (1988). Interconversion of F_{430} derivatives of methanogenic bacteria. *Ant. Leeuwen.* **54**, 207–220.

Klucas, R. V. Hanus, F. J., Russell, S. A., and Evans, H. J. (1983). Nickel: a micronutrient element for hydrogen-dependent growth of *Rhizobium japonicum* and for expression of urease activity in soybean seeds. *Proc. Natl. Acad. Sci. USA* **80**, 2253–2257.

Lancaster, J. (1988). *Bioinorganic Chemistry of Nickel*. VCH Publishers, Deerfield Beach, FL.

Lee, M.H., Mulrooney, S.B., and Hausinger, R.P. (1990). Purification, characterization, and in vivo reconstitution of *Klebsiella aerogenes* urease apoenzyme. *J. Bacteriol.* **172**, 4427–4431.

Lindahl, P. A., Kojima, N., Hausinger, R. P., Fox, J. A., Teo, B. K., Walsh, C. T., and Orme-Johnson, W. H. (1984). Nickel and iron EXAFS of F_{420}-reducing hydrogenase from *Methanobacterium thermoautotrophicum*. *J. Am. Chem. Soc.* **106**, 3062–3065.

Lissolo, T., Choi, E. S., LeGall, J., and Peck, H. D., Jr. (1986). The presence of multiple intrinsic membrane nickel-containing hydrogenases in *Desulfovibrio vulgaris* (Hildenborough). *Biochem. Biophys. Res. Commun.* **139**, 701–708.

34   Biological Utilization of Nickel

Ljungdahl, L. G. (1986). The autotrophic pathway of acetate synthesis in acetogenic bacteria. *Ann. Rev. Microbiol.* **40**, 415–450.

Lohmeyer, M., and Friedrich, C. G. (1987). Nickel transport in *Alcaligenes eutrophus*. *Arch. Microbiol.* **149**, 130–135.

McDonald, J. A., Vorhaben, J. E., and Campbell, J. W. (1980). Invertebrate urease: purification and properties of the enzyme from a land snail, *Otala laceta*. *Comp. Biochem. Physiol.* **66B**, 223–231.

Mackay, E. M., and Pateman, J. A. (1980). Nickel requirement of a urease-deficient mutant in *Aspergillus nidulans*. *J. Gen. Microbiol.* **116**, 249–251.

Mackay, E. M., and Pateman, J. A. (1982). The regulation of urease activity in *Aspergillus nidulans*. *Biochem. Genet.* **20**, 763–776.

Mackerras, A. H., and Smith, G. D. (1986). Urease activity of the cyanobacterium *Anabaena cylindrica*. *J. Gen. Microbiol.* **132**, 2749–2752.

Meyer, O., and Schlegel, H. G. (1983). Biology of aerobic carbon monoxide oxizing bacteria. *Ann. Rev. Microbiol.* **37**, 277–310.

Meyer-Bothling, L. E., Polacco, J. C., and Cianzio, S. R. (1987). Pleiotrophic soybean mutants defective in both urease isozymes. *Mol. Gen. Genet.* **209**, 432–438.

Mobley, H.L.T., and Hausinger, R. P. (1989). Microbial ureases: significance, regulation, and molecular characterization. *Microbiol. Rev.* **53**, 85–108.

Mohan, P. M., Rudra, M.P.P., and Sastry, K. S. (1984). Nickel transport in nickel-resistant strains of *Neurospora crassa*. *Curr. Microbiol.* **10**, 125–128.

Mucha, H., Keller, E., Weber, H., Lingens, F., and Trosch, W. (1985). Sirohydrochlorin, a precursor of factor F_{430} biosynthesis in *Methanobacterium thermoautotrophicum*. *FEBS Lett.* **190**, 169–173.

Mulrooney, S.B., and Hausinger, R.P. (1990). Sequence of the *Klebsiella aerogenes* urease genes and evidence for accessory proteins facilitating nickel incorporation. *J. Bacteriol.* **172**, 5837–5843.

Muth, E., Morschel, E. and Klein, A. (1987). Purification and characterization of an 8-hydroxy-5-deazaflavin-reducing hydrogenase from the archaebacterium *Methanococcus voltae*. *Eur. J. Biochem.* **169**, 571–577.

Nieboer, E., Tom, R.T., and Sanford, W.E. (1988). Nickel metabolism in man and animals. In H. Sigel and A. Sigel (Eds.), *Metal ions in biological systems*, Vol. 23. Marcel Dekker, New York, pp. 91–121.

Nielsen, F. H., and Ollerich, D. A. (1974). Nickel: a new essential trace element. *Fed. Proc.* **33**, 1767–1772.

Jaenchen, R., Gilles, H. H., Diekert, G., and Thauer, R. K. (1982). Zur kenntnis des faktors F430 aus methanogenen bakterien: struktur des porphinoiden ligand-systems. *Helv. Chim. Acta* **65**, 828–865.

Pfaltz, A., Livingston, D. A., Jaun, B., Diekert, G., Thauer, R. K., and Eschenmoser, A. (1985). Factor F_{430} from methanogenic bacteria: on the nature of isolation artefacts of F_{430} and the conformational stereochemistry of the ligand periphery of hydroporphinoid nickel (II) complexes. *Helv. Chim. Acta* **68**, 1338–1358.

Pfaltz, A., Kobelt, A., Huster, R., and Thauer, R. K. (1987). Biosynthesis of coenzyme F430 in methanogenic bacteria. Identification of $15,17^3$-*seco*-F430-17^3-acid as an intermediate. *Eur. J. Biochem.* **170**, 459–467.

Polacco, J. C. (1977). Is nickel a universal component of plant ureases? *Plant Sci. Lett.* **10**, 249–255.

Polacco, J. C., and Havir, E. A. (1979). Comparison of soybean urease isolated from seed and tissue culture. *J. Biol. Chem.* **254**, 1701–1707.

Ragsdale, S. W., Clark, J. E., Ljungdahl, L. G., Lundie, L. L., and Drake, H. L. (1983).

Properties of purified carbon monoxide dehydrogenase from *Clostrium thermoaceticum*, a nickel, iron–sulfur protein. *J. Biol. Chem.* **258**, 2364–2369.

Ragsdale, S. W., Wood, H. G., and Antholine, W. E. (1985). Evidence that an iron–nickel–carbon complex is formed by reaction of CO with the CO dehydrogenase from *Clostridium thermoacetcium*. *Proc. Natl. Acad. Sci. USE* **82**, 6811–6814.

Rayner-Canham, G. W., van Roode, M., and Burke, J. (1985). Nickel and cobalt concentrations in the tunicate *Halocynthia pyriformis*: evidence for essentiality of the two metals. *Inorg. Chim. Acta* **106**, :37–L38.

Rees, T. A. V., and Bekheet, I. A. (1982). The role of nickel in urea assimilation by algae. *Planta* **156**, 385–387.

Richardson, D. H. S., Beckett, P. J., and Nieboer, E. (1980). Nickel in lichens, bryophytes, fungi, and algae. In J. O. Nriagu (Ed.), *Nickel in the Environment*. Wiley, New York, pp. 367–406.

Rouviere, P. E., and Wolfe, R. S. (1988). Novel biochemistry of methanogenesis. *J. Biol. Chem.* **263**, 7913–7916.

Shiemke, A. K., Hamilton, C. L., and Scott, R. A. (1988a). Structural heterogeneity and purification of protein-free F_{430} from the cytoplasm of *Methanobacterium thermoautotrophicum*. *J. Biol. Chem.* **263**, 5611–5616.

Shiemke, A. K., Scott, R. L., and Shelnutt, J. A. (1988b). Resonance Raman spectroscopic investigation of axial coordination in *M. thermoautotrophicum* methyl reductase and its nickel tetrapyrrole cofactor F_{430}. *J. Am. Chem. Soc.* **110**, 1645–1646.

Soeder, C. J., and Engelmann, G. (1984). Nickel requirement in *Chlorella emersonii*. *Arch. Microbiol.* **137**, 85–87.

Sprott, G. D., Shaw, K. M., and Jarrell, K. F. (1983). Isolation and chemical composition of the cytoplasmic membrane of the archaebacterium *Methanospirillum hungatei*. *J. Biol. Chem.* **258**, 4026–4031.

Sprott, G. D., Shaw, K. M., and Beveridge, T. J. (1987). Properties of the particulate enzyme F_{420}-reducing hydrogenase isolated from *Methanospirillum hungatei*. *Can. J. Microbiol.* **33**, 896–904.

Stults, L. W., Sray, W. A., and Maier, R. J. (1986). Regulation of hydrogenase biosynthesis by nickel in *Bradyrhizobium japonicum*. *Arch. Microbiol.* **146**, 280–283.

Stults, L. W., Mallick, S., and Maier, R. J. (1987). Nickel uptake in *Bradyrhizobium japonicum*. *J. Bacteriol.* **169**, 1398–1402.

Sumner, J. B. (1926). The isolation and crystallization of the enzyme urease. *J. Biol. Chem.* **69**, 435–441.

Tabillion, R., and Kaltwasser, H. (1977). Energieabhangige Ni-aufnahme bei *Alcaligenes eutrophus* stamm H 1 und H 16. *Arch. Microbiol.* **113**, 145–151.

Takakuwa, S. (1987). Nickel uptake in *Rhodopseudomonas capsulatus*. *Arch. Microbiol.* **149**, 57–61.

Takishima, K., Suga, T., and Mamiya, G. (1988). The structure of jack bean urease. The complete amino acid sequence, limited proteolysis and reactive cysteine residues. *Eur. J. Biochem.* **175**, 151–165.

Teixeira, M., Fauque, G., Moura, I., Lespinat, P. A., Berlier, Y., Prickril, B., Peck, H. D., Jr., Xavier, A. V., LeGall, J., and Moura, J. J. G. (1987). Nickel–[iron–sulfur]–selenium-containing hydrogenases from *Desulfovibrio baculatus* (DSM 1743). *Eur. J. Biochem.* **167**, 47–58.

Thauer, R. K., Diekert, G., and Schonheit, P. (1980). Biological roles of nickel. *Trends Biochem. Sci.* **11**, 304–306.

Torres, V., Ballesteros, A., and Fernandez, V. M. (1986). Expression of hydrogenase activity in barley (*Hordeum vulgare* L.) after anaerobic stress. *Arch. Biochem. Biphys.* **245**, 174–178.

van Baalen, C., and O'Donnell, R. (1978). Isolation of a nickel-dependent blue-green algae. *J. Gen. Microbiol.* **105,** 351–353.

van der Zwaan, J. W., Albracht, S. P. J., Fontijn, R. D., and Mul, P. (1987). On the anomalous temperature behaviour of the EPR signal of monovalent nickel in hydrogenase. *Eur. J. Biochem.* **169,** 377–384.

Walker, C. D., Graham, R. D., Madison, J. T., Cary, E. E., and Welch, R. M. (1985). Effects of Ni deficiency on some nitrogen metabolites in cowpeas (*Vigna unguiculata* L. Walp). *Plant Physiol.* **79,** 474–479.

Walsh, C. T., and Orme-Johnson, W. H. (1987). Nickel enzymes. *Biochemistry* **26,** 4901–4906.

Waugh, R., and Boxer, D. H. (1986). Pleiotropic hydrogenase mutants of *Escherichia coli* K12: growth in the presence of nickel can restore hydrogenase activity. *Biochimie* **68,** 157–166.

Webb, M. (1970). Interrelationships between the utilization of magnesium and the uptake of other bivalent cations by bacteria. *Biochim. Biophys. Acta* **222,** 428–439.

Winkler, R. G., Polacco, J. C., Eskew, D. L., and Welch, R. M. (1983). Nickel is not required for apourease synthesis in soybean seeds. *Plant Physiol.* **72,** 262–263.

Wood, H. G., Ragsdale, S. W., and Pezacka, E. (1986). The acetyl-CoA pathway of autotrophic growth. *FEMS Microbiol. Rev.* **39,** 345–362.

Wu, I. F., and Mandrand-Berthelot, M.-A. (1986). Genetic and physiological characterization of new *Escherichia coli* mutants impaired in hydrogenase activity. *Biochimie* **68,** 167–179.

3

OCCUPATIONAL EXPOSURES TO NICKEL

Evert Nieboer

Department of Biochemistry and Occupational Health Program,
McMaster University, Hamilton, Ontario, Canada, L8N 3Z5

1. INTRODUCTION

Because excess respiratory cancer has been linked unequivocally to certain nickel-refining operations, most of the published exposure data refer to this industrial sector. Nevertheless, some hygiene information is available for the working environments encountered in nickel-using industries. A brief description of the most common nickel-producing processes and the uses of nickel products precedes the details of the reported nickel exposures.

Nickel and Human Health: Current Perspectives, Edited by Evert Nieboer and Jerome O. Nriagu.
ISBN 0-471-50076-3   © 1992 John Wiley & Sons, Inc.

2. PRODUCTION OF NICKEL FROM ORES*

Nickel is mined either as sulfide or lateritic ores. In the former case, it is usually present as pentlandite [(Ni, Fe)$_9$S$_8$] with ore grades of 1–3% nickel. Major deposits occur in Canada (in the provinces of Ontario and Manitoba) and Russia (at Norilsk and Pechenga), with others in Australia, South Africa, and Zimbabwe. Nickel minerals found in laterite deposits are often divided roughly into silicate ores, which refers to hydrous nickel silicates (including members of the serpentine, chlorite, and clay mineral groups), or limonitic ores, which refer to nickeliferous hydrated ferric oxide. Mature laterite deposits may contain up to 2.5% nickel and occur in New Caledonia, Colombia, Cuba, the Dominican Republic, Greece, Guatemala, Indonesia, the Philippines, and Russia.

2.1. Recovery of Nickel from Sulfide Ores

The first step in the recovery process is concentration (also referred to as ore benefication, mineral dressing, or less accurately, milling). The ore is ground to liberate the individual mineral grains and a nickel-rich concentrate ($\approx$ 8–12% Ni) is largely separated from the host rock and from other sulfide minerals by selective flotation. Pyrrhotite (Fe$_7$S$_8$), a common constituent of sulfide nickel ores, can be removed by magnetic separation.

Either a series of pyrometallurgical procedures or hydrometallurgical steps are used to remove iron, sulfur, and rock from the nickel concentrate (see Fig. 3.1). Roasting (650–800°C), smelting (1200°C), and converting (1150–1250°C) are employed in the pyrometallurgical stream to produce "matte," containing nickel, sulfur, and usually some copper. Copper and nickel sulfides (Cu$_2$S and Ni$_3$S$_2$) are then separated by cooling the matte slowly and applying benefication techniques or leaching the cooled matte. The nickel subsulfide product of matte separation is roasted to nickel oxide in fluidized beds (1100–1225°C), while the matte leachate is purified by chemical means. Further refining is achieved by one of three methods: the nickel carbonyl (Mond) process, electrolysis, or electrowinning (see Fig. 3.1).

In earlier operations that were associated with very high risks of respiratory cancer, high-temperature oxidation (1650°C) of Ni$_3$S$_2$ on sintering machines was employed to produce nickel oxide. These operations were notorious for their extreme dustiness Doll (1990).

The pure hydrometallurgical stream in Figure 3.1 is an alternative to the roasting methods. In this procedure, nickel is removed by pressure leaching, often involving ammonia. The nickel-rich leachate is purified and then reduced with hydrogen gas to produce nickel powder.

---

* Much of the material in this section and Section 3 is derived from a number of authoritative reviews: Boldt and Queneau (1967); Morgan (1979); Duke (1980a,b); Warner (1984); Mastromatteo (1986, 1988); and Tyroler and Landolt (1988).

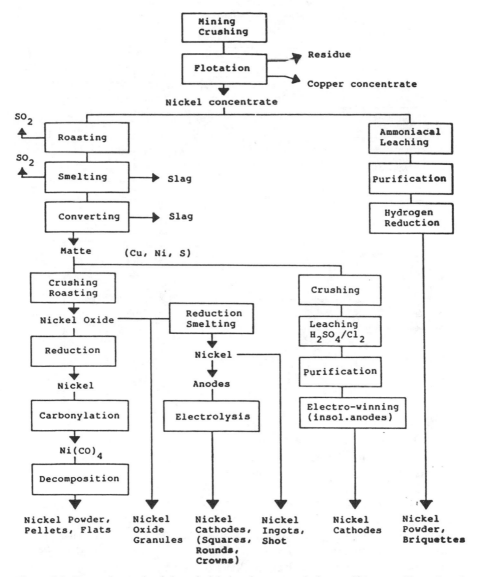

Figure 3.1. Extraction and refining of nickel and compounds from sulfide ores. (Reproduced with permission from Mastromatteo, 1988.)

2.2. Recovery of Nickel from Laterite Ores

Mining of silicate or oxide nickel ores involves surface (earth-moving) operations. Nickel is extracted from them by hydrometallurgical or pyrometallurgical methods or a combination of the two. This is illustrated in Figure 3.2. Sulfuric acid leaching at 230–260°C removes over 95% of the nickel and cobalt, leaving the iron behind. Subsequently, the nickel and cobalt are

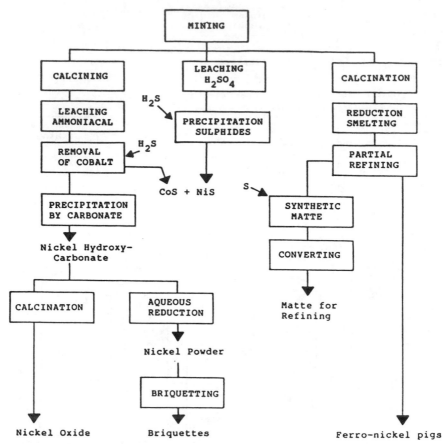

Figure 3.2. Extraction and refining of nickel and compounds from oxide and silicate ores. (Reproduced with permission from Mastromatteo, 1988.)

precipitated as the sulfides, and this product is further refined employing some of the procedures outlined in Figure 3.1. The combined pyrometallurgical–hydrometallurgical stream involves the selective reduction of the ore to metallic nickel at about 750°C, followed by ammoniacal leaching, purification, and isolation of either nickel oxide or nickel powder. Two routes are employed in the pyrometallurgical refining of nickel as indicated in Figure 3.2. Either ferronickel or synthetic high-grade nickel sulfide matte is the product. The latter is refined further by the methods already alluded to in Section 2.1.

3. APPLICATION AND USES OF NICKEL PRODUCTS

The nickel-producing industry sells nickel in various forms: nickel metal (>95% purity), ferronickel (20–50% nickel), nickel oxides, and nickel salts (hydrated crystals of the sulfate and chloride). Most of the nickel production is used in the making of alloys: stainless steel (which usually contains 8–10% Ni and 20% Cr and is employed in components of chemical and food processing equipment, commercial transport vehicles, and consumer durables) and nickel alloys with copper (corrosion-resistant, marine equipment; 66% Ni), chromium (high-temperature applications; 42–73% Ni), chromium–iron (cryogenic applications; 25–42% Ni), and aluminum (permanent magnets). Nickel cast iron is another important product and is employed for manufacturing industrial equipment and automobiles. Nickel and nickel–copper alloys are very common in the making of coins, and nickel-welding products are also produced. Nickel metal and nickel compounds are also used in the electroplating of consumer goods and industrial materials and for the manufacture of nickel–cadmium batteries. As Raney nickel, the metal is used as a catalyst in the hydrogenation of fats and oils. Other uses of nickel and its compounds as catalysts are known, such as in the petroleum, rubber, and chemical industrial sectors.

4. OCCUPATIONAL EXPOSURES TO NICKEL

Two extensive industrywide surveys of average ambient nickel concentrations encountered in the workplace are available. Warner (1984) and Nieboer et al. (1984) summarize data for nickel-producing and nickel-using operations. Mastromatteo (1986, 1988) provides a convenient abridgement of the Warner data. Representative (average) concentrations are summarized in Tables 3.1 and 3.2. Current exposure limits are compiled in Table 3.3 for comparison. It is clear from the data provided in Tables 3.1 and 3.2 that nickel exposures in both the nickel-producing and nickel-using industries on occasion exceed the recommended exposure limits (Table 3.3). Since there is no indication that the refining and manufacturing practices have altered significantly since the surveys referred to were conducted, comparable exposures may be anticipated today. Taking the ACGIH threshold limit values (TLVs) of 1 mg/m^3 (nickel and water-insoluble compounds) and 0.1 mg/m^3 (nickel salts; water-soluble compounds) as a guideline, as is done in most countries (Table 3.3), a number of jobs or processes can be designated as potential problem areas. Although group averages fall below the prescribed exposure limits, the concentration ranges shown indicate that measurements for individuals involved in the following components of the pyrometallurgical refining of sulfide ores sometimes become elevated: roasting, smelting, converting, anode casting, and especially the handling and grinding of nickel sulfide matte. By comparison, the producing of ferronickel from laterite ore appears to be less dusty. It should also be noted that the extremely high estimates for the old sintering plants given in Table 3.1 confirm the anecdotal assessment by retired workers concerning the dusty working conditions that prevailed.

Table 3.1 Occupational Exposures in Nickel-producing Industry

Operation or Occupation[a]	Exposure Details			Concentration[c] (mg Ni/m³)		Source
	Compound	Sampling Mode[b]	Date of Survey	Mean(s)	Range	
Underground miners (S)	Ore	P	1984[d]	0.025 (0.005)[e]	—	Warner, 1984
Ore beneficiation (S)	Ore and concentrate	P	1984[d]	0.03–0.13 (geom.)	—	Warner, 1984
Roasting, smelting, and converting (S)	Ni sulfides	A	1930, 1961	0.03–0.08 (geom.)	≤1.00	Warner, 1984
Fluid bed roaster (S)	Ni sulfides	P	1975, 1981[f]	0.57	0.08–1.57	Nieboer et al., 1984
Smelter (S)	Ni sulfides	P	1975, 1981[f]	0.18	0.04–0.41	Nieboer et al., 1984
Matte separation (S)	Ni sulfides	P	1975	0.43	0.2–1.2	Nieboer et al., 1984
Matte grinding (S)	Ni sulfides	P	1984[d]	0.11–0.47	0.01–2.9	Warner, 1984
Anode casting	Crude Ni	P	1980	0.34 (geom.)	<2.5	Nieboer et al., 1984
Discontinued calcining and sintering (S)	Ni sulfides and oxides	E	Before 1963	—	5–100	Inco, 1976; Warner, 1985
Ferronickel and synthetic matte (L)	Ore, Ni, Ni sulfides	P	1984[d]	(0.045)[g]	0.05–1.00	Warner, 1984
Ferronickel production (L)	Ore, Ni	P	1984[d]	0.01–0.19	0.004–0.42	Warner, 1984
Hydrometallurgical operations (S) Matte leaching	Mostly nickel salts	P	1984[d]	0.10	0.005–1.6	Warner, 1984
Electrolyte purification	Mostly nickel salts	P	1984[d]	0.13–0.22[h]	0.01–0.47[h]	Warner, 1984

42

Operation	Form		Date			Reference
Electrowinning	Mostly nickel salts	P	1984[c]	0.19	0.08–0.40	Warner, 1984
Electrowinning	Mostly nickel salts	P	1981	0.20	0.08–0.26	Nieboer et al., 1984
Electrowinning	Mostly nickel salts	A	1982	—	<0.05	Nieboer et al., 1984
Electrolysis (electrorefining)	Mostly nickel salts	P	1974, 1980	—	0.1–0.5	Nieboer et al., 1984
Electrolysis (electrorefining)	Mostly nickel salts	P	1984[d]	0.03–0.05[h,i]	0.001–0.24[h]	Warner, 1984
Hydrometallurgical operations (L)	Ni/Ni salts	P	1978	0.02–0.68	0.01–1.1	Nieboer et al., 1984
Nickel carbonyl refinery	Ni/NiO	P	1982	0.31	0.02–5.0	Nieboer et al., 1984
Nickel powder packaging	Ni	P	1980	0.19	0.03–0.99	Nieboer et al., 1984
		A	1975	0.90	0.03–3.0	Nieboer et al., 1984
Nickel powder briquetting	Ni	P	1978	0.72	0.01–2.1	Nieboer et al., 1984
Nickel carbonyl refinery	Ni(CO)$_4$	A	1975	0.04 (15 ppb)	<0.12 (50 ppb)	Nieboer et al., 1984; Mastromatteo, 1986

[a] Abbreviations: S, sulfide ore; L, laterite ore.

[b] Abbreviations: A, area sampling; P, personal sampling; E, retrospective estimate.

[c] Concentrations of nickel (time-weighted averages) refer to total nickel unless specified otherwise; geom., geometric mean.

[d] Date of publication of the data.

[e] Sampled for particles < 10μm.

[f] Not much change was evident in a 1981 survey.

[g] Sampled for particles < 30μm.

[h] Measured as water-soluble nickel.

[i] Corresponding total nickel was 0.05–0.18.

Table 3.2 Occupational Exposures in Nickel-using Industry

| Operation or Occupation | Exposure Details | | | Concentration[b] (mg Ni/m^3) | | Source |
	Compound	Sampling Mode[a]	Date of Survey	Mean	Range	
Stainless steel production	Ni	P	1984[c]	0.014–0.13	0.001–0.19	Warner, 1984
High-nickel-alloy production	Ni	P	1984[c]	0.01–0.30	0.001–4.4	Warner, 1984
Foundry operations	Ni	P	1984[c]	0.0–0.31	0.01–0.9	Warner, 1984
Electroplating	Ni salts	A	1984[c]	0.004–0.011	0.002–0.016	Warner, 1984
Nickel salt production	Ni, NiO, Ni salts	P	1976	0.15	0.009–0.59	Nieboer et al., 1984
Nickel–cadmium battery manufacture	Ni, Ni(OH)$_2$	P	1984[c]	0.38	0.02–1.91	Warner, 1984
Nickel catalyst production	Ni, NiSO$_4$	P	1984[c]	0.37 / 0.003[d]	0.19–0.53 / 0.002–0.009[d]	Warner, 1984

[a] Abbreviations; A, area sampling; P, personal sampling.
[b] Concentrations of nickel (time-weighted averages) refer to total nickel unless specified otherwise.
[c] Date of publication of data.
[d] Measued as water-soluble nickle.

Table 3.3   Occupational Exposure Limits for Nickel[a]

Location	Nickel Metal, Sparingly Soluble Compounds (mg Ni/m³)	Soluble Nickel Compounds (mg Ni/m³)	Nickel Carbonyl (mg Ni/m³)
Czechoslovakia	0.05	0.05	0.0034 (1.4 ppb)[f]
Denmark	0.5	1	—
Federal Republic of Germany	0.5[b]	0.05	0.24 (100 ppb)[f]
United Kingdom	1	0.1	0.12 (50 ppb)[f]
Japan	1	1	0.0024 (1 ppb)[f]
Netherlands	1	0.1	—
Norway	0.1	0.1	—
Soviet Union	0.05[c]	0.005[c]	0.0005[c]
Sweden	0.5[d]	0.1[e]	0.0024 (1 ppb)[f]
United States (ACGIH)	1	0.1	0.12 (50 ppb)[f]
United States (OSHA)	1	1	0.0024 (1 ppb)[f]
United States (NIOSH)	0.015[g]	0.015[g]	0.0024[h] (1 ppb)[f]

[a] Abbreviations: ACGIH, American Conference of Governmental and Industrial Hygienists; NIOSH, National Institute of Occupational Safety and Health; OSHA, U.S. Occupational Safety and Health Administration. Adapted from a compilation by Grandjean (1986) and based primarily on the listing provided by IRPTC (1987).
[b] Respirable, technical guideline only.
[c] Aerosols, ceiling value.
[d] Nickel metal only; for nickel subsulfide, 0.01.
[e] Nickel compounds.
[f] Refers to $Ni(CO)_4$ in units of nanoliters per liter: 1 ppb (v/v) = 0.007 mg $Ni(CO)_4$/m^3 = 0.0024 mg Ni/m^3.
[g] Recommended TLV.

Electrowinning and electrorefining processes have also been responsible for elevated soluble nickel levels (see Table 3.1). Mixed exposure to both soluble and insoluble nickel is characteristic of unit operations within the electrorefining process such as cementation (Warner, 1984). No doubt, this duality is reflected in some of the total nickel levels reported in Table 3.1. Interestingly, the ambient nickel levels measured by area sampling in one plant dropped from 0.1 mg/m^3 in 1973–1974 to ≤0.05 mg/m^3 in 1982 when switching to electrowinning from electrorefining (Nieboer et al., 1984). Although this suggests that with adequate containment measures electrowinning is potentially a cleaner operation, high production schedules (i.e., high electrical operating currents) result in average levels such as 0.20 mg/m^3, double the ACGIH TLV.

In all nickel powder production plants, the packaging (or briquetting) of nickel powder is extremely dusty. Ambient levels as high as 5 mg/m^3 have, on occasion, been observed (Nieboer et al., 1984). The nickel oxide reduction step (Morgan, 1979) of the nickel carbonyl process also appears to generate considerable dust (see Table 3.1). By contrast, actual exposure to nickel carbonyl gas appears to be minimal (mean of 15 ppb and an upper limit below 50 ppb), although the available data on which this judgement is based are scarce.

The ACGIH guidelines are also surpassed in some nickel-using industries, particularly in high-nickel alloy production, nickel–cadmium battery manufacture, and nickel salt production (see Table 3.2). It is noteworthy that nickel electroplating, by contrast to electrowinning, generates very low ambient nickel concentrations.

The data reported in a recent industry-wide re-evaluation of past occupational exposures to nickel complements the summary in Table 3.1 (Doll, 1990).

5. CONCLUDING REMARKS

The environmental data reviewed in this chapter indicate that considerable exposures to nickel and its compounds occur in both the nickel-producing and nickel-using industries. It will be shown in Chapter 4 that the reported ambient nickel levels are responsible for elevation in workers of nickel in serum (plasma) and urine. As might be expected, the magnitude of the scaling factor that relates ambient to body fluid nickel concentrations depends strongly on the physicochemical properties of the nickel compounds (Nieboer et al., 1984). Because of this substantial impact on nickel body burden and the known carcinogenic risk associated with certain nickel-refining operations (Doll, 1990), it seems prudent to continue to pursue improvements in industrial hygiene and engineering practices to reduce occupational exposure.

ACKNOWLEDGMENT

I am indebted to Professor Jorma Rantanen, Director General, Institute of Occupational Health, Helsinki, Finland, for the opportunity to spend an enjoyable and stimulating sabbatical period (a writing holiday!) at the Institute. I acknowledge, with thanks, the kind support of his staff, especially the excellent secretarial services provided by Raija Salmenius. Editorial input by Dr. Stuart Warner, INCO Ltd., Toronto, is gratefully noted.

REFERENCES

Boldt, J. R., Jr., and Queneau, P. (1967). *The Winning of Nickel.* Longmans, Toronto.
Doll, R., Editor-in-Chief (1990). Report of the International Committee on Nickel Carcinogesis in Man. *Scand. J. Work Environ. Health* **16**, 1–82.

Duke, J. M. (1980a). Nickel in rocks and ores. In J. O. Nriagu (Ed.), *Nickel in the Environment*. Wiley, New York, pp. 27–50.

Duke, J. M. (1980b). Production and uses of nickel. In J. O. Nriagu (Ed.), *Nickel in the Environment*. Wiley, New York, pp. 51–65.

Grandjean, P. (1986). *Health Effects Document on Nickel*. Ontario Ministry of Labour, Toronto.

International Nickel Company (INCO) (1976). Nickel and its inorganic compounds (including nickel carbonyl). Unpublished report submitted to NIOSH by INCO (U.S.) October 1976.

International Register of Potentially Toxic Chemicals (IRPTC) (1987). IRPTC Legal File, 1986. IRPTC, United Nations Environment Programme, Geneva, Switzerland.

Mastromatteo, E. (1986). Nickel. *Am. Ind. Hyg. Assoc. J.* **47,** 589–601.

Mastromatteo, E. (1988). Nickel and its compounds. In C. Zenz (Ed.), *Occupational Medicine*, 2nd ed. Year Book Medical Publishers, Chicago, pp. 597–608.

Morgan, L. G. (1979). Manufacturing processes. Refining of nickel. *J. Soc. Occup. Med.* **29,** 33–35.

Nieboer, E., Yassi, A., Jusys, A. A., and Muir, D.C.F. (1984). *The Technical Feasibility and Usefulness of Biological Monitoring in the Nickel Producing Industry, Final Report*. Special Document, McMaster University, Hamilton, Ontario, Canada. (Available from the Nickel Producers Environmental Research Association, Durham, NC.)

Tyroler, G. P., and Landolt, C. A. (1988). *Extractive Metallurgy of Nickel and Cobalt*. Metallurgical Society, Warrendale, PA.

Warner, J. S. (1984). Occupational exposure to airborne nickel in producing and using primary nickel products. In F. W. Sunderman, Jr. (Ed.-in-Chief), *Nickel in the Human Environment*. IARC Sci. Publ. No. 53. Oxford University Press, Oxford, pp. 419–437.

Warner, J. S. (1985). Estimating past exposures to airborne nickel compounds in the Copper Cliff sinter plant. In S. S. Brown and F. W. Sunderman, Jr. (Eds.), *Progress in Nickel Toxicology*. Blackwell Scientific, Oxford, pp. 203–206.

4

ABSORPTION, DISTRIBUTION, AND EXCRETION OF NICKEL

Evert Nieboer

Department of Biochemistry and Occupational Health Program, McMaster University, Hamilton, Ontario, Canada

William E. Sanford

Department of Physical Sciences, Mohawk College of Applied Arts and Technology, Hamilton, Ontario, Canada

Brian C. Stace

Department of Biochemistry, University of Surrey, Guildford, Surrey, England

Nickel and Human Health: Current Perspectives, Edited by Evert Nieboer and Jerome O. Nriagu.
ISBN 0-471-50076-3 © 1992 John Wiley & Sons, Inc.

1. INTRODUCTION

At the outset, a general survey of the absorption, distribution, and excretion of nickel is presented. This provides the framework for a dicussion of the dynamics of its renal clearance. An assessment of the ultrafilterable fraction of nickel in serum and the dependence of the fractional clearance of nickel on urine flow rate is shown to lead to a systematic approach to correcting for urinary dilution effects in biological monitoring. Effects of environmental and pathological factors on nickel in body fluids are also reviewed, including a summary of the observed correlations between ambient nickel levels in the workplace and concentrations in body fluids. Throughout this chapter, human work will be emphasized with animal work inserted where appropriate. Detailed description of the kidney and its functional unit, the nephron, is reviewed in Chapter 11.

2. ABSORPTION, DISTRIBUTION, AND EXCRETION

2.1. Absorption

Inhalation and ingestion are the major routes of nickel intake in humans. Although absorption through the skin (percutaneous) does not appear to be a major source of nickel in humans, it is of great importance in nickel dermatitis (Maibach and Menné, 1989).

The most important route of nickel uptake, especially in an industrial setting, is respiratory absorption. This has been documented for workers in nickel refineries (including nickel carbonyl plants) and electroplating shops (Doll, 1990). Current industrial exposures are generally well below the American Conference of Governmental and Industrial Hygienists (ACGIH) guidelines of 1 mg/m^3 for sparingly water-soluble compounds (e.g., nickel metal, oxides, and sulfides) and 0.1 mg/m^3 for water-soluble salts. By comparison, ambient nickel levels are very low, even in urban centers (Barrie, 1981; EPA, 1986; Valerio et al., 1989). A typical mean in 1982 for U.S. cities was 0.008 µg/m^3, with 99% of the data falling below 0.030 µg/m^3 (EPA, 1986).

The daily nickel intake is estimated as 100–300 µg/day (see Chapter 16; Pennington and Jones, 1989). Nielsen (Chapter 16) argues that above-average consumption of food items with naturally high nickel contents may result in considerably higher absorption. Foods that are relatively rich in nickel include cocoa, soya beans, other dried legumes, nuts, and certain grains. Sunderman et al. (1989) have assessed gastrointestinal absorption of nickel sulfate added to drinking water and food (to scrambled eggs as part of a standard American breakfast). Subjects were given 12–18 µg Ni/kg body weight after a 12-hr (overnight) fast. Absorbed nickel averaged 27 ± 17% of the dose ingested in water, compared to 0.7 ± 0.4% of the same dose in the food. The toxicokinetic model employed (Section 2.3; Chapter 5) did not include parameters for possible biliary elimination and intestinal reabsorption, which may be significant (see Section 2.3). On the basis of their modeling and on perusal of published dietary intakes and known fecal excretion levels, Sunderman et al. (1989) conclude that about 1% of dietary nickel is absorbed. Because of the high level of absorption from drinking water, nickel from this source may be more important than traditionally thought (Bennett, 1984; Nieboer et al., 1988b). Nickel concentrations in drinking water have been estimated as ≤5 μg/L (EPA, 1986; Nieboer et al., 1988b).

2.2. Distribution in Body Fluids and Tissues

Reliable methods for the detection of nickel and for contamination control during specimen collection, handling, storage, and analysis have assigned much lower nickel levels to serum and urine than earlier studies. Recently reported values in serum are 0.46 ± 0.26 μg/L (Sunderman et al., 1984), 0.28 ± 0.24 μg/L (Linden et al., 1985; Sunderman et al., 1986a,b; Sunderman et al., 1989), 0.44 ± 0.30 μg/L (Sanford, 1988) and 0.14 ± 0.09 μg/L (Nixon et al., 1989). Comparable normal, whole-blood levels are 1.26 ± 0.33 μg/L (Sunderman et al., 1984) and 0.34 ± 0.28 μg/L (Linden et al., 1985; Sunderman et al., 1986a,b). The drop in the whole-blood concentrations from 1.26 to 0.34 μg/L has been assigned by Sunderman and co-workers to a switch to heparin that contained no detectable nickel. Although lower serum and plasma nickel values are now commonly reported, urine concentrations appear to hold near 2 μg/L. In separate studies three years apart, Sunderman et al. (1986b, 1989) report urinary nickel levels to be 2.0 ± 1.5 μg/L (range 0.5–6.0 μg/L) and 2.0 ± 0.9 μg/L (range 0.4–3.1 μg/L) for nonexposed healthy adults. For workers occupationally exposed to nickel and its compounds, values of plasma or serum concentrations up to 35 μg/L and urine levels up to 400 μg/L have been reported (Nieboer et al., 1984; Sunderman et al., 1986a). Recently, in an industrial accident involving the contamination of drinking water with Ni(II) sulfate and chloride, serum nickel concentrations ranged from 13 to 1340 μg/L and urinary levels from 150 to 12,000 μg/g creatinine in 15 workers (Sunderman et al., 1988; also see Chapter 5). These are by far the highest levels recorded by any laboratory recognized for performing reliable nickel analysis. The doses of nickel consumed by these workers were estimated to be 0.5–2.5 g.

Nickel is transported in the body via plasma where it is bound to both high-molecular-mass proteins (e.g., albumin) and low-molecular-mass carriers (e.g., proteins and amino acids; see Section 3 for more details). The binding of Ni^{2+} to these protein and amino acid ligands allows its rapid transport throughout the body. Although the distribution of nickel in the human body is of great importance, relatively little is known in comparison to animal studies. The nickel content in human tissue has been limited to autopsy specimens. Most of the early data appears to be of questionable analytical reliability because of inadequate contamination control (e.g., Schroeder et al., 1962; Tipton and Cook, 1963; Tipton et al., 1965). A number of studies have reported that the nickel contents of some of the autopsy tissues were below the detection limit available at the time. Nevertheless, these earlier reports illustrate the presence of nickel in brain, lung, kidney, liver, heart, trachea, aorta, spleen, skin, testes, and intestine. Generally, the levels were found to be less than 600 μg/kg dry tissue. Bone, lung, and intestines appear to exhibit the highest levels (Sumino et al., 1975). The ubiquity of nickel in tissues has proven to be correct. Based on a minimum of seven subjects, Rezuke et al. (1987) reported the following nickel concentrations in normal human tissues, expressed as mean ± SD in micrograms per kilogram of dry weight: lung, 173 ± 94; thyroid, 141 ± 83; adrenal, 132 ± 84; kidney, 62 ± 43; heart, 54 ± 40; liver, 50 ± 31; brain, 44 ± 16; spleen, 37 ± 31; and pancreas, 34 ± 25. Levels in bone seem to be higher than lung (Nomoto, 1974; Sumino et al., 1975). Other than lung, the quoted levels for these tissues seem to be consensus values (Rezuke et al., 1987; Seemann et al., 1985). Concentrations for the lung are dependent on age (Kollmeier et al., 1985; Rezuke et al., 1987) and on the portion of the lung sampled (Raithel et al., 1987). Presumably, the age dependence indicates that air pollution is the source. The major contribution to ambient nickel originates from the combustion of fossil fuels, and the predominant forms in air are nickel sulfate (water soluble), nickel oxide, and complex metal oxides containing nickel (EPA, 1986). Animal studies have shown that the green Ni(II) oxide is relatively inert in the lung, in contrast to water-soluble nickel salts, which are rapidly cleared (Nieboer et al., 1988b). As might be anticipated, elevation of nickel in lungs of nickel refinery workers has been clearly demonstrated (Andersen and Svenes, 1989). The geometric mean for 39 workers was 38 μg/g dry weight with 99% confidence limits (CLs) of 18–81 μg/g, compared to 0.67 μg/g and 99% CLs of 0.48–0.95 μg/g for 16 control subjects; the tissue was taken from the right lower lobe.

Nickel also occurs in significant amounts in sweat, milk, saliva, hair, and nasal mucosa (Sunderman et al., 1986a) as well as in nails (Gammelgaard and Andersen, 1985).

By comparison to the limited amount of human work, considerable information is available on the distribution of nickel in animal tissues. For example, Ni(II) chloride administered parenterally to rodents and rabbit shows the largest accumulation in lung, kidney, liver, and endocrine glands, with relatively little nickel found in neural tissues (e.g., Parker and Sunderman, 1974; Clary, 1975; Mushak, 1980).

2.3. Excretion

Many studies have been completed on the kinetics of distribution and clearance of injected nickel compounds in experimental animals. All show a relatively short half-life for water-soluble Ni(II) salts of 24 hr or less. In rats and rabbits, Onkelinx et al. (1973) determined a two-compartment clearance model with fast and slow components. One day after intravenous injection, 78% of the dose was excreted in the rabbit; in the rat it took about 72 hr. In both species, $^{63}Ni^{2+}$ was rapidly cleared from the plasma or serum in the first two days after injection and subsequently disappeared at a much slower rater (three to seven days). Most of the nickel was cleared by the kidney; during the three days after injection of rats, 78% of the dose was excreted in the urine and 15% was excreted in feces presumably via bile.

The toxicokinetic model for nickel absorption, distribution, and elimination in humans referred to in Section 2.1, and described fully in Chapter 5, is really an extension of the two-compartment model for rodents referred to above (Onkelinx et al., 1973; Onkelinx and Sunderman, 1980). The major storage compartments in humans are the serum and the tissues. The elimination half-time into urine for both nickel absorbed from drinking water and food was 28 ± 9 hr (range, 17–48 hr). Interestingly, all rate constants are first order (i.e., for nickel uptake from drinking water into serum, transfer from serum to tissues and vice versa, and transfer from serum to urine) except uptake from the solid food, which was zero order. This means that for the latter process, the absorption rate is independent of dose; by contrast, the rates of all the other processes do depend on the nickel concentration.

Biliary excretion of nickel has been known to occur in calves with a nickel-supplemented diet (O'Dell et al., 1971) and in rabbits (9.2% of intravenous $^{63}Ni^{2+}$ injected dose) (Onkelinx et al., 1973). Marzouk and Sunderman (1985) have found that biliary excretion of nickel in rats injected with single subcutaneous doses only amounted to approximately 0.3% of the total dose over a 24-hr period. As already indicated, bile may be an important excretion route of nickel for humans. Bile specimens obtained during the postmortem examination of gallbladders from five non-occupationally-exposed subjects averaged 2.3 ± 0.8 μg Ni/L (Rezuke et al., 1987). This level appears to be comparable to normal urinary nickel levels of 2.2 ± 1.2 mg/L (or 2.6 ± 1.4 μg/day) (Sunderman et al., 1986a,b). Providing there is no significant intestinal reabsorption of the bile nickel in the gastrointestinal tract, Rezuke et al. (1987) estimate that the net biliary excretion of nickel in nonexposed persons averages 2–5 μg/day, since 1–2 L/day of bile are excreted into the human intestine. By comparison, on average, 1.5 L/day of urine are voided. Consequently, these data suggest a net excretion of approximately 6 μg Ni/day for the combined biliary and urinary routes.

Ishihara and Matsushiro (1986) point out that metals in feces have four origins: unabsorbed fraction; and excretion into hepatic bile, pancreatic juice, and by the intestinal wall (desquamation of epithelial cells and intestinal secretion). Human studies suggest that pancreatic excretion of nickel into the duodenum might occur (Ishihara et al., 1987). However, specimen contamination in the study assessing this might have been a complicating factor, judging by the very high urinary levels for Cd, Cr, and Ni reported by these authors (Ishihara and Matsushiro, 1986). If pancreatic nickel excretion is confirmed, then it would constitue another important excretionary route since the normal flow is substantial (1–2 L/day). For this to be significant, intestinal reabsorption must not occur.

It may be concluded from the above discussion that renal excretion is a major route in man and animals. Perhaps this is the reason why urinary nickel is a reliable indicator of environmental exposure (see Section 6 for more details). Urine is also the most convenient body fluid for biological monitoring because its collection is noninvasive and its nickel concentration is considerably high than in plasma, serum, or whole blood, which facilitates analysis (see Sections 5 and 6).

3. BIOCHEMICAL ASPECTS OF NICKEL DISTRIBUTION AND EXCRETION

It is common for metals to have only a fraction of the total amount in the serum available for renal filtration by the glomerulus. For example, serum calcium is recognized to have three distinct fractions: (a) protein-bound Ca^{2+}, (b) low-molecular-mass complexes, and (c) free Ca^{2+} (Lambie and Harris, 1985). These authors state that the total fraction of ultrafilterable calcium (α_{Ca}) (free-ion and low-molecular-mass complexes) is 0.60. By comparison, the ultrafilterable fraction of chromium (α_{Cr}) has been estimated as 0.055 (Lim et al., 1983; Nieboer and Jusys, 1988).

The preferred high-molecular-mass ligand for nickel appears to be human albumin and possibly a nickel metalloprotein believed to be an α_2-macroglobulin, also referred to as "nickeloplasmin" (Sunderman, 1977; Nomoto, 1980; Scott and Bradwell, 1983; Nomoto and Sunderman, 1989). The low-molecular-mass component has been shown to be a Ni(II)–amino acid complex by in vitro [63]Ni addition experiments. Of the 22 amino acids tested, the main Ni^{2+}-binding amino acid at physiological pH was found to be L-histidine (Lucassen and Sarkar, 1979). The exact distribution between these various fractions has not been firmly established. The metalloprotein nickeloplasmin has not been well characterized, and there is some doubt about its importance in the distribution of Ni^{2+} in serum. Work by Nomoto et al. (1971), Sunderman et al. (1972), and Nomoto and Sunderman (1989) determined the association of nickel with albumin, nickeloplasmin, and ultrafilterable components of rabbit and human sera. Their work involved atomic absorption spectrometry measurements on the fractions of serum that had been separated by ultrafiltration and column chromatography. It was found that approximately 40% of Ni(II) present in rabbit serum was associated with albumin, 44% with the nickeloplasmin fraction, and 16% with the ultrafilterable fraction; whereas approximately 34% of the nickel in human serum was associated with albumin, 26% with nickeloplasmin, and 40% was ultrafilterable. Asato et al. (1975) found the amount of [63]Ni^{2+} (introduced by intravenous injection) associated with the ultrafiltrable fraction of rabbit serum to be 15%. Nomoto (1980) separated α_2-macroglobulin by affinity column chromatography from the serum of healthy humans who had no industrial exposure to nickel. Using atomic absorption spectrometry, he found that 43% of the total nickel content was bound to $\alpha_2 - 1$ microglobulin. By comparison, Lucassen and Sarkar (1979) found that only about 0.1% of the total Ni^{2+} in human serum was bound to the nickeloplasmin, with 95.7% bound to albumin and 4.2% to L-histidine in the ultrafiltcrable fraction. They added $^{63}NiCl_2$ to 2.0 mL of serum and diluted to a constant volume of 2.5 mL before ultracentrifugation. The values reported are thus for Ni(II) added to serum and do not necessarily represent the endogenous nickel distribution. Caution in interpretation is thus warranted since the added nickel resulted in concentrations about 35,000-fold higher than those observed naturally in serum. The mentioned in vitro distribution studies involving 22 amino acids were also carried out under these artificial conditions. Recently, Sanford et al. (1992a) determined the actual endogenous low- and high-molecular-mass fractions of nickel for sera from six electrolytic refinery workers. The low-molecular-mass fraction was determined by nickel analysis of serum by electrothermal atomic absorption spectrometry (EAAS), ultracentrifugation of serum, followed by EAAS nickel analysis of the supernatant. The nickel content of sera from these refinery workers averaged 8 ± 5 µg/L. The value of 24% for the ultrafilterable fraction ($\alpha_{Ni} = 0.24 \pm 0.06$) is substantially higher than the 4.2% observed by Lucassen and Sarkar (1979). It is also larger than the 15% reported for rabbit serum by Asato et al. (1975) and less than the 40% ultrafilterable fraction attributed to human serum by Sunderman et al. (1972) and Nomoto et al. (1971). It has been shown that there are large species variations in the proportions of ultrafilterable and protein-bound serum nickel (Hendel and Sunderman, 1972). New Zealand rabbit sera are known to have a larger concentration of nickel than human sera, which may explain the different distribution between high- and low-molecular-mass components. The species distribution work done by Sunderman et al. (1972) and Nomoto et al. (1971) on human sera also involved the measurement of endogenous nickel. However, the nickel concentrations observed for the sera from healthy unexposed subjects (2.3 µg/L) was considerably higher than current background values (e.g., 0.46 µg/L). Consequently, the value of $\alpha_{Ni} = 0.40$ found by these researchers may reflect some inherent analytical uncertainties.

It appears that the most reliable estimate of α_{Ni} for humans to date appears to be that of Sanford et al. (1992a), namely 0.24.

4. FRACTIONAL RENAL CLEARANCE OF NICKEL

Until recently, there was only one detailed study of the renal excretion of nickel by humans. Mertz et al. (1970) looked at the renal excretion of nickel under varying conditions of diuresis in adults of both sexes with normal and impaired kidney function. Great variability in nickel excretion was observed within the individual experimental groups that could not be related to different functional states of the kidneys. However, nickel excretion was shown to be dependent on urine flow rates.

In the recent study by Sanford et al. (1992a), the renal clearance of nickel was investigated in 26 electrolytic refinery workers. Multivoid 24-hr urine collections were obtained as well as serum samples at the beginning and end of this sampling period. Nickel clearance, the volume of serum from which *ultrafilterable nickel* is removed in 1 min, was calculated using the formula

$$C_{Ni} = U_{Ni}V/\alpha_{Ni}S_{Ni} \tag{1}$$

where U_{Ni} is the nickel concentration in 24 hr collection; V the flow rate in milliliters per minute based on the 24-hr volume; α_{Ni} the ultrafilterable fraction of nickel in serum (0.24 ± 0.06); and S_{Ni} the average nickel concentration of the two serum specimens. Note that $\alpha_{Ni}C_{Ni}$ represents the hypothetical volume from which all nickel is removed. It might be designated as the *apparent nickel clearance*. When divided by the creatinine clearance ($C_{creat} = U_{creat}V/S_{creat}$), the clearance of nickel yields the fractional clearance of nickel, C_{Ni}/C_{creat} (Fig. 4.1). The observed nickel clearance of 32 ± 11 mL/min (1.73

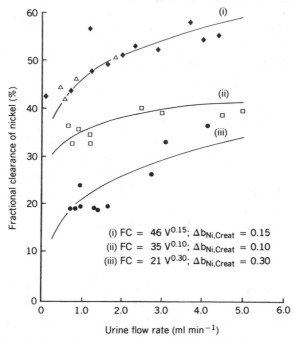

Figure 4.1. Fractional clearance of nickel in percent ($(C_{Ni}/C_{creat}) \times 100$) as a function of urinary flow rate V. Grouping is arbitrary for the four donors (see different symbols) whose data align on the same curve. [Reproduced with permission from Sanford et al. (1992a).]

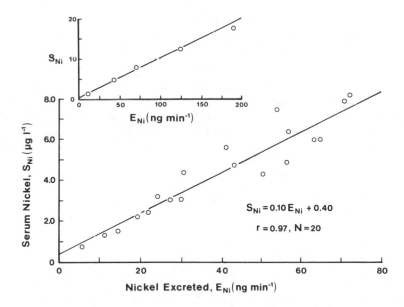

Figure 4.2. Relationship between serum nickel concentrations and nickel excretion ($VU_{Ni} = E_{Ni}$). Here S_{Ni} is the average of the nickel in serum before and after the 24-hr collection period for U_{Ni}. The donors were working in an electrolytic nickel refinery. Insert illustrates that the regression line includes data points above 80 ng Ni/min. [Reproduced with permission from Sanford et al. (1992a).]

m^2) with $N = 27$ was considerably less than the corresponding creatinine clearance of 109 ± 28 mL/min (1.73 m^2) with $N = 30$. The apparent nickel clearance, therefore, is 7.7 mL/min (1.73 m^2), which is in good agreement with average values of 5.8 and 8.3 mL/min (1.73 m^2) reported recently by Sunderman et al. (1988) (also see Chapter 5). Because creatinine clearance is a measure of glomerular filtration rate (GFR) (Duarte et al., 1980; Schuster and Seldin, 1985), any substance that has a clearance lower than the GFR is reabsorbed in the kidney (Green, 1976). Since the fractional clearances of nickel were substantially less than 100% (Fig. 4.1), absorption of nickel must occur in the human kidney. As is evident from Figure 4.1, the fractional clearance of nickel tends to increase rapidly at low urine flow rates and levels off somewhat at high flow rates. The shape of the fractional clearance plots for nickel has a close resemblance to that for the fractional clearance of urea (Smith, 1964; Green, 1976). Tubular reabsorption of urea is by passive diffusion down a concentration gradient created by the reabsorption of water associated with the active reabsorption of NaCl (Lambie and Harris, 1985). Therefore, by analogy to urea, it is reasonable to postulate that the reabsorption of nickel in the human kidney may have a large passive component.

Equation (1) predicts a linear relationship between S_{Ni} and VU_{Ni} (identical to nickel excretion rate E_{Ni}), with $(\alpha_{Ni}C_{Ni})^{-1}$ as the slope. This is observed as indicated in Figure 4.2. Employing $\alpha_{Ni} = 0.24$, the value of C_{Ni} obtained from the slope is 42 mL/min [$\equiv$ 36 mL/min (1.73 m^2)]. Similar relationships with slopes near 0.1 have been observed for plots of S_{Ni} (or plasma nickel, P_{Ni}) versus U_{Ni} for spot urine samples [as summarized by Nieboer et al. (1984) and Sunderman et al. (1986a)]. This suggests that for a group of individuals V on average assumes a value near 1 mL/min. Surprisingly, this linear relationship is maintained ($r = 0.97$, $N = 103$) for the extremely wide range of S_{Ni} and U_{Ni} concentrations reported by Sunderman et al. (1988) (see Section 2.2); the observed slope was 0.14, as expected for creatinine-adjusted U_{Ni} since U^0_{creat} is 1.3–1.4 g/L (Sanford, 1988; Nieboer et al., 1984).

5. ADJUSTMENT OF URINARY NICKEL CONCENTRATIONS FOR RATE OF URINE FLOW

One possible approach to deciding what type of normalization is most appropriate for a particular analyte is that proposed by Sunderman et al. (1986a,b). They plotted U_{Ni} versus U_{creat} or against specific gravity for specimens collected over a 24-hr period. The objective was to determine which of urinary creatinine or specific gravity most closely resembled urinary nickel. They concluded that normalization by either creatinine or specific gravity has approximately equal validity for nickel elimination in urine. When similar plots were made with data for which U_{Ni} varies over a much wider range than those studied by Sunderman et al. (1986a,b), it was found that these relationships were not necessarily linear (Sanford, 1988). This led Sanford et al. (1992b) to develop a more systematic and internally rigorous mathematical approach to this problem.

   Sanford et al. (1992b) collected 24-hr multivoid urines for 26 male volunteers who were occupationally exposed to water-soluble nickel salts at two Canadian electrolytic refining operations. Individual urine voids were collected directly into separate 500-mL polyethylene screw-top bottles. Nickel was determined by direct extraction into a 4-methylpentane-2-one (MIBK) layer and quantified by EAAS. A single-channel Technicon Auto Analyzer employing the Jaffé reaction was used to measure urinary creatinine, while a hand-held temperature-compensated refractometer was used to measure specific gravity (ρ). Evaluation of plots of log V versus log U_{Ni}, log U_{creat}, or log $U_{\rho-1}$ led to Equation (2), which was originally proposed by Araki (1980):

$$U_i V^{b_i} = U^\circ_i \tag{2}$$

Here U_i represents the urinary concentration of i (e.g., nickel, creatinine, or total solutes measured as $\rho - 1$, V is the urine flow rate in milliliters per minute, U°_i the urinary concentration of i at a flow rate of 1 mL/min, and b_i the volume exponent or power coefficient. Equation (2) was employed to derive Equations (3) and (4), which have general validity for urine spot samples:

$$\frac{U_1}{U_2} = \frac{U_1^\circ}{U_2^\circ} V^{\Delta b} \tag{3}$$

$$\frac{U_1}{U_2} = \frac{U_1^\circ}{U_2^\circ} V^{\overline{\Delta b} \pm z\text{SD}} = \frac{U_1^\circ}{U_2^\circ} V^{\Delta b'} \tag{4}$$

In these equations U_1 represents the urinary nickel concentration; U_2 denotes either urinary total solutes ($U_{\rho-1}$) or creatinine concentrations (U_{creat}), with U_1° and U_2° the corresponding values at $V = 1$ mL/min; Δb denotes the difference between b_2 and b_1, with, for example, $b_2 = b_{\text{creat}}$ or $b_{\rho-1}$ and $b_1 = b_{\text{Ni}}$; $\overline{\Delta b}$ is the average value of Δb for a group of individuals; SD the standard deviation for Δb; and z the z-statistic. If the use of the z-statistic is valid, then Equation (4) applies to spot urine specimens for individuals. It can be modified to include the confidence limits for Δb based on the t-statistic distribution for use with groups of individuals:

$$\frac{U_1}{U_2} = \frac{U_1^\circ}{U_2^\circ} V^{\overline{\Delta b} \pm t(\text{SD}/\sqrt{n})} = \frac{U_1^\circ}{U_2^\circ} V^{\Delta b''} \tag{5}$$

In Equations (4) and (5), $V^{\Delta b'}$ or $V^{\Delta b''}$ denotes the uncompensated residual flow factor. Effectively, the normalization procedure that is recommended ensures that urinary data are referred to a urine flow rate of 1 mL/min. The uncompensated residual flow factor, $V^{\Delta b}$ in Equation (3), has a value of unity when $\Delta b = 0$ or $V = 1$. The first condition may also be considered a reference state corresponding to $V = 1$ mL/min. It is clear from Equation (3) that if $\Delta b \rightarrow 0$, normalization becomes independent of the urine flow rate V. This emphasizes the need for selecting the appropriate reference metabolite.

Sanford et al. (1992b) conclude from their study of electrolytic nickel refinery workers that specific gravity adjustment of urinary nickel concentrations in spot samples, when combined with appropriate flow rate restrictions (expressed in terms of specific gravity range limits), provides the most suitable estimate of $U_{\text{Ni}}^\circ/U_{\rho-1}^\circ$ at a specified error and confidence interval. Specific gravity is preferred over creatinine adjustment because the corresponding $\overline{\Delta b}$ values are 0.004 ± 0.24 and 0.31 ± 0.20, respectively (Sanford et al., 1992b). Compilations of boundary values for U_{creat} and $U_{\rho-1}$ corresponding to acceptable error levels at the 95% confidence limit are provided by these authors. For example, if the specific gravity values of spot urine voids in a group of 20 individuals are between 1.010 and 1.039, then the uncompensated residual flow component of specific gravity–adjusted urinary nickel concentrations does not deviate from unity by more than ±10% at the 95% confidence level. Similarly, if the specific gravity of a spot urine sample of an individual donor falls between 1.013 and 1.032, then the uncompensated component does not deviate from unity by more than $\pm 30\%$ at the 95% confidence level (i.e., $z = 2$).

The question of whether to normalize urinary nickel levels, and if so, with what, should not be left to the discretion of the individual investigator. Sanford et al. (1992b) illustrate the need for a more rigorous approach. Because the approach taken by these investigators has general validity, it provides a suitable strategy for both routine applications in clinical chemistry laboratories and biological monitoring of other metals and organic compounds. In summary, the choice of an endogenous substance for normalization is essentially determined by the need to match the renal excretion pattern (i.e., its dependence on urinary flow rate) of the xenobiotic with that of the reference metabolite.

6. EFFECTS OF ENVIRONMENTAL AND PATHOLOGICAL FACTORS ON NICKEL IN BODY FLUIDS

As indicated in Section 2.2, elevated urinary and blood nickel concentrations occur in individuals occupationally exposed to nickel and its compounds (also see Chapter 11). The data in Figures 4.3 and 4.4 illustrate that the

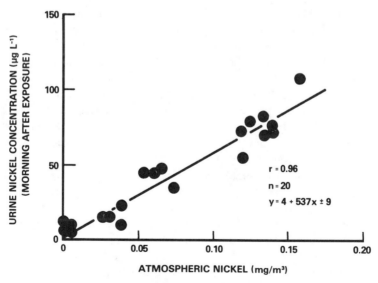

Figure 4.3. Scatter diagram showing the linear relationship between ambient nickel levels and urinary nickel concentrations of four electroplaters exposed to solutions of NiSO$_4$ and NiCl$_2$ and one unexposed employee monitored daily for a week. Urine specimens were collected in the morning of the following day before work at 0700 hr. Atmospheric nickel levels were measured by personal samplers utilizing a cellulose ester filter, Millipore, 0.8 mm, at a flow rate of 2.0 L/min. [Reproduced with permission from Nieboer et al., 1984; based on the data of Tola et al. (1979).]

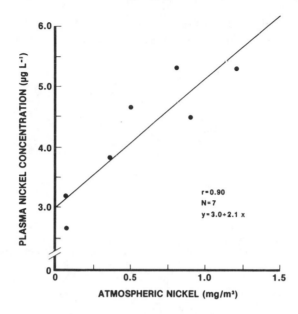

Figure 4.4. Scatter diagram showing the linear relationship between ambient nickel levels and plasma nickel concentrations for seven individuals working in a Ni_3S_2 matte crushing department. [Reproduced with permission from Nieboer et al. (1984); based on data provided by Andersen, Falconbridge Nikkelverk A/S (Andersen, 1983).]

absorption of nickel is dependent on the chemical composition (and no doubt the particle size) of the nickel-containing substance to which exposure occurs. This is demonstrated further by the calculations summarized in Table 4.1. It provides concentrations of nickel in plasma (serum) and urine corresponding to exposure at the threshold limit values (TLVs) enforced or recommended by a number of U.S. agencies. The conversion factors employed are based on observed linear relationships between ambient and body fluid nickel levels (Nieboer et al., 1984; Sunderman et al., 1986a).

Environmental exposure is also known to influence the nickel present in serum or urine (McNeely et al., 1972; Hopfer et al., 1989). Both of the referenced studies involved hospital workers in Sudbury, Ontario, Canada, where extensive mining and refining of nickel occurs. The control groups were similar hospital employees in Hartford, Connecticut, USA. The most recent data (Hopfer et al., 1989) show a reduced enhancement compared to the earlier work. In part, this can be explained in terms of "improved techniques of specimen collection and enhanced sensitivity and specificity of nickel analysis." Hopfer et al. (1989) summarize their data as follows: "Serum nickel concentration in Sudbury residents averaged 0.6 ± 0.3 µg/L (range 0.2–1.3; $N = 22$), which was significantly greater than the corresponding mean of 0.2 ± 0.2 µg/L (range <0.05 to 1.0) in Hartford, USA, residents ($p < 0.05$ by Mann–Whitney U-Test; $N = 43$)." By comparison, "nickel concentrations

Table 4.1   Body Fluid Nickel Concentrations Expected for OSHA Threshold Limit Values and NIOSH and ACGIH Guidelines[a]

Type of Exposure[b]	Agency[c]	TLV or Guideline (mg/m^3)	Urine[d] $(\mu g/L)$	Plasma (P) or Serum[e] $(\mu g/L)$
Soluble	OSHA	1.0	700	62
Insoluble	OSHA	1.0	87[f]	14[f]
Nickel carbonyl	OSHA	0.0024(1 ppb)	6	2
Soluble	NIOSH	0.015	13	2.9
Insoluble	NIOSH	0.015	3.3[f]	2.2[f]
Soluble	ACGIH	0.10	72	8
Insoluble	ACGIH	1.0	87[f]	14[f]
Nickel carbonyl	ACGIH	0.12 (50 ppb)	206	22

[a] Reproduced from Nieboer et al. (1984). The reader is referred to this document and Sunderman et al. (1986a) for details about the observed relationships between body fluid nickel and ambient nickel levels.

[b] In the soluble exposure category workers were exposed largely to water-soluble salts; exposures to insoluble nickel included significant exposures to sulfides or oxides of nickel, or powdered metal (Doll, 1990).

[c] OSHA, TLV set by U.S. Occupational Safety and Health Administration; NIOSH, TLV recommended by the U.S. National Institute of Occupational Safety and Health; ACGIH, TLV recommended by the American Conference for Governmental Industrial Hygienists.

[d] Calculated employing the relationships $U_{Ni} = 700$ Ni_{air}(soluble) $+ 2$ and $U_{Ni} = 85$ Ni_{air}(insoluble) $+ 2$. These generalized relationships were formulated after a detailed and critical appraisal of the published literature (Nieboer et al., 1984). For $Ni(CO)_4$ the relationship, although very approximate, was taken as $U_{Ni} = 1700$ $Ni_{air} + 2$.

[e] Calculated employing the relationships $P_{Ni} = 60$ Ni_{air}(soluble) $+ 2$ and $P_{Ni} = 12$ Ni_{air}(insoluble) $+ 2$. These generalized relationships were formulated after a detailed and critical appraisal of the published literature (Nieboer et al., 1984). For $Ni(CO)_4$ the relationship, although very approximate, was taken as $P_{Ni} = 170$ $Ni_{air} + 2$.

[f] Excludes workers exposed to Ni_3S_2 matte during crushing and those exposed to nickel hydroxide. Expected body fluid nickel concentrations on the basis of the available data are expected to be lower for the former and higher for the latter than the values recorded in this table.

in *random* urine specimens from the Sudbury residents averaged 2.4 ± 2.1 $\mu g/g_{creat}$ (range 0.3 to 7.6), which was slightly (but not significantly) greater than the corresponding mean of 1.5 ± 1.5 $\mu g/g_{creat}$ (range <0.5 to 4.0) in Hartford residents." Interestingly, the nickel levels in tap water were 0.4 ± 0.2 $\mu g/L$ in Hartford, Connecticut, compared to 109 ± 46 $\mu g/L$ in Sudbury ($p < 0.01$; $N = 5$).

Hypernickelemia occurs in a number of diseased states or associated treatments. Two main areas of concern are its occurence in patients with myocardial infarction (MI) and those requiring hemodialysis. The earliest report of pathophysiological fluctuations in serum nickel of patients with MI may now be considered unreliable, since the serum levels of nickel observed in control patients were extremely high (0.5–2 mg/L; D'Alonzo et al., 1963). Presumably, sample contamination was the source of the error (Cecutti and Nieboer, 1981). Nevertheless, Sunderman and colleagues have repeatedly confirmed the original conclusion by D'Alonzo et al. (1963) (Sunderman et al., 1970; McNeely et al., 1972; Sunderman, 1977; Leach et al., 1985). In the most recent of these studies (Leach et al., 1985), hypernickelemia (>1.2 μg/L) developed in three-fourths of patients with acute MI and about one-half of those with unstable angina, with peak nickel concentrations at 12–48 h after admission of 3.0 ± 3.4 μg/L ($N = 54$) and 1.4 ± 0.9 μg/L ($N = 33$), respectively. The origin of the nickel is not known, although hypernickelemia is also seen in patients with cerebral stroke and thermal burns (McNeely et al., 1971). Another recent study linking posttraumatic myocardial damage to endogenous nickel release must be discounted since it reports normal serum nickel levels between 40 and 80 μg/L (Szabo et al., 1985), corresponding to at least 100–300 times the currently accepted level. Values up to 2105 μg/L are reported after injury in patients without occupational exposure or an iatrogenic nickel source. A careful reassessment of this work is important as there is genuine concern about the potential cardiotoxicity of nickel (Nieboer et al., 1988a).

Serum nickel concentrations are consistently elevated in hemodialysis patients. Concentrations of 7.0 ± 2.4 μg/L ($N = 30$) and 6.2 ± 1.8 μg/L ($N = 42$) have been reported for patients in Sudbury, Ontario, and Hartford, Connecticut, respectively (Hopfer et al., 1989). Comparable values of 7.4 μg/L ($N = 16$) have been observed by Drazniowsky et al. (1985) and 6.4 ± 3.4 μg/L ($N = 27$) by Nixon et al. (1989). Nickel toxicity apparently developed in hemodialysis patients when the dialysis fluid became severely contaminated with nickel leached from a nickel-plated water heater (Webster et al., 1980). The contaminated dialysate was found to contain 250 μg Ni/L, compared to ≤1 μg/L for properly functioning dialysis units (Hopfer et al., 1989; Nixon et al., 1989). Interestingly, the clinical symptoms of the hemodialysis patients treated with the contaminated dialysis fluid resembled those observed for the electroplating workers who accidently drank nickel sulfate (see Section 2.2), namely, nausea, vomiting, weakness, headache, and palpitations. However, quality assurance and contamination control measures are not mentioned in the study by Webster et al. (1980), and this weakens the postulated link between the clinical symptoms and the measured serum nickel levels.

The absence of suitable contamination control and quality assurance measures (see Nieboer and Jusys, 1983; Nixon et al., 1989), and perhaps method insensitivity, also seem to confound reports on the transient hypernickelemia in pregnant women following delivery (Anonymous, 1984). In the crucial study by Rubányi et al. (1982), nonpregnant, healthy control subjects (hospital employees) are reported to have serum nickel levels of 122 ± 25 μg/L, which exceed the currently accepted value by at least 300-fold. Consequently, it is difficult to interpret the 15-fold increase observed during the immediate postpartum period.

A number of additional causes of hypernickelemia have been reported. Linden et al. (1985) examined patients with stainless steel hip prostheses of the metal-to-plastics types and conclude that elevation of serum nickel does not develop in the "apparent absence of corrosion, local complications, or systemic conditions, such as renal insufficiency." On careful scrutiny of the published data, Linden et al. (1985) conclude that "hypernickelemia in patients with metal prostheses appears to be indicative of corrosion, allergic, or inflammatory complications, or the presence of contributory systemic conditions such as renal insufficiency." In addition, analytical limitations are also evident in some of the studies reviewed.

Hopfer et al. (1987) have observed in patients with chronic alcoholism that nickel concentrations in serum, whole blood, and urine were significantly increased (17- to 39-fold) after the administration of the drug disulfiram (Antabuse): "In serum and whole blood, the Ni concentrations reached a plateau after two weeks of treatment; in urine the Ni concentrations increased progresively during the initial four months of treatment." These findings were interpreted to indicate enhanced intestinal absorption of dietary nickel. The ability of disulfiram and its metabolite diethyldithiocarbamate (DDC) to induce hypernickelemia and nickeluresis is discussed further in Chapters 20 and 22 in relation to their use in the chelation therapy of nickel carbonyl poisoning.

Sunderman (1983) has illustrated the potential of including hypernickelemia in conjunction with the administration of contaminated intravenous fluids. For example, nickel levels around 200 μg/L have been reported in human serum albumin solutions (Leach and Sunderman, 1985). Sunderman (1983) recommends "that the maximum permissible level of nickel be set at 5 μg L^{-1} for common i.v. solutions, and 10 μg L^{-1} for solutions that contain albumin or amino acids (e.g., histidine) which avidly bind Ni^{2+}.

REFERENCES

Andersen, I. (1983). Falconbridge Nikkelverk, Kristiansand, S., Norway. Personal communication.

Andersen, I., and Svenes, K. B. (1989). Determination of nickel in lung specimens of thirty-nine autopsied nickel workers. *Int. Arch. Environ. Health* **61**, 289–295.

Anonymous (1984). Transient hypernickelemia following delivery. *Nutrition Rev.* **42**, 158–159.

Araki, S. (1980). Effects of urinary volume on urinary concentrations of lead, δ-aminolaevulinic acid, coproporhyrin, creatinine, and total solutes. *Br. J. Ind. Med.* **37**, 50–54.

Asato, N., van Soestbergen, M., and Sunderman, F. W., Jr. (1975). Binding of $^{63}Ni(II)$ to ultrafilterable constituents of rabbit serum *in vivo* and *in vitro*. *Clin. Chem.* **21**, 521–527.

Barrie, L. A. (1981). Atmospheric nickel in Canada. In *Effects of Nickel in the Canadian Environment*. NRCC Document No. 18568. National Research Council of Canada, Ottawa, pp. 55–76.

Bennett, B. G. (1984). Environmental nickel pathways to man. In F. W. Sunderman Jr. (Ed.-in-Chief), *Nickel in the Human Environment*. IARC Scientific Pub. No. 53. Oxford University Press, Oxford, pp. 487–495.

Cecutti, A., and Nieboer, E. (1981). Nickel metabolism and biochemistry and effects of nickel on animals and humans. In *Effects of Nickel in the Canadian Environment*. NRCC Document No. 18568. National Research Council of Canada, Ottawa, pp. 193–260.

Clary, J. J. (1975). Nickel chloride–induced metabolic changes in the rat and guinea pig. *Toxicol. Appl. Pharmacol.* **31**, 55–65.

D'Alonzo, C. A., Pell, S., and Fleming, A. J. (1963). The role and potential role of trace metals in disease. *J. Occup. Med.* **5**, 71–79.

Doll, R., Editor-in-Chief (1990). Report of the International Committee on Nickel Carcinogenesis in Man. *Scand. J. Work Environ. Health* **16**, 1–82.

Drazniowsky, M., Parkinson, I. S., Ward, M. K., Channon, S. M., and Kerr, D.N.S. (1985). A method for the determination of nickel in water and serum by flameless atomic absorption spectrophotometry. *Clin. Chim. Acta* **145**, 219–226.

Duarte, C. G., Elveback, L. R., and Liedtke, R. R. (1980). Creatinine. In C. G. Duarte (Ed.), *Renal Function Tests*. Little, Brown, Boston, pp. 1–28.

Environmental Protection Agency (EPA) (1986). *Health Assessment Document for Nickel and Nickel Compounds*. EPA/600/8-83/012 FF. U.S. Environmental Protection Agency, Research Triangle Park, NC.

Gammelgaard, B., and Andersen, J. R. (1985). Determination of nickel in human nails by adsorption differential-pulse voltammetry. *Analyst* **110**, 1197–1199.

Green, J. H. (1976). *An Introduction to Human Physiology*, 4th ed. Oxford University Press, New York, pp. 134–147.

Hendel, R. C., and Sunderman, F. W., Jr. (1972). Species variations in the proportions of ultrafilterable and protein-bound serum nickel. *Res. Commun. Chem. Pathol. Pharmacol.* **4**, 141–146.

Hopfer, S. M., Linden, J. V., Rezuke, W. N., O'Brien, J. E., Smith, L., Walters, F., and Sunderman, F. W., Jr. (1987). Increased nickel concentrations in body fluids of patients with chronic alcoholism during disulfiram therapy. *Res. Commun. Chem. Pathol. Pharmacol.* **55**, 101–109.

Hopfer, S. M., Fay, W. P. and Sunderman, F. W., Jr. (1989). Serum nickel concentrations in hemodialysis patients with environmental exposure. *Ann. Clin. Lab. Sci.* **19**, 161–167.

Ishihara, N., and Matsushiro, T. (1986). Biliary and urinary excretion of metals in humans. *Arch. Environ. Health* **41**, 324–330.

Ishihara, N., Koizumi, M., and Yoshida, A. (1987). Metal concentrations in human pancreatic juice. *Arch. Environ. Health* **42**, 356–360.

Kollmeier, H., Witting, C., Seeman, J., Wittig, P., and Rothe, R. (1985). Increased chromium and nickel content in lung tissue. *J. Cancer Res. Clin. Oncol.* **110**, 173–176.

Lambie, A. T., and Harris, P. J. (1985). Kidney. In J. M. Forrester (Ed-in-Chief), *A Companion to Medical Studies*, Vol. 1, 3rd ed. Blackwell Scientific, London, pp. 31.1–31.33.

Leach, C. N., Jr., and Sunderman, F. W., Jr. (1985). Nickel contamination of human serum albumin. *N. Engl. J. Med.* **313**, 1232.

Leach, C. N., Jr., Linden, J. V., Hopfer, S. M., Crisostomo, M. C., and Sunderman, F. W., Jr. (1985). Nickel concentrations in serum of patients with acute myocardial infarction or unstable angina pectoris. *Clin. Chem.* **31**, 556–560.

Lim, T. H., Sargent, T., and Kusubov, N. (1983). Kinetics of trace element chromium(III) in the human body. *Am. J. Physiol.* **244** (*Regulatory Integrative Comp. Physiol.* **13**), R445–R454.

Linden, J. V., Hopfer, S. M., Gossling, H. R., and Sunderman, F. W., Jr. (1985). Blood nickel concentrations in patients with stainless-steel hip prostheses. *Ann. Clin. Lab. Sci.* **15**, 459–463.

Lucassen, M., and Sarkar, B. (1979). Nickel(II)-binding constituents of human blood serum. *J. Toxicol. Environ. Health* **5**, 897–905.

McNeely, M. D., Nechay, M. W., and Sunderman, F. W., Jr. (1972). Measurements of nickel in serum and urine as indices of environmental exposure to nickel. *Clin. Chem.* **18**, 992–995.

McNeely, M. D., Sunderman, F. W., Jr., Nechay, M. W., and Levine, H. (1971). Abnormal concentrations of nickel in serum in cases of myocardial infarction, stroke, burns, hepatic cirrhosis, and uremia. *Clin. Chem.* **17**, 1123–1128.

Maibach, H. I., and Menné, T. (Eds.) (1989). *Nickel and the Skin: Immunology and Toxicology.* CRC Press, Boca Raton, FL.

Marzouk, A., and Sunderman, F. W., Jr. (1985). Biliary excretion of nickel in rats. *Toxicol. Lett.* **27**, 65–71.

Mertz, D. P., Koschnick, R., and Wilk, G. (1974). Renale ausscheidungsbedingungen von nickel beim menschen. *Z. Klin Chem. U. Klin. Biochem.* **8**, 387–380.

Mushak, P. (1980). Metabolism and systemic toxicity of nickel. In J. O. Nriagu (Ed.), *Nickel in the Environment*. Wiley, New York, pp. 499–523.

Nieboer, E., and Jusys, A. A. (1983). Contamination control in routine ultratrace analysis of toxic metals. In S. S. Brown and J. Savory (Eds.), *Chemical Toxicology and Clinical Chemistry of Metals*. Academic, London, pp. 3–16.

Nieboer, E., and Jusys, A. A. (1988). Biological chemistry of chromium. In J. O. Nriagu and E. Nieboer (Eds.), *Chromium in the Natural and Human Environments*. Wiley, New York, pp. 21–79.

Nieboer, E., Yassi, A., Jusys, A. A., and Muir, D. C. (1984). *The Technical Feasibility and Usefulness of Biological Monitoring in the Nickel Producing Industry*. Special Document. McMaster University, Hamilton, Canada. (Available from the Nickel Producers Environmental Research Association, Durham, NC.)

Nieboer, E., Rossetto, F. E., and Menon, C. R. (1988a) Toxicology of nickel compounds. In H. Sigel and A. Sigel (Eds.), *Nickel and Its Role in Biology, Metal Ions in Biological Systems*. Vol. 23, Marcel Dekker, New York, pp. 359–402.

Nieboer, E., Tom, R. T., and Sanford, W. E. (1988b). Nickel metabolism in man and animals. In H. Sigel and A. Sigel (Eds.), *Nickel and Its Role in Biology, Metal Ions in Biological Systems*. Vol. 23, Marcel Dekker, New York, pp. 91–121.

Nixon, D. E., Moyer, T. P., Squillace, D. P., and McCarthy, J. T. (1989). Determination of serum nickel by graphite furnace atomic absorption spectrometry with Zeeman-effect background correction: values in normal population and a population undergoing dialysis. *Analyst* **114**, 1671–1674.

Nomoto, S. (1974). Determination and pathophysiological study of nickel in humans and animals. II. Measurement of nickel in human tissues by atomic absorption spectrometry. *Shinshu Igaku Zasshi* **22**, 39–44.

Nomoto, S. (1980). Fractionation and quatitative determination of alpha-2-macroglobulin-combined nickel in serum by affinity column chromatography. In S. S. Brown and F. W. Sunderman, Jr. (Eds.), *Nickel Toxicology*. Academic, London, pp. 89–90.

Nomoto, S., and Sunderman, F. W., Jr. (1989). Presence of nickel in alpha-2-macroglobulin isolated from human serum by high performance liquid chromatography. *Ann. Clin. Lab. Sci.* **18**, 78–84.

Nomoto, S., McNeeley, M. D., and Sunderman, F. W., Jr. (1971). Isolation of a nickel α_2-macroglobulin from rabbit serum. *Biochemistry* **10**, 1647–1651.

O'Dell, G. D., Miller, W. J., Moore, S. L., King, W. A., Ellers, J. C., and Jurecek, H. (1971). Effect of dietary nickel level on excretion and nickel content of tissues in male calves. *J. Anim. Sci.* **32**, 769–773.

Onkelinx, C., and Sunderman, F. W., Jr. (1980). Modelling of nickel metabolism. In J. O. Nriagu (Ed.), *Nickel in the Environment*. Wiley, New York, pp. 525–545.

Onkelinx, C., Becker, J., and Sunderman, F. W., Jr. (1973). Compartmental analysis of the metabolism of ^{63}Ni(II) in rats and rabbits. *Res. Commun. Chem. Pathol. Pharmacol.* **6**, 663–676.

Parker, K., and Sunderman, F. W., Jr. (1974). Distribution of ^{63}Ni in rabbit tissues following intravenous injection of ^{63}NiCl$_2$. *Res. Commun. Chem. Pharmacol.* **7**, 755–762.

Pennington, J. A., and Jones, J. W. (1989). Molybdenum, nickel, cobalt, vanadium and strontium in total diets. *J. Am. Dietet. Assoc.* **87**, 1644–1650.

Raithel, H. J., Ebner, G., Schaller, K. H., Schellmann, B., and Valentin, H. (1987). *Am. J. Ind. Med.* **12**, 55–70.

Rezuke, W. N., Knight, J. A., and Sunderman, F. W., Jr. (1987). Reference values for nickel concentrations in human tissues and bile. *Am. J. Ind. Med.* **11**, 419–426.

Rubányi, G., Birtalan, I., Gergely, A., and Kovách, A.G.B. (1982). Serum nickel concentration in women during pregnancy, during parturition, and post partum. *Am. J. Obstet. Gynecol.* **143**, 167–169.

Sanford, W. E. (1988). The renal clearance of nickel in man: implications for biological monitoring. Ph.D. thesis. University of Surrey, United Kingdom.

Sanford, W. E., Stace, B. C., and Nieboer, E. (1992a). Renal clearance of nickel and biochemical indices of kidney function in electrolytic refinery workers. *Scand. J. Work Environ. Health*, submitted.

Sanford, W. E., Stace, B. C., and Nieboer, E. (1992b). Illustration for nickel of a systematic approach to urinary dilution corrections. *Scand. J. Work Environ. Health*, submitted.

Schroeder, H. A., Balassa, J. J., and Tipton, I. H. (1962). Abnormal trace metals in man— nickel. *J. Chronic Dis.* **15**, 51–65.

Schuster, V. L., and Seldin, D. W. (1985). Renal clearance. In D. W. Seldin and G. Giebisch (Eds.), *The Kidney: Physiology and Pathophysiology*. Raven, New York, pp. 365–395.

Scott, B. J., and Bradwell, A. R. (1983). Identification of the serum binding proteins for iron, zinc, cadmium, nickel and calcium. *Clin. Chem.* **29**, 629–633.

Seemann, J., Wittig, P., Kollmeier, H., and Rothe, G. (1985). Analytical measurements of Cd, Pb, Zn, Cr and Ni in human tissues. *Lab. Med.* **9**, 294–299.

Smith, H. W. (1964). *The Kidney*. Oxford University Press, New York.

Sumino, K., Hayakawa, K., Shibata, T., and Kitamura, S. (1975). Heavy metals in normal Japanese tissues. *Arch. Environ. Health* **30**, 487–494.

Sunderman, F. W., Jr. (1977). A review of the metabolism and toxicology of nickel. *Ann. Clin. Lab. Sci.* **7**, 377–398.

Sunderman, F. W., Jr. (1983). Potential toxicity from nickel contamination of intravenous fluids. *Ann. Clin. Lab. Sci.* **13**, 1–4.

Sunderman, F. W., Jr., Crisostomo, M. C., Reid, M. C., Hopfer, S. M., and Nomoto, S. (1984). Rapid analysis of nickel in serum and whole blood by electrothermal atomic absorption spectrophotometry. *Ann. Clin. Lab. Sci.* **14**, 232–241.

Sunderman, F. W., Jr., Decsy, M. I., and McNeely, M. D. (1972). Nickel metabolism in health and disease. *Ann. N.Y. Acad. Sci.* **199**, 300–312.

Sunderman, F. W., Jr., Aitio, A., Morgan, L. G., and Norseth, T. (1986a). Biological monitoring of nickel. *Toxicol. Ind. Health* **2**, 17–78.

Sunderman, F. W., Jr., Hopfer, S. M., Crisostomo, M. C., and Stoeppler, M. (1986b). Rapid analysis of nickel in urine by electrothermal atomic absorption spectrophotometry. *Ann. Clin. Lab. Sci.* **16**, 219–230.

Sunderman, F. W., Jr., Dingle, B., Hopfer, S. M., and Swift, T. (1988). Acute nickel toxicity in electroplating workers who accidently ingested a solution of nickel sulfate and nickel chloride. *Am. J. Ind. Med.* **14,** 257–266.

Sunderman, F. W., Jr., Hopfer, S. M., Sweeney, K. R., Marcus, A. H., Most, B. M., and Creason, J. (1989). Nickel absorption and kinetics in human volunteers. *Proc. Soc. Exp. Biol. Med.* **191,** 5–11.

Sunderman, F. W., Jr., Nomoto, S., Pradhan, A. M., Levine, H., Bernstein, S. H., and Hirsch, R. (1970). Increased concentrations of serum nickel after acute myocardial infarction. *N. Engl. J. Med.* **283,** 896–899.

Szabo, K., Balogh, I., and Gergely, A. (1985). Endogenous nickel release in injured patients: a possible cause of myocardial damage. *Br. J. Accident Surg.* **16,** 613–620.

Tipton, I. H., and Cook, M. J. (1963). Trace elements in human tissue. Part II. Adult subjects from the United States. *Health Phys.* **9,** 103–145.

Tipton, I. H., Schroeder, H. A., Perry, H. M., Jr., and Cook, M. J. (1965). Trace elements in human tissue. Part III. Subjects from Africa, the Near and Far East and Europe. *Health Phys.* **11,** 403–451.

Tola, S., Kilpio, J., and Virtamo, M. (1979). Urinary and plasma concentrations of nickel as indicators of exposure to nickel in an electroplating shop. *J. Occup. Med.* **21,** 184–188.

Valerio, F., Brescianini, C., and Lastraioli, S. (1989). Airborne metals in urban areas. *Int. J. Environ. Anal. Chem.* **35,** 101–110.

Webster, J. D., Parker, T. F., Alfrey, A. C., Smythe, W. R., Kubo, H., Neal, G., and Hull, A. R. (1980). Acute nickel intoxication by dialysis. *Ann. Int. Med.* **92,** 631–633.

5

TOXICOKINETICS OF NICKEL IN HUMANS

F. William Sunderman, Jr.

*Departments of Laboratory Medicine and Pharmacology,
University of Connecticut School of Medicine,
Farmington, Connecticut 06030*

1. NICKEL KINETICS IN HUMANS EXPOSED TO INHALATION OF NICKEL COMPOUNDS

Sunderman et al. (1986) reviewed the reported studies of nickel elimination kinetics in industrial workers occupationally exposed to inhalation of nickel-containing compounds or alloys. Most investigators have assumed first-order kinetics for clearance of nickel from plasma and excretion of nickel in urine. In electroplating workers exposed to inhalation of soluble nickel compounds, Tossavainen et al. (1980) found that the half-time for reduction of plasma nickel concentrations was 20–34 hr and the half-time for elimination of nickel in urine was 17–39 hr. In mold makers in the glass industry exposed to inhalation of insoluble nickel compounds, Raithel et al. (1982) reported that

Nickel and Human Health: Current Perspectives, Edited by Evert Nieboer and
Jerome O. Nriagu.
ISBN 0-471-50076-3 © 1992 John Wiley & Sons, Inc.

the half-time for urinary elimination of nickel was 30–50 hr. The half-time for urinary elimination of nickel compounds inhaled by welders was estimated by Zober et al. (1984) to average 53 hr. These reports suggest relatively short biological half-times for nickel in exposed workers. They do not exclude protracted retention of some nickel in storage deposits, from which nickel could be slowly released into plasma and excreted in urine, as suggested by the reports of (a) persistent elevations of urine nickel concentrations in nickel refinery workers after plant closure for six months (Morgan and Rouge, 1984), (b) elevated nickel concentrations in urine specimens from welders of high-nickel alloys after four weeks of vacation (Akesson and Skerfving, 1985), and (c) elevated nickel concentrations in specimens of urine and plasma and in nasal mucosal biopsies obtained from retired workers years after cessation of employment in a nickel refinery (Torjussen and Andersen, 1979; Boysen et al., 1984). In regard to inhalation exposures of workers to nickel-containing dusts, scant attention has been given to the influence of particle size in determining the sites of nickel deposition and the kinetics of its elimination from the respiratory tract or the possible nonlinearity of nickel elimination kinetics at high levels of nickel exposure, owing to impaired macrophage function and inhibited ciliary activity (Sunderman et al., 1986).

2. NICKEL KINETICS IN HUMANS EXPOSED TO ORAL INGESTION OF NICKEL COMPOUNDS

Several groups of dermatologists have investigated the gastrointestinal absorption of nickel in healthy human volunteers and patients with nickel dermatitis who ingested capsules containing 0.6–6 mg of nickel as nickel sulfate (Spruit and Bongaarts, 1977; Kaaber et al., 1978; Cronin et al., 1980; Christensen and Lagesson, 1981; Gawkrodger et al., 1986). These studies showed that nickel concentrations in serum or plasma generally peak at 2.5–3 hr posttreatment and diminish to baseline by 72 hr and that elimination of nickel in urine is greatest during the initial 12 hr and gradually returns almost to baseline within 72 hr posttreatment. Christensen and Lagesson (1981) reported that the half-time for reduction of plasma nickel concentrations in volunteers after oral intake of nickel sulfate was approximately 11 hr.

Solomons et al. (1982) showed that fasting human volunteers developed marked hypernickelemia within 4 hr after ingesting 5 mg of nickel as an aqueous solution of nickel sulfate, whereas significant hypernickelemia did not occur when the same quantity of nickel sulfate was added to standard meals, suggesting that food has a marked effect on nickel absorption from the gastrointestinal tract. Sunderman et al. (1988a) studied nickel elimination kinetics in nickel electroplating workers who accidentally drank water heavily contaminated with nickel sulfate and chloride. The nickel intake in 10 workers with acute symptoms was estimated to range from 0.5 to 2.5 g. Based on serial analyses of serum nickel concentrations in these subjects, who were hospitalized and treated for three days with intravenous fluids to induce diuresis, the mean elimination half-time for serum nickel averaged 27 hours (SD ± 7 hr).

3. COMPARTMENTAL ANALYSIS OF NICKEL KINETICS IN HUMAN VOLUNTEERS

Sunderman et al. (1989) performed compartmental analysis of nickel absorption, distribution, and elimination kinetics for the first time in human subjects. The investigation was based on two experiments in which 10 healthy human volunteers ingested nickel sulfate in drinking water (experiment 1) or added to food (experiment 2). Nickel was analyzed by electrothermal atomic absorption spectrophotometry (Sunderman et al., 1988b) in serial specimens of serum, urine, and feces collected during two days before and four days after each subject received an oral nickel dose (12, 18, or 50 μg Ni/kg body weight). In experiment 1, each subject fasted 12 hr before and 3 hr after drinking the dose of nickel sulfate dissolved in water; in experiment 2, the respective subjects fasted 12 hr before consuming a standard American breakfast that contained the identical dose of nickel sulfate added to scrambled eggs.

Data for nickel concentrations in serum and urine were analyzed by a linear, compartmental, toxicokinetic model (see Fig. 5.1), adapted from an earlier model of nickel kinetics in rats and rabbits (Onkelinx et al., 1973; Onkelinx and Sunderman, 1980). The model included two inputs of nickel: (a) the single oral dose of $NiSO_4$, administered in water or food at 7 a.m. on the third day of each experiment and (b) the baseline dietary ingestion of nickel. For each subject and each experiment, the following kinetic parameters were estimated: (1) a first-order rate constant (k_{01}) for intestinal absorption of nickel from the oral dose of $NiSO_4$; (2) a pseudo-zero-order

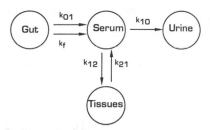

Figure 5.1. Schematic diagram of compartmental model of nickel metabolism [k_f, zero-order rate constant for fractional absorption of dietary nickel; k_{01}, first-order rate constant for intestinal absorption of nickel from oral dose of $NiSO_4$; k_{12}, first-order rate constant for nickel transfer from serum (compartment 1) to tissues (compartment 2); k_{21}, first-order rate constant for nickel transfer from tissues to serum; k_{10}, first-order rate constant for nickel excretion in urine] (Sunderman et al., 1989).

rate constant (k_f) for fractional absorption of dietary nickel, based upon analyses of urine and feces during the control (predose) periods; (3) a first-order rate constant (k_{10}) for urinary elimination of nickel; (4) two first-order rate constants (k_{12} and k_{21}) for transfer of nickel between the compartments; and (5) the mass fraction (F) of nickel absorbed from the oral dose of NiSO$_4$. The mathematical model was based upon three simultaneous equations as follows:

$$C = \frac{FDk_{01}}{V_1} [A_1 e^{-L_1 t} + A_2 e^{-L_2 t} + A_3 e^{-k_{01} t}] + C_0 \tag{1}$$

$$\text{AUC}(t) = \frac{FDk_{01}}{V_1} \left[\frac{A_1}{L_1} (1 - e^{-L_1 t}) + \frac{A_2}{L_2} \right. \tag{2}$$

$$\left. (1 - e^{-L_2 t}) + \frac{A_3}{k_{01}} (1 - e^{-k_{01} t}) \right] + C_0 t$$

$$A_u(t) = k_{10} V_1 \, \text{AUC}(t) \tag{3}$$

where C = serum nickel concentration (µg/L) at time t after the oral dose of NiSO$_4$

D = dose of nickel (µg) ingested as NiSO$_4$ in water or food

t = time (hr) after the oral dose of NiSO$_4$

C_0 = baseline concentration of serum nickel [$= f(k_f)$] established during the predose control period

V_1 = plasma volume (L), derived from standard nomograms based on sex, height, and weight (Nadler et al., 1962)

A_1 = an algebraic constant derived for the first exponential term of Equation (1), $= (L_1 - k_{21})/(k_{01} - L_1)(L_1 - L_2)$

A_2 = an algebraic constant derived for the second exponential term of Equation (1), $= (k_{21} - L_2)/(k_{01} - L_2)(L_1 - L_2)$

A_3 = an algebraic constant derived for the third exponential term of Equation (1), $= -(A_1 + A_2)$

L_1, L_2 = hybrid rate constants, which are functions of the intercompartmental rate constants (k_{10}, k_{12}, k_{21})

$\text{AUC}(t)$ = cumulative area under the serum nickel concentration curve to time t

$A_u(t)$ = cumulative urinary elimination of nickel (µg) from to time t

Computations were performed on a Compaq Desk-Pro computer using an iterative, least-squares program, PCNONLIN, to fit the equations to the data (Metzler and Weiner, 1986) and a polyexponential curve-stripping program, CSTRIP, to obtain the initial parameters (Sedman and Wagner, 1976). Goodness of fit was assessed by the Spearman correlation coefficients r for calculated versus observed values for serum nickel concentration and cumulative urinary excretion of nickel. Renal nickel clearance was calculated as the product of the urinary elimination rate constant (k_{10}) and the plasma volume. The elimination half-time ($T_{1/2}$) for absorbed nickel was calculated by dividing the natural logarithm of 2 by L_2 using observations when the mass fraction of absorbed nickel exceeded 1.2% of the administered dose.

The compartmental model developed by Sunderman et al. (1989) provided excellent goodness of fit for the paired data sets from both experiments. Correlation coefficients r for the calculated versus the observed data were 0.983 for serum nickel concentrations and 0.9996 for cumulative urinary eliminations of nickel. Parameters for nickel absorption, distribution, and elimination in the human volunteers are listed in Table 5.1. Absorbed nickel

Table 5.1   Parameters for Nickel Kinetics in Human Volunteers[a]

Parameters[b]	Experiment 1, $NiSO_4$ in water[c]	Experiment 2, $NiSO_4$ in food[c]	P[d]
Mass fraction of Ni dose absorbed from gastrointestinal tract (F, %)	27 ± 17	0.7 ± 0.4	<0.001
Rate constant for alimentary absorption of Ni from $NiSO_4$ dose (k_{01}, hr^{-1})	0.28 ± 0.11	0.33 ± 0.24	NS
Rate constant for alimentary absorption of dietary Ni intake (k_f, μg/hr)	0.092 ± 0.051	0.105 ± 0.036	NS
Rate constant for Ni transfer from compartment 1 to 2 (k_{12}, hr^{-1})	0.38 ± 0.17	0.37 ± 0.34	NS
Rate constant for Ni transfer from compartment 2 to 1 (k_{21}, hr^{-1})	0.08 ± 0.03	[e]	
Rate constant for urinary elimination of Ni (k_{10}, hr^{-1})	0.21 ± 0.05	0.15 ± 0.11	NS
Renal clearance of Ni (C_{Ni}, mL/min/1.73 m^2)	8.3 ± 2.0	5.8 ± 4.3	NS
Renal clearance of creatinine (C_{Creat}, mL/min/1.73 m^2)	97 ± 9	93 ± 15	NS
Ni clearance as percentage of creatinine clearance (C_{Ni}/C_{Creat}, ×100)	8.5 ± 1.8	6.3 ± 4.6	NS

[a] From Sunderman et al. (1989).
[b] With symbols and units.
[c] Data are listed as mean ± SD.
[d] P value computed by analysis of variance (ANOVA); NS, not significant.
[e] Indeterminate value, owing to the small mass of Ni absorbed from the alimentary tract and transferred from compartment 1 (serum) into compartment 2 (tissues).

averaged 27 ± 17% (mean ± SD) of the dose ingested in water versus 0.7 ± 0.4% of the same dose ingested in food (a 40-fold difference); rate constants for nickel absorption, transfer, and elimination were not significantly influenced by the oral vehicle; the elimination half-time for absorbed nickel averaged 28 ± 9 hr. Fecal elimination of nickel during four days posttreatment, corrected for each subject's baseline fecal elimination of nickel, averaged 76 ± 19% of the dose ingested in water versus 102 ± 20% of the dose ingested in food. Total recovery of the nickel dose in urine plus feces during four days post-treatment averaged 102 ± 8% in experiment 1, which did not differ significantly from the corresponding value of 104 ± 21% in experiment 2.

The subjects' renal clearances of nickel did not differ significantly after ingestion of $NiSO_4$ in water versus food (Table 5.1). The renal clearance of nickel did not correlate with the mass fraction of the nickel dose that was absorbed, but it did correlate directly with the renal clearance of creatinine ($r = 0.856$, $P < 0.01$). The equation of the regression line was $y = -4.76 + 0.134x$, where y is the nickel clearance and x is the creatinine clearance, expressed as milliliters per minute.

Sunderman et al. (1989) found that optimal fit of the mathematical model to the experimental data was obtained when baseline gastrointestinal uptake of nickel from the daily diet was treated as a pseudo-zero-order process (k_f) and uptake of the administered $NiSO_4$ dose was treated as a first-order process (k_{01}). This approach was justified by considering the pseudo-zero-order constant k_f to represent the steady-state summation of multiple first-order nickel inputs from foods and beverages consumed throughout the day. The model did not include parameters for possible biliary elimination and intestinal reabsorption of nickel, since the data gave no indications of an enterohepatic cycle, such as secondary peaks of nickel concentrations in serum or urine, and since biliary excretion of nickel has been shown to be quantitatively insignificant in rats (Marzouk and Sunderman, 1985). However, it is possible that biliary excretion of exogenous nickel could be significant in humans, since bile specimens obtained postmortem from nonexposed persons contained 2.3 ± 0.8 µg Ni/L (Rezuke et al., 1987).

The study of Sunderman et al. (1989) confirmed the finding of Solomons et al. (1982) that dietary constituents markedly reduce the bioavailability of Ni^{2+} for alimentary absorption: approximately one-quarter of nickel ingested in drinking water after an overnight fast was absorbed from the human intestine and excreted in urine, compared to only 1% of nickel ingested in food. The reduced absorption of nickel from food presumably reflects chelation or reduction of Ni^{2+} by dietary constituents, consistent with the findings of Solomons et al. (1982) that certain dietary constituents (e.g., ascorbic acid) and beverages, (e.g., milk, coffee, tea, orange juice) reduce the bioavailability of orally administered nickel in humans and the findings of Foulkes and McMullen (1986) that skimmed milk and Zn^{2+} reduce the absorption of nickel in isolated segments of perfused rat jejunum.

4. RELATIONSHIPS BETWEEN MATHEMATICAL MODELS OF NICKEL KINETICS IN HUMANS AND EXPERIMENTAL ANIMALS

As previously mentioned, a compartmental model of nickel kinetics in rats and rabbits, described by Onkelinx et al. (1973) and Onkelinx and Sunderman (1980), was adapted for study of human subjects by Sunderman et al. (1989). Toxicokinetic models of nickel distribution and elimination in rats after exposure to nickel compounds by intratracheal or inhalation routes have been reported by English et al. (1981), Medinsky et al. (1987), and Menzel et al. (1987). Menzel (1988) proposed several ways in which a physiologically based model of nickel kinetics in rats could be used to plan animal experiments involving nickel exposures from oral ingestion or dermal contact and hypothetically to extrapolate nickel kinetics from animals to man. Sunderman et al. (1989) discussed possible applications of their mathematical model of nickel kinetics in humans to describe the absorption, tissue distribution, and elimination of nickel compounds in experimental animals. Such comparisons of nickel kinetics in animals with those in humans could reduce the uncertainties of toxicological risk assessments of nickel exposures that rely upon interspecies extrapolations.

REFERENCES

Akesson, B., and Skerfving, S. (1985). Exposure in welding of high nickel alloy. *Int. Arch. Occup. Environ. Health* **56**, 111–117.

Boysen, M., Solberg, L. A. Torjussen, W., Poppe, S., and Høgetveit, A. C. (1984). Histological changes, rhinoscopical findings and nickel concentrations in plasma and urine in retired nickel workers. *Acta Otolaryngol.* **97**, 105–115.

Christensen, O. B., and Lagesson, V. (1981). Nickel concentrations in blood and urine after oral administration. *Ann. Clin. Lab. Sci.* **11**, 119–125.

Cronin, E., Di Michiel, A., and Brown, S. S. (1980). Oral challenge in nickel-sensitive women with hand eczema. In S. S. Brown and F. W. Sunderman, Jr. (Eds.), *Nickel Toxicology*. Academic, London, pp. 149–152.

English, J. C., Parker, R.D.R., Sharma, R. P., and Oberg, S. G. (1981). Toxicokinetics of nickel in rats after intratracheal administration of a soluble and insoluble form. *Am. Ind. Hyg. Assoc. J.* **42**, 486–492.

Foulkes, E. C., and McMullen, D. M. (1986). On the mechanism of nickel absorption in the rat jejunum. *Toxicology* **38**, 35–42.

Gawkrodger, D. A., Cook, S. W., Fell, G. S., and Hunter, J.A.A. (1986). Nickel dermatitis: the reaction to oral nickel challenge. *Br. J. Dermatol.* **115**, 33–38.

Kaaber, K., Veien, N. K., and Tjell, J. C. (1978). Low nickel diet in the treatment of patients with chronic nickel dermatitis. *Br. J. Dermatol.* **98**, 197–201.

Marzouk, A. B., and Sunderman, F. W., Jr. (1985). Biliary excretion of nickel in rats. *Toxicol. Lett.* **27**, 65–71.

Medinsky, M. A., Benson, J. M., and Hobbs, C. H. (1987). Lung clearance and distribution of ^{63}Ni in F344 rats after intratracheal instillation of nickel sulfate solution. *Environ. Res.* **43**, 168–178.

Menzel, D. B. (1988). Planning and using PB-PK models: an integrated inhalation and distribution model for nickel. *Toxicol. Lett.* **43**, 67–83.

Menzel, D. B., Deal, D. L. Tayyeb, M. I., Wolpert, R. L., Boger, J. R., III, Shoaf, C. R., Sandy, J., Wilkinson, K., and Francovitch, R. J. (1987). Pharmacokinetic modeling of the lung burden from repeated inhalation of nickel aerosols. *Toxicol. Lett.* **38**, 33–43.

Metzler, C. M., and Weiner, D. L. (1986). 'PCNONLIN' and 'NONLIN84': Software for the statistical analysis of nonlinear models. *Am. Stat.* **40**, 1–52.

Morgan, L., and Rouge, P.J.C. (1984). Biological monitoring in nickel refinery workers. In F. W. Sunderman, Jr. (Ed.-in-Chief), *Nickel in the Human Environment*. International Agency for Research on Cancer, Lyon, pp. 507–520.

Nadler, S. B., Hidalgo, J. U., and Bloch, T. (1962). Prediction of blood volume in normal human adults. *Surgery* **51**, 224–232.

Onkelinx, C., and Sunderman, F. W., Jr. (1980). Modelling of nickel metabolism. In J. O. Nriagu (Ed.), *Nickel in the Environment*. Wiley-Interscience, New York, pp. 525–545.

Onkelinx, C., Becker, J., and Sunderman, F. W., Jr. (1973). Compartmental analysis of the metabolism of ^{63}Ni(II) in rats and rabbits. *Res. Commun. Chem. Pathol. Pharmacol.* **6**, 663–676.

Raithel, H. J., Schaller, K. H., Mohrmann, W., Mayer, P., and Henkels, U. (1982). Untersuchungen zur Ausscheidungskinetik von Nickel bie Beschaftigten in der Glas-und Galvanischen Industrie. In T. M. Fliedner (Ed.), *Bericht uber die 22 Jahrestagung der Deutschen Gesellschaft fur Arbeitsmedizin*. Komb, Belastungen Arbeitsplatz, pp. 223–228.

Rezuke, W. N., Knight, J. A., and Sunderman, F. W., Jr. (1987). Reference values for nickel concentrations in human tissues and bile. *Am. J. Ind. Med.* **11**, 419–426.

Sedman, A. J., and Wagner, J. G. (1976). 'CSTRIP': A Fortran-IV computer program for obtaining initial polyexponential parameter estimates. *J. Pharm. Sci.* **65**, 1006–1010.

Solomons, N. W., Viteri, F., Shuler, T. R., and Nielsen, F. H. (1982). Bioavailability of nickel in man: effects of foods and chemically-defined constituents on the absorption of inorganic nickel. *J. Nutr.* **112**, 39–50.

Spruit, D., and Bongaarts, P.J.M. (1977). Nickel content of plasma, urine and hair in contact dermatitis. *Dermatologica* **154**, 291–300.

Sunderman, F. W., Jr., Aitio, A., Morgan, L. G., and Norseth, T. (1986). Biological monitoring of nickel. *Toxicol. Ind. Health* **2**, 17–78.

Sunderman, F. W., Jr., Dingle, B., Hopfer, S. M., and Swift, T. (1988a). Acute nickel toxicity in electroplating workers who accidently ingested a solution of nickel sulfate and nickel chloride. *Am. J. Ind. Med.* **14**, 257–266.

Sunderman, F. W., Jr., Hopfer, S. M., and Crisostomo, M. C. (1988b). Nickel analysis by atomic absorption spectrometry. *Methods Enzymol.* **158**, 382–391.

Sunderman, F. W., Jr., Hopfer, S. M., Sweeney, K. R., Marcus, A. H., Most, B. M., and Creason, J. (1989). Nickel absorption and kinetics in human volunteers. *Proc. Soc. Exper. Biol. Med.* **191**, 5–11.

Torjussen, W., and Andersen, I. (1979). Nickel concentrations in nasal mucosa, plasma, and urine in active and retired nickel workers. *Ann. Clin. Lab. Sci.* **9**, 289–298.

Tossavainen, A., Nurminen, M., Mutanen, P., and Tola, S. (1980). Application of mathematical modelling for assessing the biological half-times of chromium and nickel in field studies. *Br. J. Ind. Med.* **37**, 285–291.

Zober, A., Weltle, D., and Schaller, K. H. (1984). Untersuchungen zur Kinetik von Chrom und Nickel in biologischen Material wahrend einworchigen Lichtbogenschweissens mit Chrom-Nickel-haltigen Zusatzwerkstoffen. *Schweissen Schneiden* **10**, 461–464.

6

ANALYSIS OF NICKEL IN BIOLOGICAL MATERIALS BY ISOTOPE DILUTION GAS CHROMATOGRAPHY/ MASS SPECTROMETRY

Suresh K. Aggarwal, Michael Kinter, Michael R. Wills, John Savory, and David A. Herold

Departments of Pathology, Biochemistry and Medicine, University of Virginia Health Sciences Center, Charlottesville, Virginia, 22908

Nickel and Human Health: Current Perspectives, Edited by Evert Nieboer and Jerome O. Nriagu.
ISBN 0-471-50076-3   © 1992 John Wiley & Sons, Inc.

1. INTRODUCTION

There has been an increasing interest in the quantitative determination of nickel in biological matrices driven by investigations of its clinical, nutritional, and toxicological effects. The advantages and limitations of the various quantitative methods used for nickel in biological materials have been highlighted in reviews by Stoeppler (1980) and Sunderman (1980). To summarize these reviews, an electrothermal atomic absorption spectrometry (EAAS) method has been accepted as a reference method for the determination of nickel in serum and urine by the International Union of Pure and Applied Chemistry (IUPAC) (Brown et al., 1981). Nickel has also been measured by differential pulse absorption voltametry (DPAV) with a dimethylglyoxime-sensitized mercury electrode (Flora and Nieboer, 1980). The other techniques that are capable of detecting microgram-per-liter concentrations of nickel, the analytical sensitivity requirement for nickel analysis in biological materials, are particle-induced X-ray emission spectrometry and gas chromatography with electron capture detection (Sunderman, 1980).

The IUPAC Subcommittee on Environmental and Occupational Toxicology of Nickel (Brown et al., 1981), however, has identified the need to develop a definitive method for the analysis of nickel in biological materials. Using the classification scheme established by the International Federation of Clinical Chemistry (Buttner et al., 1976), isotope dilution gas chromatography/mass spectrometry (GC/MS) is considered a definitive technique that represents the ultimate in quality. Since isotope dilution employs an ideal internal standard, that is, an enriched isotope of the same element, its accuracy does not depent upon consistent, quantitative sample preparation and external standardization. With the advent of bench-top mass spectrometers coupled to capillary column gas chromatographs, isotope dilution GC/MS techniques could be adopted by many laboratories. Specifically, the advantages of GC/MS over the other mass spectrometric techniques of thermal ionization mass spectrometry, spark source mass spectrometry, fast atom bombardment mass spectrometry, and inductively coupled plasma mass spectrometry are (i) fewer isobaric interferences; (ii) smaller mass discrimination effects; (iii) the possibility of using several ions, molecular and/or fragment, for quantitation; (iv) higher sample throughput; and (v) the wider availability of general-purpose organic mass spectrometers.

Isotope dilution GC/MS methods have been reported for chromium by Veillon et al. (1979) and for selenium by Reamer and Veillon (1981) using trifluoroacetylacetone and 4-nitro-o-phenylenediamine as the chelating agents, respectively. Historically, however, there have been three problems in the development of GC/MS for other metals of biological interest. These problems are (i) the lack of suitable chelating agents; (ii) poor accuracy in the isotope ratio measurements; and (iii) cross contamination due to memory effect during mass spectrometric analysis of samples of different isotope ratios.

We have addressed these problems for the quantitation of nickel by stable isotope dilution GC/MS. Lithium bis(trifluorethyl)dithiocarbamate

[Li(FDEDTC)] has been found to be a suitable chelating agent for nickel. Isotope ratios have been measured with good accuracy, and it has been shown that there is negligible memory effect using Ni(FDEDTC)$_2$. Briefly, the method entails spiking the samples with a known amount of an enriched stable isotope of nickel (^{62}Ni) and digesting with HNO_3–H_2O_2 to destroy the organic matter. The Ni(FDEDTC)$_2$ chelate is then formed at a pH of 3 and is extracted into methylene chloride. Aliquots of the organic extract are injected into the GC/MS and the isotope ratios are measured. The total amount of nickel in the sample is calculated from the change in the isotope ratio due to the addition of ^{62}Ni spike to the sample.

2. EXPERIMENTAL

2.1. Instrumentation

The mass spectrometer used was a double-focusing, reverse-geometry instrument (model 8230, Finnigan MAT, San Jose, CA) with SpectroSystem 300 data system. The instrument was operated in the electron ionization (EI) mode using 70-eV electrons with a source temperature of 200°C, the conversion dynode at 5000 V, and the secondary electron multiplier at 2200 V. Data for isotope ratios were acquired in the selected ion monitoring (SIM) mode using voltage peak switching at a sampling rate of 2 Hz yielding 20 measurement cycles across the approximately 10-sec chromatographic peaks.

A Variant 3700 gas chromatograph equipped with a DB-1 (J. W. Scientific, Rancho Cordova, CA) dimethylpolysiloxane bonded-phase fused silica capillary column, 10 m × 0.32 mm, with a 0.25-μm film thickness was used. Samples were injected using an on-column injector (OCI-3, Scientific Glass Engineering, Austin, TX) at an oven temperature of 100°C followed by a 15°C/min ramp to 300°C. High-purity helium was used as a carrier gas.

2.2. Reagents

Enriched ^{62}Ni was obtained in the elemental form from Oak Ridge National Laboratory (Oak Ridge, TN). The ^{62}Ni spike solution was prepared by dissolving the metal in dilute HNO_3 with heating. The isotopic composition of this spike solution was determined experimentally by preparing the Ni(FDEDTC)$_2$ chelate. The spike solution was then calibrated by reverse-isotope dilution GC/MS using a standard solution of nickel (Certified Atomic Absorption Standard, Fisher Scientific Co., Fairlawn, NJ) as the primary standard. The urine Standard Reference Material 2670 (SRM 2670) was obtained from the National Bureau of Standards (NBS, Gaithersburg, MD) and prepared according to their directions.

The chelating agent Li(FDEDTC) was prepared by the Sucre and Jennings (1980) method using bis(trifluoroethyl)amine (PCR Research Chemicals, Gainesville, FL), n-butyllithium, and CS_2 (Aldrich Chemical Co., Milwaukee,

WI) in an inert atmosphere at $-70°C$. A 0.02-mol/L solution of Li(FDEDTC) in distilled, deionized water (DDW) was used as the chelating reagent. The double subboiling quartz distilled HNO_3 (NBS), hydrogen peroxide (50% Fisher Scientific Co.), and ammonia solution (Ultrex, J. T. Baker Chemical Co., Phillipsburg, NJ) used in this work were checked for the presence of nickel by EAAS and were found to be free of nickel (<1 μg/L).

2.3. Digestion Procedure

A known volume, 0.5 mL, of the reconstituted urine SRM 2670 was mixed with a weighed amount of the ^{62}Ni spike solution in a Teflon beaker. The amount of nickel in the spike solution added to the urine sample was optimized to obtain an isotope ratio in the mixture corresponding to the geometric mean of the isotope ratios in the sample and the spike. This was done to minimize the errors in the concentration values due to random errors in the isotope ratio measurements. The solution in the Teflon beaker was heated gently on a hot plate to reduce the volume to about 100 μL. The solution was digested twice with 200 μL of HNO_3 and evaporated to dryness, and the residue was dissolved in 200 μL of 2% HNO_3. Approximately 200 μL of 50% hydrogen peroxide was added and the solution was heated gently at about 50°C on a hot plate and inspected periodically. At each inspection, the contents were mixed and the beaker tapped gently to disperse frothing. The digestion with $HNO_3–H_2O_3$ was performed four to five times until a white residue remained on complete evaporation of the solution. The total time required was 3–4 hr. This dried residue was then dissolved in 200 μL of 2% HNO_3 and the solution left at room temperature for 1–2 hr to facilitate complete dissolution. The solution was then diluted to about 1 mL using HNO_3 of pH 3 and the pH was adjusted to 3 using a dilute ammonia solution. Derivatization with Li(FDEDTC) was carried out by adding 1 mL of pH 3 acetate buffer followed by 200 μL of the chelating reagent. The mixture was vortexed for about 2 min and the $Ni(FDEDTC)_2$ extracted into methylene chloride. The methylene chloride containing the $Ni(FDEDTC)_2$ was transferred to a plastic bullet with a conical bottom and evaporated. The residue was dissolved in 20 μL of methylene chloride and about 1 μL injected into the GC/MS. During preliminary studies on nickel determination in bovine serum samples, the samples were deproteinized using HNO_3 and heating after the addition of the ^{62}Ni spike solution and prior to the above-mentioned digestion procedure.

2.4. Gas Chromatography/Mass Spectrometry

The mass spectrum of $Ni(FDEDTC)_2$ was obtained by exponential mass scanning at a resolution of about 1000. The isotope ratios were measured after optimizing the source and the collector slits to obtain flat-topped peaks and employing voltage peak switching in the SIM mode. Isotope ratios were calculated from the integrated areas of the chromatographic peaks in the respective mass chromatograms. As the measurements were performed without using a lock mass, it was found necessary to determine each day the m/z value of the peak maximum by obtaining a histogram of the ion current at various m/z values about the calculated m/z value of each ion. This was done by measuring the ion current at intervals of 0.2 amu, up to 1 amu on either side of the calculated m/z value for the two ions of highest abundance. Subsequently, the m/z values given for the various ions of interest in the data acquisition corresponded to these experimentally determined m/z values of that ion. Further, during the measurements, the ion current was also monitored at m/z values of ± 0.2 amu on either side of the experimentally determined position of the peak maximum. This served as a check for any shift in the mass calibration during the analysis and was necessary due to slight irregularities in the exponential mass scan line. For example, the calculated m/z values for the various isotopes of nickel using the Ni(FDEDTC)$_2$ chelate and monitoring the molecular ion were 569.87 (^{58}Ni), 571.87 (^{60}Ni), 572.87 (^{61}Ni), 573.87 (^{62}Ni), and 575.87(^{64}Ni). The experimentally determined positions for the maxima of the two highest abundance peaks of ^{58}Ni and ^{62}Ni were found at the m/z 569.7 and 573.7, respectively, on a given day. These subtle changes in the peak position would appear to be insignificant, particularly considering the mass resolution used, but were found to be critical for the measurement of accurate isotope ratios.

2.5. Contamination Control

It is extremely important to minimize nickel contamination from the apparatus, reagents, personnel, and the laboratory environment. Stable isotope dilution GC/MS has the potential of providing precise and accurate values at levels below the microgram-per-liter level. The overall blank will therefore define the sensitivity of the method and limit the applicability of this technique at extremely low levels. The analysis should be performed in a scrupulously clean laboratory equipped with laminar flow hoods to prevent nickel contamination from dust. The laboratory ware should be acid leached and cleaned with DDW. All reagents used should be checked for nickel contamination and, if detected, must be purified of nickel. Lastly, powder-free gloves must be worn to avoid nickel contamination by perspiration from the hands of the analyst.

3. RESULTS AND DISCUSSION

The mass spectrum of Ni(FDEDTC)$_2$ and a reconstructed ion chromatogram obtained for the analysis of a standard solution containing 10 ng of nickel are shown in Figures 6.1 and 6.2. The presence of a molecular ion in the mass spectrum in Figure 6.1 as well as the symmetrical and sharp chro-

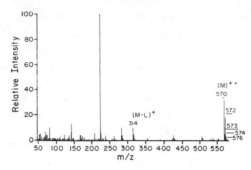

Figure 6.1. Electron ionization mass spectrum of Ni(FDEDTC)$_2$.

matographic peaks seen in Figure 6.2 indicate the adequate thermal stability of the nickel chelate at nanogram levels under the experimental conditions used. Among the various ions containing the nickel isotopes, the ions clustered around m/z 570, which correspond to the molecular ion, were of the highest intensity and were selected for the nickel isotope ratio measurements and the isotope dilution experiments.

Table 6.1 presents the results obtained for isotope ratio measurements of a sample of natural nickel using Ni(FDEDTC)$_2$. The overall precision of the isotope ratio measurements was evaluated by determining the within-run precision as well as the between-run precision. This was done to evaluate the effects of small variations in the mass spectrometer operating parameters that may occur from one day to another. It is seen that an overall precision of 1–7% is obtained for the isotope ratio measurements. The accuracy of the isotope ratio measurements was determined by comparing the experimentally measured ratios with the theoretically computed values using the reported natural abundances of the nickel isotopes in the elemental form and the contributions of the isotopes of carbon, nitrogen, and sulfur to the/ different masses in the Ni(FDEDTC)$_2$ molecular ion cluster. Accuracy values

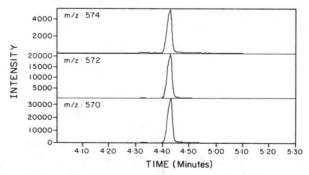

Figure 6.2. Reconstructed ion chromatogram for the analysis of a standard solution of natural nickel using Ni(FDEDTC)$_2$.

Table 6.1   Precision and Accuracy in Isotope Ratio Measurements

	Isotope Ratio		Accuracy[a]	Precision
	Natural	Observed	(%)	(%)
572/570	0.5686	0.5738	0.9	1.2
573/570	0.0973	0.1013	4.1	4.9
574/570	0.1406	0.1458	3.7	7.2
576/570	0.0293	0.0311	6.2	5.7

[a]Accuracy values were calculated as (observed − natural) × 100/natural.

[calculated as (observed-natural) × 100/natural] of 0.9–6.2% are obtained for different isotope ratios, and these are within the experimental error limits of the method.

Memory effect observed when measuring varying isotope ratios was evaluated using two different approaches. The first approach involved the sequential analysis of a standard solution of natural nickel and a solution of the ^{62}Ni spike measuring the m/z 574/570 isotope ratios. The results are shown in Figure 6.3. Some carryover is seen when these samples of extremely different isotopic composition, isotope ratios differing by a factor of 400, are analyzed sequentially. The expected isotope ratios in this case can, however, still be obtained after rejecting the data of three to four injections. This is demonstrated in Figure 6.3 for the natural nickel solution after multiple injections of the ^{62}Ni-enriched solution. The second approach was to observe the extent of memory effect in samples that should be more commonly encountered in isotope dilution experiments, namely solutions in which the isotope ratio is

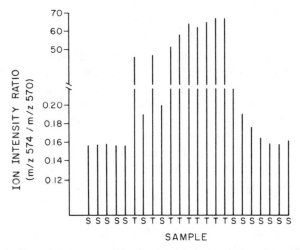

Figure 6.3. Evaluation of the cross contamination between samples of very different isotopic compositions: S, standard solution of natural nickel; T, ^{62}Ni-enriched spike solution.

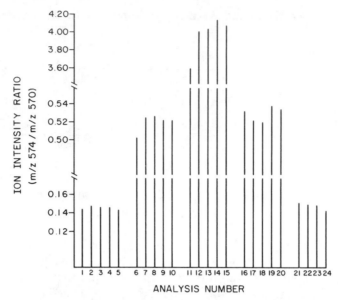

Figure 6.4. Evaluation of the cross contamination between samples of slightly different isotopic compositions. Analyses numbers 1–5 and 21–24 are of a sample of natural nickel (S), analyses numbers 6–10 and 16–20 are of a synthetic sample with an m/z 574/570 ratio of 0.5 (SM-1), and analyses numbers 11–15 are of a synthetic sample with a m/z 574/570 ratio of 4 (SM-2).

less than 10 and preferably approaching 1. Two synthetic mixtures with m/z 574/570 isotope ratios of 0.5 and 4 were prepared. The samples of natural nickel and the two samples of synthetic isotope ratios were analyzed in sequence and the results shown in Figure 6.4. It can be seen that in these samples, which differ in isotope ratio by a factor of about 25, no memory effect was observed.

The suitability of GC/MS using the Ni(FDEDTC)$_2$ chelate for accurately determining isotope ratios different from those of natural samples was also evaluated. In this experiment, five synthetic mixtures differing in isotope ratio by a factor of 10 were prepared by mixing weighed aliquots of the standard solution and ^{62}Ni-enriched spike solution. The isotope ratios determined from these mixtures were used to calculate the nickel concentration in the spike solution in a reverse-isotope dilution experiment. For comparison, the concentration of nickel in the spike solution was also determined by EAAS. From the results given in the Table 6.2, it can be seen that the concentration values obtained by isotope dilution GC/MS are in good agreement with the value determined by the EAAS method. More importantly, the results obtained for the series of different isotope ratios yield consistent concentrations for the spike solution. This indicates that this GC/MS method is providing accurate results over a range of nonnatural isotope ratios. The importance of this observation is simply that both natural and altered isotope

Table 6.2 Determination of Different Isotope Ratios

Synthetic Mixture	574 (^{62}Ni)/ 570 (^{58}Ni)	C_{SP} Using 574/570	C_{SP} Using 574/572
A	0.5	4.276	3.966
B	1	4.088	4.086
C	2	4.149	3.966
D	3	4.068	4.094
E	5	4.098	4.104
Mean		4.136 ± (2.0%)	4.043 ± (1.8%)

Note: The concentration in the spike solution, C_{SP}, was determined by EAAS = 4.35 μg Ni/g solution.

ratios can be accurately measured making this method appropriate for accurate quantitative methods.

The results of the quantitative analysis of nickel in the urine reference material using isotope dilution GC/MS are given in the Table 6.3. The nickel concentration values of 83.7 and 79.6 μg/L determined using the isotope ratios m/z 574/570 and m/z 574/572, respectively, are in good agreement with each other as well as with the NBS recommended value of 70 μg/L (no error limits specified). Figure 6.5 shows a typical reconstructed mass chromatogram for this analysis. As can be seen, no other GC peaks are observed, indicating the good selectivity of the analysis. This is an especially important consideration in the development of definitive and reference methods. Also, an excellent signal-to-noise ratio is observed indicating good sensitivity for the analysis.

The results obtained on the determination of nickel in the serum of Virginia cows are summarized in the Table 6.4. Once again, the concentration values of about 3 μg/L were obtained using two different sets of isotope ratios and

Table 6.3 Nickel in Urine Reference Material

Experiment Number	Concentration (μg/L) From 574/570	Concentration (μg/L) From 574/572
1	81.1	76.4
2	78.3	73.0
3	77.4	76.0
4	90.1	86.9
5	88.3	87.1
6	86.9	83.8
Mean	83.7 (2.6%)	79.5 (3.7%)

Note: For SRM-2670 (normal level). Recommended Ni concentration by NBS = 70 ± ? μg/L.

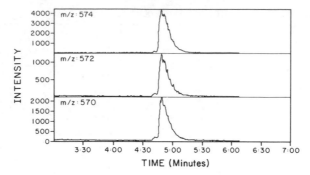

Figure 6.5. Reconstructed ion chromatogram for the analysis of a standard urine reference material, SRM 2670, by GC/MS using Ni(FDEDTC)$_2$.

are in good agreement with each other. These data indicate that the digestion procedure, which was altered to include deproteinization of the serum, can accommodate other biological matrices.

ACKNOWLEDGMENTS

Funding for the purchase of the high-resolution mass spectrometer was obtained from the National Institutes of Health, Division of Research Resources Shared Instrumentation Grant Program, grant number 1-S10-RRO-2418-01. Additional funding from the John Lee Pratt Fund of the University of Virginia and grant ESO 4464 of the National Institute of Environmental Health Sciences is also gratefully acknowledged. One of us (S.K.A.) thanks the Division of Experimental Pathology, Department of Pathology, University of Virginia Health Sciences Center, for a postdoctoral fellowship and the authorities at Bhabha Atomic Research Center, Trombay, Bombay-400 085, India, for granting leave. The authors thank James Nicholson for providing the electrothermal atomic absorption spectrometric results.

Table 6.4   Concentration of Nickel in Bovine Serum by Isotope Dilution GC/MS (μg/L)

	Using 574/570	Using 574/572
	3.05	3.31
	2.67	2.65
	3.60	3.76
	3.26	3.42
	3.63	2.69
Mean	3.24 ± 12%	3.16 ± 15%

REFERENCES

Brown, S. S., Nomoto, S., Stoeppler, M., and Sunderman, F. W., Jr. (1981). IUPAC reference method for analysis of nickel in serum and urine by electrothermal atomic absorption spectrometry. *Pure Appl. Chem.* **53,** 773–781.

Buttner, J., Borth, R., Boutwell, J. H., and Broughton, P.M.G. (1976). Provisional recommendation on quality control in clinical chemistry. *Clin. Chem.* **22,** 532–540.

Flora, C. J., and Nieboer, E. (1980). Determination of nickel by differential pulse polarography at a dropping mercury electrode. *Anal. Chem.* **52,** 1013–1020.

Reamer, D. C., and Veillon, C. (1981). Determination of selenium in biological materials by stable isotope dilution gas chromatography–mass spectrometry. *Anal. Chem.* **53,** 2166–2169.

Stoeppler, M. (1980). Analysis of nickel in biological materials and natural waters. In J. O. Nriagu (Ed.), *Nickel in the Environment.* Wiley-Interscience, New York, pp. 661–821.

Sucre, L., and Jennings, W. (1980). Lithium di(trifluoroethyl) dithiocarbamate: an alternative reagent for the preparation of di(trifluoroethyl)dithiocarbamate metal chelates. *Anal. Lett.* **13,** 497–501.

Sunderman, F. W., Jr. (1980). Analytical Biochemistry of Nickel. *Pure Appl. Chem.* **52,** 527–544.

Veillon, C., Wolf, W. R., and Guthrie, B. E. (1979). Determination of chromium in biological materials by stable isotope dilution. *Anal. Chem.* **51,** 1022–1024.

7

URINARY EXCRETION OF NICKEL IN NICKEL–CHROMIUM ELECTROPLATERS

M. A. White

Institute of Naval Medicine, Alverstoke Gosport, United Kingdom PO12 2DL

A. M. Boran

Institute of Occupational Health, University of Birmingham, Birmingham, United Kingdom

Nickel and Human Health: Current Perspectives, Edited by Evert Nieboer and Jerome O. Nriagu.
ISBN 0-471-50076-3   © 1992 John Wiley & Sons, Inc.

1. INTRODUCTION

The toxic nature of inorganic nickel compounds is well documented (Mastromatteo, 1986). Workers in the electroplating industry are exposed to aerosols of soluble nickel salts, primarily nickel chloride and sulfate. Although there is an increased risk of respiratory cancer from long-term exposure to insoluble nickel compounds (Chovil et al., 1981), the risk associated with exposure to soluble nickel salts is unresolved. There is also some suggestion of an increased risk of gastrointestinal cancer (Mastromatteo, 1986). Nickel compounds are known to be potent sensitizing agents, and contact dermatitis is common among nickel workers (Bencko et al., 1986). Several incidences of occupational asthma have also been identified in electroplaters exposed to soluble nickel salts (Davies, 1986; Malo et al., 1982).

Biological monitoring is becoming increasingly important in the assessment of occupational exposure to toxic metals. Several studies have shown that the concentration of nickel in plasma, urine, and hair may be used as an index of occupational exposure (Hyoi et al., 1986; Sunderman et al., 1986). In particular, urine nickel measurements appear to be particularly well suited to the monitoring of occupational exposure to soluble nickel salts (Norseth and Piscator, 1979). Recently, direct graphite furnace atomic absorption spectrometry (GFAAS) methods for the determination of urinary nickel have been described (Sunderman et al., 1986; Kiilunen et al., 1987) that may be considered more appropriate for biological monitoring work than existing methods employing chelation and solvent extraction.

In the study presented here, we have monitored urine nickel excretion in electroplaters from two platting shops. Urine nickel has been measured by GFAAS using a direct method in order to assess the suitability of such a method for biological monitoring of occupational exposure and to validate further the use of urine nickel measurements in assessing occupational exposure.

2. STUDY DETAILS

Two electroplating shops, A and B, carrying out a variety of plating and finishing processes were selected for the study. In both shops decorative nickel and chromium plating were the principal plating processes undertaken. Two types of nickel plating were employed. Large objects were plated in open (conveyor) baths, while small objects were barrel plated in hexagonal barrels rotating slowly in the plating bath. Ventilation in the shops was regulated by extractor fans.

In shop A, pre- and postshift urine nickel levels were determined in 44 workers at the end of a typical working week. In shop B, pre- and postshift urine nickel levels were determined daily, for one week, in a group of 10 workers. Airborne nickel concentrations in workshop B were also measured on the first, third, and last day of the week. A group of zinc galvanizers

($n = 50$) and nonoccupationally exposed individuals acted as controls ($n = 23$).

3. METHODOLOGY

3.1. Airborne Measurements

Air samples were collected onto filter membranes fitted to personal samplers and static samplers placed above the plating baths. Filters were analyzed for nickel by atomic absorption spectrophotometry using the method recommended by the Health and Safety Executive (HSE MDHS 42).

3.2. Urine Measurements

A Perkin-Elmer 5000 instrument equipped with an HGA 500 graphite furnace, AS-40 autosampler, and deuterium continuum source background correction was used for the analyses. All reagents were of the highest purity available. Double-distilled deionized water was used throughout. All glassware and disposable plasticware were soaked for 24 hr in 20% HNO_3 and rinsed with ultrapure water before use. Rubber gloves with talc were worn for all sample-handling steps to avoid nickel contamination by sweat. We have found no contamination problems when using these gloves for all biological monitoring work within the department.

A matrix-matched calibration curve was prepared by spiking a human urine pool with aqueous nickel standard to give final concentrations of 0–60 μg/L. Samples and standards were diluted $1 + 2$ (v/v) with 0.6% HNO_3 for analysis. The furnace program and instrument settings are given in Table 7.1. Urine quality control samples (Lanonorm, Behring, and Seronorm urine,

Table 7.1 Instrument Parameters and Furnace Settings for Atomic Absorption Spectrometer

Instrument Parameters

Lamp current, 25 mA	Slit, 0.2 nm
Wavelength, 232.0 nm	Mode, AA-BG
Sample volume, 25 μL	

Furnace Settings

Step	Temperature (°C)	Ramp Time (sec)	Hold Time (sec)	Gas Flow (mL/min)
Dry	120	30	10	300
Char	1100	70	40	300
Atomize	2500	0	5	30
Clean	2700	1	4	300

Nygaard, United Kingdom) were included with each analytical run, and a random selection of samples were chosen for an interlaboratory comparison.

Urine samples were voided directly into acid-washed 500-mL polypropylene containers. An aliquot of each sample was transferred to a plastic 30-mL universal container and acidified with nitric acid. Samples were stored at 4 or −20°C until analyzed.

4. RESULTS

The detection limit (3 × SD of blank) for the method was 0.6 µg/L, or 1.8 µg/L in undiluted urine samples. The detection limit is not as low as that reported for the method of Sunderman et al. (1986) owing to the increased dilution factor and the use of gas flow at atomization. Within-run precision and run-to-run precision were 5% and 10.5%, respectively. Accuracy of the method was determined by analyzing quality control urines with known reference values and by interlaboratory comparison. The results are shown in Table 7.2.

Urinary nickel was elevated in the entire group of platers (mean 15.6 µg/L, range <2–76 µg/L, $n = 54$), compared with galvanizers (mean 6.9 µg/L, range <2–20 µg/L, $n = 50$) and unexposed individuals (mean 5.5 µg/L, range <2–9 µg/L, $n = 23$). In calulating the mean nickel concentrations in the above groups, values at or below the limit of detection were excluded as a definitive concentration could not be assigned to these samples. The concentrations of nickel determined are similar to previously reported values for platers (Sunderman et al., 1986). The percentage of dispersion of urine

Table 7.2 Assessment of Performance of Analytical Method

Measured value Lanonorm 2	30 µg/L ± 3.5
Assigned value	31 µg/L
Measured value Seronorm urine	45 µg/L ± 5
Assigned value	40 µg/L

Interlaboratory Comparison

Urine Sample	Our Result (µg/L)	HSE Result[a] (µg/L)
1	16.8	16
2	24.9	25.2
3	24.2	22.7
4	5.0	9.0
5	15.8	18.4
6	4.9	4.3
7	75.8	85.7

[a] Urine nickel analyses were performed by the Occupational Medicine Department Health and Safety Executive (HSE).

nickel levels in both platers and galvanizers is plotted in Figure 7.1. It clearly shows the increased exposure to nickel encountered in the plating industry.

For monitoring studies of electroplaters, urine nickel was expressed as micrograms per gram of creatinine to correct for variable urine dilution. However, many postshift samples were very dilute (<2.7 mmol creatinine/ L), approaching the level at which a repeat determination on a different sample is recommended (Sunderman et al., 1986). In the 44 platers from shop A, urine nickel rose from 7.5 μg/g creatinine preshift (range <2–18.5) to 10.3 μg/g creatinine postshift (range <2–44.5). In the 10 platers from shop B monitored over five consecutive days, nickel excretion tended to increase during the work day and also throughout the week, rising from 8 μg/g creatinine (2.5–21.6 μg/g) at the start of the week to 17.2 μg/g creatinine (5–44 μg/g) after the last shift of the week.

The patterns of nickel excretion in open-bath platers and barrel platers are shown in Figures 7.2 and 7.3, respectively. Urine nickel levels were highest for open-bath platers, while levels in barrel platers were significantly lower ($0.02 > P > 0.01$). Also, measurements taken in shop B indicated a correlation between urine nickel and exposure to airborne nickel (Fig. 7.4). Concentrations of nickel above the plating baths ranged from 7 to 70 μg/m^3, the highest levels being recorded toward the end of the week. These levels are similar to previously reported values (Mastromatteo, 1986) and fall below the HSE and American Conference of Governmental and Industrial Hygienists (ACGIH) exposure limit of 0.1 mg/m^3 for soluble nickel salts. Personal monitoring of platers indicated a greater degree of exposure to nickel in open-bath plating than in barrel plating, which is consistent with the urinary nickel data. As the surface area of the tank open to the atmosphere is reduced in

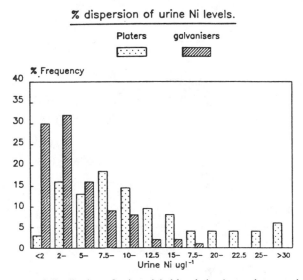

Figure 7.1. Percentage of distribution of urine nickel levels in electroplaters and zinc galvanizers.

Open Bath Platers

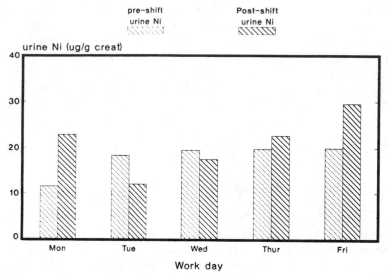

Figure 7.2. Patterns of nickel excretion in four open-bath platers. Values plotted are means of duplicate determinations.

Barrel Platers

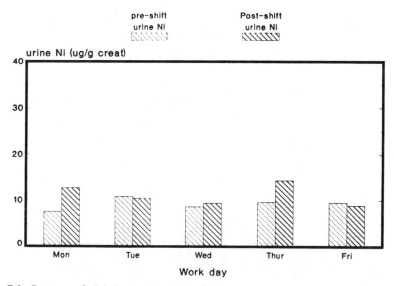

Figure 7.3. Patterns of nickel excretion in five barrel platers. Values are means of duplicate determinations.

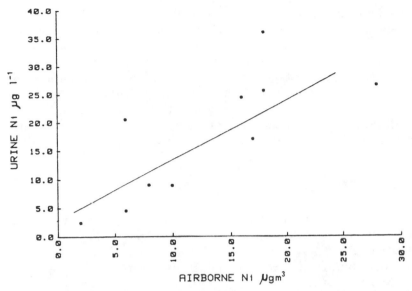

Figure 7.4. Correlation graph of urine Ni levels and exposure to airborne Ni in 10 electroplating workers. Exposure was monitored by personal sampler; $y = 3.4 + 1.04\,x$, $r = 0.74$.

the barrel plating process, the concentration of nickel above the bath may be considerably less than that above an open plating bath, and this may account for the lower nickel exposure and urine nickel levels observed in barrel platers.

5. DISCUSSION

Our monitoring studies indicate that the direct determination of urine nickel is suitable for the biological monitoring of occupationally exposed workers. A direct method considerably reduces the risk of sample contamination during preparation for analysis, which has been reported to be one of the major sources of analytical error (Versiek et al., 1987). Furthermore, it is very cost effective in terms of analyst time. The detection limit of this method is perfectly adequate for detecting low exposure levels to the metal. However, the limit could be further improved by lowering the urine dilution factor or reducing the gas flow at atomization provided the resulting background signal did not exceed the correction capabilities of the instrument.

The pattern of nickel excretion observed in this study is similar to that reported by Tola et al. (1979) for a group of four platers. However, the urine nickel concentrations and airborne levels observed are somewhat lower than those reported by the latter group and are more consistent with recent measurements (Sunderman et al., 1986). Few anomalous values were determined in the monitoring study. Concomitant measurement of urinary chromium indicated contamination during sample collection to be the likely cause of the anomalies. The dilute nature of many of the postshift urine samples suggested that preshift urines taken toward the end of the working week may be most suitable as spot samples for monitoring exposure.

The degree of correlation between urine nickel levels and airborne nickel levels found in this study is similar to that found in earlier studies (Hogetveit et al., 1978; Tola et al., 1978) and suggests that urine nickel measurements are well suited for the biological monitoring of nickel platers and associated workers, who are primarily exposed to soluble nickel salts.

ACKNOWLEDGMENTS

The authors thank Surgeon Captain J. M. Young, R.N.; Surgeon Commander M. D. Simmons, R.N.; J. N. Marks (Institute of Naval Medicine); Professor J. M. Harrington (University of Birmingham); Dr. R. Braithwaite (Regional Toxicology Unit, Dudley Rd. Hospital, Birmingham); and N. Smith (Health and Safety Executive).

REFERENCES

Bencko, V., Wagner, V. Wagnerova, M., and Zavazal, V. (1986). Human exposure to nickel and cobalt: biological monitoring and immunobiochemical response. *Environ. Res.* **40,** 399–410.

Chovil, A., Sutherland, R. B., and Halliday, M. (1981). Respiratory cancer in a cohort of nickel sinter plant workers. *Br. J. Ind. Med.* **30,** 327–333.

Davies, J. E. (1986). Occupational asthma caused by nickel salts. *J. Soc. Occup. Med.* **36,** 29–31.

Hogetveit, A. C., Barton, R. T., and Kostol, C. O. (1978). Monitoring nickel exposure in refinery workers. *Ann. Occup. Hyg.* **21,** 113–120

Hyoi, N., Imahori, A., Shiobara, S., and Fukushima, I. (1986). Trace element concentrations in hair and blood of electroplating workers. *Jpn. J. Ind. Health* **28,** 200–201.

Kiilunen, M., Jarvisalo, J., Makitie, O., and Aitio, A. (1987). Analysis, storage stability and reference values for urinary chromium and nickel. *Int. Arch. Occup. Health* **59,** 43–50.

Malo, J-L., Cartier, A., Doepner, M., Nieboer, E., Evans, L., and Dolovich, J. (1982). Occupational asthma caused by nickel sulphate. *J. Allergy Clin. Immunol.* **69,** 55–59.

Mastromatteo, M. (1986). Nickel. *Am. Ind. Hyg. Assoc. J.* **47,** 589–601.

Norseth, T., and Piscator, M. (1979), Nickel. In L. Friberg, G. F. Nordberg, and V. B. Vouk (Eds.), *Handbook on the Toxicology of Metals.* Elsevier/North Holland Biomedical, Amsterdam, pp. 541–553.

Sunderman, F. W., Jr., Hopfer, S. M., Christomo, M. C., and Stoeppler, M. (1986). Rapid analysis of nickel in urine by electrothermal atomic absorption spectrophotometry. *Ann. Clin. Lab. Sci.* **16,** 219–229.

Tola, S., Kilpio, J., and Virtamo, M. (1979). Urinary and plasma concentrations of nickel as indicators of exposure to nickel in an electroplating shop. *J. Occupat. Med.* **21,** 184–188.

Versieck, J., Vanballenberghe, L., De Kessel, A., and Van Renterghem, D. (1987). Accuracy of biological trace element determinations. *Biol. Trace Element Res.* **12,** 45–55.

8

URINARY AND HAIR NICKEL AS EXPOSURE INDICES

Zhijie Yang, Baoqi Gang, and Shanzhong Sun

School of Public Health, Harbin Medical University, Harbin 150001, China

1. INTRODUCTION

A great number of reports on the toxic effects of nickel and its compounds, especially their carcinogenic effects on nickel workers, have been issued. Because of nonspecific toxic effects and the long latency period of cancer, many authors have attempted to use nickel in body fluids as an index in biological monitoring. Høgetveit et al. (1980), Tola et al. (1979), and others have observed a correlation between nickel in plasma (or urine) and that in ambient air and that measurement of nickel in plasma was more reliable than in urinary nickel. The use of plasma nickel as an exposure index in biological monitoring was therefore recommended. However, Bernacki et al. (1978)

Nickel and Human Health: Current Perspectives, Edited by Evert Nieboer and Jerome O. Nriagu.
ISBN 0-471-50076-3   © 1992 John Wiley & Sons, Inc.

reported that nickel levels in urine were higher than that in serum. Because of this, and since collection of blood samples is invasive and more difficult, they favor the use of urinary nickel. In the studies referred to, exposure was mainly to water-soluble nickel salts. For exposure to water-insoluble nickel compounds, only a few reports on biological monitoring are available. Therefore, we proceeded to measure both ambient and urinary nickel for exposure to dust in a nickel smelter using pyrometallurgical processes. The purpose of this chapter is to clarify the dependence of the nickel contents of hair and urine on ambient nickel concentrations, on exposure duration, as well as on other factors (e.g., smoking).

2. SUBJECTS AND METHODS

Thirty healthy male workers were randomly selected as subjects from electric furnace and converter sections of a nickel smelter, where exposure was predominantly to water-insoluble nickel compounds. The following samples were collected and measurements made.

1. *Airborne Nickel.* Workers were outfitted with a personal sampler with a 0.8-μm Micropore membrane filter and operated at a flow rate of 0.5 L/min during an 8-hr work shift.
2. *Urinary Nickel.*   Urine specimens were collected before changing into work clothes just prior to commencing work in the morning. Replicate samples were analyzed (25 mL urine for each). At the same time, the creatinine content of the urine samples was determined (alkaline picric acid method). Nickel concentrations in urine were expressed in micrograms per gram of creatinine.
3. *Nickel Content of Hair.* Hair (0.5 g) was analyzed and the results were expressed in micrograms per gram of hair.
4. *Nickel Exposure Duration.* It was assessed by questionnaire.
5. *Smoking Habits.* Workers who smoked continuously were classified as smokers; those who smoked occasionally as nonsmokers.

The concentration of nickel in hair and urine and on the air filters were determined by dimethylglyoxime-sensitized pulse polarography as described by Flora and Nieboer (1980), employing an ammonia–ammonium chloride –sulfosalicylic acid buffer instead of ammonium tartrate–citrate. The acid digestion of hair, urine, and the filters were completed with a mixture of ultrapure acids (nitric, sulfuric, and perchloric acids).

Partial correlation analysis of five variables, namely airborne nickel concentration, nickel exposure duration, smoking habit, and nickel concentration in urine and hair, was carried out with a Z-80 microcomputer.

3. RESULTS

In the analysis of the data, five results were discarded because (1) the creatinine content in urine was less than 0.5 g/L or higher than 2.0 g/L or (2) replicate urine specimens were not obtained. Consequently, only data from 25 workers aged 21–40 years were left. As indicated in Table 8.1, personal exposure levels of airborne nickel were in the range 0.0034–1.2760 mg/m^3; nickel concentrations in urine were 3.645–36.470 μg/g creatinine; nickel content in hair was 0.46–18.16 μg/g; nickel exposure duration was 14–109 months; and 20 of the 25 workers examined were smokers (smoking duration 5–24 years).

Table 8.1 Smoking Habits, Exposure Duration, and Observed Nickel Levels in Urine, Hair, and Air

Worker Number	Urine Ni (μg/g creatinine)	Hair Ni (μg/g)	Air Ni (mg/m^3)	Exposure Duration (months)	Smoking[a]
1	11.315	2.58	0.0184	45	1
2	16.685	0.46	0.0063	83	1
3	16.815	2.75	0.0123	73	0
4	21.195	2.83	0.0034	45	1
5	17.330	2.00	0.0249	45	1
6	13.040	2.83	0.0110	30	1
7	10.520	2.13	0.0486	80	1
8	8.610	1.71	0.0464	35	1
9	11.030	1.29	0.0494	40	1
10	6.100	3.42	0.2190	14	1
11	39.200	4.33	1.2760	26	0
12	13.680	7.17	0.2989	26	1
13	36.470	12.67	0.0731	40	1
14	11.810	9.25	0.0570	40	1
15	3.645	3.79	0.0517	40	0
16	16.020	4.75	0.0547	108	0
17	29.890	16.33	0.0307	40	1
18	13.650	16.17	0.0324	109	0
19	25.370	3.21	0.0268	109	1
20	6.185	2.25	0.0255	40	1
21	6.335	4.38	0.1542	100	1
22	21.855	2.25	0.0184	40	1
23	8.585	18.16	0.2607	40	1
24	11.915	2.83	0.3569	28	1
25	13.230	2.96	0.1748	19	1

[a] 1, smoking; 0, no smoking.

Table 8.2   Partial Correlation Coefficient (r) for Measured
Variables

	Urine Ni	Hair Ni	Air Ni	Exposure Duration	Smoking
Nickel in urine	1.00	0.23	0.42[a]	0.16	0.09
Nickel in hair	0.23	1.00	−0.08	−0.01	−0.09

[a]Significant at 95% level.

The partial correlation coefficients r for five variables are listed in Table 8.2. It is evident that the r value for the correlation between nickel concentrations in urine and airborne nickel was the highest ($r = 0.42$); that for hair nickel and airborne nickel was low ($r = 0.08$). Both urinary and hair nickel were poorly correlated with smoking and nickel exposure duration.

4. DISCUSSION

As to the question of which index is actually most appropriate for biological monitoring purposes, most authors hold the view that the nickel concentration in urine or in plasma can be used as an exposure index, especially when the exposure is to water-soluble salts. This is not so clear for exposures to nickel-containing dusts. Aitio et al. (1985) have reported urinary excretion of chromium and nickel in six cast cleaners of a steel foundry. The alloys handled contained on average 13–23% chromium and 4–11% nickel. Air specimens were collected with personal samplers. The concentrations of nickel in the air were between 0.05 and 1.10 mg/m³, and the concentrations of acid-soluble chromium ranged from 0.06 to 0.79 mg/m³. No relationship was seen between the urinary nickel concentration and the daily period of exposure or for the time-weighted average concentration of nickel in the air. Further, Schaller et al. (1985) measured the nickel concentrations in urine and in blood of 20 arc welders, and the external exposure was estimated by personal air sampling. The nickel content of the electrodes was 8–10%, but in special cases it amounted to 65%. For a short duration, pure nickel electrodes were also in use. The median value of the external nickel exposure was 5.8 µg/m³. No correlation was observed between nickel concentration in the air and nickel concentration in urine or in plasma. However, the authors considered that urine analyses are adequate as a basis for quantifying the internal exposure. In the Jinchuan region of China, a survey conducted by Gansu Province Sanitary-antiepidemic Station in a nickel smeltery showed that nickel concentrations in the air were 0.4–5.1 mg/m³, and the hair nickel content of workers was higher than that of the controls (unpublished data). Of the three biological materials (blood, urine, and hair), the content of nickel in urine or hair may be more practical because the collection of blood specimen is rather difficult. But which of these two is the more appropriate? The data obtained in the present study suggest that in 25 workers exposed to insoluble nickel compounds in the range 0.0034–1.2760 mg/m^3, the best correlation occurred between nickel in urine and that in air. By contrast, the nickel content in hair was little related to airborne nickel.

It has been reported that nickel is metabolized rapidly, at least when exposure occurs to water-soluble salts (Norseth, 1986). Perhaps a similar explanation applies for the men in the present study as there was no dependence on exposure duration. Alternatively, an equilibrium situation might have occurred in which lung clearance of nickel is balanced by its deposition (Chapter 33). Some authors have reported that tobacco contains nickel (EPA, 1986). Workers may roll cigarettes with nickel-contaminated hands in the workplace. Both situations might be expected to enhance nickel uptake. However, we observed no correlation between nickel in urine or hair and smoking habits.

In conclusion, the present study indicates that urinary nickel was most closely related to airborne levels of insoluble nickel compounds. By contrast, the body burden of nickel was little influenced by duration of exposure and smoking habits. Therefore, nickel concentration in urine may be considered a suitable index of current exposure in the situation studied.

REFERENCES

Aitio, A., Tossavainen, A., Gustafsson, T., Kiilunen, M., Haappa, K., and Jarvisalo, J. (1985). Urinary excretion of nickel and chromium in workers at a steel foundry. In S. S. Brown and F. W. Sunderman, Jr. (Eds.), *Progress in Nickel Toxicology*. Blackwell, Oxford, pp. 149–152.

Bernacki, E. J., Parsons, G. E., Roy, B. R. Mikac-Devic, M., Kennedy, C. D., and Sunderman, F. W., Jr. (1978). Urine nickel concentrations in nickel-exposed workers. *Ann. Clin. Lab. Sci.* **8**, 184–189.

Environmental Protection Agency (EPA) (1986). Health assessment document for nickel and nickel compounds. EPA/600/8-83/012FF, Final Report. U.S. EPA, Research Triangle Park, NC, pp. 4.8–4.9.

Flora, C. J., Nieboer, E. (1980). Determination of nickel by differential pulse polarography at a dropping mercury electrode. *Anal. Chem.* **52**, 1013–1020.

Høgetveit, A. C., and Barton R. T. (1976). Preventive health program for nickel workers. *J. Occup. Med.* **18**, 805–809.

Høgetveit, A. C., Barton, R. T., and Andersen, I. (1980). Variation of nickel in plasma and urine during the work period. *J. Occup. Med.* **22**, 597–600.

Norseth, T. (1986). Nickel. In L. Friberg, G. F. Nordberg, and V. B. Vouk (Eds.), *Handbook on the Toxicology of Metals*, Vol. 2. Elsevier, Amsterdam, pp. 462–481.

Schaller, D. H., Zober, A., and Valentin, H. (1985). External and internal nickel exposure of arc welders during one working week. In S. S. Brown and F. W. Sunderman, Jr. (Eds.), *Progress in Nickel Toxicology*. Blackwell, Oxford, pp. 153–156.

Tola, S., Kilpiö, J., and Virtamo, M. (1979). Urinary and plasma concentrations of nickel as indicators of exposure to nickel in an electroplating shop. *J. Occup. Med.* **21**, 184–188.

9

NICKEL CONTENT OF FINGERNAILS AS MEASURE OF OCCUPATIONAL EXPOSURE TO NICKEL

Bente Gammelgaard

*Royal Danish School of Pharmacy,
DK-2100 Copenhagen, Denmark*

Kurt Peters and Torkil Menné

Gentofte Hospital, DK-2900 Hellerup, Denmark

Nickel and Human Health: Current Perspectives, Edited by Evert Nieboer and
Jerome O. Nriagu.
ISBN 0-471-50076-3   © 1992 John Wiley & Sons, Inc.

1. INTRODUCTION

The concentrations of nickel in plasma or urine are often used as a measure of the body burden of the element and have been shown along with scalp hair to be elevated in occupationally exposed persons (Spruit and Bongaarts, 1977). Elevated urine or blood nickel levels as a measure of occupational exposure have the implicit requirement of absorption by either the lungs or the digestive tract. These indices do not effectively assess external dermal exposures.

Contact dermatitis caused by nickel hypersensitivity is a common disease, especially among women in the industrialized part of the world. In Denmark, hand eczema associated with nickel allergy is the single most common skin disorder, which leads to permanent disability compensation (Menné, 1983). People suffering from contact dermatitis caused by nickel may have been occupationally exposed to nickel via the skin on the hands by handling nickel compounds or nickel-releasing objects. Only small amounts of nickel are required to sensitize an individual, and such exposures may not elevate urine and blood nickel levels. Perhaps nickel concentrations in fingernails may reflect such external exposure more accurately. Our hypothesis is that apart from the contaminating surface layer, nickel is impregnated in nails. Perhaps, it is chemically bound to the nail substance or is trapped as particulates in small surface pits (i.e., interstitial spaces).

Nail analysis would appear to have several advantages. The normal nickel levels in nails are in the microgram-per-gram range (Alder and Batoreau, 1983; Lindemayr, 1984; Ostapzuk et al., 1985), which is three orders of magnitude higher than the levels in blood or urine. This allows the analysis of small samples and reduces the risk of contamination during collection, handling, and analysis. The sampling procedure is simple, and there are no storage problems during the period between sampling and analysis.

The purpose of this work is threefold. The first objective is to define a normal range of nickel in human fingernails from people who are neither hypersensitive to nickel nor occupationally exposed to the metal; second, we want to examine if nickel nail concentrations are significantly elevated in occupationally exposed people; finally, we want to examine if there is any correlation between the nickel concentrations in nails and the duration of the exposure.

2. MATERIALS AND METHODS

2.1. Subjects

Fingernail parings were taken from volunteers in a hospital, a lock factory, an electrical relay and switch factory, and 12 nickel-plating plants of various sizes. Each individual was assigned to one of four different groups, numbered

0–3 in order of expected exposure to nickel. The groups are characterized as follows:

Group 0: Expected Exposure—None. This group consists of people whose occupational environment is not expected to contain nickel in amounts exceeding the exposure in everyday life. Examples of these occupations are clerks, nurses, hospital orderlies, students, and doctors. Group 0 can be thought of as a random sample of the population of people not occupationally exposed, or the "normal" population.

Group 1: Expected Exposure—Weak. This group consists of people from the hospital cleaning staff. The majority of the group works 8 hr a day and are in contact with water and a limited assortment of cleaning materials. Protective rubber gloves and 5% carbamide cream are available as required.

Group 2: Expected Exposure—Moderate. The group represents a random sample of a model population expected to be moderately exposed to nickel. These people are occupied with the assembling of nickel-plated objects in electrical switches, relays, and locks. A few process nickel containing metals by grinding and milling.

Group 3: Expected Exposure—Heavy. This group represents a random sample of heavily exposed individuals. These people are employed in electrolytic nickel plating. They work with galvanic baths in heavily contaminated rooms and in immediate contact with nickel salts, sometimes in acidic solutions. Protective gloves are used in some processes.

The following conditions resulted in exclusion from the investigation: established hypersensitivity to nickel, psoriasis, sustained eczema on the fingers, regular use of nail polish, and treatment with disulfiram during the last year (disulfiram increases the excretion of nickel from the body). Age, sex, and the duration of the exposure to nickel were noted for each person. The final material consisted of 254 nail specimens. The distribution among the four groups is seen in Table 9.1

2.2. Sampling Procedure

Volunteers were asked to wash their hands thoroughly with water and ordinary soap, followed by drying with a clean towel or a paper tissue. Some nail samples were still black or brown pigmented after this cleaning. Nails were cut from as many fingers as possible, preferentially from the right hand from right-handed persons and vice versa. A pair of scissors made of stainless steel was used. In control experiments, the scissors were shown not to release nickel. Samples were stored in acid-cleaned plastic containers until the analysis.

Table 9.1   Distribution of Nail Samples among Four Exposure Groups

Group	Exposure	Number	Male–Female Ratio
0	None	95	36:59
1	Weak	25	1:24
2	Moderate	83	22:61
3	Heavy	51	40:11

2.3.   Sample Pretreatment and Analysis

Before weighing the samples, they were washed according to a standardized washing procedure. Samples were first treated with a 0.1% solution of the surfactant Triton X-100 in an ultrasonic bath for 15 min. Then the clippings were washed five times with ultrapure water, followed by five 1-min treatments with ultrapure water in an ultrasonic bath. Finally, the samples were dried at 70°C overnight in an oven. The washed nail samples were digested by wet ashing with a mixture of concentrated nitric and sulfuric acid, and the nickel contents were determined by adsorption differential pulse voltammetry (ADPV) as previously described (Gammelgaard and Andersen, 1985).

3.   RESULTS AND DISCUSSION

The results of the analyses arranged in histogram format are shown in Figure 9.1. It can be seen that the data are not normally distributed but are asymmetric. Consequently, the random samples are better characterized by their medians rather than by the commonly used parameter, mean ± standard deviation (SD). Nevertheless, the means are given to make it possible to compare these results with previously published values.

Group 0, the random sample of people not occupationally exposed to nickel, or the normal material, shows a mean of 1.19 ± 1.61 ppm (mean ± SD) and a median ± standard error (SE) of 0.49 ± 0.13 ppm (Table 9.2). Lindemayr (1984) found a mean of 6.50 ± 3.50 ppm (range 4–210 ppm) in fingernails from 4 unexposed persons. The analytical method was flame atomic absorption spectrometry (FAAS). Using ADPV, Gammelgaard and Andersen (1985) found a mean of 2.14 ± 1.20 ppm (range 0.65–5.08 ppm) in fingernails from 14 unexposed persons and a mean of 0.78 ± 0.42 ppm (range 0.24–1.31 ppm) in toenails from 5 unexposed persons. The difference between the nickel contents in toenails and fingernails was significant. Ostapzuk et al. (1985) found a mean of 1.38 ppm (range 0.34–3.21 ppm) in fingernails from 5 unexposed persons and a mean of 0.29 ppm (range 0.18–0.39 ppm) in toenails from 5 unexposed persons. By comparison, they found a mean

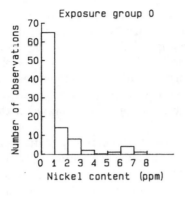

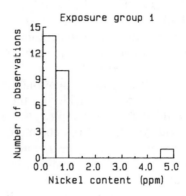

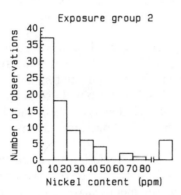

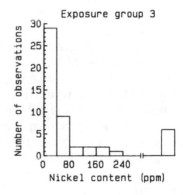

Figure 9.1. Nail nickel content among the four exposure groups: 0, no exposure; 1, weak; 2, moderate; 3, heavy.

Table 9.2   Median ± SE and Mean ± SD for Nickel Content of Nails

Group	Exposure	Median (ppm)		Mean (ppm)	
1	Weak	0.41 ± 0.21	NS	0.61 ± 0.92	NS
0	None	0.49 ± 0.13		1.19 ± 1.61	
2	Moderate	13.8 ± 5.6	p < .001	29.2 ± 56.7	p < .001
3	Heavy	29.9 ± 18.0	p < .05	123.5 ± 288.8	p < .001

Note: ppm corresponds to micrograms per gram; statistical significance was assessed relative to group 0; NS, not statistically significant.

of 1.85 ppm in fingernails from 19 persons and 0.39 ppm in toenails from 15 persons with stainless steel implants. The difference between nickel levels in toenails and fingernails can be interpreted as the difference between endogenously deposited nickel and exogenous accumulation in unexposed people. All the investigations mentioned were performed without previous washing of the samples. Considering the differences in washing protocols and the limited number of samples, there is a fairly good agreement between the previously reported results and our study.

Unexpectedly, the mean and median in group 1, the cleaning staff, were lower than the corresponding values in group 0, the normal material, although the differences are not significant ($p > 0.05$) (Table 9.2). From the outset, group 1 was expected to be weakly exposed, as people occupied with cleaning are reported to have a high incidence of nickel allergy (Clemmensen et al., 1981). According to the current findings, the cleaning staff appears not to be exposed to nickel.

The mean of the results from the moderately exposed group occupied with the assembling of nickel-plated objects is 29.2 ± 56.7 ppm, and the mean of the heavily exposed group occupied with electrolytic plating is 123.5 ± 288.8 ppm. These results are significantly different from those for the normal population ($p < 0.001$). By comparison, Lindemayr (1984) found a mean of 54.22 ± 70.15 ppm (range 4–210 ppm) in fingernails from nine female hairdressers with hand eczema.

The data in Table 9.2 and Figure 9.2 show the association between the mean nickel concentrations in fingernails and the intensity of the exposure in the occupational environment. The differences in means as well as medians for groups are significant ($p < 0.05$). These data lend themselves to the

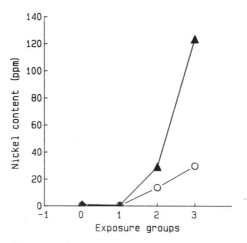

Figure 9.2. Association between nickel concentrations in fingernails and intensity of nickel exposure: (▲) mean; (○) median.

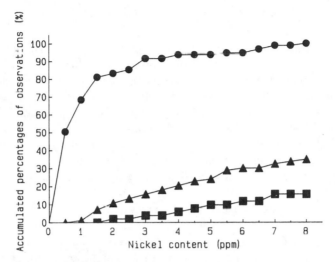

Figure 9.3. Accumulated observation frequencies for exposure groups 0, 2, and 3. Group 1 data are not shown as they fall near that for group 0. (●) Group 0; (▲) group 2; (■) group 3.

following interpretation (working hypothesis): The heavier the nickel exposure is in the occupational environment, the higher are the nickel levels in the fingernails; conversely, the more the nickel concentration in a person's fingernails deviates from the background values, the larger is the possibility that the person is occupationally exposed to nickel.

Figure 9.3 shows the accumulated frequencies of the observations in the normal and the two exposed groups. It is seen that 68% of the observations in the unexposed group 0 are below 1 ppm, while only 1.2% of the observations in group 2 and 0% of group 3 are below 1 ppm. All the observations in the sample of unexposed people are below 8 ppm, while only 35% of the observations in group 2 and 16% of the observations in group 3 are below 8 ppm. From these results it is to be concluded that if a person has a nail nickel content less than 1 ppm, it is unlikely that he or she is exposed to nickel. If the nickel content in the nails is larger than 8 ppm, it is improbable that the individual is not exposed to nickel. In the range between 1 and 8 ppm it may be assumed that the probability of environmental exposure increases with increasing nickel levels.

Correlation between the nickel concentration in nails and the duration of exposure in groups 2 and 3 was also examined. It was not possible to demonstrate any association, not even in the first 12 months of exposure. A few people reached very high values after only a short period of exposure, while some had relatively low values (below 20 ppm) even after exposure times of several years. This suggests that the nickel concentration in nails is primarily decided by the intensity of the exposure rather than by exposure duration.

4. POTENTIAL CLINICAL APPLICATIONS

Using the scheme suggested above, nail analysis may in principle be used to determine if a person is occupationally exposed to nickel. If a person during employment develops a hand eczema and patch testing shows a positive reaction for nickel, there is a good reason to believe that the eczema is the result of nickel hypersensitivity. If the allergy is occupationally acquired, the case should be reported as an occupational injury, and appropriate action should be taken to avoid future exposure. In Denmark, compensation for disablement may be claimed. If the allergy is not due to the occupational environment, the source of the sensitization should be sought in the home environment.

In practice, the nickel dermatitis patients can be placed into three groups:

1. *Nail Nickel Content < 1 ppm.* It is unlikely that the patient is occupationally exposed. The cause of the sensitization should be found in the home environment.
2. *Nail Nickel Content > 8 ppm.* The patient is most probably occupationally exposed to nickel. Unless the patient has a hobby with obvious nickel exposure, the case should be reported as an occupational injury and measures taken to avoid future exposure.
3. *Nail Nickel Content 1–8 ppm.* The probability of occupational exposure has to be assessed in each individual case. The probability of nickel exposure presumably increases with increasing nickel content of the nails. The suspicion of the presence of nickel in the occupational environment should be confirmed or invalidated by inspection.

The effectiveness of this protocol has not been assessed yet. However, special attention should be given to the occurrence of false positives (e.g., high nickel levels due to hobbies), false negatives (e.g., sensitization prior to accumulation of nickel in nails), suitability of the cutoff concentrations of 1 and 8 μg/g, and the relationship of the nail nickel exposure index with the concentrations of this metal in body fluids.

REFERENCES

Alder, J. F., and Batoreau, M. C. (1983). Determination of lead and nickel in epithelial tissue by electrothermal atomic absorption spectrometry. *Anal. Chim. Acta* **155**, 199–207.

Clemmensen, O. J., Menné, T., Kaaber, K., and Solgaard, P. (1981). Exposure of nickel and the relevance of nickel sensitivity among hospital cleaners. *Contact Derm.* **7**, 14–18.

Gammelgaard, B., and Andersen, J. R. (1985). Determination of nickel in human nails by adsorption differential pulse voltammetry. *Analyst* **110**, 1197–1199.

Lindemayr, H. (1984). Friseurekzem und Nickelallegie. *Hautarzt* **35**, 292–297.

Menné, T. (1983). Nickel allergy. Thesis, University of Copenhagen.

Ostapzuk, P., Froning, M., Stoeppler, M., and Nürnberg, H. W. (1985). Square wave voltammetry: a new approach for the sensitive determination of nickel and cobalt in human samples. In S. S. Brown and F. W. Sunderman (Eds.), *Progr. Nickel Toxicol., Proc. Int. Conf. Nickel Metabolism Toxicol., 3rd, Paris 1984.* pp. 129–132.

Spruit, D., and Bongaarts, P.J.M. (1977). Nickel content of plasma urine and hair in contact dermatitis. *Dermatologica* **154,** 291–300.

10

NICKEL CONCENTRATION IN TISSUES OF FINNISH MEN WITHOUT OCCUPATIONAL EXPOSURE

M. Kiilunen, S. Anttila, A. Aitio, and V. Riihimäki

Institute of Occupational Health, Helsinki, Finland

P. Pääkkö

University of Oulu, Oulu, Finland

1. INTRODUCTION

Occupational exposure to nickel has been monitored, in addition to the traditional air measurements, by determining the concentrations of nickel in urine, and also in blood of exposed workers (Aitio, 1984; Doll et al., 1984; Nieboer et al., 1984; Sunderman, 1988; Sunderman et al., 1986). However,

Nickel and Human Health: Current Perspectives, Edited by Evert Nieboer and Jerome O. Nriagu.
ISBN 0-471-50076-3   © 1992 John Wiley & Sons, Inc.

these monitoring methods have the inherent weakness that they do not necessarily indicate levels of nickel in the target organ, which, for the carcinogenic action of nickel, are the nasal sinuses and the lung (IARC, 1987). In the present investigation, we have determined the concentrations of nickel in different organs of Finnish men not exposed to nickel occupationally; such reference values are needed in order to evaluate the levels found in organs from exposed persons. Since the lung is one of the main targets of nickel carcinogenicity, this study was targeted on the lung.

2. SUBJECTS AND SAMPLES

Tissue samples were collected from 34 male patients who had died of non-malignant diseases and were autopsied. The causes of death were ascertained, and they were as follows: 15 had died of acute cardiac infarction, 5 of generalized arteriosclerosis, 5 of pulmonary emphysema, 3 of cerebrovascular accidents, 3 of other cardiovascular diseases, and the remaining 3 of other causes of death.

To obtain a representative material as regards age and smoking habits, four groups each containing 5–16 subjects were formed as follows: (1) age under 70 with normal lungs or mild emphysema; (2) age under 70 and moderate or severe emphysema; (3) age 70 or more and no or mild emphysema; and (4) age 70 or more and moderate or severe emphysema. The average ages and standard deviations in the four groups were 68 ± 7, 68 ± 5, 73 ± 3, and 74 ± 3 years, respectively. The severity of emphysema was determined from radiographs of the excised, air-inflated lungs (continuous air inflation through a filter, pore side 0.2 μm) according to the following criteria: peripheral vascular changes, tissue defect translucencies, and changes in the shape of the lung (Sutinen et al., 1982).

The occupations of the 34 men could be grouped into following five categories: administration, commerce, and social services ($n = 5$); agriculture, forestry, and fishing ($n = 8$); transport and communication ($n = 6$); industry and construction work ($n = 9$); and unknown or no occupation ($n = 6$). The subjects were residents of the two northernmost provinces in Finland.

The occupational and smoking histories were inquired through a questionnaire that was sent to the next of kin and supplemented by a telephone contact when necessary. For seven subjects, patient histories available from hospital records were used because the relatives could not be contacted.

Specimens were taken mostly from the right lung (from the posterior or apicoposterior segment and the anterior segment of the upper lobe of the lung and from the superior segment and the basal segment of the lower lobe of the lung) and from the liver, spleen, renal cortex, and adrenal gland.

The samples were cut with glass knives and put directly to plastic bags and frozen to $-20°C$ until analysis.

3. METHODS OF ANALYSIS

To avoid the risk of contamination, great emphasis was put on sample collection, storing, and analysis. For the cutting of tissues a special glass knife was used. It was cleaned with 0.5% nitric acid and washed in distilled water just before using. All glassware used was kept 24 hr in 20% Deconex solution and rinsed at least 10 times in distilled water. Thereafter the glassware was kept overnight in 0.5% nitric acid solution (*supra pur*) and rinsed 10 times in distilled water. Water for analysis was distilled and ion exchanged fresh before using. The plastic bags used for homogenization were analyzed and found to be free of nickel contamination.

After melting, the samples were weighed. Then they were frozen in liquid nitrogen and homogenized manually in plastic bags. The homogenized samples with a weight range of 100–300 mg were used for analysis. The tissue specimens were wet digested and extracted with *n*-octanol, not with 4-methylpentane-2-one (MIBK), which deviated from the International Union of Pure and Applied Chemistry (IUPAC) reference method for nickel in serum and urine (Brown et al., 1981).

Nickel was analyzed with a Varian SpetrAA-40 atomic absorption spectrophotometer equipped with a D_2 background corrector. Platform tubes were used for analysis. The furnace heating program employed is presented in Table 10.1. Both nitrogen and argon were used as protective gases. No remarkable differences were seen between the peak-area and peak-height measurement modes, although the sensitivity was poorer in the peak-area mode. The sample volume was 20 μL. Samples were standardized against water standards. Commercial Lanonorm 1 (Behring) (3.4μg/L, SD 17.9%

Table 10.1   Graphite Furnace Program

Step Number	Temperature (°C)	Time (sec)	Gas Flow (L/min)	Gas Type	Read Command
1	240	40.0	3.0	Nitrogen	
2	260	40.0	3.0	Nitrogen	
3	305	20.0	3.0	Nitrogen	
4	400	10.0	3.0	Nitrogen	
5	1200	20.0	3.0	Nitrogen	
6	1200	5.0	3.0	Nitrogen	
7	100	5.0	3.0	Nitrogen	
8	100	10.0	3.0	Argon	
9	100	5.0	0.0	Argon	
10	2400	1.2	0.0	Argon	Yes
11	2400	2.0	0.0	Argon	Yes
12	2400	2.0	0.8	Argon	
13	2600	2.0	3.0	Argon	

between different days) and our own water control (6 μg/L, SD 13.9% between different days). Reference wheat flour (21 ± 4 ng/g dry weight) was used as a control material in every series to validate the level of analysis. The average ($n = 34$) recovery of Lanonorm 1 was 132% of the target value, but within the accepable variation specified by the manufacturer; that for the water reference of 6 μg/L recovery was 110%. The coefficient of variation in one round was about 10%.

4. RESULTS AND DISCUSSION

According to earlier reports, the nickel content in tissues of occupationally unexposed persons varies widely. In recent years, more reliable results concerning nickel concentrations in men have been made available (Zober et al., 1984; Kollmeier et al., 1987; Rezuke et al., 1987; Raithel et al., 1987, 1988).

Earlier the measurements tended to be under the detection limits of the analytical methods or the samples were contaminated during the process (Baumgardt et al., 1986). Although we did not use the method of standard additions, which is considered to be more accurate for small concentrations, it was possible to measure nickel concentrations in tissues with a nonvolatile extraction medium (n-octanol). To guarantee the purity of n-octanol, it was extracted with ammoniumpyrrolidineditiocarbamate (APDC) for at least three times before its daily use. The very low volatility of n-octanol was an important feature because the duration of the atomic absorption spectrometry (AAS) run was long. Szathmary and Daldrup (1982) compared different methods of nickel analysis and showed that the results with AAS, gas chromatography (GC), and GC/mass spectrometry (GC/MS) agreed well with each other. Unfortunately, these samples had been taken from persons who had died of nickel poisoning and thus nickel concentrations in tissues were quite high.

In our study the concentrations of nickel in different tissues ranged from 0.3 to 13.6 ng/g wet weight, and no remarkable differences were seen between the different organs. The results are shown in Table 10.2. During the treatment of the tissue samples and analytical procedures, great efforts were made to avoid contamination and to get accurate results. Our results indicate that nickel concentrations in tissues of unexposed persons are very low.

Concentrations of nickel in the range 2000–90,000 ng/g wet weight have been found in lungs of deceased nickel workers (Raithel et al., 1988; Chapter 11), which is a clear evidence of nickel accumulation in the lungs. For the general population, Rezuke et al. (1987) found 18 ± 12 ng/g (7–46 ng/g, $n = 9$) in the lung, 26 ± 15 ng/g (13–56 ng/g, $n = 10$) in the adrenal, 9 ± 6 ng/g (3–25 ng/g, $n = 10$) in the kidney, 10 ± 7 ng/g (2–21 ng/g, $n = 10$) in the liver, and 7 ± 5 ng/g (5–11 ng/g, $n = 10$) in the spleen—all wet weight results—which are in good agreement with our results. Similarly, Zober et al. (1984) detected in the lung 7.4 ng/g and in the kidney 13.9 ng/g of nickel per wet tissue weight. The results of Raithel et al. (1987, 1988) indicate somewhat higher mean concentrations and wider ranges compared to our results that may possibly be due to a higher background exposure in the normal population of Germany and Norway.

In our study in the superior and basal segments of the lower lobe as well as in the spleen, higher mean nickel concentrations were observed among current smokers ($p < 0.05$, Student's t-test). Also, in other tissues the tendency was the same but the differences were not as great as in the specimens above. Nickel concentration in the lungs were in general higher—especially among smokers—than in the other organs analyzed (Fig. 10.1). The pack-years of smoking were significantly correlated to the nickel concentration of the basal segment of the lower lobe of the lung ($p < 0.001$, $r = 0.600$, $n = 29$). For other segments of the lung, the correlations ranged between 0.1248 and 0.4068 and were not significant. Neither were there significant correlations between pack-years and nickel in other organs ($r = -0.0206$ to $r = 0.292$).

The severity of emphysema was divided into four classes: normal ($n = 5$), mild ($n = 16$), moderate ($n = 7$), and severe ($n = 6$). Table 10.3 shows the relation between the mean nickel concentration in various parts of the lung and the severity of emphysema. The results indicate no remarkable differences between the nickel concentrations in different locations. Pulmonary nickel concentrations in the lower lobe of the lung seemed to increase in relation to the severity of emphysema, but that was not statistically significant. Significant correlations were found between the nickel concentration of the different segments of the upper lobe of the lung ($r = 0.518$, $n = 34$, $p < 0.001$) as well as between the anterior segment of the upper lobe and the superior segment of the lower lobe ($r = 0.659$, $n = 34$, $p < 0.001$). Nickel concentrations were also significantly correlated between the adrenal gland and the liver ($r = 0.616$, $n = 34$, $p < 0.001$), the renal cortex and the spleen ($r = 0.557$, $n = 34$, $p < 0.001$), as well as the renal cortex and the adrenal gland ($r = 0.536$, $n = 34$, $p < 0.001$).

Our observations showed significant correlations between smoking habits and the content of nickel in the lung and the spleen. This may be an indication of accumulation of nickel in the lungs from tobacco smoke, which is known to contain nickel (Szadkowski et al., 1969; Menden et al., 1972; Nadkarni, 1974), although the nickel level in mainstream smoke appears to be low (Gutenmann et al., 1982; EPA, 1986). Raithel et al. (1987) could not detect any effect of smoking on the pulmonary content of nickel; however, he used average concentrations in lungs and may thus have missed findings specific of different lung regions. In this context it can be noted that there was also a positive correlation, although not significant, between the nickel concentrations of the lower lobe, in both the superior ($r = 0.322$, $p < 0.10$, $n = 34$) and basal ($r = 0.338$, $p < 0.05$, $n = 34$) segments, and the severity of emphysema. Since emphysema is usually caused by smoking, it is not possible to separate the two effects, that is, that of increased exposure from the nickel

Table 10.2 Nickel Concentrations[a] of Tissues in Autopsy Material Comprising 34 Occupationally Nonexposed Elderly Subjects

Subject Number	Posterior	Anterior	Superior	Basal	Liver	Spleen	Renal Cortex	Adrenal Gland
1	2.9	2.9	2.2	1.6	1.2	1.2	1.7	1.0
2	1.1	0.9	0.5	0.6	0.1	0.3	0.7	0.3
3	4.6	0.1	0.1	2.2	<0.1	0.3	0.1	0.1
4	1.2	0.8	0.5	0.5	1.0	0.7	1.1	0.5
5	5.0	0.6	1.1	0.3	<0.1	0.6	0.7	0.6
6	4.2	4.8	6.6	8.0	0.5	1.3	0.7	1.4
7	2.8	2.9	1.8	2.0	0.8	1.1	2.3	0.8
8	1.6	1.6	0.7	1.1	0.5	1.2	1.6	1.5
9	3.7	2.7	3.7	2.3	1.9	1.0	1.4	3.6
10	3.7	3.7	3.1	2.2	2.4	2.8	4.9	3.5
11	7.0	2.3	2.0	1.9	0.8	1.5	1.4	1.3
12	0.4	1.2	2.5	7.8	7.5	0.9	0.6	1.0
13	0.4	7.1	0.2	0.2	<0.1	0.1	<0.1	0.1
14	5.7	5.2	9.0	1.9	0.7	4.4	0.6	1.8
15	1.5	1.7	2.7	4.4	1.0	2.5	2.2	0.9
16	3.2	1.7	2.3	2.0	2.4	2.8	2.1	2.7

17	5.3	4.1	4.6	3.9	3.4	3.8	3.8	4.1
18	1.5	2.8	1.8	1.2	2.5	0.9	2.6	4.1
19	13.6	7.0	3.9	8.4	3.2	3.5	3.3	3.8
20	1.0	1.3	1.4	1.9	1.1	3.4	5.7	1.5
21	3.5	3.4	3.0	2.5	1.3	1.3	2.0	2.0
22	4.3	6.8	4.5	1.5	2.6	0.4	2.1	3.4
23	7.5	9.5	7.7	6.1	1.5	1.9	2.4	1.2
24	6.1	6.9	5.1	5.1	4.4	2.9	2.9	3.9
25	8.0	4.8	7.2	5.4	1.1	1.3	1.4	4.2
26	3.8	3.2	4.5	4.1	2.2	2.2	2.5	3.8
27	3.1	2.7	2.9	3.1	4.2	2.8	2.9	4.4
28	4.8	4.4	3.0	7.2	2.4	1.9	2.2	3.5
29	5.2	1.0	0.9	1.9	0.6	0.6	2.3	1.0
30	2.2	3.1	2.9	2.7	1.2	1.4	2.6	1.9
31	4.5	3.9	4.6	2.2	2.0	2.1	1.9	1.9
32	9.1	7.2	4.4	7.3	0.8	4.8	1.5	2.3
33	3.6	2.2	2.7	2.2	1.1	1.0	2.3	1.3
34	6.4	6.1	3.8	4.7	7.4	4.0	4.1	5.5
Mean ± SD	4.2 ± 2.7	3.5 ± 2.3	3.2 ± 2.1	3.2 ± 2.3	1.9 ± 1.8	1.9 ± 1.3	2.1 ± 1.3	2.2 ± 1.5

[a]Mean ± SD. All values are in nanograms per gram.

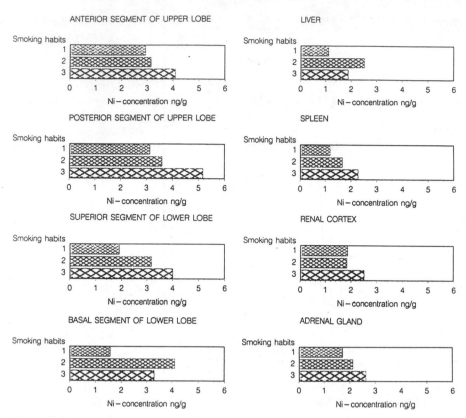

Figure 10.1. Mean nickel concentration in different tissues as a function of smoking habits: (1) nonsmokers, $n = 9$; (2) exsmokers, $n = 12$; (3) smokers, $n = 12$. The association between nickel concentration and current smoking was significant for superior and basal segments of the lower lobe and in the spleen ($p < 0.05$).

Table 10.3   Mean Nickel Concentrations[a] of Various Lung Locations in Relation to Severity of Emphysema

	Nickel Concentration (ng/g wet weight)			
Lung Location	Normal, $n = 5$	Mild, $n = 16$	Moderate, $n = 7$	Severe, $n = 6$
A	4.9 ± 5.1	4.3 ± 2.3	3.8 ± 2.4	3.7 ± 2.4
B	4.1 ± 2.9	3.6 ± 2.5	2.9 ± 1.6	3.9 ± 2.4
C	1.8 ± 1.4	3.1 ± 1.9	3.2 ± 2.2	4.5 ± 3.1
D	2.5 ± 3.4	2.8 ± 1.8	3.2 ± 1.4	5.1 ± 3.2

[a] Mean ± SD. Lung locations: A, posterior or apicoposterior segment of upper lobe; B, anterior segment of upper lobe; C, superior segment of lower lobe; D, basal segment of lower lobe.

in cigarette smoke (Raithel et al., 1987; Kollmeier et al., 1987) and impaired lung clearance mechanisms caused by emphysema. Another confounding factor might be age: Kollmeier et al. (1987) and Rezuke et al. (1987) reported an increase of pulmonary nickel concentration with age. On the other hand, Raithel and co-workers (1987) did not detect such an effect. We could not draw any conclusions regarding the effect of age because the age range of our subjects was quite narrow.

In the present study, which was based on 34 subjects only, statistically significant correlations were detected between the concentrations of nickel in different parts of the lungs. Raithel et al. (1987) found fairly large variations between samples from different locations within a lung in a very inhomogeneous material. In order to get reliable information of the pulmonary content of nickel in differently exposed groups, procurement of several tissue specimens may be advisable.

ACKNOWLEDGMENTS

We are indebted to Dr. Kaj Ahlman of Outokumpu Ltd. for useful discussions in the planning of the project. The authors also present their warmest thanks to Ms. Jaana Pirnes and Ms. Marita Sohlberg for their skillful technical assistance. The study received financial support from the Nickel Producers Environmental Research Association (NiPERA).

REFERENCES

Aitio, A. (1984). Biological monitoring of occupational exposure to nickel. In F. W. Sunderman, Jr., A. Aitio, A. Berlin, C. Bishop, E. Buringh, W. Davis, M. Gounar, P. C. Jacquignon, E. Mastromatteo, J. P. Rigaut, C. Rosenfeld, R. Saracci, and A. Sors (Eds.), *Nickel in the Human Environment.* IARC Scientific Publications No 53. International Agency of Research on Cancer, Lyon, pp. 497–506.

Baumgardt, B., Jackwerth, E., Otto, H., and Tölg, G. (1986). Beitrag zur spurenanalytischen Untersuchung von menschlichem Lungengewebe. *Fresenius Z. Anal. Chem.* **323**, 481–486.

Brown, S. S., Nomoto, S., Stoeppler, M., and Sunderman, F. W., Jr (1981). IUPAC reference method for analysis of nickel in serum and urine by electrothermal atomic absorption spectrometry. *Clin. Biochem.* **14**, 295–299.

Doll, R. (1984). Nickel exposure: a human hazard. In F. W. Sunderman, Jr., A. Aitio, A. Berlin, C. Bishop, E. Buringh, W. Davis, M. Gounar, P. C. Jacquignon, E. Mastromatteo, J. P. Rigaut, C. Rosenfeld, R. Saracci, and A. Sors (Eds.), *Nickel in the Human Environment.* IARC Scientific Publications No 53. International Agency of Research on Cancer, Lyon, pp. 3–21.

Environmental Protection Agency (EPA) (1986). Health assesment document for nickel and nickel compounds. EPA/600/8-83/012FF Final Report. U.S. EPA, Research Triangle Park, NC, pp. 4.8–4.9.

Gutenmann, W., Bache, C., and Lisk, D. (1982). Cadmium and nickel in smoke of cigarettes prepared from tobacco cultured on municipal sludge-amended soil. *J. Toxicol. Environ. Health* **10**, 423–431.

International Agency of Research on Cancer (1987). IARC monographs on the evaluation of carcinogenic risk to humans. Overall evaluations of carcinogenity: an updating of IARC monographs, volumes 1 to 42. Supplement 7. International Agency of Research on Cancer, Lyon, France, pp. 264–269.

Kollmeier, H., Seemann, J. W., Müller, K-M., Rothe, G., Wittig, P., and Schejbal, V. B. (1987). Increased chromium and nickel content in lung tissue and bronchial carcinoma. *Am. J. Ind. Med.* **11**, 659–669.

Menden, E. E., Elia, V. J., Michael, L. W., and Petering, H. G. (1972). Distribution of cadmium and nickel of tobacco during cigarette smoking. *Environ. Sci. Technol.* **6**, 830–832.

Nadkarni, R. A. (1974). Some considerations of metal content of tobacco products. *Chem. Ind.* **17**, 693–696.

Nieboer, E., Yassi, A., Jusys, A. A., Muir, D.C.F. (1984). Technical feasibility and usefulness of biological monitoring in the nickel producing industry. Nickel Producers' Environmental Research Association (NiPERA), Toronto, pp. 1–285.

Raithel, H. J., Ebner, G., Schaller, K. H., Schellmann, B., and Valentin, H. (1987). Problems in establishing normal values for nickel and chromium concentrations in human pulmonary tissue. *Am. J. Ind. Med.* **12**, 55–70.

Raithel, H. J., Schaller, K. H., Reith, A., Svenes, K. B., and Valentin, H. (1988). Investigations on the quantitative determination of nickel and chromium in human lung tissue. Industrial medical, toxicological, and occupational medical expertise aspects. *Int. Arch. Occup. Environ. Health* **60**, 55–66.

Rezuke, W. N., Knight, J. A., and Sunderman, F. W., Jr. (1987). Reference values for nickel concentrations in human tissues and bile. *Am. J. Ind. Med.* **11**, 419–426.

Sunderman, F. W., Jr. (1988). Nickel. In T. W. Clarkson, L. Friberg, G. F. Nordberg, and P. Sager (Eds.), *Biological Monitoring of Toxic Metals*. Plenum, New York and London, pp. 265–282.

Sunderman, F. W., Jr., Aitio, A., Morgan, L. G., Norseth, T. (1986). Biological monitoring of nickel. *Toxicol. Ind. Health* **2**, 17–78.

Sutinen, S., Lohela, P., Pääkkö, P., and Lahti, R. (1982). Accuracy of postmortem radiography of excised air-inflated human lungs in assessment of pulmonary emphysema. *Thorax* **37**, 906–912.

Szadkowski, D., Schultze, H., Schaller, K. H., and Lehnert, G. (1969). Zur ökologischen Bedeutung des schwermetallgehaltes von Zigaretten. *Arch. Hyg. Bacteriol.* **153**, 1–8.

Szathmary, S. C., and Daldrup, T. (1982). Zum Nachweis von Nickel in biologischen Mittels GC, GC-MS und AAS nach einer tödlichen Vergiftung. *Fresenius Z. Anal. Chem.* **313**, 48.

Zober, A., Kick, K., Schaller, K. H., Schellmann, B., and Valentin, H. (1984). Untersuchungen zum Nickel- und Chrom-Gehalt ausgewählter menschlicher Organe und Körperflüssigkeiten. *Zbl. Bakt. Hyg., I. Abt. Orig. B* **179**, 80–95.

11

RENAL TOXICITY OF NICKEL IN HUMANS

William E. Sanford

Department of Physical Sciences, Mohawk College of Applied Arts and Technology, Hamilton, Ontario, Canada

Evert Nieboer

Department of Biochemistry and Occupational Health Program, McMaster University, Hamilton, Ontario, Canada

The experimental work reported in this article was conducted by one of us (W.E.S.) in partial fulfillment of the Degree of Doctor of Philosophy, University of Surrey, United Kingdom.

Nickel and Human Health: Current Perspectives, Edited by Evert Nieboer and Jerome O. Nriagu.
ISBN 0-471-50076-3   © 1992 John Wiley & Sons, Inc.

1. INTRODUCTION

The kidneys not only form urine but also regulate the concentrations of most constituents of the extracellular fluids (homeostasis). This is achieved by excreting most of the end products of bodily metabolism while conserving those that are present at low levels or essential to the human body. This is made possible by the processes of passive glomerular filtration, passive tubular diffusion, and active tubular secretion/absorption.

The blood that enters the Bowman's capsule is "filtered" in the glomerulus, and the "primary urine" produced contains almost all of the salts, sugars, amino acids, and other small molecules that are in the blood plasma; the high-molecular-mass compounds are retained on the blood side of the glomerulus. Water and other substances needed by the body are reabsorbed (e.g., glucose, amino acids, and low-molecular-mass proteins) from the proximal convoluted tubule into the capillaries that surround the nephron. Inorganic ions (e.g., Na^+, Cl^-, HCO_3^{-1}) are reabsorbed along the entire length of the nephron. Waste products are usually not reabsorbed to any significant amount, while other substances (e.g., K^+, p-amino-hippurate, and penicillin) can be secreted into the glomerular filtrate by the tubular epithelial cells.

The kidneys, taken as a single unit, receive approximately 20–25% of the entire cardiac output. The high renal blood flow results in the delivery of relatively large amounts of toxic substances. The concentration process involved in the formation of urine can result in a nontoxic concentration of a chemical in blood becoming toxic in the kidney. As illustrated by Nieboer et al. (Chapter 4), nickel is known to be reabsorbed in the human kidney, and animal studies indicate that nickel can reach toxic levels in this organ. In the following two sections, a brief summary is provided of the nephrotoxicity of nickel and of other metals in humans. Subsequently, a synopsis is given of a new study conducted by the authors in which a number of biochemical indices of kidney function are evaluated in two groups of electrolytic nickel refinery workers.

1.1. Nephrotoxicity of Metals

As noted by Nieboer and Sanford (1985), "metal-ion-induced kidney damage is not a general cytotoxic effect but is specific for individual metals." For example, uranium as the uranyl ion (UO_2^{2+}) becomes attached to the brush-border membrane of the proximal tubule cells (Berlin and Rudell, 1986) with subsequent loss of cell function followed by cell death (Dounce, 1949; Clarkson, 1983). Cadmium is primarily excreted in the urine as a complex of the low-molecular-mass protein metallothionein (relative molecular mass of 7000) (Lauwerys, 1983; Herber, 1984; Goyer, 1983; Nieboer et al., 1988). After tubular reabsorption, this complex is believed to degrade within lysosomes with subsequent release of Cd^{2+} (Nieboer and Sanford, 1985, and references therein). Perturbation of the homeostatis of zinc, and possibly copper, may also be involved, since metallothionein that has not been induced by cadmium contains both zinc and copper. There is evidence to suggest that Cd^{2+} ion impairs cell function by interfering with the function of lysosomes and the nucleus, resulting ultimately in tubular loss of low-molecular-mass protein and suppression of amino acid transport (Fowler, 1983; Foulkes 1983). The Pb^{2+} damages and inhibits the mitochondria [the site of cellular respiration and adenosine triphosphate (ATP) production] of the proximal tubular cells. This results in an alteration of the normal reabsorptive functions (Fowler, 1983). Lilis et al. (1980) also found evidence of a reduction in the effective glomerular filtration capacity on chronic industrial exposure to lead. Gold, mercury, and platinum are other metals of concern vis-à-vis nephrotoxicity (Nieboer and Sanford, 1985; Friberg et al., 1986). From this very brief review, one can see that the major site of damage caused by metals appears to be the proximal convoluted tubule and to a lesser degree the glomerulus.

1.2. Nephrotoxic Effects of Nickel in Humans

Before presenting new experimental data, the few existing reports of nephro-toxicity in humans that are attributed to nickel exposure are reviewed. Gitlitz et al. (1975) mentions an unpublished report of proteinuria in people who chronically drank water from a well contaminated with nickel, and Sunderman and Horak (1981) reported a significant increase ($p < 0.05$) in the number of individuals with minor elevation of urinary β_2-microglobulin (β_2-m) among nickel workers. By contrast, Wall and Calnan (1980) found that there was no evidence of renal dysfunction among 17 nickel process workers in an electroforming plant during an outbreak of occupational dermatitis. It is important to note that, other than the situations mentioned and in two cases of nickel carbonyl poisoning (Sunderman, 1977), severe proteinuria or any other markers of significant renal disease have not been reported despite years of biological monitoring and observation (Nieboer et al., 1984). On the other hand, there has been a case report of immunoglobulin A (IgA) nephropathy associated with nickel sensitization to a dental alloy (Strauss and Eggleston, 1985). These authors suggest that their patient may represent an example of nickel-induced sensitization associated with IgA glomerulopathy. More recently, Sunderman et al. (1988) investigated the acute toxicity of nickel in electroplating workers who drank water contaminated with nickel sulfate and nickel chloride. Many of the 20 workers developed a wide range of symptoms such as nausea, vomiting, lassitude, and headache. Although very high serum and urinary nickel levels were evident in these workers (see Chapters 4 and 5), only three subjects had elevated urine albumin levels on the second day after exposure, which diminished by day 3 postexposure and returned to normal on day 5. From the above summary of incidence of nickel nephrotoxicity in humans, one can see that such observations are very rare in comparison to the well-documented nephropathies induced by uranium, cadmium, and lead (Nieboer et al., 1984).

By contrast to the inconclusive evidence of renal damage in humans, there is much better evidence of nickel-induced nephropathy in animal studies. For example, poisoning of dogs and cats by nickel nitrate was reported to produce acute renal injury with proteinuria and hyaline casts. Inhalation or injection of nickel carbonyl was found to cause transient azotemia, with pathological evidence of glomerular and tubular damage in various species including man (Sunderman et al., 1975; Gitlitz et al., 1975; Sunderman, 1977; Sunderman and Horak, 1981). Acute nephropathy developed in rats after single intraperitoneal injections of $NiCl_2$ with a 4-fold increase in urinary total proteins, a 2-fold increase of urine N-acetyl-D-glucosaminidase (NAG) activity, and a 20- to 25-fold increase of valine excretion (Sunderman and Horak, 1981). Mathur and Tandon (1981) also found significant elevation of urine and plasma α-amino acid levels in rats 7 days after intraperitoneal injection of $NiSO_4 \cdot 6H_2O$.

2. METHODOLOGY

Only a brief outline of the materials and methods will be given. Readers are referred to Sanford et al. (1992a,b) and Sanford (1988) for more details on the methodologies employed.

2.1. Donor Population

The subjects were 26 individuals occupationally exposed to nickel in electrolytic refining operations. Two plants were involved, with 6 workers from one (group 1) and 20 from the other (group 2).

2.2. Urine and Serum Collection Procedures

Multivoid, 24-hr urine collections were obtained, as well as serum samples at the beginning and end of this sampling period. Urine voids were made directly into an acid-washed 500-mL polyethylene (PE) bottle. Appropriate aliquots were removed from selected voids and their pH adjusted to 6–8 with 1 N NaOH. These samples were employed for the determination of β_2-m. Blood samples were collected according to the International Union of Pure and Applied Chemistry (IUPAC) nickel working method (Brown et al., 1981). After centrifugation, the serum was transferred with an acid-washed Pasteur pipet to an acid-washed polypropylene tube wrapped with parafilm for transport. Both the urine and serum samples were transported to McMaster University as soon as possible by car (group 1) or by air after freezing (group 2). Samples were kept frozen until they were analyzed for nickel. Precautions were taken to minimize contamination during collection, handling, and analysis of the samples (Nieboer and Jusys, 1983).

2.3. Analytical Procedures

Urinary nickel was directly extracted into 4-methylpentane-2-one (MIBK) at pH $\approx$ 7 with ammonium pyrolidine dithiocarbamate (APDC) without prior acid digestion; the analytical measurement was by electrothermal atomic absorption spectrometry (EAAS) (Nieboer et al., 1984). The determination of nickel in serum involved acid precipitation with HNO_3 of the protein fraction and extracting neutralized, protein-free supernatant with APDC/MIBK. This method constitutes an adaptation of the protocols reported by Adams (1980) and Sunderman et al. (1984).

All urine voids were screened, in a semiquantitative manner employing dipsticks (N-Multistix reagent strips, AMES) for indications of any abnormal general health effects and kidney function. All urinary samples that showed trace amounts of protein on the dipsticks were analyzed for total protein by the Bio-Rad procedure (Bradford, 1976). Specific gravity was measured by a hand-held refractometer (temperature compensated). The β_2-m was measured in both the serum and urine employing Phadebas β_2-micro and RIA commercial kits from Pharmacia. A single-channel Technicon Auto Analyzer with a dialysis block was employed to measure the urinary and sera creatinine levels by the Jaffé reaction.

3. RESULTS

Tables 11.1 and 11.2 summarize the combined results of the biochemical indices of nephrotoxicity for both groups 1 and 2. These data show that these indices are in the normal range in all but a few cases. Most of the dipstick tests were negative, while only trace levels were indicated in the positive responses. Two workers had abnormally low creatinine clearances (29 and 41 mL/min) adjusted for a body surface area of 1.73 m^2. Three workers were found to have elevated urinary β_2-m in individual voids. However, when averaged over three or more voids, the average values were below 200 μg/g creatinine and within the normal 24-hr excretion range. Similar excursions were observed for a small number of voids in the case of total protein (see footnote g, Table 11.1).

4. DISCUSSION

4.1. Indices of Nephrotoxicity

Two markers of renal status often measured to assess the nephrotoxicity of cadmium are urinary levels of total protein and β_2-m. Normally, only trace amounts of β_2-m and total protein are present in the urine. Excretion of β_2-m is noticeably increased in nephrotubular disorders. This change can be due to exposure to metals such as cadmium (Lauwerys, 1983) and gold-based

Table 11.1 Combined Laboratory Test Results, Groups 1 and 2

Substance[a]	Observed Range[b]	Normal Range	Units
Urine			
Creatinine[c]	14–27	21–26[d]	mg/kg day
(24, 29)	(22 ± 4)		
Total protein	28–143	<150[e]	mg/day
(26, 26)	(99 ± 37)	(77 ± 26)[f]	
Total protein[g,h]	12–253	<95[i]	mg/g creatinine
(26, 84)	(45 ± 29)		
β_2-Microglobulin[h]	ND–178	<250[j]	μg/L
25, 56)	(76 ± 35)		
β_2-Microglobulin[h,k]	ND–442	<200[l]	μg/g creatinine
(25, 56)	(83 ± 81)	(60 ± 40)[f]	
Specific gravity	1.003–1.032	1.002–1.030[m]	—
(26, 239)	(1.018 ± 0.0008)		
Serum			
Creatinine[n]	0.73–1.36	0.6–1.20[e]	mg/dL
(26, 64)	(1.03 ± 0.16)	(0.99 ± 0.12)[f]	

Note: None of the observed body fluid concentrations differed significantly from the normal values ($P > 0.05$), tested relative to the midpoint of the normal range and/or mean ± SD when employing Student's t-test.

[a] The first number in parentheses denotes number of donors and the second, number of specimens.

[b] Numbers in parentheses are values of $\bar{x} \pm SD$.

[c] Excluding two donors with low creatinine clearances in group 2 (see footnote n).

[d] From Di Giorgio, 1974.

[e] Clinical Chemistry Laboratory, Adult Reference Ranges, Chedoke-McMaster Medical Centre.

[f] From Sunderman and Horak, 1981; $N = 18$ or 28.

[g] One void in the case of two donors exceeded 95 mg/g creatinine; when combined with other voids for the same donor, the amount of total protein was less than 95 mg/g creatinine.

[h] Spot urine samples.

[i] From Herber, 1984.

[j] From Pharmacia β_2-m instruction book.

[k] Three donors had individual voids with β_2-microglobulin levels above 200 μg/g (i.e., 258, 283, 442, and 265 μg/g). When the results for all voids analyzed for β_2-m were combined, the average was less than 200 μg/g in each case.

[l] From Lauwerys, 1983.

[m] From Faulkner and King, 1976.

[n] The creatinine clearance was 119 ± 36 mL/min (all values) and 124 ± 31 mL/min (range 84–211 mL/min) when excluding the two low values. The creatinine clearances adjusted to 1.73 m^2 were 104 ± 33 mL/min (all values) and 109 ± 28 (range 70–190 mL/min) when excluding the two low values.

Table 11.2   Combined Dipstick Test Results, Groups 1 and 2

Test	Negative k	Negative n	Positive[a] k	Positive[a] n
Ketones	26	217	5	11
Bilirubin	26	215	9	13
Glucose	25	213	7	15
Nitrite	26	220	4	8
Blood	26	226	1	2
Urobilinogen	26	208	11	20
Protein	25	153	20	75

Note: k, donors; n, total analyses. Donor ages [mean $\pm$ SD (range)]: Group 1, 48 $\pm$ 12 years (31–59 years); Group 2, 45 $\pm$ 8 years (32–57 years).
[a] Only trace levels were indicated in all cases.

antirheumatic compounds (Merle et al., 1980) or to urinary tract infections (Schardijn et al., 1979) or renal parenchymal disease (Sherman et al., 1983). In renal diseases, increased serum β_2-m levels are considered to reflect alterations in glomerular filtration rate (GFR) more accurately than do serum creatinines (Wibell et al., 1973; Shea et al., 1981). The occurrence of both reduced creatinine clearance and increased serum creatinine is considered indicative of reduced GFR.

Usually, the total protein excreted in the urine from normal donors rarely exceeds 150 mg/day. Increased excretion of urinary protein is an early sign of renal damage. Proteins cross the glomerular filter by a complex process that depends on the endothelial pore size, electrical charge, and binding characteristics of the protein molecule as well as the absolute values of the GFR (Lauwerys, 1984). Proteins with relative molecular mass below 40,000–60,000 pass easily through the glomerular filter and are reabsorbed by the proximal tubular cells of the nephron. When there is glomerular damage, the glomerular permeability is usually increased, and therefore larger quantities of high-molecular-mass proteins enter the glomerular filtrate and appear in the urine (Lauwerys, 1984). Low-molecular-mass proteinuria is usually not increased because their reabsorption has not changed. In case of tubular dysfunction, the amount of protein filtered through the glomeruli is not increased. However, in this instance the low-molecular-mass proteins (and perhaps some high) normally filtered are not reabsorbed to the same degree. Hence larger quantities of the low-molecular-mass proteins appear in the urine (Lauwerys, 1984).

It is judicious to screen urine for indicators of general health effects and abnormal kidney function. Below is a brief summary of the dipstick tests used in this study and their significance for kidney function or general health [the appropriate biochemical/physiological information is found in a number of sources such as Green (1976), Leaf and Cotran (1980), Matta and Wilbraham (1981), and Vander et al. (1990)]:

Ketones. Low incidence of ketonuria is an indication that a study population is essentially free of renal complications such as papillary necrosis or nephrotic syndrome associated with diabetes mellitus (Leaf and Cotran, 1980).

Bilirubin. Any blockage or failure of the bilirubin excretory pathway (e.g., blockage of bile ducts by a gallstone) will result in its accumulation in the blood, causing the condition known as jaundice and excretion in the urine rather than in the feces.

Glucose. If present in the urine, it is usually an indication of impaired glucose metabolism associated with diabetes mellitus or an indication of poor reabsorption of glucose in the kidney.

Nitrite. A positive test for nitrite indicates significant bacteriuria that may indicate pyelonephritis, cystitis, and urethritis.

Blood. Hematuria is associated with many kidney disorders such as acute nephritis, calculi, renal carcinoma, or chronic kidney infection.

Urobilinogen. A small amount in urine is normal, and increased levels are associated with hepatic disease while decreased levels indicate biliary obstruction.

4.2. Interpretation of Results

The trace levels and transient nature of the results from the dipstick screening of urine specimens indicate that there was no evidence of complicating factors in kidney function in the workers examined, such as bacterial infection or any renal indication of other major health perturbations in the populations examined.

Examination of the measured biochemical indices of renal function indicates that in comparison to a known nephrotoxic metal such as cadmium (Elinder et al., 1985), the changes in renal function observed for the nickel workers may be designated as minimal. In particular, the excretion of urinary protein for two donors (one from each group) corresponded to a single void with total protein above 95 mg/g creatinine, with one of them exceeding 200 mg/g creatinine. When these voids were combined with other voids from the same donor, the "average" fell below 95 mg/g creatinine. Thus the occurrence of protein in the urine was considerably lower than 250 mg/g creatinine, which is usually considered to be the cutoff point for normal kidney function (Lauwerys, 1983; Herber, 1984). Similarly, it was observed that four urine voids (from three donors) had β_2-m concentrations greater than 200 µg/g creatinine. As before, when these values for each donor were averaged with the values for the remaining voids, the 24-hr "average" was less than 200 µg/g creatinine. By comparison, urinary concentrations of β_2-m exceeding 1000 or even 5000 µg/L are often observed among affected cadmium workers (Kjellstrom, 1986). Thus, in the present study both urinary total protein and β_2-m do not appear to be elevated.

For the two workers with low creatinine clearance, there were no other biochemical indices supporting the presence of kidney damage; that is, serum creatinine and β_2-m were not increased, and total protein and β_2-m in urine were not elevated. Consequently, the reduced clearance may well indicate incomplete collection of the 24-hr specimen rather than renal dysfunction. Other errors can also occur in the measurement of creatinine clearances, namely, faulty timing, unusual exercise during the collection period, and improper hydration of the patient which could result in negative error due to retention of urine in the bladder (Faulkner and King, 1976).

4.3. Comparison of Results with Other Studies

It is of interest to compare our results with those of Sunderman and Horak (1981). The data in Table 11.3 attest to the good agreement between the two

Table 11.3   Urinary β_2-Microglobulin in Nickel-exposed Workers

Group Examined[a]	N	Urinary Nickel, µg/L (mean ± SD)	Urinary β_2-Microglobulin, µg/L (mean ± SD)
Controls[b]	18	2.0 ± 1.5[c]	84 ± 61
Electroplaters[b]	18	6.1 ± 6.0	68 ± 5
Ni refinery workers[b]	19	<100	89 ± 58
Ni refinery workers[b]	11	>100	193 ± 153
Ni refinery workers,[d] group 1	9	28 ± 6[e]	62 ± 14
Ni refinery workers,[d] group 2	47	60 ± 30[e]	80 ± 37

[a] Refers to males only.
[b] From Sunderman and Horak (1981); controls were healthy males without occupational exposure to nickel; spot specimens appear to have been collected.
[c] Although not measured for the indicated control group, the nickel concentration does correspond to healthy adult persons not exposed to nickel occupationally (Sunderman et al., 1984).
[d] From the present study.
[e] Values obtained from plots of the logarithm of urinary nickel *versus* the logarithm of void volume and correspond to a urine flow rate of 1 mL/min (see Chapter 4).

studies. For urinary nickel concentrations below 100 μg/L, there is no evidence of β_2-microglobulinuria, while mild elevation occurred in the group with nickel levels above 100 μg/L. In this group, five out of eleven ($p < 0.05$) had β_2-m values exceeding the normal upper limit (taken as 240 μg/L); the highest observed concentration was 450 μg/L. There is no information to assess whether these elevations represent temporary excursions in the spot voids collected or whether they denote a more chronic effect. Nevertheless, the observed β_2-microglobulinuria is mild, since concentrations greatly exceeding 1000 μg/L are common among cadmium workers with kidney dysfunction (Kjellstrom, 1986) or patients with renal parenchymal disease (Sherman et al., 1983).

As already indicated, Sunderman et al. (1988) followed a group of workers who accidentally ingested a solution of nickel sulfate and nickel chloride. The nickel doses consumed by workers with symptoms of acute toxicity were estimated to be in the range 0.5–2.5 g. On day 1 postexposure, serum nickel levels ranged from 13 to 1340 μg/L, while urinary concentrations were 150–12,000 μg/g creatinine. These constitute some of the highest concentrations reported in a study in which the analytical data are reliable. Three of the subjects had urine albumin concentrations on day 2 postexposure that exceeded the upper limit of normal (taken as 16 mg/g creatinine). The elevated albumin levels (27, 40, and 68 mg/g creatinine) "diminished on day 3 and returned to normal by day 5, suggesting mild transient nephrotoxicity." Serum and urine nickel levels also dropped rapidly in this time period. Again, these data concur with the conclusion that nickel is a mild nephrotoxin.

ACKNOWLEDGMENT

The experimental work reported in this chapter was undertaken with the financial support of the Nickel Producers Environmental Research Association (NiPERA).

REFERENCES

Adams, D. B. (1980). The routine determination of nickel and creatinine in urine. In S. S. Brown and F. W. Sunderman, Jr. (Eds.), *Nickel Toxicology*. Academic, London, pp. 99–102.

Berlin, M., and Rudell, B. (1986). Uranium. In L. Friberg, G. F. Nordberg, and V. B. Vouk (Eds.), *Handbook on the Toxicology of Metals*, 2nd ed. Vol 2. Elsevier, Amsterdam, pp. 623–637.

Bradford, M. M. (1976). A rapid and sensitive method for the quantitation of microgram quantities of protein utilizing the principle of protein-dye binding. *Anal. Biochem.* **72**, 248–254.

Brown, S. S., Nomoto, S., Stoeppler, M., and Sunderman, F. W., Jr. (1981). IUPAC reference method for analysis of nickel in serum and urine by electrothermal atomic absorption spectrometry. *Clin. Biochem.* **14**, 295–299.

Clarkson, T. W. (1983). Molecular targets of metal toxicity. In S. S. Brown and J. Savory (Eds.), *Chemical Toxicology and Clinical Chemistry of Metals*. Academic, London, pp. 211–226.

DiGiorgio, J. (1974). Nonprotein nitrogenous constituents. In R. J. Henry, D. C. Cannon, and J. W. Winkelman (Eds.), *Clinical Chemistry Principles and Technics*. Harper/Row, Hagerstown, MD, pp. 503–563.

Dounce, A. L. (1949). The mechanism of action of uranium compounds in the animal body. In C. Voegtlin and H. C. Hodge (Eds.), *Pharmacology and Toxicology of Uranium Compounds*. McGraw-Hill, New York, pp. 951–991.

Elinder, C.-G. Edling, C., Lindberg, E., Kagedal, B., and Vesterberg, O. (1985). β_2-Microglobulinuria among workers previously exposed to cadmium: follow-up and dose–response analysis. *Am. J. Ind. Med.* **8**, 553–564.

Faulkner, W. R., and King, J. W. (1976). Renal function. In N. W. Tietz (Ed.), *The Fundamentals of Clinical Chemistry*. W. B. Saunders, Philadelphia, pp. 975–1014.

Foulkes, E. C. (1983). Tubular sites of action of heavy metals, and the nature of their inhibition of amino-acid reabsorption. *Fed. Proc.* **42**, 2965–2968.

Fowler, B. A. (1983). Role of ultrastructural techniques in understanding mechanisms of metal-induced nephrotoxicity. *Fed. Proc.* **42**, 2957–2964.

Friberg, L., Nordberg, G. F., and Vouk, V. B. (Eds.) (1986). *Handbook on the Toxicology of Metals*, 2nd ed., Vol 2. Elsevier/North-Holland, Amsterdam.

Gitlitz, P. H., Sunderman, F. W., Jr., and Goldblatt, P. J. (1975). Aminoaciduria and proteinuria in rats after a single intraperitoneal injection of Ni(II). *Toxicol. Appl. Pharmacol.* **34**, 430–440.

Goyer, R. A. (1983). Metal–protein complexes in detoxification processes. In S. S. Brown and J. Savory (Eds.), *Chemical Toxicology and Clinical Chemistry of Metals*. Academic, London, pp. 199–209.

Green, J. H. (1976). *An Introduction to Human Physiology*, 4th ed. Oxford University Press, New York.

Herber, R.F.M. (1984). Beta-2-microglobulin and other urinary proteins as an index of cadmium-nephrotoxicity, *Pure Appl. Chem.* **56**, 957–965.

Kjellstrom, T. (1986). Renal effects. In L. Friberg, C.-G. Elinder, T. Kjellstrom, and G. F. Nordberg (Eds.), *Cadmium and Health: A Toxicological and Epidemiological Appraisal*, Vol. 2. CRC Press, Boca Raton, FL, pp. 22–109.

Lauwerys, R. R. (1983). *Industrial Chemical Exposure: Guidelines for Biological Monitoring*. Biomedical Publications, Davis, CA.

Lauwerys, R. R. (1984). Assessment of early nephrotoxicity, In A. Aitio, V. Riihimaki, and H. Vainio (Eds.), *Biological Monitoring and Surveillance of Workers Exposed to Chemicals*. Hemisphere Publishing, New York, pp. 315–320. (Distributed outside USA by McGraw-Hill.)

Leaf A., and Cotran, R. S. (1980). *Renal Pathophysiology*, 2nd ed. Oxford University Press, New York.

Lilis, R., Fischbein, A., Valciukas, J. A., Blumberg, W., and Selikoff, J. (1980). Kidney function and lead: relationships in several occupational groups with different levels of exposure. *Am. J. Ind. Med.* **1**, 405–412.

Mathur, A. K., and Tandon, S. K. (1981). Effect of nickel(II) on urinary and plasma concentrations of α-amino acids in rats. *Toxicol. Lett.* **9**, 211–214.

Matta, M. S., and Wilbraham, A. C. (1981). *Atoms Molecules and Life: An Introduction to General Organic and Biological Chemistry*. Benjamin/Cummings, Menlo Park, CA.

Merle, L. J., Reidenberg, M. M., Camacho, M. T., Jones, B. R., and Drayer D. E. (1980). Renal injury in patients with rheumatoid arthritis treated with gold. *Clin. Pharmacol. Ther.* **28**, 216–222.

Nieboer, E., and Jusys, A. A. (1983). Contamination control in routine ultratrace analysis of toxic metals. In S. S. Brown and J. Savory (Eds.), *Chemical Toxicology and Clinical Chemistry of Metals*. Academic, London, pp. 3–16.

Nieboer, E., and Sanford, W. E. (1985). Essential, toxic and therapeutic functions of metals (including determinants of reactivity). In E. Hodgson, J. R. Bend, and R. M. Philpot (Eds.), *Reviews in Biochemical Toxicology*, Vol. 7. Elsevier, New York, pp. 205–245.

Nieboer, E., Yassi, A., Jusys, A. A., and Muir, D.C.F. (1984). *The Technical Feasibility and Usefulness of Biological Monitoring in the Nickel Producing Industry*. Special document, McMaster University, Hamilton, Ontario Canada. Available from the Nickel Producers Environmental Research Association, Durham, N.C.

Nieboer, E., Jusys, A. A., Tom, R. T., Brouwer, E., and Dempster, J. J. (1988). *Radio-immunological Determination of Metallothionein in Human Body Fluids*. Special research report, McMaster University, Hamilton, Ontario, February 1, 1988.

Sanford, W. E. (1988). The renal clearance of nickel in man: implications for biological monitoring. Ph.D. thesis. University of Surrey, Guildford, United Kingdom.

Sanford, W. E., Stace, B. C., and Nieboer, E. (1992a). Illustration for nickel of a systematic approach to urinary dilution corrections. *Scand. J. Work Environ. Health*, submitted.

Sanford, W. E., Stace, B. C., and Nieboer, E. (1992b). Renal clearance of nickel and biochemical indices of kidney function in electrolytic refinery workers. *Scand. J. Work Environ. Health*, submitted.

Schardijn, G., Statius van Eps, L. W., and Swaak, A.J.G. (1979). Urinary β_2-microglobulin in upper and lower urinary tract infections. *Lancet* **1**, 805–807.

Shea, P. H., Maher, J. F., and Horak, E. (1981). Prediction of glomerular filtration rate by serum creatinine and β_2-microglobulin. *Nephron* **29**, 30–35.

Sherman, R. L., Drayer, D. E., Leyland-Jones, B. R., and Reidenberg, M. M. (1983). *N*-acetyl-β-glucosaminidase and β_2-microglobulin. *Arch. Int. Med.* **143**, 1183–1185.

Strauss, F. G., and Eggleston D. W. (1985). IgA nephropathy associated with dental nickel alloy sensitization. *Am. J. Nephrol.* **5**, 395–397.

Sunderman, F. W., Jr. (1977). A review of the metabolism and toxicology of nickel. *Ann. Clin. Lab. Sci.* **7**, 377–398.

Sunderman, F. W., Jr., and Horak, E. (1981). Biochemical indices of nephrotoxicity, exemplified by studies of nickel nephropathy. In S. S. Brown and D. S. Davies (Eds.), *Chemical Indices and Mechanisms of Organ-Directed Toxicity*. Pergamon, Oxford, pp. 55–67.

Sunderman, F. W., Jr., Coulston, F., Eichhorn, G. I., Fellows, J. A., Mastromatteo, E., Reno, H. T., and Samitz, M. H. (1975). *Nickel. A Report of the Committee on Medical and Biologic Effects of Environment Pollutants*. U.S. National Academy of Science, Washington.

Sunderman, F. W., Jr., Crisostomo, M. C., Reid, M. C., Hopfer, S. M., and Nomoto, S. (1984). Rapid analysis of nickel in serum and whole blood by electrothermal atomic absorption spectrophotometry. *Ann. Clin. Lab. Sci.* **14**, 232–241.

Sunderman, F. W., Jr., Dingle B., Hopfer, S. M., and Swift, T. (1988). Acute nickel toxicity in electroplating workers who accidentally ingested a solution of nickel sulfate and nickel chloride. *Am. J. Ind. Med.* **14**, 257–266.

Vander, A. J., Sherman, J. H., and Luciano, D. S. (1990). *Human Physiology: The Mechanisms of Body Function*, 5th ed. McGraw-Hill, New York.

Wall, L. M., and Calnan, C. D. (1980). Occupational nickel dermatitis in the electroforming industry. *Contact Derm.* **6**, 414–420.

Wibell, L., Evrin, P.-E., and Berggord, I. (1973). Serum β_2-microglobulin in renal disease. *Nephron* **10**, 320–331.

12

NICKEL AT THE RENAL GLOMERULUS: MOLECULAR AND CELLULAR INTERACTIONS

Douglas M. Templeton

University of Toronto, Department of Clinical Biochemistry, Banting Institute, Toronto, Canada, M5G 1L5 and the Hospital for Sick Children, Research Institute, Department of Biochemistry, Toronto, Canada, M5G 1X8

Nickel and Human Health: Current Perspectives, Edited by Evert Nieboer and Jerome O. Nriagu.
ISBN 0-471-50076-3 © 1992 John Wiley & Sons, Inc.

1. INTRODUCTION

Acute exposures to soluble Ni(II) complexes are generally well tolerated both in man and experimental animals. Nevertheless, toxic effects on a number of tissues and organs have been observed following single exposures to nickel compounds (for reviews see Sunderman, 1988; Mastromatteo, 1986; Norseth, 1984). The kidney plays a central role in the toxicokinetics of nickel, since it serves as the major organ of excretion (Sunderman, 1988; Onkelinx and Sunderman, 1980) and as a site of accumulation (Sunderman, 1988; Parker and Sunderman, 1974) of absorbed nickel, and is itself a target of nickel toxicity under some circumstances. The renal glomerulus is responsible for the ultrafiltration of circulating low-molecular-weight complexes of nickel (Nordberg, 1982) and is in turn a site of transient pathology following a single administration of $NiCl_2$ to rats (Gitlitz et al., 1975). A detailed understanding of the molecular interactions of nickel at the glomerular level is, therefore, of the utmost importance.

In this chapter, we briefly review the current state of knowledge of the toxicity of metals to the glomerulus, with emphasis on nickel, and describe certain properties of the isolated glomerulus relevant to its use in studying the in vitro toxicity of nickel. We then highlight our own work on the binding of nickel to glomerular extracellular components and on the toxicity of nickel to isolated glomeruli in suspension. Evidence is presented that supports the view that toxic effects are manifest at the extracellular level and result from an effect of ionic nickel on the assembly of the glomerular basement membrane (GBM). Some details of methodology for handling the isolated glomerulus, its cells, and extracellular matrices are given in Section 3.

2. GLOMERULAR HANDLING OF NICKEL *IN VIVO*

2.1. Glomerular Filtration of Nickel

Nickel is transported in serum primarily bound to the N-terminal tripeptide of albumin (Lucassen and Sarkar, 1979; Nomoto, 1980; Lau and Sarkar, 1984) and circulates bound also to higher molecular weight proteins such as α_2-macroglobulin (Nomoto et al., 1971). The ultrafiltrable fraction of serum nickel shows a species dependence, comprising 41% of total serum nickel in man and 27% in the rat, the latter representing 1.8 µg/L of ultrafiltrable ion (Hendel and Sunderman, 1972). This fraction is primarily responsible for the excretion of nickel (Asato et al., 1975; Onkelinx and Sunderman, 1980; Nordberg, 1982) and is generally believed to represent amino acid or oligopeptide complexation (Asato et al., 1975; Lucassen and Sarkar, 1979; Abbracchio et al., 1982). The nature of these ligands strongly influences the cellular uptake and toxicity of ionic nickel (Abbracchio et al., 1982; Costa and Heck, 1985). Following glomerular filtration of its low-molecular-weight complexes, nickel undergoes extensive tubular reabsorption (99%; Nordberg, 1982).

Numerous authors have separated a variety of low-molecular-weight nickel complexes from the renal cytosolic compartment (Parker and Sunderman, 1974; Oskarsson and Tjälve, 1979, Sunderman et al., 1981, 1983; Herlant-Peers et al., 1982, 1983; Templeton and Sarkar, 1985). In general, these ligands differ from those thought to occur in circulation and indicate that a change in the chemical form of nickel occurs either during the processes of glomerular filtration and tubular reabsorption or in the cytosol itself. Observing that a significant fraction of the cytosolic nickel was bound to glycosamino-glycanlike oligosaccharides, thought to occur as turnover products of the GBM (Schurer et al., 1980), Templeton and Sarkar (1985, 1986) speculated that ligand exchange may occur at the level of glomerular filtration.

In summary, features relevant to the handling of nickel by healthy glomeruli are the extent of the plasma ultrafiltrable fraction, the effect of low-molecular-weight ligands on potential interactions with glomerular cells, and the possibility of binding or ligand exchange interactions in the glomerular extracellular matrix.

2.2. Glomerular Toxicity of Metals

A number of metals and their compounds induce nephropathies characterized by high-molecular-weight proteinurias or glomerulonephritic changes in the kidney. In contrast to an extensive literature dealing with toxicity to the renal tubule, a systematic description of metal-induced glomerular injury is lacking. Glomerular pathology is almost invariably overshadowed by tubular damage and in general occurs at higher levels of exposure than the latter and as a more chronic effect. High-molecular-weight proteinurias are a common feature of metal-induced glomerular damage. Structural and biochemical alterations in the GBM and its supporting cells are manifest as an increased permeability to albumin and, if progressive, to larger proteins such as transferrin and immunoglobulin G (IgG). Glomerular filtration rate (GFR) may be increased or decreased. Well-documented clinical syndromes involving glomerular damage as one component follow poisoning with Hg, Pb, Au, and U (Humes and Weinberg, 1986). These metals act at various levels to cause derangement of plasma filtration, with consequent loss of the size selectivity contributed to the process by the GBM. Nickel is notably absent from a recent discussion of metal-induced acute toxic nephropathies in humans (Humes and Weinberg, 1986), although its inclusion among experimental glomerulopathies is well justified (Gitlitz et al., 1975; Sunderman and Horak, 1981). Observations of glomerular consequences of elemental exposures are summarized in Table 12.1 and include both histological and functional alterations. No attempt is made at completeness, our present aim being to illustrate the range of effects.

Several mechanisms should be considered in any attempt to understand the glomerulus as a target of metal poisoning in general and nickel toxicity in particular.

1. Vascular effects of metals may lead to changes in GFR through direct mechanical or pharmacological effects on the vessels. For example, vaso-contriction caused by vanadium administration to cats as $NaVO_3$ decreases GFR (Larsen and Thomsen, 1980).

Vasoactivity may also be mediated through renin-angiotensin or prosta-glandin pathways (Oken, 1976; Oken et al., 1982). Alterations in renocortical blood flow might also trigger ischemic damage (Hook, 1981). Mechanical blockage of the glomerular capillaries with cell debris leads to edematous destruction at the tamponed site, as observed following hemorrhage and hemolysis due to arsine poisoning (Arnold, 1988).

2. Disturbances in glomerular filtration may occur as a consequence of changes in fluid flow in the tubules. Tubular occlusion may cause back pressure and edematous changes, and backflow of solutes will reduce GFR as is suggested to occur in uranyl nitrate poisoning (Blantz, 1975). A glomerulo-tubular filtration feedback loop has been hypothesized (Oken, 1976). Increased GFR is an early accompaniment of irreversible lead-induced interstitial fibrosis (Goyer, 1988).

Table 12.1 Selected Examples of Involvement of Glomerulus in Metal-related Nephropathy

Element	Chemical Form	Species	Nature of Exposure	Glomerular Effects	Reference
Ag	$AgNO_3$ solution	Man	Chronic oral	Deposition of Ag_2Se in GBM; normal renal function	Aaseth et al., 1981
		Rat		Ag granules in GBM	Walker, 1973
Al	?Antacid	Man	Chronic oral (dialysis patient)	Al-dense deposits in GBM, identified by X-ray probe	Smith et al., 1982
As	AsH_3 gas	Man	Acute inhalation	Blockage of glomerular capillaries with cell fragments	Arnold, 1988
	AsO_4^{3-}	Dog	Intravenous infusion	Glomerular sclerosis and necrosis	Tsukamoto et al., 1983
Au	Organothiols	Man	Iatrogenic	Au-containing vacuoles in glomerular epithelium; prominent loss of foot processes	Watanabe et al., 1976
				Immune complex deposition; possible aggravation of rheumatoid glomerulonephritis	Watanabe and Watanabe, 1987
		Rat	Intraperitoneal injection	Au filaments in mesangium	Kaizu et al., 1987
				Au deposition in lysosomes of glomerular visceral epithelium	Kaizu et al., 1987

(Table continues on p. 140.)

Table 12.1 (*Continued*)

Element	Chemical Form	Species	Nature of Exposure	Glomerular Effects	Reference
Bi	Organo salts	Man	Acute oral	Nephropathy with glomerular and tubular damage; decreased GFR	Thomas et al., 1988; Urizar and Vernier, 1966
Cd	$CdCl_2$	Rat	Intraperitoneal injection	Cd–metallothionein staining in glomerular mesangial and visceral epithelial cells	Banerjee et al., 1982
	Dust	Man	Chronic occupational	High-molecular-weight (glomerular-type) proteinuria	Lauwerys et al., 1984
				Autoimmune response to GBM laminin	Bernard et al., 1984
	Solution (unspecified)	Rat	Chronic oral	Glomerular epithelial cell swelling and fusion and glomerular swelling in presence of tubulointerstitial disease	Aughey et al., 1984
Cu	$CuSO_4$ solution	Sheep	Oral	Damage to glomerular endothelium; deposition of putative Fe and Cu aggregates along GBM; impaired glomerular function	Gooneratne and Howell, 1983
	Dietary	Man	Wilson's disease	Cu deposits in glomerular cell nuclei; decreased GFR; glomerular protein leakage	Reynolds et al., 1966

140

Element	Form	Species	Route	Effect	Reference
Hg	Organic and inorganic	Various	Various	Membranous (immune complex) glomerulonephritis; glomerular protein leakage	Humes and Weinberg, 1986; Clarkson and Shaikh, 1982
	$HgCl_2$	Rat	Subcutaneous injection	Filtration failure due to increased preglomerular and decreased efferent resistance	Wolfert et al., 1987
Ni	$NiCl_2$	Rat	Intraperitoneal injection	Reversible fusion of foot processes and nephropathy at doses too low to produce tubular damage	Gitlitz et al., 1975
Pb	Various	Man	Chronic occupational	Increased GFR an early sign of irreversible tubulointerstitial disease; glomerulosclerosis and peritubular fibrosis later	Goyer, 1988
Te	Inorganic	Rat	Oral and inhalational	Slight to severe damage to tubules and glomeruli	Alexander et al., 1988
U	Nitrate	Rat	Intravenous injection	Reduction in single nephron GFR, 1° due to decreased ultrafiltration coefficient; also 2° to backflow of solutes from damaged tubules	Blantz, 1975
	Acetate	Rabbit	Intravenous injection	Reversible reduction of glomerular endothelial fenestrae; loss of epithelial foot processes; depressed glomerular function	Kobayashi et al., 1984
V	$NaVO_3$	Cat	Oral	Decreased GFR 2° to vasoconstriction	Larsen and Thomsen, 1980

3. Direct toxic effects on glomerular cells are possible with those metals taken up during filtration. Gold has been found in epithelial cells of patients receiving chrysotherapy (Watanabe et al., 1976), and cadmium–thionein has been localized immunohistochemically in both epithelial and mesangial cells following parenteral administration of $CdCl_2$ to rats (Banerjee et al., 1982). Copper deposits were found in the glomerular epithelium of a patient with Wilson's disease (Reynolds et al., 1966). The consequences of these accumulations are at present unknown but may be involved in the glomerular component of the nephrotoxicity produced by each metal. Glomerular epithelial cells frequently respond to insult by retraction of podocytes (*fusion of foot processes*), while damage to mesangial cells could eventually lead to replacement of the mesangium and glomerulosclerosis (Lovett and Sterzel, 1986). Smooth muscle mesangial cells (see Section 6) are contractile in nature and are involved in regulation of glomerular capillary flow (Brenner et al., 1986), so hemodynamic sequelae of mesangial cell toxicity may also be expected.

4. Changes in the GBM may result in increased permeability in the absence of obvious structural disruption. Decreased single-nephron GFR was found to result in part from a decreased glomerular permeability coefficient in rats treated with uranyl nitrate prior to any histological changes in the glomerulus (Blantz, 1975). Architectural disruption and GBM thickening are seen in a number of nephrotoxic phenomena. Membranous glomerulonephritis resulting from immune complex deposition along the GBM is common and is a prominent feature of, for example, mercury (Humes and Weinberg, 1986) and gold (Duke et al., 1982) nephrotoxicity. In some cases the autoimmune response may be directed against the GBM itself, as, for example, the production of antilaminin antibodies during chronic cadmium exposure (Bernard et al., 1984). Since the GBM serves also as an anionic electrostatic barrier (see Section 4.1.1), deposition of metal cations could lead to charge reduction and proteinuria. Lauwerys and co-workers (1984) have speculated that such a mechanism may underlie the high-molecular-weight proteinuria seen in those chronically exposed to cadmium in the workplace. Direct evidence for such a mechanism is lacking, although metal deposition along the GBM has been demonstrated in some instances. Sheep dying with hemolysis resulting from copper toxicosis had electron-dense deposits along the GBM, which were thought to represent aggregates of iron and copper with amino acids and other low-molecular-weight ligands (Gooneratne and Howell, 1983). Silver granules clearly mark an anionic lattice in the laminae rara of the GBM of mice chronically fed silver in the drinking water (Walker, 1973; Naeser and Rastad, 1982). X-ray emission and electron diffraction were used to demonstrate the presence of Ag_2Se in the GBM of a patient with argyria (Aaseth et al., 1981), and X-ray microprobe revealed the aluminum content of dense deposits in the GBM of a patient with renal failure secondary to a dysproteinemia (Smith et al., 1982). Accumulation of gold particles in the visceral epithelium and along the subepithelial GBM of a patient receiving chrysotherapy have

been considered as an immediate cause of altered glomerular permeability (Watanabe et al., 1976). Different modes of binding to the GBM polyions are open to small metal cations, small organic cations, and cationic macromolecules. Electrostatic consequences of metal binding to the GBM are considered below (Section 4.4).

It may be useful to distinguish between nephropathies that result from the direct toxic effects of metals on glomerular cells or extracellular structures and those that are secondary to systemic or postglomerular renal toxicity. Mercury intoxication results in immune complex deposition along the GBM (Clarkson and Shaikh, 1982; Humes and Weinberg, 1986). Insignificant amounts of mercury cross the GBM due to the low ultrafiltrable fraction of mercuric ion in plasma, but damage to the tubule resulting from transtubular secretion appears to be the source of circulating antigenic material. In contrast, while membranous glomerulonephritis and immune complex deposition are also hallmarks of gold nephropathy, gold probably causes an additional deterioration in glomerular function concomitant with deposition in epithelial cells (Watanabe et al., 1976; Kaizu et al., 1987). Localization by X-ray dispersive analysis of gold in the lysosomes of epithelial cells correlates with vacuolization in those cells (Oochi et al., 1987). As noted above, some workers (Lauwerys et al., 1984) have attributed a direct glomerular component to chronic cadmium nephropathy. Supporting this interpretation, Aughey et al. (1984) described glomerular swelling and fusion of foot processes, while the existence of metallothionein in glomerular cells following cadmium exposure (Banerjee et al., 1982) implies uptake of cadmium by those cells. Unfortunately, sufficient information to make the mechanistic distinction is unavailable for most metals. A motivation for studying the isolated glomerulus is its potential for distinguishing direct glomerular effects from those dependent on extraglomerular toxicity.

2.3. Glomerular Toxicity of Nickel

Gitlitz et al. (1975) exposed rats parenterally to $NiCl_2$ at 34 μmol/kg body weight. Significant proteinuria was present during the first two days after exposure, although higher doses of nickel were required to produce aminoaciduria. These functional changes were reversible. Forty-eight hours after a dose of 68 μmol/kg, all five rats examined showed focal fusion of glomerular epithelial foot processes, while only one animal showed micrographic evidence of tubular damage (focal tubular necrosis). These observations support a direct, reversible effect of nickel on the glomerulus as the more prominent manifestation of low-dose exposure to nickel (Sunderman and Horak, 1981). Despite its low acute nephrotoxicity, nickel may therefore be of great interest as an unusual example of a metal cation causing direct glomerular effects prior to tubular damage. Alternatively, this could be an example of a more general phenomenon underlying transient glomerulopathy in metal poisoning recognized with nickel by the careful work of Sunderman's lab.

3. ISOLATED RAT GLOMERULUS

Isolated glomeruli have been used by a number of authors for a variety of metabolic and cell biological studies (see Schlondorff, 1986 and references therein). They offer many of the advantages of cultured cells, including separation of local from systemic effects and freedom to manipulate growth conditions and exposures. Their potential for use in toxicological studies has been anticipated (Ormstad, 1982), though not widely exploited to date. Recently, we (Templeton and Chaitu, 1990) and others (Wilks et al., 1990) have studied the toxic effects of divalent metals on isolated rat glomeruli. In this section, we describe the methodology used in our laboratory for the preparation, maintenance, and assessment of rat glomeruli.

3.1. Preparation of Glomeruli

Several techniques have been used to obtain clean preparations of glomeruli, including microdissection of isolated nephrons, separation in a magnetic field following perfusion of the kidney with iron oxide (Meezan et al., 1973), and density gradient centrifugation (Mandel et al., 1973). Most investigators now use straightforward mechanical sieving (Langeveld and Veerkamp, 1981), which is favored in our laboratory.

Routinely, male Wistar rats of 200 ± 25 g are anaesthetized, and their kidneys are removed under clean conditions for collection in sterile Hank's balanced salt solution (HBSS). Under sterile airflow, kidneys are decapsulated, sectioned thinly, and pressed through a 180-μm mesh stainless steel sieve. The collected pressings are washed sequentially through sieves of 125, 106, and 90 μm, with a total of 1 L of saline per pair of kidneys. Saline is used at room temperature to reduce mechanical damage to the more rigid chilled glomeruli. Glomerular fractions are collected from the 106- and 90-μm screens in HBSS, pooled, and allowed to settle under gravity. They are transferred to RPMI 1640 medium containing penicillin G (100 U/mL) and streptomycin sesquisulfate (100 μg/mL), which has equilibrated at 37°C in a 5% CO_2 atmosphere, and counted under a phase contrast microscope. Typical yields of 30,000–40,000 glomeruli per pair of rat kidneys are obtained, with 2–3% contamintaion with tubular fragments (Fig. 12.1a). Only preparations with <5% contamination are used for overnight incubations. All preparations are given 2 hr recovery in the incubator (37°C, 5% CO_2, humidified), and the medium is then changed before undertaking any of the experiments reported here. When lower sulfate content is desired in some sulfate-labeling experiments (Section 3.2.5), Ham's F12 medium (6×10^{-6} M inorganic sulfate) is substituted for RPMI 1640 (4.2×10^{-4} M sulfate).

(a)   (b)

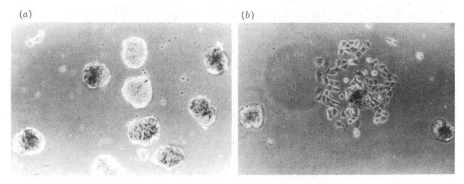

Figure 12.1. Rat glomeruli in vitro: (*a*) freshly isolated; (*b*) epithelial outgrowths after 5 days of culture. [Reproduced from Templeton and Khatchatourian (1988), with permission.]

3.2. Assessment of Viability and Metabolic Activity

3.2.1. *General*

It has been noted that, as a minimum, characterization of glomerular preparations for use in toxicological studies should include measurement of lactate dehydrogenase leakage and trypan blue exclusion (Ormstad, 1982). It is this author's experience that such information is seldom, if ever, provided. Despite a claim that 100% of glomeruli isolated in one laboratory were viable by the criterion of trypan blue exclusion (Klein et al., 1986), we find this approach problematic. Persistence of Bowman's capsule on some glomeruli, diffusion and trapping of dye in the extensive glomerular extracellular matrix, and difficulty in visualizing individual cells in the interior of the mesangium all render interpretation difficult (Templeton and Khatchatourian, 1988). Partial collagenase digestion allowed visualization of stained cells in the glomerular cores, but these cells were not identified (Templeton and Chaitu, 1990). Extensive reports of metabolic activity (Brendel and Meezan, 1973; Jim et al., 1982; Folkert et al., 1984) do not bear upon the question of fractional glomerular viability, or more correctly, the proportion of normally functioning cells in a given glomerulus. For these reasons we prefer to characterize the glomerular preparations by several different metabolic activities.

Perhaps a major factor accounting for the low number of toxicological investigations using isolated glomeruli is the relative quiescence of adult material. Cells from glomeruli of mature rats have low replicative rats (Pabst and Sterzel, 1983) and tissue from younger animals is to be preferred for establishing primary cultures of glomerular cells (Foidart et al., 1980; Lovett and Sterzel, 1986). The first cells to grow from attached glomerular explants of young (150-g) rats are epithelial in nature, which is the only cell type to incorporate [³H]thymidine during the first three days (Nørgaard, 1987). Mesangial cells are labeled only after five days, while labeling of glomerular endothelium has not been observed. At the earlier times, during epithelial labeling, dedifferentiation, loss of podocyte structure, and overgrowth of the parietal epithelium of Bowman's capsule are already in progress (Nørgaard, 1987). These factors raise a note of caution in the use of isolated glomeruli for certian types of toxicological investigation and suggest that nongenotoxic, metabolic effects influencing the native glomerulus should be sought in the early hours after isolation.

3.2.2. Enzyme Leakage

Lactate dehydrogenase (LDH; EC 1.1.1.27) and glucose-6-phosphate dehydrogenase (G6PDH; EC 1.1.1.49) were measured as the rates of oxidation of reduced nicotinamide-adenine dinucleotide (NADH) in the presence of pyruvate and reduction of nicotinamide-adenine dinucleotide phosphate (NADP) in the presence of glucose-6-phosphate, respectively. A parallel burst in released activities of both enzymes are seen during the initial recovery period and are easily measured in aliquots of culture medium when glomeruli are suspended at a density of 5000 mL^{-1}. Because glomerular G6PDH represents about 40% of the total renal activity of this enzyme (Bonner et al., 1982), it may be a more specific indicator of glomerular cell leakiness.

3.2.3. 2-Deoxyglucose Uptake and Release

2-Deoxyglucose is taken up by cells in an insulin-dependent manner and converted to 2-deoxyglucose-6-phosphate, which cannot be metabolized further and accumulates intracellulary (Wick et al., 1957). Following exposure of cultured cells to 2-deoxy[^{14}C] glucose, leakage of ^{14}C back into the medium is a useful indicator of membrane integrity (Cherian, 1982). Glomeruli were let recover for 2 hr in RPMI 1640 medium, then resuspended at 5000 glomeruli/ mL in fresh medium containing 2-deoxy[^{14}C] glucose (2 μCi/mL). After 3 hr of labeling, glomeruli were washed thrice with fresh medium. At intervals, aliquots of medium were withdrawn for counting of radioactivity. The uptake of label could be greatly enhanced by performing the 3 hr exposure in glucose-free HBSS containing 2 μg/mL of insulin without apparent adverse effects on glomeruli, allowing individual measurements on as few as 2500 glomeruli. However, low levels of glycolytic enzymes in the glomerulus, including hexokinase (Guder and Ross, 1984), result in low rates of conversion of the substrate.

3.2.4. Protein Synthesis

Glomerular suspensions were incubated with [^{3}H]leucine at 1 μCi/mL for 16 hr. To measure its incorporation into total protein, aliquots of 5000 glomeruli were harvested on glass fiber filters and washed with 10 mL of cold HBSS, 10 mL of cold 10% trichloroacetic acid (TCA), and 10 mL of water, and the filters were transferred to scintillation vials for counting. To determine incorporation of [^{3}H]leucine into TCA-precipitable material in initial cell outgrowths in multiwell plastic plates, the wells were washed free of glomeruli and fresh serum-free medium was added to each well. [^{3}H]leucine was added

(2 μCi/well) and the cells were incubated for 24 hr. Cells were then washed with HBSS, fixed to the plate with methanol, and treated with TCA, and the precipitate was solubilized with NaOH for scintillation counting, according to the procedure of Wilson (1986).

3.2.5. Proteoglycan Synthesis

Production of the appropriate populations of proteoglycans involves not only protein synthesis but also extensive posttranslational modification, secretion, targeting, and matrix assembly. Therefore, proteoglycan synthesis is not only an important functional aspect of the glomerulus but also serves as a more global indicator of the physiological state of the cells. Proteoglycan synthesis was determined by isolation of proteoglycans following incubation of glomeruli with [^{35}S]sulfate (5 μCi/mL) for desired periods of time. Other authors have used low-sulfate media to increase the specific activity of the label (Klein et al., 1986; Kobayashi et al., 1983), as we have done in the experiments reported here, employing Ham's F12 medium for the sulfate-labeling studies. However, the sulfate concentration of the medium can affect the population distribution of labeled proteoglycans in some cases (Ito et al., 1982; Humphries et al., 1986), and this parameter should be considered when labeling glomerular preparations (cf. Templeton and Khatchatourian, 1988).

Following incubation, glomeruli were harvested, washed thrice with HBSS, and suspended in 4 M guanidine–HCl in sodium acetate (50 mM, pH 5.8) containing 1% Triton X-100 and protease inhibitors, at 4°C for 24 hr. This dissociative extraction solution for proteoglycans (Wagner, 1985; Heinegård and Sommarin, 1987) is generally supplemented with a battery of protease inhibitors, since a number of unidentified peptidases are involved in basement membrane turnover. We include disodium ethylenediamine tetraacetate (10 mM), benzamidine hydrochloride (5 mM), ortho-phenanthroline (3 mM), tryptamine hydrochloride (5 mM), and phenylmethylsulfonyl fluoride (1 mM). The soluble extract is then eluted from a Sepharose CL-4B column (0.9 × 30 cm) with 8 M urea in sodium acetate buffer containing 0.15 M NaCl. The pooled void volume fractions, containing the labeled proteoglycans, are loaded on a DEAE–Sephacel ion exchange column (10 mL in a disposable plastic syringe) equilibrated in the same buffer with 0.5% Triton X-100, and eluted with a 0.15–1.15 M gradient of NaCl in this buffer. This routinely gives rise to two peaks of radioactivity, corresponding to glycosaminoglycans of lower and higher charge density (Section 5.3).

To determine the composition of the two peaks derived from ion exchange (Section 3.2.5), radiolabeled material is dialysed extensively against ultrapure water (Milli Q system, Millipore) and lyophillized. The residue is then degraded either chemically with nitrous acid or enzymatically with chondroitinase ABC (EC 4.2.2.4), chondroitinase AC (EC 4.2.2.5), or heparitinase (heparin lyase II; EC 4.2.2.8) (Templeton and Castillo, 1990). Nitrous acid treatment is carried out by dissolving the lyophilized residue in 1.8 M acetic acid

(Lindahl et al., 1973), dividing in half, and adding fresh sodium nitrite to one-half and water to the other. Under these conditions, only glycosamino-glycans containing deacetylated 2-N-hexose residues (i.e., heparins and heparan sulfates) are degraded. The enzymes (all from Sigma Chemical Co., St. Louis, MO) are glycosidases that have been found in our laboratory to be free of significant protease activity. Chondroitinase ABC is specific for chon-droitin-4 and 6-O-sulfates and dermatan sulfate, while chondroitinase AC attacks only chondroitin-4- and 6-O-sulfates. Digestions with both enzymes are carried out in 50 mM Tris-HCl/50 mM sodium acetate, pH 7.5, at 37°C. Heparitinase degrades heparan sulfates and is used in 0.25 M sodium acetate containing 2.5 mM calcium acetate at pH 7.0 and 43°C. The extent of deg-radation is determined by gel filtration of these reaction mixtures, and the proportion of each glycosaminoglycan type is defined by susceptibility to the appropriate treatments (e.g., see Fig. 12.6).

3.2.6. Glutathione Levels

Total (reduced and oxidized) glutathione was measured by the glutathione reductase-dependent reduction of 5,5'-dithio-(bis-2-nitrobenzoic acid), ac-cording to Akerboom and Sies (1981). Samples of 10,000 glomeruli are con-veniently homogenized in cold 2M perchloric acid with a glass rod in the tip of an Eppendorf 1.5-mL centrifuge tube. The glutathione content of healthy glomeruli was found to be about 50 fmol/glomerulus.

3.3. Other Methods

3.3.1. Preparation of Extracellular Matrix and GBM

Extraction of the extracellular matrix with detergent has been described (Carlson et al., 1978) and is widely used today. In brief, this technique involves osmotic lysis of whole glomeruli in water followed by detergent extraction with Triton X-100, enzymatic removal of deoxyribonucleic acid (DNA), and further solubilization with sodium deoxycholate. Sodium azide is included in all steps except deoxyribonuclease (DNase) treatment. Alter-natively, sonication of glomeruli (Spiro, 1967) yields preparations of GBM and may be preferable for retention of the sulfate content of the basement membrane proteoglycans (Parthasarathy and Spiro, 1982). Both approaches rely on the relative insolubility of GBM and ease of disruption of the mesangial matrix to affect partial purification of the former. We have used sonication to produce a material of constant chemical composition (Templeton, 1987a) and now prefer the term *GBM-enriched fraction* to describe it (Templeton and Khatchatourian, 1988). In the present chapter, *isolated GBM* refers to this fraction and is retained for consistency with our earlier work. We have also attempted to disrupt the mesangium with TCA, following Krakower and Manaligod (1980), with less satisfactory results (Templeton, 1987a).

For the studies reported here, the GBM-enriched fraction was obtained by sonicating glomeruli in ice-cold 1 M NaCl using 30-sec bursts of a Branson sonifier (power setting 6/10). Sonic disruption is continued until intact glomeruli are microscopically absent, usually for a total of 3–5 min. The GBM fraction is then collected by centrifugation, washed extensively with ultrapure water, and lyophilized.

3.3.2. Attachment and Cell Growth

To achieve attachment of glomeruli and initial cell outgrowth, glomeruli are prepared and allowed to recover as described in Section 3.1. The medium is then replaced with fresh RPMI 1640 containing penicillin and streptomycin, supplemented with 20% fetal calf serum (Gibco, Grand Island, NY) and buffered with 15 mM N-2-hydroxyethylpiperazine-N'-2-ethanesulfonic acid (HEPES, Gibco). The glomeruli are transferred to multiwell tissue culture plates (Flow Laboratories, McLean, VA) at a density of 500–1000 mL^{-1} and incubated undisturbed for five days. Under these conditions, initial epithelial cell outgrowths are observed at this time (Fig. 12.1b).

4. EXTRACELLULAR MATRIX OF GLOMERULUS

4.1. Mesangial Matrix and the GBM

The glomerulus has an extensive extracellular matrix consisting of capillary GBM and less organized mesangium. Morphometric estimates place the volumes of the peripheral capillary GBM and the total mesangial compartment at about 0.01 and 0.04 μm^3/glomerulus, respectively, in the healthy 300-g Lewis rat (Hirose et al., 1982), so the GBM accounts for less than 20% of the glomeruluar extracellular volume. Little is known about the regulation of synthesis and turnover of the mesangium. Mesangial cells in culture secrete types I, III, IV, and V collagen, fibronectin, and variable proportions of sulfated proteoglycans (Lovett and Sterzel, 1986). Prolonged glomerular inflammation leads to hypercellularity, expansion of mesangial matrix, sclerosis, and glomerular obliteration. The effects of toxic metal ions on mesangial cell metabolism and cell matrix interaction are potentially an important aspect of the progression of toxic nephropathy to end-stage disease.

The more highly organized GBM has been widely studied, and details of the interplay of composition and function are emerging. Pathological changes, including accumulation and thickening of GBM, alterations of anionic site distribution, and disturbances in permeability can now be discussed in molecular terms.

4.1.1. Structure and Function of the GBM

This brief summary is based on several recent reviews (Kanwar, 1984; Abrahamson, 1986; Timpl, 1986) and references therein. The GBM consists mainly of type IV collagen, sulfated proteoglycans, laminin, and a number of other minor proteins and glycoproteins, including fibronectin and the self-associating protein nidogen. Some of these minor components are likely extrinsic, accumulating from the circulation. Proper assembly of the GBM reflects a number of specific interactions, including those between collagen and heparan sulfate proteoglycan (HSPG), collagen and laminin, laminin and HSPG, HSPG and fibronectin, and of course self-association of collagen (Yurchenco et al., 1986). Coordination of the biosynthetic processes involved in producing these constituents (Heathcoate and Grant, 1981; Yurchenco and Schittny, 1990) results in a fused basal lamina supporting capillary endothelium on one side and visceral epithelial cells on the other. The GBM is therefore a double basement membrane (Abrahamson, 1986, 1987) that requires attachment sites for both cell types and may play a role in maintaining the interdigitated structures that comprise the epithelial foot processes. It is further highly specialized for participation in plasma ultrafiltration. This process involves molecular discrimination on the basis of hydrodynamic size and shape (Brenner et al., 1986) as well as charge (Brenner et al., 1977; Kanwar, 1984). Proper glomerular function therefore requires maintenance of an appropriate electrostatic barrier as well as a controlled distribution of effective pore sizes.

Because we are interested in the filtration of metal cations, our recent work has focused on interactions with the electrostatic components of the filtration unit. Negative charges on cell sufaces and likely sialoglycoproteins (Brenner et al., 1977; Lelongt et al., 1987) contribute to the net negative charge of the unit, but the glycosaminoglycan chains of the HSPG appear to be major determinants (Kanwar et al., 1980; Assel et al., 1984). This molecule accounts for most of the proteoglycan of the GBM. A number of authors have used cationic probes to demonstrate a regular latticelike arrangement of anionic sites in the laminae rara interna and externa of the GBM, with a periodicity of about 60 nm (Kanwar et al., 1980; Mahan et al., 1986; Rohrbach, 1986). This lattice is well established by enzyme degradation (Kanwar et al., 1980) and immunolocalization (Lelongt et al., 1987) to represent HSPG, although the regular arrangement in the laminae rara is considered by some (Laurie et al., 1982; Grant and Leblond, 1988) to represent an artifact of staining and fixation for electron microscopy. Equally unresolved is the true content of HSPG in the mammalian GBM, estimates ranging from less than 1% glycosaminoglycan by weight (Parthasarathy and Spiro, 1982) to molar ratios of glycosaminoglycan to collagen around unity (Grant and Leblond, 1988), more in keeping with the composition of other basement membranes (Yurchenco et al., 1986).

Whatever the true disposition of the HSPG in the GBM, this anion now appears likely to modify the pattern of filtration of charged macromolecules, retarding polyanions such as albumin and dextran sulfates (Kanwar et al., 1984; Assel et al., 1984). Several pathological states are thought to allow increased permeability of the GBM to albumin as a consequence of a reduction in the amount of HSPG. Changes in the number or arrangement of GBM anionic sites have been discussed in relation to minimal lesion disease and its animal counterpart, puromycin aminonucleoside nephrosis (Mahan et al., 1983; but cf. Lelongt et al., 1987), congenital nephrotic syndrome (Vernier et al., 1983), immune-mediated nephritis (Melnick et al., 1981), and diabetes mellitus (Rohrbach, 1986).

Equally important, although empirically less substantiated, is the possibility that filtration of cationic molecules is enhanced by negative charges in the membrane. There is some evidence for greater transport of cationic proteins across the glomerular capillary wall (Rennke and Venkatachalam, 1977) and increased fractional clearance of diethylaminoethyl dextran (Brenner et al., 1977), as compared with neutral molecules of similar dimensions. These effects, even if attributable to a charge barrier, may be insignificant for small ultrafiltrable cationic metal complexes, which are expected to be unimpeded in their passage through the GBM. Perhaps of greater interest is the possibility that binding of cations, large or small, to the anionic components of the GBM reduces the effective charge barrier and in this manner alters the characteristics of plasma filtration.

4.1.2. *Proteoglycan Structure*

For the following discussion of charge compensation and Ni(II) binding to GBM components, it is helpful to introduce some general aspects of the structure of heparan sulfate and other proteoglycans (PGs). Good reviews abound (Comper and Laurent, 1978; Höök, 1984; Poole, 1986; Gallagher et al., 1986; Silbert, 1987). For present purposes the salient features are exemplified by the basement membrane HSPG. In the rat, this molecule consists of a protein core of 18 kD and about four heparan sulfate side chains of 25 kD each (Fujiwara et al., 1984; Kanwar et al., 1984). These glycosaminoglycan chains consist of 1,4-linked disaccharide units of alternating uronic (iduronic or glucuronic) acid and glucosamine residues. The glucosmanine residues (originating as *N*-acetyl glucosamine) are extensively de-*N*-acetylated, and variable positions of sulfation of this unit include the 6-*O* and 2-*N* sites. Thus the heparan sulfate side chain presents carboxyl and sulfate groups as potential metal coordination sites and is a polyanion with a significant charge density determined by variation in the degree of sulfation superimposed on a constant background of evenly spaced carboxyl groups. The protein core may engage in important interactions with other GBM components. The glycosaminoglycan chains contribute to membrane properties due to their physical characteristics. These include electrostatic charge and extensive hydration volumes. While we speculate here on the electrostatic consequences of charge compensation, the effects on hydration, though seldom discussed, may also be profound.

4.1.3. *Consequences of GBM Charge Reduction* in Vivo

The possibility has been raised that metal binding to the GBM contributes to cadmium nephropathy by reducing the charge barrier (Lauwerys et al., 1984), but conclusive demonstration of such an effect is lacking. Indeed, deposits of gold in the GBM were not accompanied by any change in protein excretion (Aaseth et al., 1981). With polycations, the situation is more ambiguous. Increased albumin clearance after perfusion with protamine has been attributed to a direct effect of GBM charge compensation by deposition of that polycation (Assel et al., 1984). Infusion of protamine (Seiler et al., 1977) and cationic hexadimethrine (Hunsicker et al., 1981) can produce retraction of foot processes and transient proteinuria. A similar mechanism may play a role in the proteinuria of immune complex nephritis following deposition of cationic antibodies (Batsford et al., 1980). On the other hand, binding of anti-HSPG IgG to the GBM did not produce abnormal protein excretion (Abrass and Cohen, 1988). Part of the explanation of these differences may lie with the different binding modes open to *point charge* metal cations and macromolecular polycations and the allowable degree of charge compensation for each (Section 4.4). We have investigated the details of metal binding to the isolated GBM in order to elucidate the chemistry underlying these differences. The results obtained with nickel are described below (Section 4.3).

4.2.   Binding of Nickel to GBM Components

Potential binding sites for nickel in the GBM include acidic glycoproteins and the surfaces of the adjacent cells (Brenner et al., 1977; Lelongt et al., 1987) as well as collagen and HSPG, although these possibilities have not been investigated in any systematic way. Solublized type IV collagen binds copper through $Cu(II)-N_4$ (one imidazole, three peptide nitrogens) coordination (Ferrari and Marzona, 1987), and presumably the same mechanism is available for Ni(II). The high linear charge densities of the glycosaminoglycan side chains of the HSPG make them excellent candidates for metal binding. Some important details are known about the effects of metal ion binding on the solution properties, conformations, and biological activities of the glycosaminoglycans (see Templeton, 1988, and references therein), and Ni(II) has been studied by some authors. Nickel reduces the acceleration of reaction between cationic reactants normally observed in the presence of chondroitin sulfate polyanion to an extent comparable to other divalent transition metal ions, supporting a nonspecific valence-dependent electrostatic interaction of Ni^{2+} with the glycosaminoglycan polymer (Booij, 1981). Balt et al. (1983b) found that Ni(II) was kept in solution at alkaline pH in the presence of chondroitin sulfate and used ligand field data and compositional analysis to argue for an octahedral ternary hydroxo complex of Ni(II) and glycosaminoglycan. The same group (Balt et al., 1983a) used viscometry, potentiometry, and spectroscopy to show that a number of metal ions, including Ni(II), site bound to the carboxyl groups of chondroitin according to the Irving–Williams series, with some electrostatic stabilization from sulfate groups. Nickel(II) forms octahedral complexes with D-glucosamine (Lerivrey et al., 1986), but direct comparison between glycosaminoglycan–metal complexes and reactions with glycosaminoglycan subunits is generally not possible due to involvement of the free C_1-OH group in monosaccharide coordination (Balt et al., 1983b).

4.3. Binding of Nickel to Whole GBM

As a starting point in our investigation of the interaction of nickel with glomerular components, we undertook a study of the binding of Ni(II) to the isolated GBM (Templeton, 1987a,b). Material prepared by sonication of glomeruli obtained from bovine kidneys taken fresh at slaughter bound ^{63}Ni when incubated with labeled $NiCl_2$ over a range of values of ionic strength and pH. Several lines of evidence supported the conclusion that this binding was dominated by glycosaminoglycan chains of the HSPG. These included the following:

(i) The binding was sensitive to treatment of the GBM preparation with heparitinase, which degrades the HSPG with a high degree of specificity and was found to be free of significant contamination with protease activity.

(ii) The binding was not sensitive to mild treatment with collagenase until significant amounts of membrane had been solubilized.

(iii) The binding was sensitive to treatment of the membrane with methanolic HCl, which quantitatively cleaved sulfate esters from the preparation.

(iv) The binding was competitively eradicated by the hexavalent cationic dye ruthenium red at concentrations confirmed by electron microscopy to stain selectively the anionic HSPG sites of the GBM.

Scatchard plots constructed at conditions approximating a physiological solution revealed a high affinity binding site for nickel with an association constant of about $4.5 \times 10^6 M^{-1}$ based on uronic acid content (Fig. 12.2a). A refinement in the fit was achieved by employing a Friedman–Manning isotherm to include overlapping binding sites and electrostatic effects (Fig. 12.2b), which predicted the observed negative cooperativity of binding (Templeton, 1987b).

One conclusion to be drawn from these observations is that electrostatic effects in the whole GBM can be understood in terms of ionic interactions of the glycosaminoglycans with Ni(II), and so we attempted to model these interactions with the hope of gaining further insight into the electrostatic potentials involved in plasma filtration. Some mathematical details of the Poisson–Boltzmann and Manning models of polyelectrolytes, as applied to the glycosaminoglycans of the GBM, have been given elsewhere (Templeton, 1987b,c). Here a less formal summary is provided.

McLaughlin (1977) has described the variation in the electrostatic potential with distance from a charged surface in an electrolyte solution, where a Boltzmann distribution of coions and counterions is achieved. The resulting Poisson–Boltzmann equation treats the surface as one plate of a parallel-plate capacitor, the electric double layer of the Gouy–Chapman theory providing the second plate. An approximate solution to the equation has been widely applied to the description of phospholipid membranes (McLaughlin, 1977), and extension to a cylindrical plate capacitor descriptive of a rigid

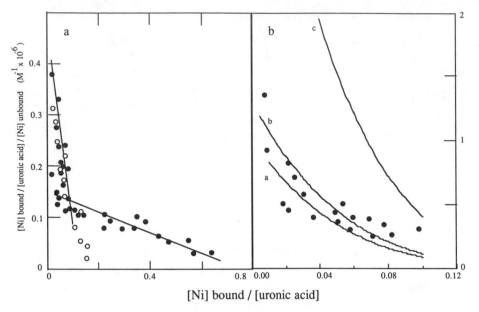

Figure 12.2. Scatchard analysis of ^{63}Ni(II) binding to isolated bovine GBM. (*a*) Plot of experimental data in Scatchard coordinates, obtained in (●) Tris-HCl buffer (5 mM, pH 7.4) or in (○) buffer containing 140 mM NaCl. (*b*) Calculated Friedman–Manning isotherms superimposed on experimental data, assuming tetracoordination and a polymer charge density of 0.9, with K_a varying as 1.0×10^6, 1.3×10^6, and $4.5 \times 10^6\ M^{-1}$ (curves *a*–*c*, respectively). [Adapted from Templeton (1987b; Fig. 6), where details can be found.]

rod–like glycosaminoglycan polymer is immediate (Templeton, 1987b,c). In this kind of model, the driving force for binding is the electrostatic attraction between polyion surface and solution counterions, and the attractions and repulsion among polyions, counterions, and coions are treated by classical electrostatics.

A complementary approach provided by Manning (1978, 1979) has been shown to be valid in a number of experimental systems. In this approach, the binding of counterions to a polyion is described fully in terms of the linear charge density of the polymer (ξ) and the valence of the counterions (*N*). The binding is expressed as the fraction *f* of the net structural charge of the polymer that is left uncompensated by counterion binding and is given by the discontinuous equation

$$ f = (\xi N)^{-1} \qquad \xi > N^{-1} $$

$$ f = 1 \qquad \xi < N^{-1} $$

where ξ is a dimensionless quantity dependent upon the spacing of monovalent charges along the polyion backbone. This rather remarkable equation predicts that when $\xi < N^{-1}$, the uncompensated charge equals the net structural charge (i.e., no binding occurs), while for counterions the reciprocal of whose valence is less than ξ, binding will proceed just until ξ is reduced to N^{-1} regardless of the nature of the ion. This would imply that Ni^{2+} (and all other divalent cations) would condense onto polyanions with charge densities greater than 0.5 and would do so to the extent necessary to reduce their charge densities to just that value. In part, this behavior can be rationalized by concluding that the thermodynamic contributions to binding are dominated by the relief of electrostatic strain inherent in the polyion. The like charges of the polyion are mutually repellent. Their partial compensation is thermodynamically favorable but does not proceed to completion because the concentration gradient of counterions between bulk and polymer domains eventually becomes limiting.

Values of ξ for the glycosaminoglycans are in the range of 0.6–2.4 (Templeton, 1987c), crucial for discrimination between monovalent and divalent cations. Several observations indicate that the Manning model provides a better description of the ionic interactions of GBM glycosaminoglycans than classical electrostatics. Put another way, HSPG in the BM shows behavior more characteristic of polyanions in dilute solution. The observations (summarized in Templeton, 1987b) include (i) comparison of the binding of $^{63}Ni^{2+}$ and 14 other metal and alkaline earth ions, (ii) the low dependence of the binding of $^{63}Ni^{2+}$ on monovalent salt, (iii) the extent of binding of the nickel ion, and (iv) its sharp ejection from the polymer domain when the charge density drops below 0.5 due to binding of trivalent lanthanide ions. The observations also require a charge density of slightly less than unity ($\xi \approx$ 0.9) for the GBM heparan sulfate molecule.

4.4. Electrostatic Consequences of Nickel Binding to the GBM

From the above discussion it is apparent that Ni(II) binds to the GBM in vitro to an extent that reduces the net charge of the HSPG by $1 - f = 1 - [1/(0.9 \times 2)]$, or 44%. If binding occurs in vivo, it should not exceed this amount, and this residual charge may still present an effective charge barrier to plasma solutes. This mechanism of binding, inherent in the polyelectrolyte properties of the HSPG, could account for the general failure of deposition of metals in the GBM to increase the loss of albumin (Section 4.1.3). Nickel would of course be in equal competition for HSPG binding sites with other divalent cations and would be relatively insensitive to changes in concentration of monovalent serum electrolytes. For these reasons, it seems unlikely that Ni(II) will exert any toxic effects on the kidney by compensating charges in the GBM. In fact, from the measured glycosaminoglycan content of the GBM, and the predicted maximum binding capacity, the extracellular matrix of one glomerulus should be able to accommodate about 20 μM of Ni(II) due to GBM HSPG alone, and this compartment may provide a degree of protection to glomerular cells in times of transient increases in circulating ion.

5. EFFECTS OF NICKEL ON THE ISOLATED GLOMERULUS

In order to observe possible direct toxic effects of Ni(II) on glomerular cells, we have examined the effects of $NiCl_2$ on several properties of rat glomerular suspensions. Because Ni(II) binds preferentially to HSPG of the GBM, we have examined the effects of nickel salts on the biosynthesis and utilization of proteoglycan. The results presented in Sections 5.1–5.4 may be summarized briefly: Ni(II) added to serum-free medium in concentrations up to the milli-molar range had no short-term effect on proteoglycan synthesis, plasma membrane leakiness, or glutathione content in glomeruli, although stimulating protein synthesis at lower concentrations. However, the appropriate incor-poration of HSPG into the GBM was markedly inhibited by 10 μM $NiCl_2$, and longer term attachment and cell outgrowth were prevented completely by concentrations above 100 μM. The latter phenomenon may be a result of failure to produce normal matrix in the presence of Ni(II).

5.1. Synthesis of Proteoglycans and Proteins

Over the first 24 hr following isolation, glomeruli incubated with [^{35}S]sulfate accumulated labeled proteoglycans in a linear fashion, both in the whole glomerulus and in the GBM (Fig. 12.3). This seems to be a good indicator of preservation of complex cellular processes, since it involves protein syn-thesis, glycosylation and Golgi transport, posttranslational modification, se-cretion, and matrix assembly. No effect on the total amount of sulfate in-corporated over 16 hr into glomerular proteoglycan was seen when $NiCl_2$ was added to the medium at concentrations ranging from 1×10^{-5} to 1×10^{-3} M. Over this same concentration range, an increase in total protein synthesis ([^{3}H]leucine incorporation) was noted (0.43 ± 0.1 and 0.45 ± 0.09 dpm/glomerulus for 10^{-4} and 10^{-3} M $NiCl_2$, respectively, vs. 0.1 ± 0.06 dpm/glomerulus for controls), but this was not reflected in an increased amount of sulfated proteoglycan.

5.2. Membrane Integrity and Oxidative Stress

Following an initial burst of LDH release, glomeruli appeared to recover and no further enzyme was released (Fig. 12.3). Similar behavior was observed for release of G6PDH activity and leakage of 2-deoxy[^{14}C]glucose. The initial rates of leakage and recovery times for each of these three markers were not altered by exposure to 1×10^{-3} M $NiCl_2$ (data not shown). Total glutathione levels were slightly higher in glomeruli from younger rats than from older, but in neither case did exposure to 1×10^{-4}–1×10^{-3} M $NiCl_2$ for 16 hr

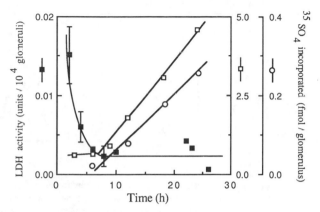

Figure 12.3. Time course of release of lactate dehydrogenase activity from glomerular suspensions (■) and incorporation of [^{35}S]sulfate into whole glomeruli (□) and a glomerular basement membrane–enriched fraction (○). [Reproduced from Templeton and Khatchatourian (1988), with permission.]

affect these levels (Table 12.2). We conclude that the isolated glomerulus is protected from oxidative stress during short-term exposure to Ni(II), as determined by membrane integrity and glutathione content. This may be due to sequestration of nickel in the extracellular matrix, with intracellular levels being insufficient to produce damage to membrane lipids (Donskoy et al., 1986) or to inhibit enzymes of glutathione metabolism (Sunderman et al., 1984; Athar et al., 1987) as seen in some tissues.

5.3. Matrix Assembly and GBM Composition

Labeled proteoglycans extracted from the glomerulus following incubation with [^{35}S] sulfate were collected by gel filtration and fractionated on DEAE–Sephacel as described (Section 3.2.5.). In both control incubations

Table 12.2   Total Glutathione Measured in Homogenates of Glomeruli after 16 hr Incubation in Presence (+Ni) or Absence (Control) of NiCl$_2$

Experiment	Conditions	Glutathione (fmol/glomerulus)
1	Control	59.4 ± 5.4
	+ Ni	64.2 ± 0.6
2	Control	46.4 ± 7.1
	+ Ni	46.9 ± 19.9

Note: In experiment 1, glomeruli were obtained from a 150-g rat, and [Ni] = 1×10^{-3} *M*. In experiment 2, a 400-g rat was used, and [Ni] = 1×10^{-4} *M*. Values are mean ± SD for three determinations.

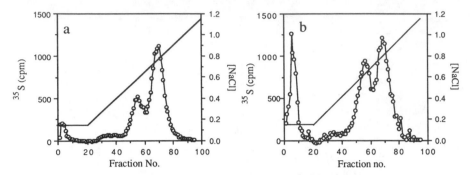

Figure 12.4. Fractionation of total glomerular proteoglycan extract on DEAE–Sephacel, as described in Section 3.2.5. Proteoglycans were ^{35}S-labeled in vitro for 16 hr in (*a*) Ham's F12 medium or (*b*) F12 medium supplemented with 1×10^{-5} *M* NiCl$_2$.

and those containing up to 1×10^{-3} *M* NiCl$_2$, the total amount of [35S]sulfate incorporated during 16 hr of exposure was constant and represented two populations of proteoglycan with lower and higher charge density (the peaks eluting at 0.55 and 0.75 *M* NaCl, respectively; Fig. 12.4). A similar profile has been observed by others for this total glomerular extract of proteoglycan labeled in vitro (Kobayashi et al. 1983; Klein et al., 1986). Both peaks can contain HSPG and chondroitin sulfate proteoglycan (CSPG), although under the conditions of this experiment, HSPG predominates in each (Templeton and Khatchatourian, 1988; Templeton and Castillo, 1990). When a GBM-enriched preparation of extracellular matrix was prepared from control glomeruli (not exposed to nickel), extracted, and fractionated, only labeled material from the first peak was present (Fig. 12.5*a*). That is, under normal circumstances, the isolated glomerulus incorporates only proteoglycan corresponding to the material of lower charge density. In contrast, glomeruli

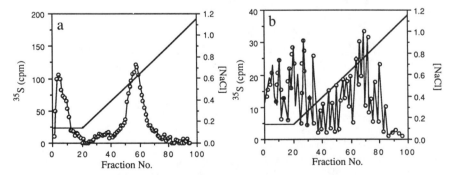

Figure 12.5. Fractionation on DEAE–Sephacel of proteoglycan extracted from a glomerular basement membrane–enriched fraction prepared from the glomerular cultures used in Fig. 12.4. Glomeruli were labeled in (*a*) control or (*b*) NiCl$_2$-supplemented medium, as described for Fig. 12.4.

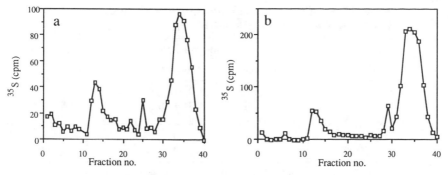

Figure 12.6. Sephadex G-50 gel filtration of nitrous acid–treated proteoglycans. Pooled fractions from the first eluting peak in Fig. 12.4 were degraded with nitrous acid as described in Section 3.2.5. After stopping the reaction with excess ammonium sulfamate, reaction mixtures were eluted from a 0.9 × 30-cm column with sodium acetate buffer (pH 5.8, 50 mM). The small peaks of ^{35}S eluting at the void volume (fractions 10–15) represent undegraded (non–heparan sulfate) proteoglycan. (*a*) Proteoglycan from control glomeruli of Fig. 12.4*a*. (*b*) Proteoglycan from Ni-exposed glomeruli of Fig. 12.4*b*.

exposed to only 1×10^{-5} M NiCl$_2$, while producing normal amounts of labeled proteoglycan, do not incorporate detectable amounts of the lower charge density proteoglycan into GBM during the time of observation (Fig. 12.5*b*). With both control and NiCl$_2$-treated glomeruli, the earliest peak to elute from DEAE–Sephacel was confirmed to contain mostly HSPG by nitrous acid degradability (Fig. 12.6), heparitinase sensitivity, and chondroitinase ABC resistance.

These experiments demonstrate an adverse effect of nickel on the incorporation of HSPG into the GBM. This may be due to complexation of Ni(II) with newly synthesized heparan sulfate chains, since there is no effect on their rate of synthesis. Then failure to incorporate the Ni(II) complex during membrane assembly must be postulated. Whatever the molecular mechanism, the observed failure of normal GBM production in the presence of NiCl$_2$ is a novel effect of this cation on tissue dynamics. It could be an important determinant of transient changes in the structure of the GBM and its attached cells and in the charge barrier to filtration during short-term exposure to nickel.

5.4. Attachment and Growth

When glomeruli were cultured as described in Section 3.3.2., they attached to the culture vessel by 24–48 hr, and cellular outgrowths appeared by four to five days (Fig. 12.1*b*). These new cells do not grow to confluence; rather they form islands of variable size around the glomerulus of origin, as in Figure 12.1*b*. These islands may grow and fuse to encompass several glomeruli or may represent only a few cells around a single glomerulus. Proliferation

Table 12.3   Effect of Incubation with $NiCl_2$ on
Epithelial Cell Outgrowth and Island Formation

[$NiCl_2$] (M)	Island Characterization		
	Extensive	Moderate	Sparse
0	1.5 ± 1.3	2.0 ± 1.6	2.3 ± 1.3
10^{-5}	1.7 ± 2.1	1.7 ± 0.6	5.7 ± 2.5
10^{-4}	0.25 ± 0.5	1.8 ± 1.0	4.8 ± 2.6
10^{-3}	0	0	0
5×10^{-3}	0	0	0

Note: Glomeruli from 200-g male Wistar rats were suspended at 500/ 2 mL per well in multiwell plates; $NiCl_2$ was included in quadruplicate wells at the concentrations indicated, and after 5 days the extent of growth was rated as described in the text (Section 5.4). Values are numbers of colonies per well (mean ± SD).

at this point is regulated by a complex and poorly understood interplay among growth-stimulating and growth-inhibitory factors secreted by epithelial and mesangial cells (Castellot et al., 1985; Lovett and Sterzel, 1986).

It is well established that these first cells are epithelial in nature (Harper et al., 1984; Striker and Striker, 1985; Oberley et al., 1986); mesangial cells take at least two to three weeks to appear, and conditions for culture of glomerular endothelium have probably not been achieved. Although these early cells have generally been assumed to originate from the visceral epithelium, there is recent compelling evidence that they represent overgrowths of the parietal epithelium lining the Bowman's capsule (Nørgaard, 1987). Whatever their origin, they arise only from attached glomeruli, and so require prior synthesis of appropriate matrix and adhesion molecules. Therefore, we decided to look at the effects of $NiCl_2$ on the attachment and potential for epithelial outgrowth of freshly isolated glomeruli.

In the preliminary studies reported here, we have cultured glomeruli in the presence of increasing concentrations of $NiCl_2$ and counted the numbers of islands of epithelial cells on day 5. We describe islands qualitatively as extensive (encompassing several glomeruli), moderate (as pictured in Fig. 12.1b), or sparse (consisting of only a few, often isolated, cells in the region of a glomerulus). The results of a typical experiment (Table 12.3) show a dramatic effect of nickel on growth of the explants. A shift to sparser colonies is seen at 10^{-5} M $NiCl_2$, large colonies are inhibited at 10^{-4} M, and all outgrowth is prevented by 10^{-3} M.

6.   CELLULAR CONSTITUENTS OF THE GLOMERULUS

Advances in the in vitro toxicology of the glomerulus will certainly be made with homogeneous cultures of glomerular cells. Such studies are presently underway in our laboratory, and details have been presented elsewhere (Templeton, 1990, 1991; Templeton and Sheepers, 1991). Here we briefly describe some of the approaches that allow investigations on the individual cell types and raise a few difficulties that must be overcome before definitive results can be obtained.

At least five types of cells have been identified in the glomerulus (Striker and Striker, 1985). These include the visceral and parietal epithelium, the glomerular capillary endothelium, and two mesangial cell types. In addition to the role of endothelium in maintaining the vascular wall, the endothelial fenestrae may prevent the passage of some very large particulates into the GBM (Tisher and Madsen, 1986). The visceral epithelium is highly differentiated and forms the interdigitated podocytes along the lamina rara externa of the GBM. It plays a central role in filtration through maintenance of the GBM and slit pore membranes and possibly through the involvement of cell surface anions, such as podocalyxin (Abrahamson, 1986) in the glomerular charge barrier. The parietal epithelium lines and maintains Bowman's capsule. The two mesangial cell types (Schlondorff, 1987) are of smooth muscle and bone marrow lineage, displaying contractile and phagocytic properties, respectively. The contractile mesangial cells are responsive to angiotensin (Brenner et al., 1986), may participate in regulation of glomerular capillary blood flow, and probably help maintain capillary wall tension (Kriz et al., 1990). The macrophagelike cells are involved in scavenging debris that constantly accumulates in a filtering structure. Visceral epithelial cells, and possibly endothelium, elaborate GBM and synthesize primarily HSPG. Mesangial cells, at least in culture, make larger amounts of CSPG (Striker et al., 1980; Foidart et al., 1980). Mesangial hypercellularity in response to inflammation leads to accumulation of CSPG-containing mesangial matrix and eventual replacement of the glomerulus with connective tissue (Lovett and Sterzel, 1986).

It is apparent that any toxic insult affecting the glomerulus could operate through a variety of cellular targets and produce a range of effects. Almost nothing is currently known about effects of metal ions on glomerular cells or of the importance of such information for understanding metal-induced nephropathies. Improved techniques of glomerular cell culture should soon change the situation. While early subculture of epithelial cells is thought to provide homogeneous lines of visceral epithelium, two problems must be overcome before definitive toxicological studies can be undertaken. First, the true nature of these cells as visceral, not parietal, epithelium must be seriously questioned (Nørgaard, 1987). Second, if visceral epithelium is obtained, a means must be found to preserve podocytic differentiation. In vivo, retraction of foot processes is often the first subtle sign of glomerular damage as, for example, following nickel exposure (Gitlitz et al., 1975). The flat polygonal epithelial cells growing in culture may already represent irreversibly dedifferentiated cells. While growth on a proper matrix is undoubtedly important for maintenance of the differentiated appearance (Lovett and Sterzel, 1986), other factors may be anticipated that will be difficult to control. These could include damage during glomerular isolation (Nørgaard, 1978, 1987), changes during the transition to replicating cells available for subculture, and cellular response to normal glomerular capillary pressures. Capillary pressures also affect the nature of the substratum on which the cells grow in vivo; the GBM may be a thixotropic gel (Simpson, 1980).

7. SUMMARY OF INTERACTIONS OF NICKEL AT THE GLOMERULUS

In this chapter we have described some interactions of ionic nickel with the renal glomerulus with the aim of providing general insights into the involvement of the glomerulus in metal-induced nephropathies in general and nickel toxicity in particular. We have also described current techniques for use of the isolated glomerulus, its extracellular matrices, and its individual cell types.

Nickel is cleared from the body mainly by glomerular filtration of low-molecular-weight complexes. Acute parenteral exposure to nickel salts can produce reversible proteinuria and structural changes in the glomerular visceral epithelium at levels that avoid disturbances in tubule function. Divalent nickel binds preferentially to the heparan sulfate proteoglycans of the isolated glomerular basement membrane in a manner described quantitatively by a model of counterion condensation about a polyelectrolyte. At saturation, this binding reduces the net negative charge of the proteoglycan by about half and may not be sufficient to affect the function of the glomerular charge barrier in vivo. The extensive extracellular matrix of the glomerulus can sequester significant amounts of cations and appears to offer some protection against toxicity to glomerular cells at relatively high concentrations of nickel. However, at lower concentrations, nickel can interfere with the incorporation of the heparan sulfate proteoglycan into the glomerular basement membrane. This novel toxic mechanism may adversely affect the filtration characteristics of the glomerulus and could conceivably operate in other tissue with extensive extracellular matrices, such as lung. Nickel prevents attachment and epithelial cell outgrowth from rat glomeruli in culture at concentrations that do not affect the short-term viability of the glomerulus. This may be a consequence of abnormal synthesis of matrix. New insights into the glomerular toxicology of nickel will almost certainly come from experiments with homogeneous cultures of glomerular cells.

ACKNOWLEDGMENTS

This work was undertaken with the financial support of the Medical Research Council of Canada and the Kidney Foundation of Canada.

REFERENCES

Aaseth, J., Olsen, A., Halse, J , and Hovig, T. (1981). Argyria—tissue deposition of silver as selenide. *Scand. J. Clin. Lab. Invest.* **41**, 247–251.

Abbracchio, M. P., Evans, R. M., Heck, J. D., Cantoni, O., and Costa, M. (1982). The regulation of ionic nickel uptake and cytotoxicity by specific amino acids and serum components. *Biol. Trace Elem. Res.* **4**, 289–301.

Abrahamson, D. R. (1986). Recent studies on the structure and pathology of basement membranes. *J. Pathol.* **149**, 257–278.

Abrahamson, D. R. (1987). Structure and development of the glomerular capillary wall and basement membrane. *Am. J. Physiol.* **253**, F783–F794.

Abrass, C. K., and Cohen, A. H. (1988). Characterization of renal injury initiated by immunization of rats with heparan sulfate. *Am. J. Pathol.* **130**, 103–111.

Akerboom, T.P.M., and Sies, H. (1981). Assay of glutathione, glutathione disulfide, and glutathione mixed disulfides in biological samples. *Methods Enzymol.* **77**, 373–382.

Alexander, J., Thomassen, Y., and Aaseth, J. (1988). Tellurium. In H. G. Siler and H. Sigel (Eds.), *Handbook on Toxicity of Inorganic Compounds.* Marcel Dekker, New York, pp. 669–674.

Arnold, W. (1988). Arsenic. In H. G. Siler and H. Sigel (Eds.), *Handbook on Toxicity of Inorganic Compounds.* Marcel Dekker, New York, pp. 79–93.

Asato, N., van Soestbergen, M., and Sunderman, F. W., Jr. (1975). Binding of ^{63}Ni(II) to ultrafiltrable constituents of rabbit serum *in vivo* and *in vitro*. *Clin. Chem.* **21**, 521–527.

Assel, E., Neumann, K.-H., Schurek, H.-J., Sonnenburg, C., and Stolte, H. (1984). Glomerular albumin leakage and morphology after neutralization of polyanions. I. Albumin clearance and sicving coefficient in the isolated perfused rat kindey. *Renal Physiol.,* **7**, 357–364.

Athar, M., Hasan, S. K., and Srivastava, R. C. (1987). Role of glutathione metabolizing enzymes in nickel mediated induction of hepatic glutathione. *Res. Commun. Chem. Pathol. Pharmacol.* **57**, 421–424.

Aughey, E., Fell, G. S., Scott, R., and Black, M. (1984). Histopathology of early effects of oral cadmium in the rat. *Environ. Health Perspect.* **54**, 153–161.

Balt, S., de Bolster, M.W.G., Booij, M., van Herk, A. M., and Visser-Luirink, G. (1983a). Binding of metal ions to polysaccharides. V. Potentiometric, spectroscopic, and viscosimetric studies of the binding of cations to chondroitin sulfate and chondroitin in neutral and acidic aqueous media. *J. Inorg. Biochem.* **19**, 213–226.

Balt, S., de Bolster, M.W.G., and Visser-Luirink, G. (1983b). Binding of metal ions to polysaccharides. II. The binding of metal ions to chondroitin sulphate in alkaline media. *Inorg. Chim. Acta* **78**, 121–127.

Banerjee, D., Onosaka, S., and Cherian, M. G. (1982). Immunohistochemical localization of metallothionein in cell nucleus and cytoplasm of rat liver and kidney. *Toxicology* **24**, 95–105.

Batsford, S., Oite, T., Takamiya, H., and Vogt, A. (1980). Anionic binding sites in the glomerular basement membrane: possible role in the pathogenesis of immune complex glomerulonephritis. *Renal Physiol.* **3**, 336–340.

Bernard, A., Lauwerys, R., Gengoux, P., Mahieu, P., Foidart, J. M., Druet, P., and Weening, J. J. (1984). Anti-laminin antibodies in Sprague Dawley and Brown Norway rats chronically exposed to cadmium. *Toxicology* **31**, 307–313.

Blantz, R. C. (1975). The mechanism of acute renal failure after uranyl nitrate. *J. Clin. Invest.* **55**, 621–635.

Bonner, F. W., Bach, P. H., and Dobrota, M. (1982). The biochemistry of the kidney. In P. H. Bach, F. W. Bonner, J. W. Bridges, and E. A. Lock (Eds.), *Nephrotoxicity: Assessment and Pathogenesis*. Wiley, Chichester, pp. 27–53.

Booij, M. (1981). Binding of metal ions to polysaccharides. I. Polyelectrolyte catalysis by chondroitin sulphate. *Inorg. Chim. Acta* **55**, 109–115.

Brendel, K., and Meezan, E. (1973). Properties of a pure metabolically active glomerular preparation from rat kidneys. II. Metabolism. *J. Pharmacol. Exp. Ther.* **187**, 342–349.

Brenner, B. M., Bohrer, M. P., Baylis, C., and Deen, W. M. (1977). Determinants of glomerular permselectivity: insights derived from observations *in vivo. Kidney Int.* **12**, 229–237.

Brenner, B. M., Dworkin, L. D., and Ichikawa, I. (1986). Glomerular ultrafiltration. In B. M. Brenner and F. C. Rector, Jr. (Eds.), *The Kidney*, 3rd ed. W. B. Saunders, Philadelphia, pp. 124–144.

Carlson, E. C., Brendel, K., Hjelle, J. T., and Meezan, E. (1978). Ultrastructural and biochemical analyses of of isolated basement membranes from kidney glomeruli and tubules and brain and retinal microvessles. *J. Ultrastruct. Res.* **62**, 26–53.

Castellot, J. J., Hoover, R. L., Harper, P. A., and Karnovsky, M. J. (1985). Heparin and glomerular epithelial cell-secreted heparin-like species inhibit mesangial cell proliferation. *Am. J. Pathol.* **120**, 427–435.

Cherian, M. G. (1982). Studies on toxicity of metallothionein in rat kidney epithelial cell culture. In E. Foulkes (Ed.), *Biological Roles of Metallothionein*. Elsevier North-Holland, Amsterdam, pp. 193–202.

Clarkson, T. W., and Shaikh, Z. A. (1982). General principles underlying the renal toxicity of metals. In P. H. Bach, F. W. Bonner, J. W. Bridges, and E. A. Lock (Eds.), *Nephrotoxicity: Assessment and Pathogenesis*. Wiley, Chichester, pp. 263–279.

Comper, W. D., and Laurent, T. C. (1978). Physiological function of connective tissue poly-saccharides. *Physiol. Rev.* **58**, 255–315.

Costa, M., and Heck, J. D. (1985). Perspectives on the mechanism of nickel carcinogenesis. In G. L. Eichorn and L. G. Marzilli (Eds.), *Advances in Inorganic Biochemistry*, Vol. 6. Elsevier, Amsterdam, pp. 285–309.

Donskoy, E., Donskoy, M., Forouhar, F., Gillies, C. G., Marzouk, A., Reid, M. C., Zaharia, O., and Sunderman, F. W., Jr. (1986). Hepatic toxicity of nickel chloride in rats. *Ann. Clin. Lab. Sci.*, **16**, 108–117.

Duke, O., Richter, M., and Panayi, G. S. (1982). Gold related nephropathy. In P. H. Bach, F. W. Bonner, J. W. Bridges, and E. A. Lock (Eds.), *Nephrotoxicity: Assessment and Pathogenesis*. Wiley, Chichester, pp. 349–355.

Ferrari, R. P., and Marzona, M. (1987). On the analogy and specificity of copper(II) binding sites in type I and type IV collagen. *Inorg. Chim. Acta* **136**, 123–126.

Foidart, J. B., Pirard, Y. S., Winand, R. J., and Mahieu, P. R. (1980). Tissue culture of normal rat glomeruli: glycosaminoglycan biosynthesis by homogeneous epithelial and mesangial cell populations. *Renal. Physiol.* **3**, 183–190.

Folkert, V. W., Yunis, M., and Schlondorff, D. (1984). Prostaglandin synthesis linked to phosphatidylinositol turnover in isolated rat glomeruli. *Biochim. Biophys. Acta* **794**, 206–217.

Fujiwara, S., Wiedemann, H., Timpl, R., Lustig, A., and Engel, J. (1984). Structure and interactions of heparan sulfate proteoglycans from a mouse tumor basement membrane. *Eur. J. Biochem.* **143**, 145–157.

Gallagher, J. T., Lyon, M., and Steward, W. P. (1986). Structure and function of heparan sulfate proteoglycans. *Biochem. J.* **236**, 313–325.

Gitlitz, P. H., Sunderman, F. W., Jr., and Goldblatt, P. J. (1975). Aminoaciduria and proteinuria in rats after a single intraperitoneal injection of Ni(II). *Toxicol. Appl. Pharmacol.* **34**, 430–440.

Gooneratne, S. R., and Howell, J. McC. (1983). Structural changes in the kidney of chronic copper poisoned sheep. In P. Brätter and P. Schramel (Eds.), *Trace Element-Analytical Chemistry in Medicine and Biology*, Vol. 2. Walter de Gruyter, Berlin, pp. 341–352.

Goyer, R. A. (1988). Lead. In H. G. Siler and H. Sigel (Eds.), *Handbook on Toxicity of Inorganic Compounds*. Marcel Dekker, New York, pp. 359–382.

Grant, D. S., and Leblond, C. P. (1988). Immunogold quantitation of laminin, type IV vollagen, and heparan sulfate proteoglycan in a variety of basement membranes. *J. Histochem. Cytochem.* **36**, 271–283.

Guder, W. G., and Ross, B. D. (1984). Enzyme distribution along the nephron. *Kidney Int.* **26**, 101–111.

Harper, P. A., Robinson, J. M., Hoover, R. L., Wright, T. C., and Karnovsky, M. J. (1984). Improved methods for culturing rat glomerular cells. *Kidney Int.* **26**, 875–880.

Heathcoate, J. G., and Grant, M. E. (1981). The molecular organization of basement membranes. *Int. Rev. Connect. Tissue Res.* **9**, 191–264.

Heinegård, D., and Sommarin, Y. (1987). Isolation and characterization of proteoglycans. *Methods Enzymol.* **144**, 319–372.

Hendel, R. C., and Sunderman, F. W., Jr. (1972). Species variations in the proportions of ultrafiltrable and protein bound serum nickel. *Res. Commun. Chem. Pathol. Pharmacol.* **4**, 141–146.

Herlant-Peers, M.-C., Hildebrand, H. F., and Biserte, G. (1982). ^{63}Ni(II)-incorporation into lung and liver cytosol of Balb/C mouse. An *in vitro* and *in vivo* study. *Zentralbl. Bakteriol., Abt. 1:Orig.* B **176**, 368–382.

Herlant-Peers, M.-C., Hildebrand, H. F., and Kerckaert, J.-P. (1983). *In vitro* and *in vivo* incorporation of ^{63}Ni(II) into lung and liver subcellular fractions of Balb/C mice. *Carcinogenesis* **4**, 387–392.

Hirose, K., Østerby, R., Nozawa, M., and Gundersen, H.J.G. (1982). Development of glomerular lesions in experimental long-term diabetes in the rat. *Kidney Int.* **21**, 689–695.

Hook, J. B. (1981). Mechanisms of renal toxicity. In S. S. Brown and D. S. Davies (Eds.), *Organ-directed Toxicity: Chemical Indices and Mechanisms*. Pergammon, Oxford, pp. 45–53.

Höök, M. (1984). Cell-surface glycosaminoglycans. *Ann. Rev. Biochem.* **53**, 847–869.

Humes, D. H., and Weinberg, J. M. (1986). Toxic nephropathies. In B. M. Brenner and F. C. Rector, Jr. (Eds.), *The Kidney*, 3rd ed. W. B. Saunders, Philadelphia, pp. 1491–1532.

Humphries, D. E., Silbert, C. K., and Silbert, J. E. (1986). Glycosaminoglycan production by bovine aortic endothelial cells cultured in sulfate-depleted medium. *J. Biol. Chem.* **261**, 9122–9127.

Hunsicker, L. G., Shearer, T. P., and Shaffer, S. J. (1981). Acute reversible proteinuria induced by infusion of the polycation hexadimethrine. *Kidney Int.* **20**, 7–17.

Ito, K., Kimata, K., Sobue, M., and Suzuki, S. (1982). Altered proteoglycan synthesis by epiphyseal cartilages in culture at low SO_4^{2-} concentration. *J. Biol. Chem.* **257**, 917–923.

Jim, K., Hassid, A., Sun, F., and Dunn, M. J. (1982). Lipoxygenase activity in rat kidney glomeruli and glomerular epithelial cells. *J. Biol. Chem.* **257**, 10294–10299.

Kaizu, K., Matsuno, K., Kodama, Y., Uriu, K., and Eto, S. (1987). Nephrotoxicity of gold: clinical and experimental studies. In S. S. Brown and Y. Kodama (Eds.), *Toxicology of Metals: Clinical and Experimental Research*. Ellis Horwood, Chichester, pp. 403–414.

Kanwar, Y. S. (1984). Biophysiology of glomerular filtration and proteinuria. *Lab. Invest.* **51**, 7–21.

Kanwar, Y. S., Linker, A., and Farquhar, M. G. (1980). Increased permeability of the glomerular basement membrane to ferritin after removal of glycosaminoglycans (heparan sulfate) by enzyme digestion. *J. Cell Biol.* **86**, 688–693.

Kanwar, Y. S., Veis, A., Kimura, J. H., and Jakubowski, M. L. (1984). Characterization of heparan sulfate-ptroteoglycan of glomerular basement membranes. *Proc. Natl. Acad. Sci. U.S.A.* **81**, 762–766.

Klein, D. J., Brown, D. M., and Oegema, T. R., Jr. (1986). Partial characterization of heparan and dermatan sulfate proteoglycans synthesized by normal rat glomeruli. *J. Biol. Chem.* **261**, 16636–16652.

Kobayashi, S., Oguri, K., Kobayashi, K., and Okayama, M. (1983). Isolation and characterization of proteohepran sulfate synthesized *in vitro* by rat glomeruli. *J. Biol. Chem.* **258**, 12052–12057.

Kobayashi, S., Nagase, M., Honda, N., and Hishida, A. (1984). Glomerular alterations in uranyl acetate-induced acute renal failure in rabbits. *Kidney Int.* **26**, 808–815.

Krakower, C. A., and Manaligod, J. R. (1980). Mesangiolysis of isolated renal glomeruli with the formation of lobular sacs or cysts. *Renal Physiol.* **3**, 226–236.

Kriz, W., Elger, M., Lemley, K., and Sakai, T. (1990). Structure of the glomerular mesangium: a biomechanical interpretation. *Kidney Int.* **38**, S2–S9.

Langeveld, J.P.M., and Veerkamp, J. H. (1981). Chemical characterization of glomerular and tubular basement membranes of various mammalian species. *Comp. Biochem. Physiol.* **68B**, 31–40.

Larsen, J. A., and Thomsen, O. Ø. (1980). Vanadate-induced oliguria and vasoconstriction in the cat. *Acta Physiol. Scand.* **110**, 367–374.

Lau, S.-J., and Sarkar, B. (1984). Comparative studies of manganese(II)-, nickel(II)-, zinc(II)-, copper(II)-, cadmium(II)-, and iron(III)-binding components in human cord and adult sera. *Can. J. Biochem. Cell Biol.* **62**, 449–455.

Laurie, G. W., Leblond, C. P., and Martin, G. R. (1982). Localization of type IV collagen, laminin, heparan sulfate proteoglycan and fibronectin to the basal lamina of basement membrane. *J. Cell Biol.* **95**, 340–344.

Lauwerys, R. R., Bernard, A., Roels, H. A., Buchet, J.-P., and Viau, C. (1984). Characterization of cadmium proteinuria in man and rat. *Environ. Health Perspect.* **54**, 147–152.

Lelongt, B., Makino, H., and Kanwar, Y. S. (1987). Status of glomerular proteoglycans in aminonucleoside nephrosis. *Kidney Int.* **31**, 1299–1310.

Lerivrey, J., Dubois, B., Decock, P., Micera, Urbanska, J., and Kozlowski, H. (1986). Formation of D-glucosamine complexes with Cu(II), Ni(II) and Co(II) ions. *Inorg. Chim. Acta* **125**, 187–190.

Lindahl, U., Bäckström, G., Jansson, L., and Hallén, A. (1973). Biosynthesis of heparin: II. Formation of sulfamino groups. *J. Biol. Chem.* **248**, 7234–7241.

Lovett, D. H., and Sterzel, R. B. (1986). Cell culture approaches to the analysis of glomerular inflammation. *Kidney Int.* **30**, 246–254.

Lucassen, M., and Sarkar, B. (1979). Nickel(II)-binding constituents of human blood and serum. *J. Toxicol. Environ. Health* **5**, 897–905.

McLaughlin, S. (1977). Electrostatic potentials at membrane–solution interfaces. *Curr. Top. Membr. Transp.* **9**, 71–144.

Mahan, J., Sisson, S., and Vernier, R. L. (1983). Decrease in anionic charge sites in lamina rara externa in aminonucleoside nephrosis. *Kidney Int.* **23**, 187 (Abstract).

Mahan, J. D., Sisson-Ross, S., and Vernier, R. L. (1986). Glomerular basement membrane anionic charge site changes in early aminonucleoside nephrosis. *Am. J. Pathol.* **125**, 393–401.

Mandel, P., Helwig, J. J., and Bollack, C. (1973). Rapid preparation of pure glomeruli and tubular fragments from rat kidney. *Exp. Cell Res.* **83**, 414–417.

Manning, G. S. (1978). The molecular theory of polyelectrolyte solutions with applications to the electrostatic properties of polynucleotides. *Quart Rev. Biophys.* **11**, 179–246.

Manning, G. S. (1979). Counterion binding in polyelectrolyte theory. *Accts. Chem. Res.* **12**, 443–449.

Mastromatteo, E. (1986). Nickel. *Am. Ind. Hyg. Assoc. J.* **47**, 589–601.

Meezan, E., Brendel, K., Ulreich, J., and Carlson, E. C. (1973). Properties of a pure metabolically active glomerular preparation from rat kidneys. I. Isolation *J. Pharmacol. Exp. Ther.* **187**, 332–341.

Melnick, G. F., Ladoulis, C. T., and Cavallo, T. (1981). Decreased anionic groups and increased permeability precedes deposition of immune complexes in the glomerular capillary wall. *Am. J. Pathol.* **105**, 114–120.

Naeser, P., and Rastad, J. (1982). Silver labeling of vascular basement membranes in streptozotocin diabetic mice. *Experientia* **38**, 937–938.

Nomoto, S. (1980). In S. S. Brown and F. W. Sunderman, Jr. (Eds.), *Nickel Toxicology*. Academic, London, pp. 89–90.

Nomoto, S., McNeely, M. D., and Sunderman, F. W., Jr. (1971). Isolation of a nickel α_2-macroglobulin from rabbit serum. *Biochemistry* **10**, 1647–1651.

Nordberg, G. (1982). Renal uptake, storage and excretion of metals. In P. H. Bach, F. W. Bonner, J. W. Bridges, and E. A. Lock (Eds.), *Nephrotoxicity: Assessment and Pathogenesis*. Wiley Heyden, Chichester, pp. 250–262.

Nørgaard, J.O.R. (1978). Retraction of epithelial foot processes during culture of isolated glomerulia. *Lab. Invest.* **38**, 320–329.

Nørgaard, J.O.R. (1987). Rat glomerular epithelial cells in culture: parietal or visceral epithelial origin? *Lab. Invest.* **57**, 277–290.

Norseth, T. (1984). Clinical effects of nickel. In F. W. Sunderman, Jr. (Ed.), *Nickel in the Human Environment*. International Agency for Research on Cancer, Lyon, pp. 395–401.

Oberley, T. D., Yang, A.-H., and Gould-Kostka, J. (1986). Selection of kidney cell types in primary glomerular explant outgrowths by *in vitro* culture conditions. *J. Cell Sci.* **84**, 69–92.

Oken, D. E. (1976). Acute renal failure caused by nephrotoxins. *Environ. Health Perspect.* **15**, 101–109.

Oken, D. E., Landwehr, D. M., and Kirschbaum, B. B. (1982). The hemodynamic basis for experimental acute renal failure. In P. H. Bach, F. W. Bonner, J. W. Bridges, and E. A. Lock (Eds.), *Nephrotoxicity: Assessment and Pathogenesis*. Wiley, Chichester, pp. 168–181.

Onkelinx, C., and Sunderman, F. W., Jr. (1980). Modelling of nickel metabolism. In J. O. Nriagu (Ed.), *Nickel in the Environment*. Wiley-Interscience, New York, pp. 525–545.

Oochi, N., Kobayashi, K., Nanishi, F., Onoyama, K., and Fujishima, M. (1987). Histopathological study of gold nephropathy. In S. S. Brown and Y. Kodama (Eds.), *Toxicology of Metals: Clinical and Experimental Research*. Ellis Horwood, Chichester, pp. 433–434.

Ormstad, K. (1982). The use of isolated kidney cells, glomeruli and subcellular fractions for assessing renal function. In P. H. Bach, F. W. Bonner, J. W. Bridges, and E. A. Lock (Eds.), *Nephrotoxicity: Assessment and Pathogenesis*. Wiley, Chichester, pp. 161–168.

Oskarsson, A., and Tjälve, H. (1979). Binding of ^{63}Ni by cellular constituents in some tissues of mice after the administration of ^{63}NiCl$_2$ and ^{63}Ni(CO)$_4$. *Acta Pharmacol. Toxicol.* **45**, 306–314.

Pabst, R., and Sterzel, R. B. (1983). Cell renewal of glomerular cell types in normal rats. *Kidney Int.* **24**, 626–631.

Parker, K., and Sunderman, F. W., Jr. (1974). Distribution of ^{63}Ni in rabbit tissues following intravenous injection of ^{63}NiCl$_2$. *Res. Commun. Chem. Pathol. Pharmacol.* **7**, 755–762.

Parthasarathy, N., and Spiro, R. G. (1982). Basement membrane glycosaminoglycans: examination of several membranes and evaluation of the effect of sonic treatment. *Arch. Biochem. Biophys.* **213**, 504–511.

Poole, A. R. (1986). Proteoglycans in health and disease: structures and functions. *Biochem. J.* **236**, 1–14.

Rennke, H. G., and Venkatachalam, M. A. (1977). Glomerular permeability: *in vivo* tracer studies with polyanionic and polycationic ferritins. *Kidney Int.* **11**, 44–53.

Reynolds, E. S., Tannen, R. L., and Tyler, H. R. (1966). The renal lesion in Wilson's disease. *Am. J. Med.* **40**, 518–527.

Rohrbach, R. (1986). Reduced content and abnormal distribution of anionic sites (acid proteoglycans) in the diabetic glomerular basement membrane. *Virchows Arch. [Cell Pathol.]* **51**, 127– 135.

Schlondorff, D. (1986). Isolation and use of specific nephron segments and their cells in biochemical studies. *Kidney Int.* **30**, 201–207.

Schlondorff, D. (1987). The glomerular mesangial cell: an expanding role for a specialized pericyte. *FASEB J.* **1**, 272–281.

Schurer, W., Fleuren, G. J., Hoedemaeker, P. J., and Molenaar, I. (1980). A model for the glomerular filter. *Renal Physiol.* **3**, 237–243.

Seiler, M. W., Rennke, H. G., Venkatachalam, M. A., and Cotran, R. S. (1977). Pathogenesis of polycation-induced alterations ("fusion") of glomerular epithelium. *Lab. Invest.* **36**, 48– 61.

Silbert, J. E. (1987). Advances in the biochemistry of proteoglycans. In J. Uitto and A. J. Perejda (Eds.), *Connective Tissue Disease: Molecular Pathology of the Extracellular Matrix*. Marcel Dekker, New York, pp. 83–98.

Simpson, L. O. (1980). Biological thixotropy of glomerular basement membrane and the implications of thixotropy in explaining basement membrane permeability. *Renal Physiol.* **3**, 272–279.

Smith, D. A., Jr., Pitcock, J. A., and Murphy, W. M. (1982). Aluminum-containing dense deposits in the glomerular basement membrane. *Am. J. Clin. Pathol.* **77**, 341–346.

Spiro, R. G. (1967). Studies on the renal glomerular basement membrane: preparation and chemical composition. *J. Biol. Chem.* **242**, 1915–1922.

Striker, G. E., and Striker, L. J. (1985). Glomerular cell culture. *Lab. Invest.* **53**, 122–131.

Striker, G. E., Killen, P. D., and Fairn, F. M. (1980). Human glomerular cells *in vitro*: isolation and characterization. *Transplant. Proc.* **12**, 88–99.

Sunderman, F. W., Jr. (1988). Nickel. In H. G. Siler and H. Sigel (Eds.), *Handbook on Toxicity of Inorganic Compounds*. Marcel Dekker, New York, pp. 454–468.

Sunderman, F. W., Jr. and Horak, E. (1981). Biochemical indices of nephrotoxicity, exemplified by studies of nickel nephropathy. In S. S. Brown and D. S. Davies (Eds.), *Organ-directed Toxicity: Chemical Indices and Mechanisms*. Pergammon, Oxford, pp. 55–67.

Sunderman, F. W., Jr., Costa, E. R., Fraser, C., Hui, G., Levine, J. J., and Tse, T. P. (1981). [63]Nickel-constituents in renal cytosol of rats after injection of [63]nickel chloride. *Ann. Clin. Lab. Sci.* **11**, 488–496.

Sunderman, F. W., Jr., Mangold, B.L.K., Wong, S.H.Y., Shen, S. K., Reid, M. C., and Jansson, I. (1983). High-performance size-exclusion chromatography of [63]Ni-constituents in renal cytosol and microsomes from [63]NiCl$_2$-treated rats. *Res. Commun. Chem. Pathol. Pharmacol.* **39**, 477–492.

Sunderman, F. W., Jr., Zaharia, O., Reid, M. C., Beliveau, J. F., O'Leary, G. P., Jr., and Griffin, H. (1984). Effects of diethyldithiocarbamate and nickel chloride on glutathione and trace metal concentrations in rat liver. *Toxicology* **32**, 11–21.

Templeton, D. M. (1987a). Interaction of toxic cations with the glomerulus: binding of Ni to purified glomerular basement membrane. *Toxicology* **43**, 1–15.

Templeton, D. M. (1987b). Meal-binding properties of the isolated glomerular basement membrane. *Biochim. Biophys. Actat.* **926,** 94–105.

Templeton, D. M. (1987c). Acceleration of the mercury-induced aquation of bromopentammine Co(III) by naturally occurring glycosaminoglycans. *Can. J. Chem.* **65,** 2411–2420.

Templeton, D. M. (1988). Acceleration of ionic reactions by naturally occurring glycosaminoglycans. II. *Inorg. Chim. Acta* **153,** 165–170.

Templeton, D. M. (1990). Cadmium uptake by cells of renal origin. *J. Biol. Chem.* **265,** 21764–21770.

Templeton, D. M. (1991). Metal-proteoglycan interactions in the regulation of renal mesangial cells. In A. Aitio, A. Aro, J. Järvisalo, and H. Vainio (Eds.), *Trace Elements in Health and Disease.* Royal Society of Chemistry, Cambridge, pp. 209–218.

Templeton, D. M., and Castillo, G. (1990). Variability of proteoglycan expression in the isolated rat glomerulus. *Biochim. Biophys. Acta* **1033,** 235–242.

Templeton, D. M., and Chaitu, N. (1990). Effects of divalent metals on the isolated rat glomerulus. *Toxicology* **61,** 119–133.

Templeton, D. M., and Khatchatourian, M. (1988). Synthesis of heparan sulfate proteoglycans by the isolated glomerulus. *Biochem. Cell Biol.* **66,** 1078–1085.

Templeton, D. M., and Sarkar, B. (1985). Peptide and carbohydrate complexes of nickel in human kidney. *Biochem. J.* **230,** 35–42.

Templeton, D. M., and Sarkar, B. (1986). Low molecular weight targets of metals in human kidney. *Acta Pharm. Tox.* **59(SVII),** 416–423.

Templeton, D. M., and Sheepers, J. (1991). Effects of Ni and Cd on proteoglycan synthesis by the isolated glomerulus and glomerular cells in culture. In P. H. Bach, N. J. Gregg, M. F. Wilks, and L. Delacruz (Eds.), *Nephrotoxicity: Mechanisms, Early Diagnosis, and Therapeutic Management.* Marcel Dekker, New York, pp. 377–382.

Thomas, D. W., Hartley, T. F., Coyle, P., and Sobecki, S. (1988). Bismuth." In H. G. Siler and H. Sigel (Eds.), *Handbook of Toxicity of Inorganic Compounds.* Marcel Dekker, New York, pp. 115–127.

Timpl, R. (1986). Recent advances in the biochemistry of glomerular basement membrane. *Kidney Int.* **30,** 293–298.

Tisher, C. C., and Madsen, K. M. (1986). Anatomy of the kidney. In B. M. Brenner and F. C. Rector, Jr. (Eds.), *The Kidney,* 3rd ed. W. B. Saunders, Philadelphia, pp. 3–60.

Tsukamoto, H., Parker, H. R., Gribble, D. H., Mariassy, A., and Peoples, S. A. (1983). Nephrotoxicity of sodium arsenate in dogs. *Am. J. Vet. Res.* **44,** 2324–2330.

Urizar, R., and Vernier, R. L. (1966). Bismuth nephropathy. *J. Am. Med. Assoc.* **198,** 187–189.

Vernier, R. L., Klein, D. J., Sisson, S., Mahan, J. D., Oegema, T. R., and Brown, D. M. (1983). Heparan sulfate-rich anionic sites in the human glomerular basement membrane: decreased concentrations in congenital nephrotic syndrome. *N. Engl. J. Med.* **309,** 1001–1009.

Wagner, W. D. (1985). Proteoglycan structure and function as related to atherosclerosis. *Ann. N.Y. Acad. Sci.* **454,** 52–68.

Walker, F. (1973). The origin, turnover and removal of glomerular basement-membrane. *J. Pathol.* **110,** 233–243.

Watanabe, I. S., and Watanabe, M. (1987). Pathology of gold nephropathy: a review. In S. S. Brown and Y. Kodama (Eds.), *Toxicology of Metals: Clinical and Experimental Research.* Ellis Horwood, Chichester, pp. 415–422.

Watanabe, I., Whittier, F. C., Moore, J., and Cuppage, F. E. (1976). Gold nephropathy: ultrastructural, fluorescence, and microanalytic study of two patients. *Arch. Pathol. Lab. Med.* **100,** 632–635.

Wick, A. N., Drury, D. R., Nakada, H. I., and Wolfe, J. B. (1957). Localization of the primary metabolic block produced by 2-deoxyglucose. *J. Biol. Chem.* **224**, 963–969.

Wilks, M. F., Kwizera, E. N., and Bach, P. H. (1990). Assessment of heavy metal toxicity in vitro using isolated rat glomeruli and proximal tubular fragments. *Renal Physiol. Biochem.* **13**, 275–284.

Wilson, A. P. (1986). Cytotoxicity and viability assays. In R. I. Freshney (Ed.), *Animal Cell Culture: A Practical Approach.* IRL Press, Oxford, pp. 183–216.

Wolfert, A. I., Laveri, L. A., Reilly, K. M., Oken, K. R., and Oken, D. E. (1987). Glomerular hemodynamics in mercury-induced acute renal failure. *Kidney Int.* **32**, 246–255.

Yurchenco, P. D., and Schittny, J. C. (1990). Molecular architecture of basement membranes. *FASEB J.* **4**, 1570–1590.

Yurchenco, P. D., Tsilibary, E. C., Charonis, A. S., and Furthmayr, H. (1986). Models for the self-assembly of basement membrane. *J. Histochem. Cytochem.* **34**, 93–102.

13

BIOCHEMICAL ASPECTS OF RENAL TOXICITY OF NICKEL

B. Sarkar and P. Predki

Research Institute, The Hospital for Sick Children, Toronto, Ontario, Canada, M5G 1X8, and the Department of Biochemistry, University of Toronto, Toronto, Ontario, Canada, M5S 1A8

Nickel and Human Health: Current Perspectives, Edited by Evert Nieboer and Jerome O. Nriagu.
ISBN 0-471-50076-3   © 1992 John Wiley & Sons, Inc.

1.   INTRODUCTION

An epidemiological study of long-term workers at INCO's nickel mines in Ontario has led Roberts et al. (1984) to conclude that a substantial risk of kidney cancer may exist along with cancers of lung, nose, and larynx. One reason for suspecting an increased risk of kidney cancer is the fact that intravenous and intrarenal injections of nickel subsulfide, a major constituent of refinery dust, is known to induce renal cancer in rats (Jasmin and Riopelle, 1976; Sunderman et al., 1979). Sunderman et al. (1979) observed a dose–response relationship to renal cancer among different strains of rat. Earlier work by Lau and Sunderman (1972) indicated that intravenous injection of nickel carbonyl in rats could result in malignant kidney carcinoma.

Toxic nephropathy with proteinuria, aminoaciduria, and reduced renal clearance has been shown to develop in rats given intraperitoneal injections of 65–85 μmol/kg of nickel chloride (Gilitz et al., 1975). Electron microscopic examination revealed alterations of glomerular epithelial cells of these rats. This nickel-induced toxicity was found to be prevented with concurrent administration of triethylenetetramine (Sunderman et al., 1976). Thus, although a correlation between nickel exposure and renal cancers has not been shown in man, animal models indicate that nephrotoxicity and renal cancer can result from exposure to various nickel compounds.

2.   KIDNEY AS A TARGET ORGAN FOR NICKEL

Nickel, along with a number of other metals, is known to be primarily excreted through urine. Distribution of nickel following intraperitoneal administration of ^{63}Ni chloride in rat clearly showed the highest concentration of nickel in kidney and urine (Sarkar, 1980). These studies were carried out over periods of 6, 18, and 24 hr and results showed that after 6 hr, there were higher concentrations of nickel in all organs. The concentration of nickel in these organs, however, declined with time. The results are consistent with those of Anke et al. (1983). Onkelinx and Sunderman (1980) have also shown that the kidney is primarily responsible for ^{63}Ni excretion. It is known that nickel induction for microsomal heme-oxygenase activity is greater in kidney than in any other organs (Maines and Kappas, 1977; Sunderman et al., 1983).

3.   SUBCELLULAR DISTRIBUTION OF NICKEL IN KIDNEY

Rats were administered intraperitoneally a dose of 102 μg nickel/kg body weight as $NiCl_2$. Radioactive $^{63}NiCl_2$ with specific activity 11.7 mCi/mg nickel was used as a tracer. The kidneys obtained after 6 hr following the nickel injection were used for subcellular fractionation. The subcellular fractionation

Table 13.1   Subcellular Distribution of ^{63}Ni at 6 hr After Administration of Nickel

	Fraction	Nickel content (μg/g wet wt/g kidney)
I	Nuclear	0.024
II	Mitochondrial	0.023 ± 0.002
III	Microsomal	0.079 ± 0.015
IV	Cytosol	0.400 ± 0.00

was carried out by differential centrifugation (Hogeboom, 1955). The nuclear fraction was isolated by centrifugation at 800 g for 10 min. The supernatant from the above fraction was further centrifuged at 5,000 g for 15 min to pellet the heavy mitochondrial fraction. The 5000 g supernatant was centrifuged at 13,000 g for 15 min, and the pellet represented the light mitochondrial fraction. The heavy and light mitochondrial fractions were combined. The above supernatant was again centrifuged at 105,000 g for 1 hr to obtain the microsomal fraction. The final supernatant is the cytosolic fraction. All homogenates and supernatants were counted for ^{63}Ni radioactivity using a 0.5-mL solution to which 0.5 mL of hydrogen peroxide (30%) was added and shaken with 10 mL of aquasol. Results are expressed in micrograms of nickel per 100 g of wet weight kidney. It is shown (Table 13.1) that in all the groups, the amount of nickel distribution was in the increasing order nuclear < mitochondrial < microsomal < cytosol. The subcellular distribution of lung and liver of Balb/c mice shows that nickel is distributed in various fractions in the order mitochondrial < microsomal < cytosol < nuclear. The changes in the distribution pattern in kidney might be due to the clearance function of this organ. It can also be attributed to the organ specificity of nickel.

4.   STUDIES OF NICKEL-BINDING SUBSTANCES IN KIDNEY

Nickel-binding studies of kidney are presented in Table 13.2. Since nickel is initially accumulated in kidney, we attempted to detect the nickel-binding constituents from kidney cytosol using Sephadex G-50 gel chromatography. Substantial amounts of nickel were seen bound to a protein of molecular weight 12,000 but could not be isolated or characterized due to lack of sufficient material (Sarkar, 1980). Sunderman et al. (1981) fractionated renal cytosol from rats treated with ^{63}Ni chloride and found that about 68% of nickel in these samples were associated with low-molecular-weight components (~2000). The remainder was bound to five macromolecular constituents, with molecular weights of ~130,000, ~70,000, ~55,000, ~30,000, and 10,000 daltons. None of these constituents were isolated or characterized. Using the soluble postmicrosomal fraction of rat kidney and fractionation by Sephadex

Table 13.2 Studies of Nickel-binding Substances from Kidney

Reference	Species	Mode of Administration Dosage (μmol/kg)	Fractionation Technique
Sarkar, 1980	Rat	Intraperitoneal, 1.4	Sephadex G-50
Sunderman et al., 1981	Rat	Intravenous, 0.1–0.5	Sephadex G-200
		Intramuscular, 5–100	Sephacryl s-300, agarose gel electrophoresis
Abdulwajid and Sarkar, 1983	Rat	Intraperitoneal, 1.7.	Sephadex G-75, DEAE–Sephadex, A-25
Templeton and Sarkar, 1985	Human	in vitro	Sephadex G-75, HPLC 1-60, HLPC C-18, reversed phase

G-75, Abdulwajid and Sarkar (1983) obtained several nickel-binding proteins in the fractionating range comprising about 80% of the nickel in the homogenate. By invoking further resolution using Sephadex G-75, high-performance liquid chromatography (HPLC) on an I-60 and C-18 reversed-phase column, Templeton and Sarkar (1985) detected a low-molecular-weight nickel-binding protein of 4000 daltons and a nickel-binding fraction containing sulfated oligosaccharides.

5. ISOLATION AND CHARACTERIZATION OF NICKEL-BINDING SUBSTANCES IN KIDNEY

5.1. Carbohydrate Fraction

About 70% of the nickel in soluble fraction exists in a region of the elution profile rich in carbohydrates. These are further resolved by chromatography on Biogel P-2 into two nickel-binding constituents having molecular weights of 500 and 900 daltons (Templeton and Sarkar, 1986a). These are oligosaccharides rich in hexosamines and uronic acids and are sulfated.

5.2. Acidic Polypeptide

The acidic polypeptide comprising approximately 30% of the low-molecular-weight cytosolic fraction was purified from both human and bovine kidneys. Kidney was homogenized at 4°C in isotonic saline and centrifuged at 100,000 g for 1 hr. The resulting cytosolic fraction was then passed through an ultrafiltration membrane with a molecular weight cutoff of 10,000 daltons. This material was fractionated through a Biogel P-2 column and the polypeptide collected. This was then concentrated and passed through a C-18 SepPak. The eluate was then passed through gel permeation and then reversed-phase HPLC columns and the purified polypeptide collected. The reversed-phase HPLC step was found to be essential to remove glutamic acid contamination present after gel permeation HPLC.

Acidic nickel-binding polypeptides isolated in this way from both human and bovine kidneys show numerous similarities. Both polypeptides, present in very low amounts in the kidney, have virtually identical amino acid compositions (Table 13.3) and an estimated molecular weight of about 4000 daltons. There is one lysine residue in each polypeptide at which tryptic cleavage into two fragments is possible. Additionally, the amino terminal residue of both polypeptides is blocked to sequencing. Scatchard analysis of nickel binding using an analytical HPLC technique (Templeton and Sarkar, 1986b) indicates that one high-affinity nickel-binding site ($K_{app} = 1.1 \times 10^5$) is present in each polypeptide.

These acidic polypeptides are unlikely to be proteolytic fragments generated during isolation, as boiling before homogenization and isolation in the presence of protease inhibitor yields identical results.

Table 13.3   Amino Acid Compositions of Nickel-binding Polypeptides from Human and Bovine Kidneys

Amino Acid	Human	Bovine
Asx	2.4	3.1
Thr	1.1	1.0
Ser	4.5	4.2
Glx	5.5	4.9
Pro	2.2	1.4
Gly	7.9	9.7
Ala	2.8	4.2
Val	1.9	1.0
Ile	1.0	0.9
Leu	1.9	1.9
His	1.1	0.9
Lys	0.7	1.1

Automated sequencing of the bovine carboxyl terminal tryptic fragment was initially unsuccessful, but a partial sequence was obtained after pyro-glutaminase treatment of the fragment. Scanning of protein sequence libraries shows no identical sequence in any known proteins. Sequencing of the bovine polypeptide is currently underway.

6. SIGNIFICANCE OF ACIDIC COMPONENTS IN METAL BINDING

6.1. Sulfated Oligosaccharides

The nickel-binding carbohydrate fraction has been found rich in sulfated hexosamines and uronic acids. Such compositions are typical of oligosaccharides of glycosaminoglycan alternating units of uronic acid and hexosamine, which are sulfated in variable degrees (Hook et al., 1984). They are anionic in nature. The available carboxyl, sulfate, and polyol functions provide ideal sites for metal binding to sulfated oligosaccharide fragments. In kidney, these components occur on cell surfaces and as heparan sulfate, an important functional component of glomerular and tubular basement membranes (Kefalides, 1981). When intact, they can act as polyelectrolytes imparting an important physiological role for kidney function. On the other hand, glycosaminoglycan fragments may occur as degradation products in the normal turnover of glomerular basement membranes (Schurer et al., 1980). Because of their anionic nature and potential chelating sites, they may play an important role in the excretion of metals.

6.2. Acidic Proteins

That a positively charged metal ion such as nickel or cadmium would bind to an anionic, acidic polypeptide is not surprising. Acidic proteins of varying molecular weights from a wide range of sources are known to bind a number of different metals. (Table 13.4).

As might be expected a number of acidic proteins are known to bind calcium. The helix–loop–helix, or *EF-hand*, binding domain initially described by Kretsinger and Nockolds (1973) is a common motif in many calcium-binding proteins. Calcium coordination in the loop region of such domains is roughly octahedral, the calcium being coordinated by water and protein oxygen atoms. While aspartic and glutamic acid side-chain carboxyl groups are the most common ligands, serine and threonine side-chain oxygen as well as main-chain carbonyl oxygen atoms may also be involved. The EF-hand calcium-binding proteins are thought to have evolved from an ancestral four-domain calcium-binding protein that itself is thought to have resulted from two gene duplications of a single calcium-binding domain gene (Kretsinger, 1976).

The EF-hand domain was initially recognized in parvalbumin. Parvalbumins are soluble proteins isolated from muscle of a variety of sources having molecular weights of around 12 kDa and isoelectric points (pIs) ranging from 3.9 to 6.6 (Gerday, 1988). Although they have three EF domains, one is nonfunctional. Thus parvalbumins bind two calcium ions with K_{app} of approximately 10^8 or two magnesium atoms with lower affinity ($K_{app} = 10^4$). Recently, parvalbumin has been detected in a number of nonmuscle tissues including kidney (Heizmann, 1988). Troponin C, an EF-type acidic calcium-binding protein with a molecular weight of 18 kDa, is known to have two classes of calcium-binding sites, calcium–magnesium and calcium-specific binding sites (Potter and Gergely, 1975). The two calcium–magnesium sites have high affinity for calcium ($K_{app} = 10^7$) and lower affinity for magnesium ($K_{app} = 10^3$) while the two calcium-specific sites bind calcium with a $K_{app} = 10^5$.

Intestinal calcium-binding proteins found in the absorptive cells of the intestinal mucosa are thought to be involved in some way in calcium transport. The acidic 9-kDa bovine intestinal calcium-binding protein has two calcium-binding EF domains (Szebenyi and Moffat, 1983).

Oncomodulin, an acidic protein with an approximate molecular weight of 12 kDa, is also a member of the EF-hand family of calcium-binding proteins (MacManus and Whitfield, 1983), having an affinity for two molecules of calcium per molecule. Interestingly, this is an oncodevelopmental protein not found in adult tissues. It is found following neoplastic transformation in

Table 13.4 Metal Binding to Acidic Proteins

Protein	Molecular Weight (kDa)	pI	Metal	Sites	K_{app}	Reference
Parvalbumin	12	3.9, 6.6	Ca(II)	2	10^8	Heizmann, 1988
Troponin C	18	—	Ca(II)	2	10^7	Potter and Gergely, 1975
			Mg(II)	2	10^3	
			Ca(II)	2	10^5	
Oncomodulin	12	—	Ca(II)	2	High	MacManus and Whitfield, 1983
S100b	21	—	Ca(II)	4	High	Baudier, 1988
			Zn(II)	2	High	
Bovine intestinal Ca-binding protein	9	—	Ca(II)	2	—	Szebenyi and Moffat, 1983
Calsequestrin	42	—	Ca(II)	40–50	10^3	Reithmeyer et al., 1987
Salivary acidic PRP AC	14	—	Ca(II)	—	Low	Anders et al., 1981
Tobacco Cd binding	2	3.15	Cd(II)	—	—	Reese and Wagner, 1987
Cr-binding compound	—	—	Cr(III)	—	—	Yamamoto et al., 1981
Lead inclusion bodies	—	—	Pb(II)	—	—	Moore et al., 1973
HMW mercury-binding component	>150	—	Hg(II)	—	—	Chandra and Cherian, 1973
Renal acidic nickel-binding polypeptide	—	—	Ni(II)	1	1.1×10^5	Templeton and Sarkar, 1986a
			Cd(II)	1	2.3×10^5	

rat and occurs extraembryonically in rat and human placenta (MacManus and Brewer, 1987).

The S100 proteins are a highly acidic water-soluble group of calcium-binding proteins composed of closely related 11-kDa alpha and beta classes that purify as noncovalent alpha–alpha, alpha–beta, and beta–beta dimers. Sequence analysis shows two potential EF hands in both the alpha and beta proteins. Bovine, human, and rat S100b (beta–beta dimer) bind zinc and calcium with high affinity (Baudier et al., 1984, 1985, 1986). While there are four high-affinity calcium-binding sites per S100b protein, calcium-binding studies on zinc-bound S100b show that the calcium and zinc sites are different (Baudier, 1988). A number of other acidic proteins such as calmodulin (Babu et al., 1985) and calcineurin-B subunit (Aitken et al., 1984) belong to this family of calcium-binding proteins.

Acidic calcium-binding proteins that do not belong to the EF-hand family are also known to exist. Calsequestrin is an acidic protein thought to be a major site of calcium storage within the muscle cells from which it is isolated (Reithmeyer et al., 1987). Rabbit calsequestrin has an approximate molecular weight of 42 kDa and has one N-linked oligosaccharide chain. It binds 40–50 atoms of calcium per molecule with a $K_{app} \approx 10^3$. Of the 360 amino acid residues present, 44 are aspartic acid and 59 are glutamic acid, and although no repeat sequence can be detected, many of these acidic residues exist in clusters within the molecule.

Salivary acidic proline-rich phosphoproteins are calcium-binding proteins thought to take part in calcium exchange in the oral cavity (Bennick et al., 1981). The two major salivary acidic proline-rich phosphoproteins, named salivary protein A and C, share an identical amino terminus that has been identified as the calcium-binding site in these molecules. Two types of weak calcium-binding sites have been identified in the amino terminal tryptic fragments of salivary proteins A and C, but the coordinating ligands have not been identified.

A number of less well-characterized acidic proteins and polypeptides also bind to various metals. A tobacco cadmium-binding polypeptide with an apparent molecular weight of 2 kDa and a pI of 3.15 has recently been described (Reese and Wagner, 1987; Grill et al., 1985). It is lacking in basic and aromatic amino acids and is highly acidic, having 30–50 residue percentage of glutamate/glutamine. A low-molecular-weight chromium-binding substance has been reported to exist as an anionic complex with chromium and is composed of amino acids and other ultraviolet-absorbable components (Yamamoto et al., 1981).

Inclusion bodies are metal–protein complexes that are characteristically formed in response to elevated levels of lead. These inclusion bodies are most abundant in liver and kidney nuclei, and amino acid composition analysis has revealed that they contain a high amount of aspartic and glutamic acids, glycine and cysteine, making them quite acidic (Moore et al., 1973). It is also thought that some of the lead is found in insoluble acidic protein fraction of nuclear proteins in these organs.

A high-molecular-weight (>150-kDa) Hg(II)-binding nonhistone protein has been isolated from rat kidney nuclei (Chandra and Cherian, 1973). This acidic protein has a high content of aspartic and glutamic acid (8.4 and 11.8 mol %, respectively) and appears to exist in vivo bound to deoxyribonucleic acid (DNA). Thus, a wide range of acidic protein–metal interactions are known to occur in nature.

In the present study, the nickel-binding polypeptide isolated from kidney is as yet of unknown origin. It has an unusually high content of aspartic acid/asparagine and glutamic acid/glutamine. Electrophoretic studies indicate that many or all of these residues are in acid form. One is led to conclude that aspartic acid and glutamic acid provide polycarboxylic chelation of nickel by this polypeptide. Initial results suggest apparent specificity of nickel for this polypeptide. Work is currently underway to determine the localization and biological function of this acidic polypeptide.

ACKNOWLEDGMENT

The research was supported by the Medical Research Council of Canada.

REFERENCES

Abdulwajid, A. W., and Sarkar, B. (1983). Nickel sequestering renal glycoprotein. *Proc. Natl. Acad. Sci. USA* **80,** 4509–4512.

Aitken, A., Klee, C. B., and Cohen, P. (1984). The structure of the B subunit of calcineurin. *Eur. J. Biochem.* **139,** 663–671.

Anke, M., Griin, M., Gröppel, B., and Kronemann, H. (1983). Nutritional requirements of nickel. In B. Sarkar (Ed.), *Biological Aspects of Metals and Metal-related Diseases.* New York, Raven, pp. 89–105.

Babu, Y. Y., Sack, J. S., Greenbough, T. J., Bugg, C. E., Means, A. R., and Cook, W. J. (1985). Three-dimensional structure of calmodulin. *Nature* **316,** 37–40.

Baudier, J. (1988). S100 proteins: structure and calcium-binding properties. In C. H. Gerday, R. Gilles, and L. Bolis (Eds.), *Calcium and Calcium-binding Proteins.* Springer-Verlag, New York, pp. 102–113.

Baudier, J., Glasser, N., Haglid, K., and Gerard, D. (1984). Purification, characterization and ion binding properties of human brain S100b protein. *Biochim. Biophys. Acta* **790,** 164–173.

Baudier, J., Labourdette, G., and Gerard, D. (1985). Rat brain S100b proteins: purification, characterization and ion-binding properties. A comparison with bovine S100b protein. *J. Neurochem.* **44,** 76–84.

Baudier, J., Glasser, N., and Gerard, D. (1986). Calcium and zinc-binding properties of brain S100 proteins: zinc regulates calcium binding in 100b protein. *J. Biol. Chem.* **261,** 8192–8203.

Bennick, A., McLaughlin, A. C., Grey, A. A., and Madapallimattam, G. (1981). The location and nature of the calcium-binding sites in salivary acidic proline-rich phosphoproteins. *J. Biol. Chem.* **256,** 4741–4746.

Chanda, S. R., and Cherian, M. G. (1973). Isolation and partial characterization of a mercury-binding nonhistone protein component from rat kidney nuclei. *Biochem. Biophys. Res. Comm.* **50,** 1013–1020.

Gerday, C. H. (1988). Souble calcium-binding proteins in vertebrte and invertebrate muscles. In C. H. Gerday, L. Bolis, and R. Gilles (Eds.), *Calcium and Calcium-binding Proteins.* Springer-Verlag, New York, pp. 23–39.

Gilitz, P. H., Sunderman, F. W., Jr., and Goldblatt, P. J. (1975). Aminoaciduria and proteinuria in rats after a single intraperitoneal injection of nickel (II). *Toxicol Appl. Pharmacol.* **34,** 430–440.

Grill, E., Winnacher, E. L., and Zenk, M. H. (1985). Phytochelatins: the principal heavy-metal complexing peptides of higher plants. *Science* **230,** 674–676.

Heizmann, C. W. (1988). Parvalbumin in non-muscle cells. In C. H. Gerday, L. Bolis, and R. Gilles (Eds.), *Calcium and Calcium-binding Proteins.* Springer-Verlag, New York, pp. 93–101.

Hogeboom, G. H. (1955). Fractionation of cell components of animal tissues. In S. P. Colowick and N. D. Kaplan (Eds.), *Methods in Enzymology,* Vol. 1, Academic, New York, pp. 16–19.

Hook, M., Kjellen, L., Johansson, S., and Robinson, J. (1984). Cell-surface glycosaminoglycans. *Ann. Rev. Biochem.* **53,** 847–869.

Jasmin, G., and Riopelle, J. L. (1976). Renal carcinomas and erythrocytosis in rats following intrarenal injection of nickel subsulfide. *Lab. Invest.* **35,** 71–78.

Kefalides, N. A. (1981). Basement membranes: structure function relationships. *Renal Physiol.* **4,** 57–66.

Kretsinger, R. H. (1976). Calcium-binding proteins. *Ann. Rev. Biochem.* **45,** 239–266.

Kretsinger, R. H., and Nockolds, C. E. (1973). Carp muscle calcium-binding protein, structure determination and general description. *J. Biol. Chem.* **248,** 3319–3326.

Lau, T. J., Hacket, R. L., and Sunderman, F. W., Jr. (1972). The carcinogenicity of intravenous nickel carbonyl in rats. *Cancer Res.* **32,** 2253–2258.

MacManus, J. P., and Brewer, C. M. (1987). Oncomodulin in health and disease. In A. W. Norman, T. C. Vanaman, and A. R. Means (Eds.), *Calcium-binding Proteins in Health and Disease.* Academic, New York, pp. 593–595.

MacManus, J. P., and Whitfield, J. F. (1983). Oncomodulin: a calcium-binding protein from hepatoma. In W. Y. Cheung (Ed.), *Calcium and Cell function,* Vol. IV. Academic, New York, pp. 411–440.

Maines, M. D., and Kappas, A. (1977). Nickel-mediated alterations in the activity of hepatic and renal enzymes of heme metabolism and heme dependent cellular activities. In S. S. Brown (Ed.), *Clinical Chemistry and Chemical Toxicology of Metals.* Elsevier/North-Holland Biomedical, Amsterdam, pp. 75–81.

Moore, J. F., Goyer, R. A., and Wilson, M. (1973). Lead-induced inclusion bodies. *Lab. Invest.* **29,** 488–494.

Onkelinx, C., and Sunderman, F. W., Jr. (1980). Modelling of nickel metabolism. In J. O. Nrigu (Ed.), *Nickel in the Environment.* Wiley, New York, pp. 525–545.

Potter, J. D., and Gergely, J. (1975). The calcium and magnesium binding sites on troponin and their role in the regulation of myofibrillar adenosine triphosphatase. *J. Biol. Chem.* **250,** 4628–4633.

Reese, R. N., and Wagner, G. J. (1987). Properties of tobacco cadmium-binding peptide(s). *Biochem. J.* **241,** 641–647.

Reithmeyer, R.A.F., Ohnish, M., Carpenter, M. R., Slupsky, J. R., Gounden, K., Fliege, L., Khanna, N. K., and MacLennan, D. H. (1987). Calsequestrin. In A. W. Norman, T. C. Vanaman and A. R. Mean (Eds.), *Calcium-binding Proteins in Health and Disease*. Academic, San Diego, pp. 62–71.

Roberts, R. S., Julian, J. A., Muir, D. C. F., and Shannon, H. S. (1984). Cancer mortality associated with the high-temperature oxidation of nickel subsulfide. In F. W. Sunderman (Ed-in-Chief), *Nickel in the Human Environment, IARC Sc. Publ. 53*, International Agency for Research on Cancer, Lyon, pp. 23–35.

Sarkar, B. (1980). Nickel in blood and kidney. In S. S. Brown and F. W. Sunderman, Jr. (Eds.), *Nickel Toxicology*. Academic, London, pp. 81–84.

Schurer, W., Fleureng, G. J., Hoedemaeker, P. J., and Molenaar, I. (1980). A model for the glomerular filter. *Renal Physiol.* **3**, 237–243.

Sunderman, F. W., Jr., Kasprzak, K., Korak, E., Gilitz, P., and Onkelinx, C. (1976). Effects of triethylenetetramine upon the metabolism and toxicity of ^{63}nickel chloride in rats. *Toxicol Appl. Pharmacol* **38**, 177–188.

Sunderman, F. W., Jr., Maenza, R. M., Hopfer, S. M., Mitchel, J. M., Allpass, P. R., and Damjanov, I. (1979). Induction of renal cancers in rats by intraperitoneal injection of nickel subsulfide. *J. Environ. Pathol. Toxicol.* **2**, 1511–1527.

Sunderman, F. W., Jr., Costa, E. R., Fraser, C., Hui, G., Levine, J. J., and Tse, T.P.H. (1981). ^{63}Nickel-constituents in renal cytosol of rats after injection of ^{63}nickel chloride. *Ann. Clin. Lab. Sci.* **11**, 488–496.

Sunderman, F. W., Jr., Reid, M. C., Bibeau, L. M., and Linden, J. V. (1983). Nickel induction of microsomal heme-oxygenase activity in rodents. *Toxicol. Appl. Pharmacol.* **68**, 87–95.

Szebenyi, D., and Moffat, K. (1983). The three-dimensional structure of the vitamin D–dependent calcium binding proteins from bovine intestine. In B. deBernard et al. (Eds.), *Calcium-binding Proteins*. Elsevier Science, New York, pp. 199–205.

Templeton, D. M., and Sarkar, B. (1985). Peptide and carbohydrate complexes of nickel in human kidney. *Biochem. J.* **230**, 35–42.

Templeton, D. M., and Sarkar, B. (1986a). Low molecular weight targets of metals in human kidney. *Acta Pharmacol. Toxicol.* **59**, 416–423.

Templeton, D. M., and Sarkar, B. (1986b). Nickel-binding to the C-terminal tryptic fragment of a peptide from human kidney. *Biochim. Biophys. Acta* **884**, 383–386.

Yamamato, A., Waoa, O., and Ono, T. (1981). A low molecular weight chromium-binding substance in mammals. *Toxicol. Appl. Pharm.* **57**, 515–523.

14

CORRELATION OF URINARY ENZYME ACTIVITY AND RENAL LESIONS AFTER INJECTION OF NICKEL CHLORIDE

Michele A. Medinsky, David M. Chico, and Fletcher F. Hahn

Inhalation Toxicology Research Institute,
Lovelace Biomedical and Environmental Research Institute,
Albuquerque, New Mexico 87185

Nickel and Human Health: Current Perspectives, Edited by Evert Nieboer and Jerome O. Nriagu.
ISBN 0-471-50076-3 © 1992 John Wiley & Sons, Inc.

1. INTRODUCTION

Previous studies in rats have shown that the major route for excretion of soluble nickel after intratracheal instillation is urine (Medinsky et al., 1987). In these studies urinary excretion of nickel is not related linearly to the amount of nickel instilled. These investigators found that with increasing amounts instilled, from 0.09 to 9.0 μmol Ni/kg body weight, the percentage of the instilled nickel excreted in urine increases, and the half-time for urinary excretion decreases. This suggests that, with higher doses, nickel is being excreted at a faster rate and to a greater extent.

Other investigators have suggested that nickel is transported in serum in association with low-molecular-weight proteins and serum albumin (Van Soestbergen and Sunderman, 1972; Glennon and Sarkar, 1982; Lucassen and Sarkar, 1979). Very little nickel is transported in the ionic form. L-Histidine is the major Ni(II)-binding amino acid in human serum. In laboratory animals, nickel–histidine complexes are believed to be excreted in urine.

The purpose of our present study was to determine if increased excretion of nickel with increasing dose resulted from structural or functional injury to the kidney. Endpoints measured in this study included histopathological lesions, urinary excretion of renal enzymes, and concentration of nickel in urine and kidneys.

2. METHODS

Female F344/N rats, 11–12 weeks of age, were used in these studies. Rats were born and raised in the barrier-maintained colony at the Inhalation Toxicology Research Institute. Test compounds were administered by intra-peritoneal injection. Rats were divided into five groups of 20 rats per group: three experimental groups that received 90, 9, or 0.9 μmol Ni/kg body weight as nickel chloride (Baker Chemical Co.); a positive control group that received 20 μmol U/kg body weight as uranyl nitrate (Sargent Welsch); and a negative control group that received 1 mL of 0.9% saline. Radiolabeled ^{63}Ni (New England Nuclear; 1 μCi per rat) was incorporated into the nickel injection solutions to facilitate detection of nickel in urine and kidney samples. Doses of 9 and 0.9 μmol Ni/kg body weight were used to compare to studies of Medinsky et al. (1987) that noted increased urinary excretion with increased dose. The 90 μmol Ni/kg body weight dose was used to compare to studies of Gitlitz et al. (1975) where nephrotoxic effects of nickel were observed.

For 5 days before injection, rats were acclimated to plastic metabolism cages and to an altered feeding cycle with feed being withheld during the dark cycle (1800 to 0600) and supplied during the light cycle (0600 to 1800). This procedure decreases the variability in enzyme levels in urine collected from rats (Berlyne, 1984). Urine was collected during the 24 hr before injection of each test compound to serve as a baseline value for renal enzyme assays and was collected at 24-hr intervals over the next 4 days from one set of five rats per exposure group.

Other sets of five rats per exposure group were sacrificed at 1 hr, 1 day, or 4 days by an intraperitoneal injection of sodium pentobarbitol (250 mg/kg body weight). Kidneys were fixed in 10% neutral-buffered formalin. Kidney sections were embedded in paraffin and processed for histopathological evaluation.

Samples of kidneys were analyzed for ^{63}Ni content by liquid scintillation spectrometry. Urine was analyzed by the methods of Amador et al. (1963) and Woolen et al. (1961) for the total amounts of lactate dehydrogenase (LDH), N-acetyl glucosaminidase (NAG), β-galactosidase (GAL), and alkaline phosphatase (AP) excreted during each of the 24-hr periods. The substrates used were p-nitrophenol (NAG and GAL), 4-nitrophenyl phosphate (AP), and pyruvate (LDH). Urine samples were also analyzed for total amount of ^{63}Ni and were subjected to ultrafiltration using Amicon Diaflo ultrafilters to determine the amount of ^{63}Ni associated with molecular weight fractions below 5000 or above 100,000. Data were analyzed for statistical significance using analysis-of-variance techniques. Results for urinary enzymes were compared to control values using Dunnett's Multiple Range Test. Significant differences were noted at $p < 0.05$.

3. RESULTS

Figures 14.1A–D show the total milli-international units (mIU) of each of the four enzymes excreted in the urine per 24-hr collection interval. At 2 days after injection, significantly elevated amounts of LDH, NAG, GAL, and AP were found in urine from rats injected with uranyl nitrate, compared to urine from control animals injected with saline. Maximum amounts of all four enzymes were found in urine collected 3 days after injection, and levels remained elevated at 4 days after injection. Only two enzymes, NAG and LDH, were elevated in the urine of rats given 90 μm of Ni/kg body weight. Increased amounts of LDH were found in urine collected 1–3 days after injection and increased amounts of NAG were present at 2 and 3 days after injection. However, by 4 days after injection, amounts of these two enzymes in urine had returned to control values. No increases in enzyme activity in urine (amount excreted per 24-hr interval) were seen after injection of the other nickel doses.

Table 14.1 summarizes the histopathological changes seen after injection of rats with uranyl nitrate, 90 μmol Ni/kg, 9.0 μmol Ni/kg, or saline. The lesions induced by uranyl nitrate, seen at 4 days after injection, were degeneration and necrosis of renal tubular epithelium in the pars recta (the P_2–P_3 portion of the proximal tubule) typical of heavy-metal poisonings (Owen, 1986). There was protein in the tubule lumen, and a few necrotic cells sloughed into the lumen of the distal portion of the tubule.

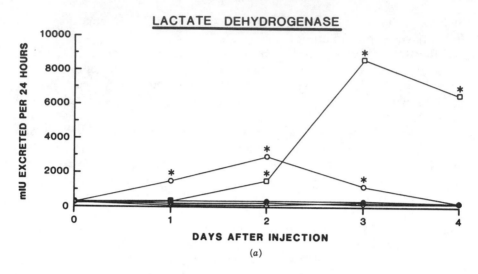

(a)

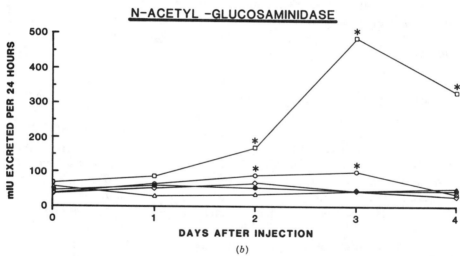

(b)

Figure 14.1. Total milli-international units (mIU) of each enzyme excreted per 24-hr collection period is indicated for days of urine collection. Day 0 represents urine collected over a 24-hr period before injection. Data points represent the mean of five animals. Stars indicate a statistically significant ($p < 0.05$) increase in urinary excretion, compared to controls, as determined by a one-way analysis of variance. (□) Uranyl nitrate; (○) 90 μm Ni/kg; (△) 9 μm Ni/kg; (◇) 0.9 μm Ni/kg; (●) saline injected controls.

B-GALACTOSIDASE

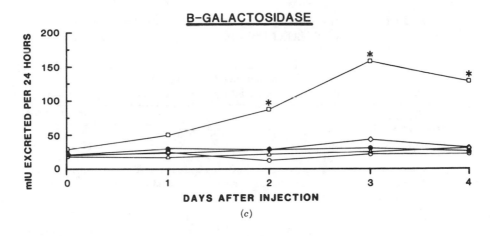

(c)

ALKALINE PHOSPHATASE

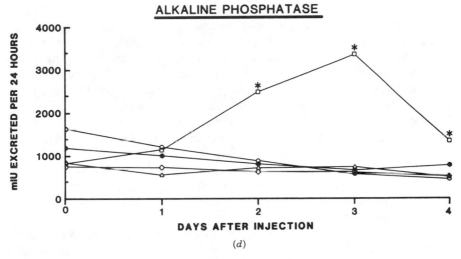

(d)

Figure 14.1. (*Continued*)

The lesion observed in the animals injected with 90 μmol Ni/kg was of a different nature, involving a different portion of the tubule. The pars convoluta was primarily affected. The major histopathological finding was a regeneration of tubular epithelium, evidenced by smaller, more numerous basophilic lining cells. Only a very few dead cells and debris were found in the lumen. Localization of the lesion in the pars convoluta corresponds to lesions observed for cadmium–metallothionein and chromate ion. No similar lesions were found in the 9 μmol Ni/kg group.

In the controls, mineralized foci in the interstitium of the medulla were found. Similar lesions were noted in other groups (Table 14.1).

Analysis of the quantities of [63]Ni retained in the kidneys of rats at various times after injection (Table 14.2) indicated that the largest quantities were

Table 14.1   Summary of Histopathological Lesions Observed 4 days after Injection of Nickel Chloride, Uranyl Nitrate, or Saline

Chemical	Diagnosis[a]			
	Normal	Mineralized Interstitial Foci	Tubular Necrosis	Tubular Regeneration
Uranyl nitrate	—	—	5/5[b]	5/5[c]
90 μmol Ni/kg	—	—	—	4/4[b]
9 μmol Ni/kg	4/5	—	—	1/5[c]
Saline	3/5	2/5[c]	—	—

[a] Numbers are ratios of number affected to number examined.
[b] Relative severity was moderate.
[c] Relative severity was minimal.

retained by the 90 μmol Ni/kg group. However, when expressed as a percentage of the injected nickel (Table 14.2), there were no significant differences in the retention of nickel in the kidneys as a function of the amount injected.

Urine was also analyzed for the percentage of injected nickel excreted over the 4-day period (Table 14.3). The percentage of the dose excreted was significantly less at the lowest injected dose, 0.9 μmol Ni/kg, compared to the two higher doses. Likewise, the percentage of the total urinary nickel that was excreted in the first 24 hr was smaller for the 0.9-μmol/kg dose. These data are consistent with previous work (Medinsky et al., 1987) that indicated that for doses up to 9 μmol/kg, both the total nickel excreted in urine and the rate at which it was excreted (measured in the present study as the fraction of the total excreted in 24 hr) increased with increasing amount of nickel administered.

For all three injected doses, 70–80% of the nickel excreted in urine was not retained by ultramembrane filters with pore size of molecular weight 5000. This nickel could be associated with small-molecular-weight peptides or amino acids or it could be present as free nickel. From 10–14% of the nickel in urine was associated with proteins exceeding 100,000 molecular weight. Association with proteins did not appear to be influenced by the amount of nickel administered.

4. DISCUSSION

The results of these studies confirmed that levels of renal enzymes excreted in urine can be a good indicator of potential kidney toxicity. Appearance of increased amounts of renal enzymes in urine can be used as a measure of kidney damage (Fowler, 1982). Increased excretion of cytosolic enzymes is generally associated with minor damage to the renal tubules, whereas increased excretion of lysosomal and membrane-bound enzymes are associated with

Table 14.2 Total Nickel in Kidneys at Different Times after Injection of Nickel Chloride[a]

μmol Ni Injected/ kg Body Weight	1 hr		24 hr		96 hr	
	Total pmol Ni in Kidneys	Percentage of Dose	Total pmol Ni in Kidneys	Percentage of Dose	Total pmol Ni in Kidneys	Percentage of Dose
0.9	5.22 ± 1.64	3.03 ± 0.98	0.937 ± 0.052	0.535 ± 0.038	0.188 ± 0.011	0.159 ± 0.010
9	40.8 ± 2.8	3.17 ± 0.24	7.60 ± 0.33	0.576 ± 0.042	1.84 ± 0.27	0.147 ± 0.020
90	666 ± 120	5.41 ± 0.94	300 ± 153	2.07 ± 0.92	13.0 ± 1.6	0.108 ± 0.014

[a]Data represent mean ± standard error, $n = 5$.

Table 14.3   Cumulative Excretion of Nickel in Urine after Administration of Various Amounts of Nickel Chloride[a]

μmol Ni Injected/ kg Body Weight	Percentage of Injected Ni Excreted	Percentage of Total Ni Excreted in First 24 hr
0.9	14.9 ± 2.37[b]	61.0 ± 5.82[b]
9.0	96.6 ± 14.4	91.3 ± 2.43
90	77.3 ± 6.58	86.5 ± 1.94

[a] Data represent mean ± standard error; $n = 5$.
[b] Significantly different from 9.0 and 90 μmol/kg groups with $p < 0.05$.

more extensive damage (Plummer et al., 1982; Price, 1982). Of the enzymes assayed in this study, both LDH and NAG are cytosolic enzymes. The GAL is a lysosomal enzyme and alkaline phosphatase is a brush-border enzyme (i.e., associated with the cell membranes of the proximal tubule cells). Only increased excretion of the two cytosolic enzymes was seen after injection with the highest dose of nickel, suggesting minor tubular damage. This observation was supported by histopathological evaluation. In contrast, after injection with uranyl nitrate, elevated urine levels of all renal enzymes in the urine were seen, and these levels were still elevated at the end of the study. Histopathological evaluation indicated necrosis of renal tubules after exposure to uranyl nitrate. Both the time course for urinary excretion of renal enzymes and histopathological lesions were consistent with the progressive nature of kidney damage caused by uranyl nitrate (Haley, 1982).

Gitlitz et al. (1975) reported toxic effects in rats injected intraperitoneally with 85 μmol $NiCl_2$/kg. In their studies the mean urinary excretion of total protein and amino acids was greatly increased during the first 2 days after injection and returned to baseline by days 3 or 4. This is consistent with the time course we observed in our studies for urinary excretion of enzymes after injection of 90 μmol Ni/kg. By 4 days after injection enzyme levels in nickel-treated rats returned to normal. In studies by Gitlitz et al. (1975) maximum urinary excretion of histidine (10 times control values) occurs the first day after injection. This was the period for maximal excretion of nickel in our studies also. These findings are compatible with the observation that a nickel–histidine complex may be involved in the renal excretion of Ni(II) (Van Soestbergen and Sunderman, 1972; Lucassen and Sarkar, 1979). The increased excretion of amino acids observed by Gitlitz et al. (1975) suggests that nickel may be inhibiting amino acid transport systems located on the membranes of the renal tubules.

We have demonstrated that at doses of nickel sufficiently high to produce kidney lesions (90 μmol Ni/kg), urinary excretion of nickel was similar to the medium dose (9 μmol Ni/kg), in which there was no renal toxicity. This suggested that damage to the kidney tubules did not alter the rate at which nickel was excreted in urine. However, at doses not resulting in nephrotoxicity there was an increase in urinary excretion of nickel with increasing dose of nickel injected (from 0.9 to 9 μmol/kg). Increased urinary excretion of nickel in the absence of nephrotoxicity suggested that mechanisms by which nickel was reabsorbed in the proximal tubules might have become saturated. Other investigators (Van Soestbergen and Sunderman, 1972) have suggested that nickel is excreted in association with the amino acid histidine. The results of our studies indicated that about 80% of the urinary nickel was associated with molecules of molecular weight of at most 5000. Thus, it is possible that low-molecular-weight nickel–histidine complexes are not being reabsorbed as readily as the free histidine molecules. Alternatively, nickel might disassociate from the histidine complex in the more acidic ultrafiltrate of the proximal tubule and be excreted as a positively charged cation.

In summary, injected nickel can produce renal lesions that are evident histologically. These nickel doses also result in elevated levels of cytosolic renal enzymes being excreted into urine. The time course for excretion of these enzymes correlates with the observed kidney damage. However, renal injury has no effect on the urinary excretion of nickel when compared to excretion following nonnephrotoxic doses.

ACKNOWLEDGMENTS

Research was sponsored by the U.S. Department of Energy Office of Health and Environmental Research under Contract No. DE-AC04-76EV01013 and conducted in facilities fully accredited by the American Association for Accreditation of Laboratory Animal Care. One of us (D.M.C.) was an Associated Western Universities Student Research Participant during the conduct of the research.

REFERENCES

Amador, E., Zimmerman, A. B., and Wacher, W.E.C. (1963). Urinary alkaline phosphatase activity. I. Elevated urinary LDH and alkaline phosphatase activities for diagnosis of renal adenocarcinomas. *J. Am. Med. Assoc.* **185,** 769.

Berlyne, G. M. (1984). Toxic nephropathies and current methods for early detection of the toxicity of the kidney. In M. A. Mehlman, C. P. Hemstreet, III, J. J. Thorpe, and N. K. Weaner (Eds.), *Renal Effects of Petroleum Hydrocarbons*, Vol. VII, Advances in Environmental Toxicology. Princeton Scientific Publishers, Princeton, NJ, pp. 173–184.

Fowler, J. S. (1982). Micro-middle and macromolecules in blood and urine. In P. H. Bach, F. W. Bonner, J. W. Bridges, and E. A. Lock (Eds.), *Nephrotoxicity, Assessment, and Pathogenesis*. Wiley, New York, pp. 66–77.

Gitlitz, F., Sunderman, W., Jr., and Goldblatt, P. J. (1975). Aminoaciduria and proteinuria in rats after a single intraperitoneal injection of Ni(II). *Toxicol. Appl. Pharmacol.* **34,** 430–440.

Glennon, J. D., and Sarkar, B. (1982). Nickel(II) transport in human blood serum: studies of nickel(II) binding to human albumin and to native-sequence peptide, and ternary-complex formation with L-histidine. *Biochem. J.* **203**, 15–23.

Haley, D. P. (1982). Morphologic changes in uranyl nitrate–induced acute renal failure in saline- and water-drinking rats. *L. Investigat.* **46**(2), 196–208.

Lucassen, M., and Sarkar, B. (1979). Nickel(II)-binding constituents of human blood serum. *J. Toxicol. Environ. Health* **5**, 897–905.

Medinsky, M. A., Benson, J. M., and Hobbs, C. H. (1987). Lung clearance and disposition of ^{63}Ni in F344/N rats after intratracheal instillation of nickel sulfate solutions, *Environ. Res.* **43**, 168–178.

Owen, R. A. (1986). Acute tubular lesions, kidney, rat. In T. C. Jones, U. Mohr, and R. D. Hunt (Eds.), *Urinary System, Monographs on Pathology of Laboratory Animals*. Springer-Verlag, Berlin, pp. 229–239.

Plummer, D. T., Ngaha, E. O., Wright, P. J., Leathwood, P. D., and Blake, M. E. (1982). The sensitivity of urinary enzyme measurements for detecting renal injury. In U. C. Dubach and U. Schmidt (Eds.), *Diagnostic Significance of Enzymes and Proteins in Urine*. Current Problems in Clinical Biochemistry, vol. 9. Hans Huber Publishers, Vienna, pp. 71–87.

Price, R. G. (1982) Urinary N-acetyl-β-D-glucosaminidase (NAG) as an indicator of renal disease. In U. C. Dubach and U. Schmidt (Eds.), *Diagnostic Significance of Enzymes and Proteins in Urine*. Current Problems in Clinical Biochemistry, vol. 9. Hans Huber Publishers, Vienna, pp. 150–163.

Van Soestbergen, M., and Sunderman, F. W., Jr. (1972). ^{63}Ni complexes in rabbit serum and urine after injection of ^{63}NiCl$_2$. *Clin. Chem.* **18**, 1478–1484.

Woolen, J. W., Heyworth, R., and Walker, P. G. (1961). Studies on glucosaminidase 3. Testicular N-acetyl-β-glucosaminidase and N-acetyl-β-galactosaminidase. *Biochem. J.* **78**, 111–116.

15

NICKEL CONTACT HYPERSENSITIVITY

Torkil Menné

Dermatologisk Afd K Kas Gentofte Niels Andersensvej 65
DK-2900 Hellerup, Denmark

1. INTRODUCTION

Nickel dermatitis was initially recognized as *Das Galvanizierekzem* in 1889. In 1925 patch testing proved nickel allergy to be the cause of dermatitis in the electroplating industry. Occupational nickel dermatitis was common in the 1920s and 1930s while consumer nickel dermatitis first appeared in the early 1930s. Five decades after recognition of nickel dermatitis as a common disease, the situation is unchanged with no accepted preventive strategy.

Nickel and Human Health: Current Perspectives, Edited by Evert Nieboer and Jerome O. Nriagu.
ISBN 0-471-50076-3   © 1992 John Wiley & Sons, Inc.

2. NICKEL SENSITIZATION

Dermatologists use the terms *allergy/sensitization* and *eczema/dermatitis* synonymously. Nickel allergy is a cellular-mediated sensitization (type IV allergy; see Chapters ■ and ■). Using two different experimental designs, Vandenberg and Epstein (1963) and Kligman (1966) sensitized 9 and 48% of subjects, respectively, within a short period of time. These limited human experimental experiences classified nickel as a medium-to-strong contact sensitizer. Most cases of primary nickel sensitization are a consequence of prolonged (hours) nonoccupational skin contact with nickel-plated objects or nickel alloys. Short-lasting skin contact with nickel as from coins, door-handles, kitchen equipment, and so on, will not give rise to sensitization but might be the cause of chronicity in individuals previously nickel sensitized. At one time, suspenders and metal buttons in blue jeans caused most cases of sensitization; today ear piercing, costume jewelry, and cheap wrist watches are the main causes.

The sensitization risk from different types of nickel coatings and alloys has only recently been evaluated in experimental studies. Menné et al. (1987) examined 11 widely used nickel alloys with respect to corrosion stability and reactivity in nickel-sensitized individuals. Alloys with a nickel release in synthetic sweat exceeding 1 $\mu g/cm^2$ per week gave a strong patch test reactivity in nickel-sensitive persons; those with a release below 0.5 $\mu g/cm^2$ per week such as nickel tin, stainless steel, and white gold showed generally were weakly reactive. Even though the study only included patients already sensitized to nickel, it gives an indication of the sensitizing capacity of different nickel alloys. Items releasing more nickel than 1 $\mu g/cm^2$ per week are known from our clinical experience to cause the majority of sensitizations; nickel releases below 0.5 $\mu g/cm^2$ per week do not give rise to primary nickel dermatitis. Occupational nickel sensitization is relatively rare and mainly starts as a primary hand eczema that spreads to the forearms. Jobs with significant occupational nickel exposure are listed in Table 15.1. Nickel dermatitis appears not to be a problem at nickel-refining plants. Perhaps immunological tolerance or desensitization occurs because of chronic exposure to nickel compounds by inhalation.

Table 15.1   Selected Occupations with Nickel Exposure

Cashier	Hairdresser
Ceramic worker	Jeweler
Electrician	Mechanic
Electroplater	Metal worker and welder

3. EPIDEMIOLOGY OF NICKEL DERMATITIS

Knowledge of the frequency of nickel allergy comes from studies performed in dermatological departments and population-based epidemiological investigations. The results from such investigations are not comparable, as patients in dermatological departments are selected because of diseased skin and are therefore expected to have an increased frequency of contact sensitivities.

Collection of data from consecutive patch-tested patients in dermatological clinics is an easy and inexpensive method of monitoring the sensitization pattern in a population. Recent international studies using this method are summarized in Table 15.2. It appears that 10–20% of females tested react to nickel, compared to 2–10% of men. Repeated longitudinal studies within the same clinic disclose an increasing frequency of nickel allergy among dermatological patients (Edman and Möller, 1982). Cheap nickel-plated wrist watches and watch straps predominantly from the far East are the main source of nickel sensitization in Nigeria (Olumide, 1985). The hot and humid climate facilitates nickel corrosion, and an increasing number of nickel-sensitized persons can be expected in the tropical areas of the world.

Population studies have disclosed an unexpected high prevalence of nickel allergy in the general population. The outcome of studies worldwide are rather uniform with a prevalence of nickel allergy between 7 and 10% in females (Table 15.3) and 1 and 2% in men. School children were included in three of the studies shown, and they exhibit the same high prevalence rate of nickel sensitization as adults. There is some evidence that the annual incidence in this age group has increased in recent years (Menné et al., 1982). The prevalence studies have also revealed that most nickel-sensitive individuals have light and intermittent problems with contact dermatitis; only a minority among the total number sensitized develop severe dermatitis leading to sick leave and permanent impairment.

4. CLINICAL PATTERN OF NICKEL DERMATITIS

4.1. Primary Sensitization

Primary nickel dermatitis occurs at skin sites in close contact with costume jewelry and clasps in clothing, illustrating a patchy, eventually symmetrical, pattern. The relation to metal contact sites is often so obvious that patients establish the diagnosis of nickel allergy themselves. Diagnostically, nickel allergy is established by patch testing, a technique where the patients are reexposed to a small quantity of nickel under occlusion on the upper back for 48 hr. A positive reaction discloses redness, edema, and eventually vesicles. The primary nickel dermatitis has a good medical prognosis, and the individual

Table 15.2 Clinical Patch-Test Studies of Nickel Sensitivity in Patients with Dermatitis

Study	Number of Patients Tested			Percentage Positive to Nickel		
	Male	Female	Total	Male	Female	Total
Europe: Fregert et al., 1969	2039	2786	4825	1.8	10.2	6.7
Kuwait: Kanan, 1969	—	—	389	24[a]	10[a]	34[a]
Belgium: Oleffe et al., 1972	184	116	300	3.8	18.1	9.3
United States: NACDG, 1973	509	691	1200	5.6	15.0	11.0
Scotland: Husain, 1977	603	709	1312	5.0	26.0	16.0
Brazil: Moriearty et al., 1978	271	265	536	2.6	11.7	7.1
Japan: Sugai et al., 1979	291	447	738	4.8	4.3	4.5
Denmark: Hammershøy, 1980	1451	1774	3225	—	—	6.4
Nigeria: Olumide; 1985	230	223	453	11.0	12.4	11.7
Italy: Angelini et al., 1986	3587	3483	7070	4.6	20.0	12.0
Eastern Europe: Schubert et al., 1987	913	1487	2400	2.1	10.5	7.3

[a]Total number.

Table 15.3 Prevalence of Nickel Allergy in Different Population Groups

Reference	Study Population	Number of Persons Investigated			Prevalence (%)	
		Male	Female	Total	Male	Female
Menné, 1978	Female nondermatological inpatients	—	213	213	—	9.4[a]
Prystowsky et al., 1979	Paid adult volunteers	460	698	1158	0.9	9.0
Peltonen, 1979	General population	478	502	980	0.8	8.0
Kieffer, 1979	Veterinary students	247	168	415	2.8	9.8
Magnusson and Möller, 1979	Orthopedic patients	106	168	274	1.0	10.0
Menné et al., 1982	General population	—	1976	1976	—	14.5[a]
Boss and Menné, 1982	Hairdresser school	—	53	53	—	20.0
Menné and Holm, 1983	Female twins	—	1546	1546	—	9.6
Larsson-Stymme and Widström, 1985	School girls	—	960	960	—	9.0
von Schmiel, 1985	Surgical inpatients and hospital staff	244	259	503	1.6	7.7
Hums, 1986	School children	135	131	266	—	7.0
van der Burg et al., 1986	Nursing school	29	188	217	—	13.0
	Hairdresser school	12	74	86	17.0	27.0

[a]Nickel allergy diagnosed by history only.

only experiences minor discomfort. However, if the condition is neglected and the patient continues with intensive nickel contact, there is a risk for chronicity and spread of the dermatitis to skin sites distant from the primary sensitization.

4.2. Secondary Nickel Dermatitis

Nickel dermatitis has a tendency to spread to the hands, elbow flexures, eyelids, and genital area. A population-based study has established a statistically significant risk for a nickel-sensitized individual for developing hand eczema (Menné et al., 1982). The dissemination of nickel dematitis can be caused by percutaneous or systemic (inhalation or oral) nickel exposure (Menné et al., 1987).

5.   TREATMENT OF NICKEL DERMATITIS

The mainstay in management of nickel-allergic individuals is instruction on how to use the "nickel spot" test (dimethylglyoxime test; Fisher, 1987), which helps to identify blatant contact with nickel. The spot-test kit contains 1% dimethylglyoxime in alcohol and 10% amonium hydroxide solution. A few drops of the two solutions are mixed on the metallic object, and if free nickel is present, a red color appears. Patients with active nickel dermatitis, particularly hand eczema, are treated with topical steroid creams for weeks or months. Short terms of systemic steroid treatment may be indicated in severe cases. Topical and systemic use of nickel-chelating drugs as well as the use of low-nickel diets must be considered only as experimental approaches to chronic nickel dermatitis. Even with effective management, only 20–30% of the patients with hand eczema and nickel allergy experience healing according to long-term follow-up studies (Fregert, 1975; Christensen, 1982).

6.   PROPHYLAXIS

Prevention of nickel sensitization is not only rational but also a realistic possibility. Using nickel alloys releasing less than 0.5 $\mu g/cm^2$ per week in synthetic sweat and/or that are negative to the dimethylglyoxime spot test for items designed for prolonged skin contact will reduce the number of nickel-sensitized individuals in the population significantly. [*Editor's note*: As of July 10, 1989, objects made from nickel-containing alloys or with nickel-containing surface coatings and releasing nickel at a rate exceeding 0.5 $\mu g/cm^2$ per week (as determined by the dimethylglyoxime spot test) may not be sold in Denmark. Articles regulated include ear ornaments, jewelry, wrist watches, glasses, and garment accessories such as buttons and zippers. Offenders are subject to a fine or detention.]

REFERENCES

Angelini, G., Vena, G. A., Fiordalisi, F., Giglio, G., and Meneghini, C. L. (1986). Allergia da contatto al nickel. Rilievi epidemiologici e clinici. *Giorn. It. Derm. Vener.* **121,** 121–126.

Boss, A., and Menné, T. (1982). Nickel sensitization from ear piercing. *Contact Dermatitis* **8,** 211–213.

Christensen, O. B. (1982). Prognosis in nickel allergy and hand eczema. *Contact Dermatitis* **8,** 7–15.

Edman, B., and Möller, H. (1982). Trends and forecasts for standard allergens in a 12-year patch test material. *Contact Dermatitis* **8,** 95–104.

Fisher, A. A. (1986). *Contact Dermatitis*, 3rd ed. Lea and Febiger, Philadelphia, pp. 772–773.

Fregert, S. 1975. Occupational dermatitis in a 10-year material. *Contact Dermatitis* **1,** 96–107.

Fregert, S., Hjorth, N., Magnusson, B., Bandman, H.-J., Calnan, C. D., Cronin, E., Malten, K., Meneghini, C. L., Pirilä, V., and Wilkinson, D. S. (1969). Epidemiology of contact dermatitis. *Trans. St. John's Hosp. Derm. Soc.* **55,** 17–35.

Hammershøy, O. (1980). Standard patch test results in 3.225 consecutive Danish patients from 1973 to 1977. *Contact Dermatitis* **6,** 263–268.

Hums, R. (1986). Nickelexposition–Nickelsensibilisierung. *Dermatol. Mon. Schr.* **172,** 697–700.

Husain, S. L. (1977). Contact dermatitis in the West of Scotland. *Contact Dermatitis* **3,** 327–332.

Kanan, M. W. (1969). Contact dermatitis in Kuwait. *J. Kwt. Med. Assoc.* **3,** 129–144.

Kieffer, M. (1979). Nickel sensitivity: relationship between history and patch test reaction. *Contact Dermatitis* **5,** 398–401.

Kligman, A. M. (1966). Identification of contact allergies by human assay. III. Maximization test: a procedure for screening and rating contact sensitizers. *J. Invest. Dermatol.* **47,** 393–409.

Larsson-Stymne, B., and Widström, L. (1985). Ear piercing—a cause of nickel allergy in schoolgirls? *Contact Dermatitis* **13,** 289–293.

Magnusson, B., and Möller, H. (1979). Contact allergy without skin disease. *Acta Derm.-Venereol.* **59** (Suppl. 89), 113–115.

Menné, T. (1978). The prevalence of nickel allergy among women. *Berufsdermatosen* **26,** 123–125.

Menné, T., and Holm, N. V. (1983). Nickel allergy in a female twin population. *Int. J. Dermatol.* **22,** 22–28.

Menné, T., and Maibach, H. I. (1987). Systemic contact allergy reactions. *Sem. Dermatol.* **6,** 108–118.

Menné, T., Borgan, Ø., and Green A. (1982). Nickel allergy and hand dermatitis in a stratified sample of the Danish female population: an epidemiological study including a statistic appendix. *Acta Derm.-Venereol.* **62,** 35–41.

Menné, T., Brandrup, F., Thestrup-Pedersen, K., Veien, N. K., Andersen, J. R., Yding, F., and Valeur, G. (1987). Patch test reactivity to nickel alloys. *Contact Dermatitis* **16,** 255–259.

Moriearty, P. L., Pereira, C., and Guimaraes, N. A. (1978). Contact dermatitis in Salvador, Bazil. *Contact Dermatitis* **4,** 185–189.

NACDG (1973). Epidemiology of contact dermatitis in North America 1972. *Arch. Dermatol.* **108,** 537–538.

Oleffe, J., Nopp-Oger, M. J., and Achten, G. (1972). Batterie européenne de tests epicu-tariés—Bilan de 300 observations. *Berufs-Dermatosen* **20,** 209–213.

Olumide, Y. M. (1985). Contact dermatitis in Nigeria. *Contact Dermatitis* 12, 241–246.

Peltonen, L. (1979). Nickel sensitivity in the general population. *Contact Dermatitis* 5, 27–32.

Prystowsky, S. D., Allen, A. M., Smith, R. W., Nonomura, J. H., Odom, R. B., and Akers, W. A. (1979). Allergic contact hypersensitivity to nickel, neomycin, ethylenediamine and benzocaine. *Arch. Dermatol.* 115, 959–962.

Schubert, H., Berova, N., Czernielewski, A., Hegyl, E., Jirásek, L., Kohánka, V., Korossy, S., Michailov, P., Nebenführer, L. Prater, E., Rothe, A., Rudzki, E., Stranski, L., Süss, E., Tarnick, M., Temesvàri, E., Ziegler, V., and Zschunke, E. (1987). Epidemiology of nickel allergy. *Contact Dermatitis* 16, 122–128.

Sugai, T., Takagi, T., Yamamoto, S., and Takahashi, Y. (1979). Age distribution of the incidence of contact sensitivity to standard allergens. *Contact Dermatitis* 5, 383–388.

Vandenberg, J., and Epstein, W. (1963). Experimental skin contact sensitization in man. *J. Invest. Dermatol.* 41, 413–416.

Van der Burg, C.K.H., Brunzeel, D. P., Vreeburg, K.J.J., von Blomberg, B.M.E., and Scheper, R. J. (1986). Hand eczema in hairdressers and nurses: a prospective study. I. Evaluation of atopy and nickel hypersensitivity at the start of apprenticeship. *Contact Dermatitis* 14, 275–279.

Von Schmiel, G. (1985). Häufigkeit von Nickel–Kontaktallergien am unausgewählten Patientengut im Raum München. *Dermatosen* 33, 92–96.

16

ORAL CHALLENGE OF NICKEL-ALLERGIC PATIENTS WITH HAND ECZEMA

Gitte Dalsgaard Nielsen

Department of Environmental Medicine, Odense University, J. B. Winsløwsvej 17, Denmark

1. INTRODUCTION

In Chapter 15, Menné documents that nickel contact hypersensitivity is a serious problem for the female population in several countries, especially if one is so unfortunate as to develop a vesicular hand eczema of the pompholyx pattern with tiny vesicles on the palms and the sides of the fingers. This chapter deals with the possible effects of dietary habits on this hand eczema.

Nickel and Human Health: Current Perspectives, Edited by Evert Nieboer and Jerome O. Nriagu.
ISBN 0-471-50076-3   © 1992 John Wiley & Sons, Inc.

Table 16.1   Nickel Content of Foods High
in Nickel[a]

Food	N	Range (μg/g)	Mean (μg/g)
Cocoa	7	8.2–12	9.8
Soya beans	3	4.7–5.9	5.2
Soya products	7	1.08–7.8	5.1
Walnuts	1	—	3.6
Peanuts	2	1.6–3.9	2.8
Oats	37	0.33–4.8	2.3
Buckwheat	3	1.3–2.8	2.0
Bitter chocolate	7	1.3–2.7	1.9
Hazelnuts	12	0.66–2.3	1.8
Oatmeal	18	0.80–4.7	1.8
Dried legumes	17	0.57–3.3	1.7
Almonds	5	1.2–1.3	1.3

[a]Modified from Flyvholm et al. (1984).

2. DAILY NICKEL INTAKE

The daily nickel intake normally ranges from 100 to 300 μg-day (Horak and Sunderman, 1973; Myron et al., 1978; Flyvholm et al., 1984). However, some food items are naturally high in nickel and can therefore lead to increased dietary nickel intakes. Table 16.1 shows food items naturally high in nickel content. Cocoa, soya beans and other dried legumes, different nuts, and oatmeal are the most important items. In Figure 16.1 the daily nickel intake has been estimated in a number of hypothetical situations, where items in the average diet are replaced by special food items or where these are added to an average diet. One hundred grams each of oatmeal, chocolate, or hazel nuts contains more nickel than the average diet, and a portion of soya bean stew contains more than three times the amount of nickel found in the total average diet. If we combine them all, a situation representing the maximum amount of nickel that a person can be expected to consume, the daily intake will be about 1200 μg nickel. However, 100 g of oatmeal, 100 g of chocolate, 100 g of hazel nuts, and a portion of soya bean stew represent quite a heavy diet.

3. ORAL CHALLENGE WITH INORGANIC NICKEL

Since 1975 several studies have added to the growing evidence that oral exposure to nickel may provoke vesicular hand eczema. Table 16.2 lists previous investigations. They have all used oral exposure with an inorganic nickel salt, nickel sulfate, in lactose capsule.

A   Average diet

B   100 g oatmeal

C   100 g chocolate

D   100 g hazel nuts

E   Soya bean stew

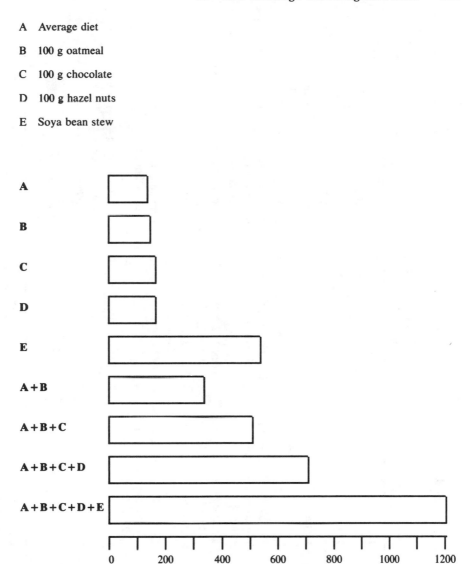

Figure 16.1. Daily nickel intake. Modified from Flyoholm et al. (1984).

In the first study, a double-blind study, a dose of 5.6 mg of nickel produced an exacerbation in 9 out of 12 subjects with hand eczema of the pompholyx type, with none reacting to the placebo. In the same subjects, the effect of external nickel exposure of the hands was recorded. Intense handling of nickel-contaminated metal objects did not induce any visible eczematous activity, thus indicating that ingestion of small amounts of nickel could be

Table 16.2 Previous Provocation Studies

Author	Design	Ni^{2+} dose (mg)	N	Results	Time after Which Examination Was Made
Christensen and Möller, 1975	Double blind	5.6	12	9 positive, none to placebo	6 and 30 hr
Kaaber et al., 1978	Single blind	2.5	28	17 positive, none to placebo	3 days
Cronin et al., 1980	Open	2.5	5	5	24 hr
		1.25	5	3	24 hr
		0.6	5	2	24 hr
Jordan and King, 1979	Double blind, randomized 4 weeks, 2 challenge weeks	0.5 (2 days) per challenge	10	Reactions did not exceed those to placebo	1 and 2 days
Burrows et al., 1981	Double blind, randomized 4 weeks, 2 challenge weeks	Week 1: 2×0.5 (2 days); Week 2: 2×1 (2 days)	22	Reactions did not exceed those to placebo	1 and 3 days
Gawkrodger et al., 1986	Double blind, randomized crossover (each patient 2 weeks challenge)	0.4 (2 days)	10	Reactions did not exceed those to placebo	3 days
		2.5 (2 days)	10	Reactions did not exceed those to placebo	3 days
		5.6	6	6 positive, 2 positive after placebo	3 days

of greater importance in maintaining the hand eczema than external contact with the metal.

This tendency was confirmed in the two subsequent studies by Kaaber et al. (1978) and Cronin et al. (1980), who used doses of 2.5 mg nickel. However, the findings were not confirmed by two double-blind studies by Jordan and King (1979) and Burrows et al. (1981) and by the three double-blind studies by Gawkrodger et al. (1986). In one of Gawkrodger's studies, six individuals ingested 5.6 mg nickel, and all six subjects either developed pompholyx or had an exacerbation of preexisting microvesicular hand dermatitis; two patients showed similar changes following placebo. In the rest of the studies, using lower doses, reaction to nickel did not exceed those to placebo (Gawkrodger et al., 1986). Common to these last three investigations is that they are all double-blind studies, where weeks with nickel exposure and weeks with placebo randomly followed each other. Further, evaluation of the eczema status was performed within less than three days after exposure, thus assuming that no delayed response occurred. Finally, the inorganic nickel doses given in these studies are at or above the upper limit of the daily intake of nickel naturally contained in the diet.

4. ORAL CHALLENGE WITH DIETARY NICKEL

Inorganic nickel in lactose capsule differs chemically and physically from nickel naturally contained in the diet. Thus a single-blind study was performed to investigate whether oral intake of nickel naturally contained in the diet can result in a worsening of the hand eczema in nickel-allergic patients with vesicular hand eczema (Nielsen et al., 1990). For this purpose a supplementary diet was prepared and elected with regard to a high natural content of nickel as possible and that one could eat without too much difficulty each day for several days. This diet consisted of oatmeal, chocolate cake, and soya bean stew, which added up to a daily nickel intake of 490 μg when analyzed. When calculated from the mean nickel contents found in the literature for the individual food items, the nickel content of the supplementary diet was estimated at 860 μg nickel, perhaps indicating that analytic technology previously resulted in erroneously high nickel concentration.

The single-blind study comprised of 12 nickel-allergic female patients aged 21–64 years; all responded to nickel when patch tested. They all had a vesicular hand eczema but were without a history of professional toxic contact dermatitis. After an initial observation period of at least 14 days, the patients were challenged at different points of time for a 4-day period with the supplementary diet.

The eczema status was evaluated by each patient on a continuous scale. In addition, the dermatologist was responsible for a relative clinical evaluation and a semiquantitative assessment of the number of vesicles in a well-defined area of the palm and of a finger. The patients were examined on day 14 or

Table 16.3   Evaluation of Eczema: Difference from Day 0[a]

Time of Evaluation	Better			Worse			Same		
	P	V	D	P	V	D	P	V	D
Day 14/21 before	4	4	1	7	5	7	1	3	4
Day 4	4	2	0	8	6	7	0	4	5
Day 11	0	2	0	12*	10**	12***	0	0	0

[a] Abbreviations: P, patients' impression; V, vesicle count; D, dermatologists' evaluation. Friedman two-way analysis of variance by ranks indicated by asterisks: (*) $P = 0.01$; (**) $P = 0.03$; (***) $P = 0.001$.
[b] Modified from Nielsen et al. (1990).

21 before start of provocation and on days 0, 4, and 11 from start of the provocation period. The patients were challenged from day 0 to day 4. Whereas the patient obviously knew when the high-nickel diet was ingested, the dermatologist was unaware of the information.

Table 16.3 shows the results of the evaluation of the eczema. When the patients' impression, the vesicle count, and the dermatologists are compared, no changes were observed from before start of challenge to day 0, whereas the eczemas of 6 patients were considered to be worse on day 4 after challenge when compared to day 0. On day 11 the eczemas of 10 out of 12 patients were unanimously considered to be worse than on day 0. The two others were possibly worse, but in one case it was impossible to count the vesicles due to scratching, and in the other the skin of the hands had deep fissures and was atrophied after many years of treatment with local steroids. When the nonparametric Friedmans test was used on the vesicle data, no change was observed from day 14/21 before start of challenge to day 0 and from day 0 to day 4, whereas a worsening was observed at day 11 ($P = 0.03$).

Two conclusions can be drawn from this investigation. First, a diet naturally high in nickel content can result in a worsening of hand eczema in nickel-allergic patients. Second, the worsening of the hand eczema is observed on day 11 after start of challenge with the supplementary diet, that is, on day 7 after finishing the 4-day provocation period. This very late reaction may be the reason why Jordan and King (1979), Burrows et al. (1981), and Gawk-rodger et al. (1986) did not observe any reactions to nickel exceeding those to placebo in their double-blind studies where weeks with nickel exposure and weeks with placebo randomly followed each other.

5. ELIMINATION STUDIES

Investigations eliminating food items high in nickel content were performed by Kaaber et al. (1978) and Veien et al. (1985). The first investigation comprised

17 patients with vesicular hand eczema; all patients had experienced aggravation following ingestion of 2.5 mg nickel. The dermatitis of 9 of the 17 patients improved during a period of six weeks on a low-nickel diet. The dermatitis of 7 of the 9 patients flared again when a normal diet was resumed. In the second investigation 204 patients with vesicular hand eczema participated (143 patients with positive patch test and 61 with negative patch test but with a positive result after oral provocation with nickel sulfate). After one to two months on a diet low in nickel content, the hand eczema disappeared in 32 patients and improved in 89 patients. The low-nickel diet used in this study was based on analytical data from 1962 when reliable methods for detecting nickel in organic material were not available. The food items not eliminated included nuts, almonds, oatmeal, and dried legumes, all of which were later shown to be high in nickel content. An improvement of the dietary recommendations could possibly further improve the effect of a nickel-restricted diet.

6. DIETARY RECOMMENDATIONS

Table 16.4 is a nickel diet instruction sheet prepared by the National Food Agency (1988) in Denmark mainly on the basis of the literature and analyses performed by Nielsen and Flyvholm (1984). This instruction sheet is available in Denmark and is used by many dermatologists when there is an indication of an orally provoked hand eczema in nickel-allergic patients. The food items are divided into four groups according to the nickel content. Group 1 consists of food items low in nickel, and group 2 consists of food items normally low in nickel but where higher contents can appear. Group 3 comprises food items with a moderate nickel content, and group 4 consists of food items with a high nickel content. Food items in group 1 can be eaten without limitations, whereas food items in group 4 should be completely avoided. The diet can be supplemented with food items from groups 2 and 3, but it is not recommended that the entire diet consist of food items from group 3.

REFERENCES

Burrows, D., Creswell, S., and Merrett, J. D. (1981). Nickel, hands and hip prostheses. *Br. J. Dermatol.* **105**, 437–444.

Christensen, O. B., and Möller, H. (1975). External and internal exposure to the antigen in the hand eczema of nickel allergy. *Contact Dermatitis* **1**, 136–141.

Cronin, E., DiMichiel, A. D., and Brown, S. S. (1980). Oral challenge in nickel-sensitive women with hand eczema. In S. S. Brown and F. W. Sunderman, Jr. (Eds.), *Nickel Toxicology*. Academic, New York, pp. 149–155.

Flyvholm, M., Nielsen, G. D., and Andersen, A. (1984). Nickel content of food and estimation of dietary intake. *Z. Lebensin. Unters. Forsch.* **179**, 427–431.

Table 16.4 Nickel Diet Instruction Sheet

Food Group	GROUP 1 Low	GROUP 2 Normally low	GROUP 3 Moderate	GROUP 4 High
Milk products	Cream, yoghurt, ice cream	Milk	—	Cheese
Meat, poultry	All meat except what is mentioned elsewhere	Liver, kidney	Sausages, liver paste, meat pie, etc.	—
Fish	All fish except what is mentioned elsewhere	Eel	Cod liver	Mussels and clams, shellfish
Egg		Egg		—
Bread, grain, and flour	White bread, rice (parboiled), spaghetti/macaroni, potato flour	Rice (polished)	Rye bread; whole grain bread; coarse bread; kernels of barley, rye, or wheat; corn meal	Buckwheat, corn flakes, oat meal, millet, bran, müssli, rice (unpolished), soybean flour
Roots and vegetables	Cucumber, asparagus, celery, pumpkin, cabbage, radish, red cabbage, celery root, Jerusalem-artichoke, Chinese cabbage	Cauliflower, kale, carrots, potatoes, onions, peppers, leeks, beet root, spinach, tomatoes	Broccoli, green beans, mushroom, potato chips, corn, parsnip, horseradish	Dried legumes (e.g., soya beans, lentils), baked beans, salad, peas

Fruits	Apple, banana, cherries, mandarin, currants, elderberries, gooseberries, grapefruit, grapes, orange, plums, raisins, rhubarb, strawberries	Peach, pear	Dates, raspberries	Pineapple, linseed, figs, peanuts, almonds, all nuts, sunflower seeds, prunes, products of dried fruits
Fats	Vegetable oil, butter	—	Vegetable margarine	Margarine
Sugar, honey and sweets	Sugar	—	Honey, caramel	Sweets, chocolate, liquorice, marzipan
Beverages	Orange and apple juice, soda and mineral water, beer and wine, coffee and tea	—	Vegetable juice	Cocoa
Besides	Yeast, salt	—	Grilled food	Juniper berry, cinnamon, caraway, soy products

Gawkrodger, D. J., Cook, S. W., Feel, G. S., and Hunter, J.A.A. (1986). Nickel dermatitis: the reaction to oral nickel challenge. *Br. J. Dermatol.* **15,** 33–38.

Horak, E., and Sunderman, F. W., Jr. (1973). Faecal nickel excretion by healthy adults. *Clin. Chem.* **19,** 429–430.

Jordan, W. P., and King, S. E. (1979). Nickel feeding in nickel sensitive patients with hand eczema. *J. Am. Acad. Dermatol.* **1,** 506–508.

Kaaber, K., Veien, N. K., and Tjell, J. C. (1978). Low nickel diet in the treatment of patients with chronic nickel dermatitis. *Br. J. Dermatol.* **98,** 197–201.

Myron, D. R. Zimmerman, T. J., Schuler, T. R., Klevay, L. M., Lee, D. E., and Nielsen, F. H. (1978). Intake of nickel and vanadium by humans. A survey of selected diets. *Am. J. Clin. Nutr.* **31,** 527–531.

National Food Agency. Levnedsmiddelstyrelsen, Sundhedsministeriet (1988). Nickel allergy. Diet instruction sheet. (In Danish).

Nielsen, G. D., and Flyvholm, M. (1984). Risks of high nickel intake with diet. In F. W. Sunderman, Jr. et al. (Eds.), Nickel in the human environment. International Agency for Research on Cancer, Lyon, IARC Scientific Publication No. 53, pp. 333–338.

Nielsen, G. D., Jepsen, L. V., Jørgensen, P. J., Grandjean, P., and Brandrup, F. (1990). Nickel-sensitive patients with vesicular hand eczema: oral challenge with a diet naturally high in nickel. *Br. J. Dermatol.* **122,** 299–308.

Veien, N. K., Hattel, T., Justesen, O., and Nørholm, A. (1985). Dietary treatment of nickel dermatitis. *Acta. Derm. Venerol. (Stockh.)* **65,** 138–142.

17

TOPICAL NICKEL SALTS: THE INFLUENCE OF COUNTERION AND VEHICLE ON SKIN PERMEATION AND PATCH TEST RESPONSE

A. Fullerton and A. Hoelgaard

Department of Pharmaceutics, Royal Danish School of Pharmacy, DK 2100 Copenhagen, Denmark

T. Menné

Department of Dermatology, Gentofte Hospital, DK 2900 Hellerup, Denmark

Nickel and Human Health: Current Perspectives, Edited by Evert Nieboer and Jerome O. Nriagu.
ISBN 0-471-50076-3 © 1992 John Wiley & Sons, Inc.

1. INTRODUCTION

Allergic contact dermatitis to nickel is verified routinely by patch testing with 5% nickel sulfate in petrolatum (pet). This test material is regarded as suboptimal, producing false-negative and irritant reactions as demonstrated by Cronin (1975) and Menné (1981). Since there is a relationship between the quantity of allergen applied, exposure time, and the immune response, an important parameter to be considered is percutaneous penetration. In order to avoid false-negative responses and to minimize simultaneously the amount of potential allergen in contact with the skin, the topical bioavailability of the allergen has to be maximized. Conditions that influence the topical bioavailability of a substance are its physical and chemical form and the nature of the vehicle in which it is applied (Barry, 1983). In this chapter, in vitro skin permeation is examined as well as uptake in the various skin layers and the epidermal binding of the nickel ion. In addition, in vivo patch testing is performed with selected test materials and compared to the standard patch test material.

2. EXPERIMENTAL DETAILS

Permeation of nickel through human skin was studied in vitro by using excised operation tissues. The subcutaneous fat was carefully removed and the skin mounted in Franz glass diffusion cells (Franz, 1975) having an available diffusion area of 1.8 cm^2. The epidermal side of the skin was sealed off with a glass stopper in order to simulate occlusion of the skin. The dermal side was bathed with saline, 37°C, acting as the recipient phase. At appropriate intervals after application of a test preparation, aliquots of the recipient phase were withdrawn for nickel analysis. The recovery of nickel in the various skin layers was determined at the end of the experiment. The stratum corneum was stripped 10 times with adhesive tape and the skin separated into the epidermis and dermis (by incubating the skin in a dessicator for 5 min at 60°C and 100% relative humidity). All diffusion experiments were conducted in duplicate to verify the consistency of the observed permeation rates. The nickel dose in the test preparations corresponded to the dose normally used in a patch test with Finn-Chamber and 5% nickel sulfate in pet (9–16 mg preparation giving 139–228 μg nickel/cm^2).

The stripping tapes, epidermis, and dermis sections were wet ashed prior to the nickel analysis. Analyses were done using a highly sensitive electrochemical method, adsorption differential pulse voltammetry (ADPV), developed by Pihlar et al. (1981). The instrumentation, conditions, and reagents used were as described by Gammelgaard and Andersen (1985). To avoid environmental contamination of the recipient phase and skin sections with nickel, the recipient chamber, all containers, utensils, and so on, were soaked in 4 M nitric acid for several days and then rinsed with copious amounts of

Milli-Q water. All analytical manipulations were carried out inside a laminar airflow clean bench.

3. RESULTS AND DISCUSSION

3.1. Counterion Effects

Nickel permeation after application of aqueous solutions of nickel chloride or nickel sulfate is shown in Figure 17.1. The profiles displayed show that nickel is capable of penetrating the skin barrier, but the nickel ion passes extremely slowly through the skin, exhibiting a lag time of about 50 hr. This long lag time may explain observations of Kalimo and Lammintausta (1984) and Menné (1981), who found the allergic and irritant reactions to nickel to

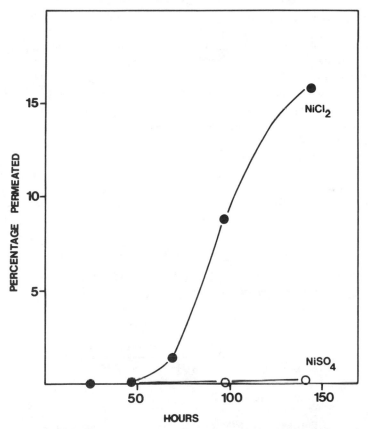

Figure 17.1. Permeation rates of nickel through human skin in vitro from aqueous solutions containing nickel chloride and nickel sulfate (250-μl solution of 1.32 mg nickel/ mL applied on 1.8 cm² skin surface). (From Fullerton et al., 1986.)

be more frequent after 48 hr than after 24 hr occlusion in patch testing. The permeation rate is considerably higher when nickel chloride is applied compared to nickel sulfate. Similar observations have been reported for the Cr(III) salts by Samitz et al. (1967). A specific interplay of counterion on the permeation rate of nickel has been investigated by adding sodium chloride to a nickel sulfate solution and sodium sulfate to a nickel chloride solution (Fullerton et al., 1986). A clear deaccelerating effect of the sulfate ion upon the permeation rate of the nickel ion was observed. In vivo patch testing has indicated similar effects. Wall and Calnan (1980) found nickel chloride to be a more reliable test substance than nickel sulfate. Kalimo and Lammintausta (1984) found that nickel chloride gives a higher number of positive allergic and irritant reactions than does nickel sulfate. The more potent nature of nickel chloride in patch testing may well be caused by the higher cutaneous bio-availability of this salt, and the choice of salt is therefore an important consideration in nickel patch testing.

3.2. Effect of Vehicles

Petrolatum is not an optimal vehicle for nickel since the allergen is insoluble in this vehicle. False-negative patch test reactions are difficult to identify but may be due to the failure of the pet vehicle to release the nickel and thus gives an insufficient cutaneous penetration of the allergen. False-positive reactions are usually irritant reactions and may be due to an uneven dispersion of nickel as crystals of varying size in the pet vehicle as proposed by Fischer and Maibach (1984).

In order to achieve good conditions for skin penetration, the vehicle should adhere tightly to the skin surface and nickel should be almost, or better still completely, dissolved in the vehicle. Wahlberg (1965) found a higher penetration rate of a metal ion from a solution than from a semisolid suspension. Table 17.1 shows the 48-hr permeation of nickel from alternative

Table 17.1   Permeation of Nickel from Alternative Patch Test Vehicles through Human Skin *In Vitro*

5% $NiCl_2$ Hydrogels[a]	48 hr Permeation (μg Ni/cm^2)
Methylcellulose 1500 (3%)	0.51
Methocel E 4M (3%)	1.65
Methocel K 4M (3%)	1.1
Methocel F 4M (4%)	0.95
Pluronic F-127 (30%)	0
Polyviol W 25/140 (10%)	0.25

[a] Numbers in parentheses are polymer concentrations.

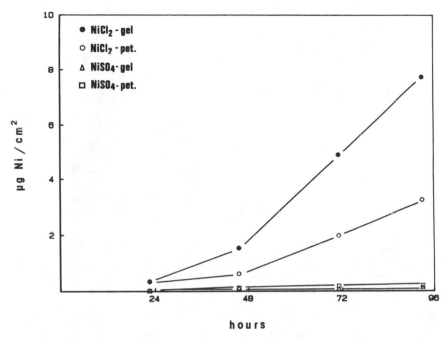

Figure 17.2. Effect of counterion and vehicle on the permeation rate of nickel through human skin in vitro. (From Fullerton and Hoelgaard, 1988a.)

patch test vehicles, such as 5% $NiCl_2$ hydrogels. The hydrogels are nonionic and are all presumed to be nonallergenic and nonirritant. They have good film-forming properties, leaving the nickel spread across the skin surface in a thin film without microscopically detectable crystals. As seen in Table 17.1, the gels made from methylcellulose and its hydroxypropyl derivatives (Methocel E 4M, K 4M, F 4M) show the highest 48-hr permeation of nickel. Significantly faster permeation rates are obtained with 5% $NiCl_2$ in these hydrogels than with a pet vehicle (Fullerton et al., 1988c). When using nickel chloride in a polyoxyethylene–polyoxypropylene–copolymer gel (Pluronic F-127), no 48-hr permeation is observed, probably because the nickel ion is kept in micelles caused by the high polymer concentration (30%). The nickel chloride gel prepared from Methocel E 4M (3%) is superior, producing the highest release rate. Furthermore, this gel is the only transparent methyl-cellulose gel and makes visual inspection possible. When comparing the standard patch test vehicle (pet) with the Methocel E 4M hydrogel containing 5% nickel chloride or nickel sulfate, permeation profiles as depicted in Figure 17.2 are obtained. The nickel permeation from the $NiSO_4$ preparation is almost unaffected by the vehicle, and it is still slow even from the gel. However, with the chloride as counterion the gel is superior compared to pet.

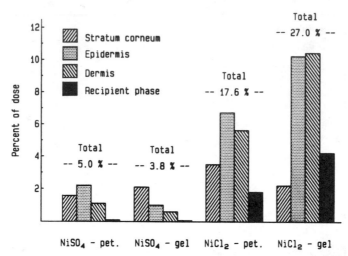

Figure 17.3. The 93-hr recovery of nickel in the stratum corneum (strippings 6–10), epidermis, dermis, and the recipient phase as percentage of the dose applied on the skin surface. (From Fullerton and Hoelgaard, 1988a.)

Figure 17.3 shows the 93-hr recovery of nickel from the stratum corneum (five last stripping tapes), epidermis, and dermis. The amount reaching the recipient phase at 93 hr is included in the diagram. A large fraction could be quantified in these skin layers: with the $NiCl_2$ gel a total of 27% of the dose applied on the skin surface is recovered, 22.8% from the skin layers and 4.2% in the recipient phase, whereas the $NiSO_4$ gel only releases a total of 3.8% of the dose to the skin layers. With $NiCl_2$ and $NiSO_4$ in pet, the recovery is 17.6 and 5.0%, respectively. The nickel content found in the skin layers matches the permeation profiles. The preparation that possesses the highest permeation rate (i.e., the $NiCl_2$ gel) also delivers the highest amount of nickel to the skin layers; consequently, the amount of nickel in the epidermis is raised along with the increasing permeation rate. Thus, the permeation studies are relevant in the evaluation of the topical bioavailability of nickel patch test preparations.

It is likely that the concentration of nickel necessary to elicit a contact allergic response is related to the actual amount of nickel in the outermost layers of the skin. Therefore, the next question is: What concentration of $NiCl_2$ in the gel is equivalent to the standard $NiSO_3$ pet material with respect to skin uptake? The same recovery procedure was peformed with lower $NiCl_2$ concentrations in the gel and compared to 5% $NiSO_4$ pet. Table 17.2 shows the stratum corneum–epidermis uptake at 96 hr. The values of the standard material lie between 1.25 and 2.5% $NiCl_2$, closest to the 1.25% $NiCl_2$ gel. Thus, it indicates that a 1.5% $NiCl_2$ gel is bioequivalent to the standard patch test material with respect to topical delivery of the nickel ion.

Table 17.2 Nickel Content in Stratum Corneum and Epidermis after Application of Various Concentrations of NiCl$_2$ in Methocel E 4M Gel and of 5% NiSO$_4$ in Petrolatum[a]

NiCl$_2$ Concentration	μg Ni/1.8 cm^2
0.63% NiCl$_2$ gel	3.1
1.25% NiCl$_2$ gel	10.1
5.0% NiSO$_4$ pet	12.5
2.5% NiCl$_2$ gel	26.0
5.0% NiCl$_2$ gel	51.5

[a]Reflects 96 hr recovery in stratum corneum and epidermis.

3.3. Binding of Nickel to Homogenized Epidermis

The observed large fraction of nickel quantified in the outmost skin layers combined with the extended lag time of the permeation profiles indicate that the epidermal binding phenomenon interferes with the diffusion of nickel across the skin. In binding studies, using homogenized human epidermis, we found considerable nickel uptake (Fullerton and Hoelgaard, 1988b). The binding of nickel is in accordance with the Freundlich adsorption isotherm with a slope of 0.55 (Fig. 17.4), which indicates that nickel is associatively bound to constituents of the epidermis. The binding characteristics of nickel chloride and nickel sulfate solutions are found to be similar. In the concentration range investigated, the lower nickel concentrations are of primary interest. More than 90% of the available nickel is in a bound state. During the diffusion of nickel across the stratum corneum, a local reservoir in the epidermis might gradually build up, and as the process of nickel permeation continues, an increasing fraction of nickel is free to permeate the skin.

The ability of three metal-chelating agents that form water-soluble chelates of nickel to remove nickel from its binding sites in the epidermis was tested with the purpose of finding agents that block nickel binding. The chelating agents tested were ethylenediaminetetraacetic acid disodium (Na–EDTA), L-histidine (L-His), and D-penicillamine (D-Pa). The results obtained are shown in Figure 17.5. The Na–EDTA is the most effective of the three agents. It removes about 70–90% of the nickel bound to the epidermis tissue in the investigated nickel concentration range. The L-His appears less effective at the lower nickel concentration, but its blocking ability increases at the high nickel concentrations, removing about 70% of nickel bound to epidermis. The D-Pa, which has been shown to protect against acute effects of nickel chloride exposure in animal experiments as reported by Horak et al. (1976), removes less than 20% of nickel from the homogenized epidermis.

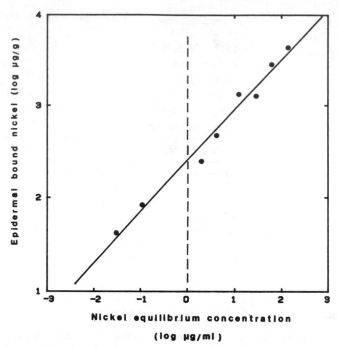

Figure 17.4. Freundlich adsorption isotherm plot of epidermal nickel binding. (From Fullerton and Hoelgaard, 1988b.)

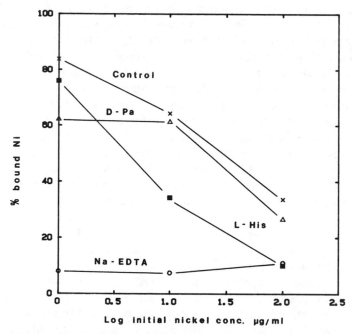

Figure 17.5. Effect of chelating agents on the binding of nickel at various initial nickel concentrations in the incubation media. The molar ratio of chelate to Ni(II) is 5. (○) Na-EDTA; (■) L-His; (△) D-Pa; (×) control. (From Fullerton and Hoelgaard, 1988b.)

In the past, some metal-chelating materials have been investigated in an attempt to demonstrate a blocking effect on patch test reactions with nickel salts (Kctcl and Bruynzeel, 1982, 1984; Kurtin and Orentreich, 1954; Samitz and Pomerantz, 1958). Of particular interest is the effect of Na–EDTA. Ketel and Bruynzeel (1982, 1984) found that the blocking effect of a 10% EDTA cream was significant in comparison to that of cream base only. This effect can be attributed either to less skin penetration of the nickel in the chelate form or to a reduced binding of nickel to proteins in the skin.

3.4. *In Vivo* Patch Testing

The $NiCl_2$ hydrogel prepared from Methocel E 4M was compared with respect to reliability and concentration equivalence to the standard patch test using 5% $NiSO_4$ in pet (Fullerton et al., 1989). Four hundred and thirty patients with known or suspected nickel contact allergy participated in patch testing, comparing different concentrations of $NiCl_2$ in hydrogel with the standard patch test material. The hydrogel base without nickel served as a control in 175 patients. The patch tests were performed with Finn Chambers fixed with Scanpore-tape on the upper part of the back. The exposure time was 48 hr, and the first reading was recorded at this time, with subsequent readings at 72 and 168 hr. The test reactions were evaluated and the readings graded according to standards accepted by the International Contact Dermatitis Research Group (ICDRG) (Wilkinson ct al., 1970).

The testing for nickel allergy was performed through five experimental series with test preparations as summarized in Table 17.3. The number of positive reactions to each of the five $NiCl_2$ gels are to be compared to the responses to 5% $NiSO_4$ pet. No statistical significant difference was obtained between the standard preparation and $NiCl_2$ gels of a concentration range of 0.5–2.0% $NiCl_2$. With 0.1% $NiCl_2$ or less, the response frequency to 5% $NiSO_4$ pet was highest. The hydrogel base without nickel produces no allergic or irritant reactions.

Table 17.3 Number of Responses from 430 Patients with Known or Suspected Contact Allergy to Experimental Battery of Nickel Salt Preparations Applied for 48 hr

Number of Patients	5% $NiSO_4$ pet	By $NiCl_2$ Concentration in Hydrogel[a]				
		0.01%	0.1%	0.5%	1%	2%
121	13	—	8	—	—	—
71	10	0	5	9	—	—
63	9	—	5	5	7	—
41	2	—	1	1	1	—
134	14	—	—	—	—	16

[a] Hydrogel base as control: no response ($n = 175$).

Table 17.4   Patient Response (Ratio) to 5% NiSO₄ Petrolatum and 0.5–2.0% NiCl₂ Hydrogels

	By $NiCl_2$ Concentrations in Hydrogel		
Response	0.5%	1.0%	2.0%
+pet/−gel	7/175	3/104	1/134
+pet/+gel	14/175	8/104	13/134
−pet/+gel	1/175	0/104	3/134

A closer evaluation of the data takes the individual responses into account. Table 17.4 shows the number of patients responding to either both nickel preparations or one of the two. The 1% $NiCl_2$ gel fails to identify three patients who respond to 5% $NiSo_4$ pet; three patients respond only to 2% $NiCl_2$ gel, while one patient responds only to 5% $NiSO_4$ pet. Thus the 2% $NiCl_2$ gel produces an increase in responses and seems to reduce false-negative reactions relative to the standard preparation.

Irritant reactions are common in patch testing with metal salts. No irritant effect was observed with 0.1% $NiCl_2$ or less in the gel. The use of 2% $NiCl_2$ gel compared to 5% $NiSO_4$ pet raised the number of irritant reactions from 6 to 8. The results indicate that the topical bioavailability of 5% $NiSO_4$ pet is normally high enough to exhibit response on patients with nickel contact allergy. If topical bioavailability is raised by using a more appropriate vehicle or nickel salt, the nickel levels in the skin become higher. However, the advantage is limited because an increase in bioavailability is followed by more irritant reactions. When the nickel ion is made more available to the skin, the risk of exceeding a threshold of irritant reactions rises. Irritant reactions are probably related to the physical amount of allergen in epidermis, and they may thus increase in number concurrently with the ability of patch test material to detect false-negative reactions.

4.   CONCLUSIONS

A specific interplay of counterion on skin permeation of the nickel ion is observed. It is found that skin penetration of nickel is considerably increased when nickel chloride is applied rather than nickel sulfate. Nickel is bound to constituents of the epidermis. This explains the extended lag time of nickel permeation profiles and the observed large fraction of nickel quantified in the outermost skin layers.

Methocel E 4M is a useful alternative patch test vehicle for water-soluble allergens. A 1.5% $NiCl_2$ hydrogel of this type appears to be the bioequivalent to the standard patch test with respect to both in vivo patient response frequency and in vitro skin permeation. Interestingly, a hydrogel prepared from Methocel E 4M with 2% $NiCl_2$ seems to be a useful patch test material in cases showing a positive history but exhibiting a negative response to 5% $NiSO_4$ in pet. However, the increase in topical bioavailability of nickel afforded by methylcellulose hydrogels is tempered by more irritant reactions. It is concluded that the topical bioavailability of 5% $NiSO_4$ pet is normally high enough to elicit a positive response from most patients with nickel contact allergy.

REFERENCES

Barry, B. W. (1983). *Dermatological Formulations*. Marcel Dekker, New York, p. 145.

Cronin, E. (1975). Patch testing with nickel. *Contact Dermatitis* 1, 56–57.

Fischer, T. I., and Maibach, H. I. (1984). Patch test allergens in petrolatum: a reappraisal. *Contact Dermatitis* 11, 285–287.

Franz, T. J. (1975). Percutaneous absorption. On the relevance of in vitro data. *J. Invest. Dermatol.* 64, 190–195.

Fullerton, A., and Hoelgaard, A. (1988a) Nickel patch test material—effect of counterion and vehicle on skin permeation in vitro. *Farmac. Sci. Ed.* 16, 53–58.

Fullerton, A., and Hoelgaard, A. (1988b). Binding of nickel to human epidermis in vitro. *Br. J. Dermatol.* 119, 675–682.

Fullerton, A., Andersen, J. R., Hoelgaard, A., and Menné, T. (1986). Permeation of nickel salts through human skin in vitro. *Contact Dermatitis* 15, 173–177.

Fullerton, A., Andersen, J. R., and Hoelgaard, A. (1988c). Permeation of nickel through human skin in vitro—effect of vehicles. *Br. J. Dermatol.* 118, 509–516.

Fullerton, A., Menné, T., and Hoelgaard, A. (1989b). Patch testing with nickel in a hydrogel. *Contact Dermatitis* 20, 17–20.

Gammelgaard, B., and Andersen, J. R. (1985). Determination of nickel in human nails by adsorption differential pulse voltammetry. *Analyst* 110, 1197–1199.

Horak, E., Sunderman, F. W., Jr., and Sarkar, B. (1976). Comparisons of antidotal efficacy of chelating drugs upon acute toxicity of Ni (II) in rats. *Res. Commun. Chem. Pathol. Pharmacol.* 14, 153–165.

Kalimo, K., and Lammintausta, K. (1984). 24 and 48 h allergen exposure in patch testing. *Contact Dermatitis* 10, 25–29.

Ketel, W. G. van, and Bruynzeel, D. P. (1982). The possible chelating effect of sodium diethyl-dithiocarbamate (DDC) in nickel allergic patients. *Dermatosen* 30(6), 198–202.

Ketel, W. G. van, and Bruynzeel, D. P. (1984). Chelating effect of EDTA on nickel. *Contact Dermatitis* 11, 311–314.

Kurtin, A., and Orentreich, N. (1954). Chelation deactivation of nickel ion in allergic eczematous sensitivity. *J. Invest. Dermatol.* 22, 441–445.

Menné, T. (1981). Nickel allergy-reliability of patch test evaluated in female twins. *Dermatosen Beruf Umwelt* 29(6), 145–160.

Pihlar, B., Valenta, P., and Nürnberg, H. W. (1981). New high-performance analytical procedure for the voltammetric determination of nickel in routine analysis of waters, biological materials and food. *Fresenius Z. Anal. Chem.* 307, 337–346.

Samitz, M. H., and Pomerantz, H. (1958). Studies of the effect on the skin of nickel and chromium salts. *Arch. Indust. Health.* 18(2), 473–479.

Samitz, M. H., Katz, S. K., and Schrager, J. D. (1967). Studies of the diffusion of chromium compounds through skin. *J. Invest. Dermatol.* **48**, 514–520.

Wahlberg, J. T. (1965). Disappearance measurements, a method for studying percutaneous absorption of isotope-labelled compounds emmiting gamma-rays. *Acta Dermatovenereol.* **45**, 397–414.

Wall, L. M., and Calnan, C. D. (1980). Occupational nickel dermatitis in the electro forming industry. *Contact Dermatitis* **6**, 414–420.

Wilkinson, D. S., Fregert, S., Magnusson, B. et al. (1970). Acta Dermatovenereol **50**, 2871.

18

THE HUMAN MAJOR HISTOCOMPATIBILITY COMPLEX IN NICKEL SENSITIVITY

Lennart Emtestam

*Departments of Dermatology and Clinical Immunology,
Karolinska Institute, Huddinge Hospital, Stockholm, Sweden*

1. **HLA Class I Molecules**
2. **HLA Class II Molecules**
3. **HLA and Disease Susceptibility**
4. **HLA and Immune Response (HLA Restriction)**
    4.1.   MHC Restrictions Specificity of Nickel-reactive T Cells
5. **HLA Association in Nickel Sensitivity**
    References

The major histocompatibility complex (MHC) was first defined in the mouse [Gorer, 1936; for review see Klein (1986)] and has so far been demonstrated in all mammalian species studied. In man this system is called HLA (human leukocyte antigen group A) and is located on the short arm of chromosome 6. The genes are arranged in haplotypes, that is, characteristic constellations of these genes. They encode polymorphic cell surface structures, HLA molecules, that have two major functions. First, they are self-markers for the

Nickel and Human Health: Current Perspectives, Edited by Evert Nieboer and Jerome O. Nriagu.
ISBN 0-471-50076-3   © 1992 John Wiley & Sons, Inc.

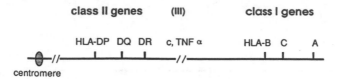

Figure 18.1. Genetic map of the human MHC region on chromosome 6.

developing T-cell immune system through the induction of tolerance and, second, the HLA molecules serve as restriction elements, which means that they participate in the cellular interactions responsible for an immune response by associating with foreign peptides, a necessary prerequisite for T-cell recognition. For a review of the current progress of MHC research, see Klein (1986). A schematic outline of the HLA complex is shown in Figure 18.1.

1. HLA CLASS I MOLECULES

Class I molecules are expressed on the surface of all nucleated cells throughout the body, and the variable antigens are divided into HLA-A, HLA-B, and HLA-C [for references see Bodmer et al. (1984)]. Each class I molecule consists of a membrane-bound heavy chain associated with β_2-microglobulin (Fig. 18.2). The crystal structure of HLA-A2 has recently been described. The region distal from the membrane is a platform of eight antiparallel β strands topped by α helices. A large groove between the α helices provides a binding site for processed foreign antigens (Bjorkman et al., 1987a,b). The class I molecules mediate presentation of fragments of antigens, such as viral antigens, to cytotoxic T cells.

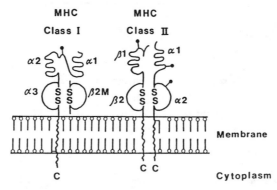

Figure 18.2. Schematic picture of the HLA molecules.

2. HLA CLASS II MOLECULES

Class II molecules are encoded by three distinct loci, called DR, DQ, and DP (Bodmer et al., 1984). Class II molecules are selectively expressed on the surfaces of some cells involved in immune responses, such as antigen-presenting cells and B lymphocytes, but can be expressed on most cells provided they are stimulated with lymphokines such as interferon-gamma (Lindahl et al., 1974; Pober et al., 1983; Basham et al., 1984; Gonwa et al., 1986; Scheynius et al., 1987). Class II molecules consist of two membrane-spanning glycosylated polypeptide chains (Walsh and Crumpton, 1977), called the α and the β chains (Giles and Capra, 1985) (Fig. 18.2).

Studies on the HLA class II genes have demonstrated that the DP locus contains two A and two B genes. However, one A and one B gene are pseudogenes and therefore cannot be expressed (Servenius et al., 1984). This means that only one DPα-β heterodimer can be formed from each haplotype. The DQ locus also contains two A and two B genes. None of these genes seem to have pseudogene characteristics, and therefore, the possibility exists for the formation of multiple class II molecules being expressed from the DQ locus (Kaufman et al., 1984; Okada et al., 1985; Jonsson et al., 1987). In addition, hybrid molecules have been demonstrated that contain an α peptide from one haplotype and a B gene product from the other haplotype (Giles et al., 1985; Lotteau et al., 1987). Therefore, as many as 16 distinct DQ molecules could conceivably exist on class II positive cells.

The DR locus is also complex. First, it contains a varying number of B genes in different haplotypes (Böhme et al., 1985). One DR B gene is found in DRw8 positive haplotypes, whereas as many as four B genes exist in DR4 and in DR7 positive haplotypes (Böhme et al., 1985). In the latter two, however, only two are functional. The DR molecule is also different from DQ and DP molecules since the A gene is invariant.

The conventional specificities DR1, DR2-DRw14 are determinants on the DRβ-1 chain, whereas the DRw52 and Rw53 supertypic specificities are determinants present on the DRβ-2 chain. The DRw8 haplotype contains only one DR B gene, which would be the only one expressed in such individuals (Andersson, 1987). This means that the DRw8 specificity is not a true "allele" to the other well-defined specificities DR1,2-w14. The HLA class II antigens are involved in the helper T-lymphocyte recognition of foreign antigens. This is discussed further below.

The DR and DQ alleles are not randomly inherited together. Certain alleles are more or less common in combination with others than expected from the gene frequencies, a phenomenon called linkage disequilibrium (Tiwari and Terasaki, 1985).

Only minute differences in primary class II gene structure could be important for disease susceptibility, as suggested by Todd et al. (1987) regarding insulin-dependent diabetes mellitus. Single amino acid substitutions in HLA peptide chains are sufficient to evoke a T-cell response in mixed lymphocyte cultures.

This is how HLA-Dw specifities have been defined. Andersson (1987) elegantly showed that very minute molecule sequence differences distinguish the five Dw subtypes of DR4. Dissection of the polymorphism of HLA class II molecules with serological techniques has provided the basis for the definition of DR and DQ alleles. Also, cellular techniques have provided the basis for Dw and DP polymorphism. However, available tissue-typing methods will not always discriminate such small but relevant changes as demonstrated by Todd et al. (1987). Characterization of the restriction fragment length polymorphism (RFLP) obtained by hybridizing HLA–gene complementary DNA (cDNA) probes to restriction-enzyme-digested and size-separated genomic DNA offers an alternative in HLA typing (Wake et al., 1982; Andersson et al., 1984; Böhme et al., 1985; Carlsson et al., 1987). Notably for the DQ locus, additional polymorphism is detected by RFLP analysis compared to serological tissue typing. With the restriction enzyme Taq I digested DNA, five DQA and eight DQB allelic patterns are found (Carlsson et al., 1987). The findings by Böhme et al. (1986) indicate that genomic hybridization may be used to identify individuals with an increased risk of acquiring insulin-dependent diabetes mellitus. By the use of cDNA probes for DQβ chains, they were able to show that DR2 and DR4 can be divided into two groups each. Only one RFLP-defined gene in each group is associated with diabetes mellitus. The biological relevance of the variability demonstrated in RFLP is also described by Paulsen et al. (1985) for a chlamydial antigen HLA-DR restriction element detected by RFLP but not by other tests.

3. HLA AND DISEASE SUSCEPTIBILITY

Many diseases with various etiologies have been shown to be more or less strongly associated with certain HLA alleles, both class I and class II (see Tiwari and Terasaki, 1985). The mechanisms involved are not fully understood, but there are two major functions of HLA structures that could form the basis for all disease associations. First, the HLA molecules serve as self-markers. Second, they regulate immune response through MHC restriction. As an example, foreign antigens might resemble MHC products and lead to a low response due to tolerance through eliminated self-reactive T-cell clones. Human cytomegalovirus encodes a glycoprotein that is similar to HLA class I molecules and has been suggested to be relevant for its virulence (Beck and Barell, 1988).

The association of a particular HLA type with a given disease does not necessarily mean that the risk for an individual carrying this HLA type to develop the disease is high. There are at least two possible reasons for the variations of the penetrance of HLA-linked diseases. First, the HLA genotype provides the hereditary factors for the disease. Second, external factors, such as viral infection, medication, or other unknown factors, are directly responsible for the precipitation of the disease process.

4. HLA AND IMMUNE RESPONSE (HLA RESTRICTION)

Antigen-presenting cells, usually macrophages, are needed for immune responses (Mosier, 1967). The specific T lymphocytes recognize the antigen in connection with products of the MHC on the surface of the antigen-presenting cells; thus the responses are restricted by MHC molecules (Rosenthal and Shevach, 1973; Zinkernagel and Doherty, 1974; reviewed by Benacerraf, 1981). The identity of an HLA molecule restricting a certain T-cell response can be studied by the ability of allogeneic macrophages to stimulate specific T lymphocytes. It has been shown that allogeneic macrophages have to share HLA molecules with the responding cell donor in order to be able to present a foreign antigen. Using monocytes from peripheral blood, Bergholtz and Thorsby (1978) demonstrated that the proliferative response to purified protein derivative (PPD) was restricted by HLA-D region products as defined by homozygous typing cells. Later, the same group reported that the Langerhans cell–dependent T-cell response to PPD and herpes simplex virus was HLA-D restricted (Braathen and Thorsby, 1982). Over the last few years many investigators have tried to define the major restricting elements of the class II loci for responses against different foreign substances. A bulk of evidence has accumulated favoring HLA-DR products as being the principal restriction elements for the response of helper T lymphocytes (Lamb et al., 1982; Thorsby et al., 1982; Ottenhoff et al., 1985), but the possibility of DQ- and DP-restricted T-lymphocyte responses also exists (Eckels et al., 1983; Qvigstad et al., 1984; Ball and Stastny, 1984a,b; Mellins et al., 1987). In our laboratory, we have focused on defining the HLA restriction specificities for responses against nickel since we assumed that a test system using a hapten would work as a model (Al-Tawil et al., 1985; Emtestam, 1988).

4.1. MHC Restriction Specificity of Nickel-reactive T Cells

In earlier studies it was demonstrated that other HLA molecules than HLA-DR molecules might serve as restriction elements for nickel-reactive T lymphocytes (Al Tawil et al., 1985). Therefore, we have further studied the class II restriction by purifying activated T-cell blasts in vitro on Percoll discontinuous density gradients. The ability of different allogeneic panel cells to function as antigen-presenting cells was investigated. Panel cell donors were carefully selected for positivity to a broad variation of DR, DQ, and DP types. Earlier definitions of class II polymorphism relied heavily on serological determination of class II molecules. However, we have defined a new level of class II polymorphism by characterizing HLA class II gene RFLPs (Carlsson et al., 1987). The panel of cells used for restriction experiments were tested and the results of positive restimulation in the presence of nickel related to their class II genotypes were recorded.

In our studies using heterogeneous T-cell cultures the nickel-sensitive donor was DR2,w10,DQw1 (Emtestam et al., 1988). All cell donors that were DR2 positive stimulated the nickel-specific T lymphocytes in the presence of nickel. Seven additional cell donors that were DR2 negative but shared DQw1 with the nickel-sensitive donor could induce a secondary response. This indicated that a DQ product could function as a restriction element alone or in addition to a DR element. However, one cell, which was DR2 negative but DQw1 positive did not restimulate. The level of polymorphism that could be defined using complementary DNA probes for DQ and DR genes did not provide additional information as to the identity of the preferential class II restriction element in the present test model. Blocking experiments with monoclonal antibodies did not give conclusive evidence as to the locus identity of restriction in this test system.

To sum up the results of the work with heterogeneous T-cell cultures, we found that cell lines from one patient could readily be restimulated by cells that shared either certain DR and/or DQ serological specificities with the responding cell population. Patterns of restimulation in the different responding cell–antigen presenting cell combinations did not agree with any earlier defined DR and/or DQ specificity. However, our data were clearly compatible with the notion that products of loci other than DR could serve as restriction elements for nickel reactivity in vitro.

The enriched T cells in this kind of heterogeneous culture contain a population of T lymphocytes with T-cell antigen receptors recognizing different sets of HLA class II elements. In further studies, we established clones of cells from one additional nickel-allergic patient to study the fine specificity of HLA restriction in nickel reactivity (Emtestam et al., 1989). Cells were cloned by limiting dilution after primary stimulation and selection of nickel-specific blasts. The clones, all shown to carry the CD4 markers, were restimulated once weekly with autologous antigen-presenting cells in the presence of nickel salt. Conditioned medium with exogenous interleukin 2 (IL-2) was added twice weekly. All established clones were completely blocked by monoclonal antibodies directed against DR antigens but were unaffected by antibodies against DQ and/or DP, demonstrating their DR specificity (Fig. 18.3).

Other groups have also produced metal salt–specific T-cell clones or lines. In the report by Löfström and Wigzell (1986), a cobalt-specific T-cell line was not restimulated by either of three tested allogeneic cells but only by autologous antigen-presenting cells. In addition, nickel-reactive T-cell clones have been presented by Sinigaglia et al. (1985), and they were found to be DR restricted in experiments with four clones and six different donors of antigen-presenting cells. Reports have been made on the total IL-2-responsive T-cell clones obtained from dermal lymphocytic infiltrate induced in nickel patch tests from nickel-allergic individuals; 15% were nickel specific (Kapsenberg et al., 1987). The clones were DR restricted as shown in blocking experiments with anti-DR antisera, which was also the case in the two studies

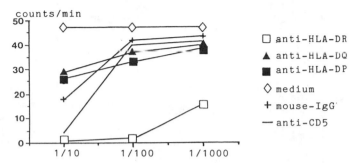

Figure 18.3. Restimulation of nickel-specific T-cell clone MCE2 by autologous antigen-presenting cells performed in the presence of monoclonal antibodies and control reagents. Results are expressed as mean counts per minute. The dilutions are shown on the x axis.

referred to above. However, the possibility of restriction by DQ products in metal reactivity was not excluded in the mentioned studies, since neither HLA-DQ typing nor blocking with locus specific anti-DR or anti-DQ monoclonal antibodies were performed.

The specificity of the clones produced from our nickel-allergic patients were all distinct, and some interesting new findings were observed (Tables 18.1 and 18.2). Thus, among the clones carefully tested, one was DR specific

Table 18.1   Hapten Specificity and MHC Restriction Specificity of the T-Cell Clone MCE2

Cell Donor	DR Type	Reagent Added	cpm
Autologous			
AN	3,4	Ni (nickel sulfate)	40,702
		Me (medium)	508
		Co (cobalt chloride)	326
Allogeneic			
BR	3,4	Ni	37,800
		Me	222
CG	2,7	Ni	288
		Me	134
HN	2,12	Ni	682
		Me	360
JM	4,8	Ni	1,984
		Me	240
OR	1	Ni	270
		Me	178

The clone MCE2: 10^4 cells per well in triplicate with (a) 50 μg/mL PHA: 116 counts per minute (CPM); (b) 6.25 μg/mL of nickel sulfate: 346 cpm; and (c) medium: 97 cpm.

Table 18.2 HLA Class II Restriction Specificity of Different Nickel-specific T-Cell Clones

Cell Donor	Class II Serological Definition			RFLP Typing DQB[a]	Clone Designation[b]			
	DR	DRw	DQw		MGD7	MGD8	MBD7	MCE2
Autologous								
AN	3,4	52,53	2,3	III/IV	pos	pos	pos	pos
Allogeneic								
BR	3,4	52,53	2,3	III/IV	neg	pos	pos	pos
CG	2,7	53	1,2	II/IV	neg	neg	neg	neg
HN	2,12	52	1	II/V	neg	neg	neg	neg
SH	1,4	53	1,3	I/V	neg	pos	neg	neg
LM	1,13	52	1	I/II	neg	neg	neg	neg
JW	3,7	52,53	2	III/IV	neg	neg	pos	neg
AF	2,12	52	1,3	II/V	neg	neg	neg	neg
TS	2,7	53	1,2	II/IV	ND	neg	ND	neg
OR	1	—	1	I	ND	ND	ND	neg
HS	1,4	53	1,3	I/IV	ND	pos	ND	pos
DM	3,7	52,53	2	III/IV	ND	neg	ND	neg

[a] For nomenclature see Carlsson et al. (1987).
[b] Positive response (pos) is defined as more than 40% of the response in the autologous combination; neg, negative response; ND, not done.

but was only restimulated by autologous cells. We have not yet used allospecific antibodies to identify if a hybrid HLA molecule is responsible for restriction of this clone. Of three other clones, one was DR3 specific and two were DR4 restricted, but with different fine specificity. One of these was tested in detail, and careful typing of panel cells with serological tools including the TA 10 monoclonal antibody (Carlsson et al., 1987), RFLP typing using cDNA probes and cellular typing with homozygous typing cells (Dw types; Jörgensen and Lamm, 1974) did not in any instance reveal a level of polymorphism that adequately and precisely correlated to the findings in this presentation (Table 18.3). This indicates that the restriction element is part of some, but not all, DR4 molecules and that this element has not yet been characterized as a determinant capable of stimulating an allogeneic proliferative response. Since DR4 haplotypes contain four DRB genes (Böhme et al., 1985) and express two DR polypeptides, DR4 and DRw53, it is possible that the restriction element is present on either of these two molecules. Blocking of restimulation with specific antisera could also elucidate this problem.

The clone designated MCE2 was also restimulated by one allogeneic cell in the absence of nickel, implying that the nickel with self-specificity mimics an allogeneic determinant, presumably of class II type (Table 18.3). The fact that self + X can mimic certain allogeneic determinants is not new (Sredni and Schwartz, 1980; Braciale et al., 1981). On the contrary, it is frequently observed in studies on the fine specificity of T-cell clones (Ashwell et al., 1986; Nisbet-Brown et al., 1987). To our knowledge the first such clone was a class I reactive cytotoxic mouse T-cell clone, specific for male H-Y carrying cells expressing H-2D^b, which reacted equally well with male and female cells of the D^d genotype (von Boehmer et al., 1979).

Other research groups have studied the splits of DR4 in detail (Holbeck et al., 1985; Cairns et al., 1985; Gregersen et al., 1987), and in a communication by Andersson (1987), sequence analyses of the DRβ chain genes of the different Dw types of DR4 showed that only minor sequence differences between the distinct Dw types were present in DR4-positive individuals. Since the specificity of the clone designated MCE2 did not fit with the Dw typing, we have defined yet another level of polymorphic variability within the HLA class II products that is biologically relevant since it can be recognized at the T-cell receptor level. It would be interesting to perform sequence analysis of the different panel cells that have now been recognized to be DR incompatible.

As could be expected, these data also demonstrate that the genomic variability that can be detected with RFLP analysis does not always reveal the total polymorphism of these genes. It is possible that the use of other restriction enzymes could circumvent this problem. However, it is also possible that such a strategy would reveal nonsense variability within intronic DNA sequences that are not relevant to T-cell-recognizable variability.

In summarizing the results of our works with cloned T cells, specificity analysis revealed that the nickel-specific lymphoid clones were restricted by

Table 18.3 HLA Class II Restriction Specificity of Nickel-specific Clone MCE 2

	Class II Serological Definition				RFLP Typing[a]		Cellular Typing[b]		Response[c]
	DR	DRw	DQw	TA10	DQB	DRB	Dw	DP	
Autologous	3,4	52,53	2,3	ND	III/IV	IV/XII	ND	ND	pos
Allogeneic									
BR	3,4	52,53	2,3	—	III/IV	IV/XII	ND	ND	pos
CG	2,7	53	1,2	—	II/IV	II/XIII	ND	ND	neg
HN	2,12	52	1	—	II/V	II/VII	2,5	2,4	neg
SH	1,4	53	1,3	+	I/V	I/XII	1	2,4	neg
LM	1,13	52	1	ND	I/II	I/VIII	1,18	1,5	neg
JW	3,7	52,53	2	—	III/IV	IV/XIII	ND	ND	neg
JM	4,8	52,53	3	—	IV	V/XII	8	2,3	neg
TS	2,7	53	1,2	—	II/IV	II/XIII	ND	ND	neg
OR	1	–	1	—	I	I	1	1,4	neg
HS	1,4	53	1,3	—	I/IV	I/XII	1,14	2,3	pos
DM	3,7	52,53	2	ND	III/IV	IV/XIII	7	NTR	neg
AF	2,12	52	1,3	—	II/V	II/VII	2,5	3,4	neg
EM	4,13	52,53	1	—	IV/VII	X/XII	4,19	3,5	AR
BC	4,13	52,53	1,3	+	V/VII	X/XII	19	3,4	neg
ST	4,8	52,53	3	—	IV	V/XII	4,8	NTR	neg
GLa	2,4	53	1,3	+	II/V	II/XII	2,4	1,4	pos
BC1	4,13	52,53	1,3	+	II/V	VIII/XII	ND	NTR	pos
GöL	4,13	52,53	1,3	—	IV/VII	X/XII	4,19	4	pos
IJ	4,8	52,53	3	+	IV/V	V/XII	8	2,3	neg
SSP	4,13	52,53	1,3	+	II/IV	IX/XII	4,18	3	pos

[a]DQB (BamHI) and DRB (TaqI) genotypes; for nomenclature see Carlsson et al. (1987). NTR, no typing response.

[b]ND, not done.

[c]Positive response is defined as more than 40% of the response in the autologous combination; neg, negative response; AR, alloreactivity, i.e., high proliferative response in the absence of nickel.

either DR3- or DR4-like molecules, which is consistent with the fact that the nickel-allergic donor was DR3, DR4 positive. These experiments were first initiated with the belief that we could find DQ-restricted T-cell clones due to our earlier findings that DQ could restrict nickel responses. We wanted to assess whether the genomic variability with regard to DQA and B genes that had been established with cDNA probes would reveal the total relevant variability of class II restriction. We did not find any DQ-restricted clones; instead our results demonstrated the enormous complexity of the human class II genes and their products and indicated that, as yet, there are no adequate methods to assess whether cells from two individuals are class II incompatible or not other than by direct testing of immunological function.

5. HLA ASSOCIATION IN NICKEL SENSITIVITY

Genetic control of delayed type sensitivity has been reported in animal models. Since 1978 several studies on HLA association in nickel sensitivity have been published. In two studies, significant increases of the frequencies of HLA-B8 and HLA-B21 in nickel-sensitized individuals were found (Lidén et al., 1978; Kapoor-Pillarisetti et al., 1981), but this has not been confirmed by others. The strongest evidence for a genetic influence in humans was found by Menné and Holm (1983) in an investigation on female twins where an increased concordance rate among monozygotic twins compared to dizygotic was shown.

We have studied 33 patients fulfilling the in vitro criteria for nickel sensitivity (Olerup and Emtestam, 1988). Complementary DNA probes for the genes of HLA-DR and HLA-DQ were applied for the demonstration of RFLP. We found a significant association between a Taq I (endonuclease recognizing four bases) HLA-DQA allelic fragment of 4.5 kb and nickel allergy (corrected $p < 0.05$) (Table 18.4). The etiologic fraction of this RFLP-defined DQA

Table 18.4 Phenotypic Frequencies (%) of Allelic Taq I HLA-DQA Restriction Fragments in Patients with Allergic Contact Sensitivity to Nickel and Healthy Controls

TaqI DQA Allelic Restriction Fragments (kb)	Patients with Nickel Reactivity ($n = 33$)	Controls ($n = 100$)
2.5	27	29
6.5	42	49
8.0	9	22
4.5	67[a]	31
5.7	39	46

[a] $\chi^2 = 11.7$, $P_{corr} < 0.05$.

allele was calculated to 0.52, indicating that the found DQA allele might represent approximately 50% of the etiology of nickel allergy. The relative risk was 4.5. Thus, this is the strongest genetic marker for nickel sensitivity yet published. There was no significant differences between high, intermediate, and low responders in the lymphocyte proliferation assay for any of the studied HLA class II genes.

The DQA genotype is in linkage disequilibrium with DR3-DQw2, DRw11-DQw3, and DRw12-DQw3 haplotypes (Carlsson et al., 1987). In turn, HLA-DR3 in linkage disequilibrium with HLA-B8, indicating that our findings might have some correlation to the results of the earlier report by Lidén et al. (1978), where they demonstrated an increase of the B8 antigen frequency. The amino acid sequences of the amino terminal domain of the DQα chain are identical in DR3-DQw2 and DR5-DQw3 haplotypes (see Todd et al., 1987). This favors the hypothesis that the discovered HLA-DQA genotype association in nickel allergy is of biological relevance.

In conclusion, we have found a strong association with a Taq I HLA-DQA allelic restriction fragment in patients with nickel sensitivity. Thus, at least some of the genetic predisposition to become sensitized to nickel seems to be linked to the MHC in humans. The association of this newly defined HLA-DQ genotype to nickel sensitivity in combination with the results of our work with heterogeneous and cloned nickel-specific T-cell populations indicate that further studies on the DQ locus might give additional information on the immunological specificity of nickel sensitivity. Data from our research mentioned above using heterogeneous nickel-specific T lymphocyte were compatible with the idea that other loci than DR could serve as restriction elements for nickel reactivity. Therefore, clones of another nickel-allergic patient were produced with the advantage of having identical sets of HLA molecules compared to the mixed heterogeneous lymphocytes. Specificity analysis and blocking experiments with monoclonal antibodies revealed that the clones were restricted by DR-like molecules. In other words, it seems that the nickel-specific lymphocytes are restricted in one situation by DR-like molecules and in another by DQ-like molecules. This differential reactivity is most likely relevant in vivo. The experiments also indicate that the cellular-defined HLA restriction elements recognized by T cells cannot be classified properly with available HLA class II typing methods, and the results of these experiments document an additional HLA class II polymorphism that has not been previously described.

REFERENCES

Al-Tawil, N. G., Marcusson, J. A., and Möller, E. (1985). HLA-class II restriction of the proliferative T lymphocyte responses to nickel, cobalt and chromium compounds. *Tissue Antigens* 25, 163–172.

Andersson, G. (1987). *Concerted evolution of human class II major histocompatibility genes.* Ph.D. Thesis, Uppsala.

Andersson, M., Böhme, J., Andersson, G., Möller, E., Thorsby, E., Rask, L., and Peterson, P. A. (1984). Genomic hybridization with class II transplantation antigen cDNA probes as a complementary technique in tissue typing. *Human Immunol.* **11**, 57–67.

Ashwell, J. D., Chen, C., and Schwartz, R. H. (1986). High frequency and nonrandom distribution of alloreactivity in T cell clones selected for recognition of foreign antigen in association with self class II molecules. *J. Immunol.* **136**, 389–395.

Ball, E. J., and Stastny, P. (1984a). Antigen-specific HLA-restricted human T-cell lines. I. An MT3-like restriction determinant distinct from HLA-DR. *Immunogenetics* **19**, 13–26.

Ball, E. J., and Stastny, P. (1984b). Antigen-specific HLA-restricted human T-cell lines. II. A GAT-specific T-cell line restricted by a determinant carried by an HLA-DQ molecule. *Immunogenetics* **20**, 547–564.

Basham, T., Smith, W., Lanier, L., Morhenn, V., and Merigan, T. (1984). Regulation of expression of class II major histocompatibility antigens on human peripheral blood monocytes and Langerhans cells by interferon. *Human Immunol.* **10**, 83–93.

Beck, S., and Barrell, B. G. (1988). Human cytomegalovirus encodes a glycoprotein homologous to MHC class-I antigens. *Nature* **331**, 269–272.

Benacerraf, B. (1981). Role of MHC gene products in immune regulation. *Science* **212**, 1229–1238.

Bergholtz, B. O., Thorsby, E. (1978). HLA-D restriction of the macrophage-dependent response of immune human T lymphocytes to PPD in vitro: inhibition by anti-HLA-DR antisera. *Scand. J. Immunol.* **8**, 63–73.

Bjorkman, P. J., Saper, M. A., Samraoui, B., Bennett, W. S., Strominger, J. L., and Wiley, D. C. (1987a). Structure of the human class I histocompatibility antigen, HLA-A2. *Nature* **329**, 506–512.

Bjorkman, P. J., Saper, M. A., Samraoui, B., Bennett, W. S., Strominger, J. L., and Wiley, D. C. (1987b). The foreign antigen binding site and T cell recognition regions of class I histocompatibility antigens. *Nature* **329**, 512–518.

Bodmer, W. F., Albert, E. D., Bodmer, J. G., Dausset, J., Kissmeyer-Nielsen, F., Mayr, W., Payne, R., van Rood, J. J., Trnka, Z., and Walford, R. L. (1984). Nomenclature for factors of the HLA system 1984. *Tissue Antigens* **24**, 73–80.

Böhme, J., Andersson, M., Andersson, G., Möller, E., Peterson, P. A., and Rask, L. (1985). HLA-DR beta genes vary in number between different DR specificities, whereas the number of DQ beta genes is constant. *J. Immunol.* **135**, 2149–2155.

Böhme, J., Carlsson, B., Wallin, J., Möller, E., Persson, B., Peterson, P. A., and Rask, L. (1986). Only one DQ-beta restriction fragment pattern of each DR specificity is associated with insulin-dependent diabetes. *J. Immunol.* **137**, 941–947.

Braathen, L. R., and Thorsby, E. (1982). Studies of human epidermal Langerhans cells. IV. HLA-D restriction of Langerhans cell-dependent antigen activation of T lymphocytes. *Scand. J. Immunol.* **15**, 55–61.

Braciale, T. J., Andrew, M. E., and Braciale, V. L. (1981). Simultaneous expression of H-2-restricted and alloreactive recognition by a cloned line of influenza virus-specific cytotoxic T lymphocytes. *J. Exp. Med.* **153**, 1371–1376.

Cairns, J. S., Curtsinger, J. M., Dahl, C. A., Freeman, S., Alter, B. J., and Bach, F. H. (1985). Sequence polymorphism of HLA DR beta 1 alleles relating to T-cell-recognized determinants. *Nature* **317**, 166–168.

Carlsson, B., Wallin, J., Böhme, J., and Möller, E. (1987). HLA-DR-DQ haplotypes defined by restriction fragment analysis. Correlation to serology. *Human Immunol.* **20**, 95–113.

Eckels, D. D., Lake, P., Lamb, J. R., Johnson, A. H., Shaw, S., Woody, J. N., and Hartzman, R. J. (1983). SB-restricted presentation of influenza and herpes simplex virus antigens to human T-lymphocyte clones. *Nature* **301**, 716–718.

Emtestam, L. (1988). Studies on epidermal Langerhans cells and on HLA factors in nickel sensitivity. Thesis, Stockholm.

Emtestam, L., Marcusson, J. A., and Möller, E. (1988). HLA class II restriction specificity for nickel reactive T-lymphocytes. *Acta Dermatovenereol.* **68**, 395–401.

Emtestam, L., Carlsson, B., Marcusson, J. A., Wallin, J., and Möller, E. (1989). Specificity of HLA restricting elements for human nickel reactive T cell clones. *Tissue Antigens* **33**, 531–541.

Giles, R. C., and Capra, J. D. (1985). Biochemistry of MHC class II molecules. *Tissue Antigens* **25**, 57–68.

Giles, R. C., DeMars, R., Chang, C. C., and Capra, J. D. (1985). Allelic polymorphism and transassociation of molecules encoded by the HLA-DQ subregion. *Proc. Natl. Acad. Sci. USA* **82**, 1776–1780.

Gonwa, T. A., Frost, J. P., and Karr, R. W. (1986). All human monocytes have the capability of expressing HLA-DQ and HLA-DP molecules upon stimulation with interferon-gamma. *J. Immunol.* **337**, 519–524.

Gregersen, P. K., Goyert, S. M., Song, Q., and Silver, J. (1987). Microheterogeneity of HLA-DR4 haplotypes: DNA sequence analysis of LD"KT2" and LD"TAS" haplotypes. *Human Immunol.* **19**, 287–292.

Holbeck, S. L., Kim, S. J., Silver, J., Hansen, J. A., and Nepom, G. T. (1985). HLA-DR4-associated haplotypes are genotypically diverse within HLA. *J. Immunol.* **135**, 637–641.

Jonsson, A.-K., Hylidg-Nielsen, J.-J., Servenius, B., Larhammar, D., Andersson, G., Jörgensen, F., Peterson, P. A., and Rask, L. (1987). Class II genes of the human major histocompatibility complex. Comparisons of the DQ and DX alpha and beta genes. *J. Biol. Chem.* **262**, 8767–8777.

Jörgensen, F., and Lamm, L. U. (1974). MLC—A micro-modification of the mixed leucocyte culture technique. *Tissue Antigens* **4**, 482–494.

Kapoor-Pillarisetti, A., Mowbray, J., Brostoff, J., and Cronin, E. A. (1981). HLA dependence of sensitivity to nickel and chromium. *Tissue Antigens* **17**, 261–264.

Kapsenberg, M. L., Res, P., Bos, J. D., Schootemijer, A., Teunissen, M.B.M., and Schooten, W. V. (1987). Nickel-specific T lymphocyte clones derived from allergic nickel-contact dermatitis lesions in man: heterogeneity based on requirement of dendritic antigen-presenting cell subsets. *Eur. J. Immunol.* **17**, 861–865.

Kaufman, J. F., Auffray, C., Korman, A. J., Shackelford, D. A., and Strominger, J. (1984). The class II molecules of the human and murine major histocompatibility complex. *Cell* **36**, 1–13.

Klein, J. (1986). *Natural History of the Major Histocompatibility Complex.* Wiley, New York.

Lamb, J. R., Eckels, D. D., Lake, P., Johnson, A. H., Hartzman, J., and Woody, J. N. (1982). Antigen-specific human T lymphocyte clones: induction, antigen specificity, and MHC restriction of influenza virus-immune clones. *J. Immunol.* **128**, 233–238.

Lidén, S., Beckman, L., Cedergren, B., Göransson, K., and Nyquist, H. (1978). HLA antigens in allergic contact dermatitis. *Acta Dermatovener (Stockh.)*, Suppl. 79, pp. 53–55.

Lindahl, P., Leary, P., and Gresser, I. (1974). Enhancement of the expression of histocompatibility antigens of mouse lymphoid cells by interferon in vitro. *Eur. J. Immunol.* **4**, 779–784.

Löfström, A., and Wigzell, H. (1986). Antigen specific human T cell lines specific for cobalt chloride. *Acta Derm. Venereol. (Stockh.)* **66**, 200–206.

Lotteau, V., Teyton, L., Burroughs, D., and Charron, D. (1987). A novel HLA class II molecule (DR alpha-DQ beta) created by mismatched isotype pairing. *Nature* **329**, 339–341.

Mellins, E., Woelfel, M., and Pious, D. (1987). Importance of HLA-DQ and -DP restriction elements in T-cell responses to soluble antigens: Mutational analysis. *Human Immunol.* **18**, 211–223.

Menné, T., and Holm, N. V. (1983). Nickel allergy in a female twin population. *Int. J. Dermatol.* **22**, 22–28.

Mosier, D. E. (1967). A requirement for two cell types for antibody formation in vitro. *Science* **158**, 1573–1575.

Nisbet-Brown, E., Lee, J.W.W., Letarte, M., Falk, J. A., and Gelfand, E. W. (1987). Major histocompatibility complex (MHC) restricted antigen recognition: high frequency of human T-cell clones recognizing novel MHC class II determinants. *Human Immunol.* **19**, 41–52.

Okada, K., Prentice, H. L., Boss, J. M., Levy, D. J., Kappes, D., Spies, T., Raghupathy, R., Mengler, R. A., Auffray, C., and Strominger, J. L. (1985). SB subregion of the human major histocompatibility complex: gene organization, allelic polymorphism and expression in transformed cells. *EMBO J.* **4**, 739–748.

Olerup, O., and Emtestam, L. (1988). Allergic contact dermatitis to nickel is associated with a TaqI HLA-DQA allelic restriction fragment. *Immunogenetics* **28**, 310–313.

Ottenhoff, T.H.M., Elferink, D. G., Hermans, J., and de Vries, R.R.P. (1985). HLA class II restriction repertoire of antigen-specific T cells. I. The main restriction determinants for antigen presentation are associated with HLA-D/DR and not with DP and DQ. *Human Immunol.* **13**, 105–116.

Paulsen, G., Qvigstad, E., Gaudernack, G., Rask, L., Winchester, R., and Thorsby, E. (1985). Identification, at the genomic level, of an HLA-DR restriction element for cloned antigen-specific T4 cells. *J. Exp. Med.* **161**, 1569–1574.

Pober, J. S., Gimbrone, M. A., Jr., Cotran, R. S., Reiss, C. S., Burakoff, S. J., Fiers, W., and Ault, K. A. (1983). Ia expression by vascular endothelium is inducible by activated T cells and by human gamma-interferon. *J. Exp. Med.* **157**, 1339–1353.

Qvigstad, E., Moen, T., and Thorsby, E. (1984). T-cell clones with similar antigen specificity may be restricted by DR, MT(DC), or SB class II HLA molecules. *Immunogenetics* **19**, 455–460.

Rosenthal, A. S., and Shevach, E. M. (1973). Function of macrophages in antigen recognition by guinea pig T lymphocytes. I. Requirement for histocompatible macrophages and lymphocytes. *J. Exp. Med.* **138**, 1194–1212.

Scheynius, A., Tjernlund, U., Johansson, C., Alm, G., and Van der Meide, P. (1987). A simple in vitro technique for studies on induction of class II transplantation antigens on keratinocytes. *J. Immunol. Methods* **102**, 59–63.

Servenius, B., Gustafsson, K., Widmark, E., Emmoth, E., Andersson, G., Larhammar, D., Rask, L., and Peterson, P. A. (1984). Molecular map of the human HLA-SB (HLA-DP) region and sequence of an SB-alpha (DP-alpha) pseudogene. *EMBO J.* **3**, 3209–3214.

Sinigaglia, F., Scheidegger, D., Garotta, G., Scheper, R., Pletscher, M., and Lanzavecchia, A. (1985). Isolation and characterization of Ni-specific T cell clones from patients with Ni-contact dermatitis. *J. Immunol.* **135**, 3929–3932.

Sredni, B., and Schwartz, R. H. (1980). Alloreactivity of an antigen-specific T-cell clone. *Nature* **287**, 855–857.

Thorsby, E., Berle, E., and Nousiainen, H. (1982). HLA-D region molecules restrict proliferative T cell responses to antigen. *Immunological Rev.*, **66**, 9–56.

Tiwari, J. L., and Terasaki, P. I. (1985). *HLA and Disease Associations.* Springer-Verlag, New York, Berlin, Heidelberg, Tokyo.

Todd, J. A., Bell, J. I., and McDevitt, H. O. (1987). HLA-DQ beta gene contributes to susceptibility and resistance to insulin-dependent diabetes mellitus. *Nature* **329**, 599–604.

von Boehmer, H., Hengartner, H., Nabholz, M., Lernhardt, W., Schreier, M. H., and Haas, W. (1979). Fine specificity of a continuously growing killer cell clone specific for H-Y antigen. *Eur. J. Immunol.* **9**, 592–597.

Wake, C. T., Long, E. O., and Mach, B. (1982). Allelic polymorphism and complexity of the genes for HLA-DR beta-chain—direct analysis by DNA-DNA hybridization. *Nature* **300**, 372–374.

Walsh, F. S., and Crumpton, M. J. (1977). Orientation of cell-surface antigens in the lipid bilayer of lymphocyte plasma membrane. *Nature* **269**, 307–311.

Zinkernagel, R. M., and Doherty, P. C. (1974). Restriction of in vitro T cell-mediated cytotoxicity in lymphocytic choriomeningitis within a syngeneic or semiallogeneic system. *Nature* **248**, 701–702.

19

NICKEL AND THE IMMUNE SYSTEM: CURRENT CONCEPTS

Stephen Nicklin

Department of Immunotoxicology, British Industrial Biological Research Association, Carshalton, Surrey, United Kingdom

Gitte D. Nielsen

Department of Environmental Medicine, Odense University, Odense, Denmark

Nickel and Human Health: Current Perspectives, Edited by Evert Nieboer and Jerome O. Nriagu.
ISBN 0-471-50076-3   © 1992 John Wiley & Sons, Inc.

1. INTRODUCTION

A wealth of literature now exists that catalogues the discovery and precise chemistry of nickel and its compounds. There are also numerous reviews detailing aspects of isolation, refining, specific industrial applications, and occupational and environmental effects. For further information interested readers are referred to accompanying chapters in this volume and specialist publications and literature surveys obtainable through the U.S. Environmental Protection Agency (EPA, 1986) and the Nickel Producers Environmental Research Association (Nieboer et al., 1984a).

This chapter explores various immunological effects associated with nickel exposure. It is clearly not feasible within the confines of a short text to fully review all of the possible interactions that may occur between individual nickel compounds and the various components that constitute the mammalian immune system. Discussion is therefore restricted to those interactions that have most attracted the attention of immunotoxicologists in recent years.

2. CURRENT IMMUNOLOGICAL CONCEPTS

Despite the massive amount of data relating to the general biological effects and organ toxicity of nickel (Nieboer et al., 1988a,b), it is only relatively recently that attention has focused on the possible adverse effects that nickel compounds may have on the immune system and immunologically related responses. Accumulating evidence based on both human and experimental animal studies now links nickel exposure with various immunotoxic processes. These are broadly divisible into immunosupressive effects and the induction of various nickel-specific hypersensitivity states. The former occur as a consequence of nickel-dependent immunotoxicity, whereas the latter involve hapten recognition of nickel-modified serum/tissue determinants. However, before discussing these effects in detail, it is necessary to introduce the reader to the salient features and function of the normal mammalian immune system.

2.1. Mammalian Immune System

Over the years a detailed analysis of the immune processes of man and a variety of experimental animals has revealed the mammalian immune system to be a finely tuned and precisely regulated mechanism involving multiple interactions between ranks of regulatory and effector T cells (thymus-matured lymphocytes), antibody-secreting B cells (bone marrow/peripherally matured) lymphocytes and a variety of antigen processing/presenting accessory cells of macrophage origin, including Langerhans cells in the skin, alveolar macrophages in the lungs, monocytes in the blood, dendritic/marginal zone and interdigitating cells in the spleen and lymph nodes, and the specialized follicle-associated and subepithelial zone macrophages in the Peyer's patches and lamina propria of the intestinal tract. These cellular elements are seen to act either independently or in concert with a spectrum of other cell types including polymorphonuclear leukocytes, natural killer cells, basophils, eosinophils, and mast cells. Together these cells provide an integrated surveillance mechanism that not only functions as an effective and versatile defence against invading parasitic or pathogenic microorganisms but also provides a mechanism whereby the host's own body tissues may be continuously monitored for neoplastic events, a concept usually referred to as immune surveillance [see Herberman (1985) and Roitt et al. (1986) for general review].

In most instances, macrophages are the first cells to encounter antigen entering the tissues (the site of entry dictating the type of cell involved). The antigen is usually internalized, processed, and then returned to the cell membrane to be presented to antigen-receptive B cells and T cells (Blaese, 1975). The B cells are the precursor cells for antibody production and are responsible for the humoral immune response. The T cells are responsible for cell-mediated reactions and are involved in both effector and control functions. Following interaction with antigen, B cells and T cells are stimulated to undertake different maturation pathways. In both instances, a proportion of cells, although primed for action, remain as a population of long-lived memory cells that are capable of an anamnestic reaction following subsequent reexposure to antigen (Miller, 1978).

2.2. B Cells and Humoral Immunity

The B cell originates from stem cells in the bone marrow through a maturation process that is independent of antigen stimulation. Before birth progenitor cells are found in the liver and spleen; after birth the bone marrow becomes the principal source. The B cell differentiates from a stem cell to an early, immature cell and finally to a late, mature cell. Following contact with antigen, resting B cells undergo a limited series of divisions that terminate in the antibody-secreting plasma cell. The B cells constitute a small percentage of the circulating lymphocyte pool but make up a significant proportion of the cells in the spleen and lymph nodes. Plasma cells are normally absent from the circulation and are found in the lymph nodes, spleen, and bone marrow. They are diffusely distributed through the lamina propria of the gut and the respiratory interstitia.

It is apparent that each B cell has the ability to make only one class of antibody, and molecules of the antibody are presented on the membrane of the cell where they function as antigen receptors. Different B cells present different antibodies, so that the B-cell population as a whole presents antibodies with a wide spectrum of different specificities. A given antigen will combine only with lymphocytes carrying antibody with which it is a good fit. On interaction, such cells are stimulated to differentiate and divide: Some become memory cells, while others differentiate into plasma cells that synthesize and secrete antibody with the same specificity as that on the surface of the parent B cell. Collectively, antibodies are referred to as immunoglobulins. Structurally, each immunoglobulin molecule consists of equal numbers of heavy and light polypeptide chains (general formula $(H_2L_2)_n$, where $n = 1$, . . . , 5) held together by noncovalent forces and interchain disulfide bridges. Essentially each antibody is bifunctional; one region of the molecule is concerned with binding to the antigen, while the opposite region mediates binding to host cells or initiates complement activation. The antigen-combining end of each chain contains extensive variations in amino acid sequence, and it is the three-dimensional structure of these hypervariable regions that determines antigen specificity (cf. the lock-and-key analogy).

Immunoglobulins have been assigned five principal classes based on the primary polypeptide structure of their respective heavy chains; these are IgG, IgA, IgM, IgD, and IgE. Briefly, IgG is the major component of normal serum accounting for some 70–75% of the total immunoglobulin pool. This immunoglobulin class is evenly distributed between the intra- and extravascular pools and is the major antibody responsible for opsinization, complement activation, and antitoxin/antivirus activity. Immunoglobulin M accounts for about 10% of the immunoglobulin pool. The molecule has a pentameric structure and is by and large confined to the intravascular pool. This molecule represents the predominant early response immunoglobulin directed against antigenically complex infectious organisms/pathogens. Immunoglobulin A represents 15–20% of the human serum immunoglobulin pool. Immunoglobulin A can cross epithelial barriers and is the predominant antibody in external seromucus secretions such as saliva, respiratory/gastrointestinal secretions, genitourinary tract secretions, tears, colostrum, and milk where it neutralizes toxic agents and protects mucosal surfaces from invasion by viruses and infectious microorganisms. Immunoglobulin D accounts for less than 1% of the total plasma immunoglobulin but is known to be present in significant amounts on the membrane of many circulating B cells. The precise biological function of IgD remains controversial, although evidence suggests that it plays a role in lymphocyte maturation/differentiation. Immunoglobulin E is present at very low concentrations in the serum but is found bound to the surface of mast cells and basophils where, following interaction with antigen, it initiates protection against parasitic infection. In the Western world, however, IgE-dependent mast cell/basophil activation is more commonly associated with immediate-type hypersensitivity (see below). The interested reader is directed to the literature (Hahn, 1982) for further details.

2.3.   T Cells and Cell-mediated Immunity

Maturation of T cells is now reasonably well characterized. Precursor T cells originating in the haemopoietic tissues of the embryonic liver and in the bone marrow of adults migrate to the thymus. There they are triggered to undergo both proliferation and differentiation. The process of differentiation is marked by morphological changes and the expression of characteristic T-cell markers (Katz, 1977; Owen and Jenkinson, 1982). Extensive proliferation is thought to be one mechanism favoring increased genetic variation within the T-cell reportoire, which produces cells with different immune reactivity patterns. The vast majority of the cells are however surplus to requirements, and these die within the thymus. Selected cells, approximately 10% of the resident population (a figure also under debate!), emigrate from the thymus. In the neonate these cells "seed" the spleen and peripheral lymph nodes. In adult life emigration continues, though at a reduced rate. The T cells circulate between the blood and lymph and back again to the blood via the thoracic duct, thereby maintaining surveillance within the body tissues as a whole, and even though the thymus involutes postpuberty, the organ continues to maintain basal mitosis and "tops up" the T-cell system as required. Following antigen stimulation, T lymphocytes differentiate into a variety of antigen-specific effector cells that are seen to be involved in all aspects of immunity: as direct killer cells in cytotoxic processes attacking virally infected cells, allografts, tumors, etc., as indirect effector cells triggering and modulating delayed-type hypersensitivity reactions such as contact sensitivity (see below for further detail), or as internal regulatory *helper* (Th) or *suppressor* (Ts) cells controlling the magnitude and time course of both humoral and cell-mediated responses (Roitt et al., 1986; Adorni, 1987).

2.4. Immunoregulation and Homeostasis

Activation, regulation, and feedback control of immune responses clearly depends on precise cell–cell communications. The latter occurs either by direct contact between the relevant cells or via the release of biological mediators, either locally into the cellular microenvironment or systemically into the circulation. A number of these secreted cell products have been isolated and characterized and include prostaglandins, leukotriene products of lipoxygenase metabolism, cytokines such as the interferons and interleukins, and a variety of less well defined thymic hormones (Fabris et al., 1983; Möller, 1982). The facility for cooperative cellular interaction between the various components of the immune system provides a complex homeostatic mechanism that can both direct and regulate the immune mechanism. In addition to this network of intercellular communication, the integrity of the system may be further modified via both the endocrine/neurological axis and by nutritional/environmental factors (Fabris et al., 1983; Adorni, 1987).

Normal immune competence is therefore the result of adaptive/reactive cellular and biochemical changes designed to ensure the survival of the species in an ever-changing and potentially hostile environment. Unfortunately the inherent complexity of the system is also its "Achilles heel," since even minor perturbations within the immune network can produce dramatic and potentially damaging changes in immune reactivity. This imbalance can be manifested in a variety of ways, including the induction of aberrant or in-appropriate responses to self-determinants resulting in autoimmune disease or an exaggerated response to antigen leading to hypersensitization. Direct or indirect suppression of the effector phase functions of the immune response may also leave the host more susceptible to infectious agents and may even produce an increased risk of cancer following the loss of immune surveillance (Herberman, 1985; Miller and Nicklin, 1986; CEC, 1987).

To date there is no direct evidence that links nickel exposure with auto-immune disease. However, there is a strong association between nickel and hypersensitivity reactions and some evidence that suggests nickel can modify normal immune competence.

2.5. Hypersensitivity Reactions

Whereas the normal immune response is designed to benefit the host, there are a number of instances when immune reactions do not provide protection but are actually damaging and occasionally fatal. These adverse effects are known as allergic, or more correctly *hypersensitivity reactions*. These reactions are by definition immunologically mediated and elicited following blind challenge. Coombs and Gell described four principal types of hypersensitivity reaction (types I, II, III, and IV), but in reality these types do not necessarily occur in isolation from each other. In many instances responses may indeed be mixed. It should be stressed, however, that these reactions are by and large no more than exaggerated or uncontrolled expressions of the normal immune repertoire acting in such a way as to cause tissue damage [see Stanworth (1985) for current concepts in hypersensitivity]. Types I–III are all antibody-mediated reactions, while type IV is a cell-mediated reaction involving both T cells and macrophages. The essential basis of these reactions is summarized below. Nickel-related effects are susequently considered separately in more detail.

Type I, or immediate, hypersensitivity occurs when immune recognition of a given antigen results in the formation of cytophilic IgE antibodies. These have the capacity to bind to and specifically sensitize mast cells and basophils. When the immunizing antigen is next encountered, it reacts with the cell-bound antibody and initiates a degranulation process that results in the immediate release of a range of preformed vasoactive components including histamine and 5-hydroxytryptamine. In addition, activation of a number of enzyme pathways culminates in the production and release of prostaglandin and leucotriene mediators. The combination of these events produces an acute inflammation and the rapid onset of respiratory, circulatory, and dermal symptoms. Reactions of this type include insect venom/drug sensitivity, hay fever, food allergy, perennial rhinitis, occupational asthma, and atopic eczema (Stanworth, 1973; Roitt et al., 1986).

Type II reactions, also referred to as antibody-dependent cytotoxic hyper-sensitivity, occur when an immune response produces antibodies with specificity for cell surfaces or foreign antigens/novel determinants bound to host cells. Antibody binding to the target cell then initiates a number of reaction pathways that may result in direct phagocytosis through opsonic (Fc) or immune (C3) adherence, nonphagocytic cytolysis mediated by killer cells with receptors for IgFc or enzymic lysis following complement activation. Reactions of this type include haemolytic disease of the newborn and drug-induced autoimmune haemolytic anaemia, thrombocytopenia, leukopenia, and hyperacute graft rejection. Type II reactions are in essence a reflection of normal immune processes directed against novel or "unexpected" self-determinants [see Roitt et al. (1986) for further reading].

Type III reactions arise from the combination of free antigen with circulating antibody and the subsequent formation of immune complexes (microprecipitates) in and around small blood vessels. This deposition initiates platelet agglutination, activates the complement cascade, and triggers tissue-damaging inflammatory processes. Furthermore, C3a and C5a, by-products of complement activation, have anaphylactoid properties and can activate mast cells directly. The clinical manifestations of this type of reaction depend upon the nature of the antigen involved and where the immune complexes lodge but may include urticaria, pyrexia, arthralgia, lymphadenopathy, myocarditis, vasculitis, and occasionally nephritis. Examples of type III hypersensitivity reactions include serum/vaccine sickness, insulin sensitivity, late-phase occupational asthma, adverse food/drug reactions, eczema, plus a variety of other diverse symptoms (WHO, 1977; Williams, 1980).

Type IV reactions, also referred to as cellular or delayed-type hypersensitivity, are encounted in many allergic reactions (both alone and in combination with those above) to bacteria, viruses, fungi, and transplanted tissues (allografts) and in contact dermatitis resulting from sensitization to certain simple chemicals. Type IV reactions are by definition slowly evolving responses involving specifically sensitized lymphocytes, regulatory helper and suppressor cells, and Langerhans cells. Initiation is independent of serum antibody and is mediated and transferable by specifically sensitized T lymphocytes. Initial sensitization occurs when incoming antigen is intercepted by antigen-processing macrophages, phagocytosed, and subsequently presented to reactive T cells. These undergo proliferation resulting in the generation of a clone of specifically sensitized lymphocytes. These cells carry on their surface-specific receptors for the sensitizing agent and, following subsequent encounter with the same antigen, release a series of immunopharmacologically active mediators that localize a mixed-cellular inflammatory response at the site of challenge.

Four types of delayed-hypersensitivity responses are recognized, and of these, the Mantoux reaction, the Jones-Mote reaction, and contact hypersensitivity all occur within 72 hr of antigen challenge. By contrast, the fourth type, granulomatous reactions, may develop over a period of weeks or months. The responses are to some extent mixed in that some reactions overlap or occur sequentially after a single antigen challenge. The Mantoux reaction occurs following the injection of tuberculin into the skin of an individual in whom previous infection with mycobacterium has induced a state of cellular immunity. The reaction is characterized by erythema and induration, which appears some hours later and reaches a maximum after 24–48 hr. Histologically the lesion is characterized by initial perivascular cuffing by monocytes followed by a more extensive exudation of monocytes and polymorphonuclear leucocytes. The latter migrate out from the lesion leaving behind a predominantly mononuclear cell infiltrate of lymphocytes and macrophages. Jones-Mote hypersensitivity represents a comparable form of sensitivity in protein-sensitive individuals; in this instance, parenteral challenge results in subepidermal basophil infiltration, and skin swelling is maximal after 24 hr. Thereafter, the reaction gradually subsides.

Contact hypersensitivity is an eczematous/inflammatory skin reaction that occurs some 24–48 hr after contact with the sensitizing agent. In Europe and Scandinavia, some of the most common sensitizing agents are nickel and chromium and related salts, acrylates, and chemicals used in or associated with the rubber industry; in the United States poison ivy and poison oak represent additional hazards. Such low-molecular-weight chemicals are referred to as haptens, which by definition are nonantigenic in the free state but have the potential to interact with dermal/serum proteins in such a way as to render them antigenic. The conjugate is then treated as a foreign sensitizing agent and is phagocytosed, processed, and presented to reactive T cells by Langerhans cells. Contact hypersensitivity lesions show a characteristic mononuclear cell infiltration at 6–8 hr after challenge, which peaks at 12–16 hr. Cellular infiltration is accompanied by epidermal oedema and microvesicle formation, localized tissue damage, and inflammation. Granulomatous hypersensitivity is clinically one of the more debilitating forms of hypersensitivity and is responsible for a number of chronic tissue reactions. Granulomas result from the presence of a persistent agent within macrophages, usually a microorganism (e.g., *Mycobacterium tuberculosis*) or some inert particle (talc or zirconium oxide) the cell is unable to destroy or break down (the latter examples being distinguishable from the former by the lack of lymphocytic infiltration). In some instances granuloma formation also appears to be linked with the continued presence/production of immune complexes, as, for example, in allergic alveolitis. The typical immunologically mediated granuloma has a core of epithelioid cells and macrophages surrounded by a cuff of lymphocytes all enclosing a single, centralized lesion. In some diseases such as tuberculosis, this central area may contain an inner zone of necrosis; in others such as talc or zirconium granulomas necrosis may be absent. Examples of granuloma-associated diseases include tuberculosis, leprosy, sarcoidosis, schistosomiasis, and foreign body reactions to talc, zirconium, and beryllium (nickel perhaps). For additional references and detail the reader should consult Turk (1980).

3. NICKEL-ASSOCIATED HYPERSENSITIVITY REACTIONS

3.1. Biochemical Aspects

Over the last few decades, considerable evidence links exposure to nickel salts with a number of human hypersensitivity states. Whereas the vast majority of cases relate to contact dermatitis (type IV hypersensitivity), occupational exposure to nickel has also been linked with respiratory sensitivity. Three basic types of asthmatic responses have been reported: (1) a rapid-onset attack associated with acute bronchospasm; (2) a late-response reaction appearing 6–12 hr after exposure; and (3) a mixed or combined response. As discussed above, the immediate asthmatic response is antibody mediated (type I hypersensitivity) and results from the interaction between antigen and antigen-specific IgE bound to mast cell. This association induces a conformational change in the mast cell membrane that triggers degranulation and the release of pharmacologically active mediators. The late response is less well defined but appears to be the consequence of an antigen–antibody immune complex–mediated inflammatory reaction, involving a series of reaction pathways including platelet activation, arachidonic acid metabolism, and cellular accumulation in response to chemotactic stimuli (a type III response).

These immunological reactions are all antigen specific, with sensitization being dependent upon a direct interaction between "nickel" and antigen-responsive T cells. Nickel as an ion is too small to function as an antigen. Therefore, in order to initiate hypersensitization, it must first interact with a carrier molecule, usually a protein, to form a hapten (Ni)–protein conjugate. In this respect, nickel has been shown to interact with a number of carriers. Coordination chemists have known for some time that the Ni(II) ion has an apparently unique ability to form stable complexes with deprotonated amide bonds. The chemical nature of such interactions has been studied in some detail, and the interested reader is directed to two recent reviews written by Andrews et al. (1988) and Martin (1988). Here we restrict our attention to the primary interaction between nickel and serum proteins.

Nickel in serum exists primarily as a chelate complex of serum albumin (Nieboer et al., 1988a). The Ni(II)-binding site has been identified and consists of a square-planar chelate structure formed by the three N-terminal amino acids (aspartic acid–alanine–histidine); in detail, the donor atoms are comprised of the N-terminus α-amino nitrogen, the intervening deprotonated amide nitrogens, and the 3-nitrogen of the imidazole ring of residue 3. Nuclear magnetic resonance (NMR) evidence suggests that the carboxylic side-chain residue of the N-terminus aspartic acid interacts axially with the metal center. This is consistent with spectrometric analysis that implies the existence of an octahedral form in equilibrium with a square-planar complex. Nickel is also chelated by the amino acid histidine and to a lesser extent by various ill-defined glycoprotein molecules (Andrews et al., 1988; Martin, 1988: Nieboer et al., 1988a). Whereas it has been established that antibodies with specificity for the Ni–human serum albumin (HSA) complex are present in the serum of some nickel-sensitive asthmatics (see next section), it is unclear as to whether the Ni–HSA interaction is a necessary prerequisite for sensitization or merely designed to facilitate renal clearance. Equally there is no information on the nature of the protein carrier involved in the induction of type IV hypersensitivity.

3.2. Nickel-associated Contact Dermatitis

Epidemiological investigations have shown a gradually increasing number of nickel-sensitive individuals since World War II, with more women affected than men (see Chapter 15). The higher prevalence in women (nearly 10-fold) is believed to most likely reflect a higher daily skin exposure to everyday objects containing nickel, such as jewelry made of German Silver (containing 10–20% nickel), nickel-plated spectacle frames, watches, buttons, clasps, buckles, zippers, and hair grips, not to mention the increasing practice of single/multiple ear rings or studs as an adornment (now popular among both sexes). Additional nonoccupational sources of nickel exposure include coins, tools, cutlery, medical instruments, surgical implants, and also certain cosmetics and detergent preparations. Although nickel salts are used in a number of industrial applications and occupational outbreaks have been reported (Wall and Calnan, 1980), nickel dermatitis is not currently believed to represent a major industrial problem (Nieboer et al., 1984a). It is however significant that many of the women who present dermatitis link their condition with wet working practices. Thus extended exposure of the hands to detergents as well as wet or moist working conditions tend to weaken the skin barrier and provide an easier route of entry for solubilized nickel ions. Further, preexisting irritant dermatitis, pressure, friction, and sweating all will clearly act as predisposing factors. These effects must however be superimposed upon a suitable genetic predisposition as indicated by differences between the pairwise concordance rates in mono- and dizygotic twins (Menné and Holm, 1983). To date, however, nickel sensitivity has not been clearly linked with any particular HLA background (Menné and Holm, 1983; Karvonen et al., 1984; but see Chapter 15).

Classically, determination of nickel dermatitis utilizes patch testing and the visual evaluation of the site following occluded exposure to a nickel salt solution or a nickel-containing object. The optimal nickel concentration in a patch test solution is currently set at 5.0% as nickel sulfate (see Chapter 15). Patch testing, however, is not to be undertaken lightly. Reactions may be ambiguous in that they can reflect primary irritation rather than preexisting immunity and there is always the possibility of sensitizing a normal individual. Intradermal testing as described by Epstein (1956) has also been advocated, but this procedure offers no overall advantage over the conventional method. An increasingly attractive technique, however, is that of in vitro lymphocyte transformation. Over the last 10 years or so in vitro transformation techniques for the rapid screening of nickel sensitivity have become advocated. Although somewhat controversial initially [see NAS (1975) for details], more recent studies have established the validity of the lymphocyte transformation test as a possible alternative to patch testing (Svejgaard et al., 1978; Braathen, 1980; Sinigalia et al., 1985; Res et al., 1987; Botham, 1988). The comparable value of the so-called leucocyte migration inhibition assay (based on Ni-induced inhibitory mediator released from antigen-primed lymphocytes) however remains to be demonstrated conclusively (Thulin, 1976; Polak, 1983).

The immunological processes associated with the induction of nickel dermatitis are clearly complex, and animal models of nickel hypersensitivity are still to be validated (see below). Consequently, interpretations of the immunological processes must be tempered accordingly. As with other contact-sensitizing systems, the hapten Ni(II) penetrates the skin and conjugates with some as-yet-undefined dermal (possibly serum) protein(s) in such a way as to render them antigenic. Several investigators have indicated that Langerhans cells (and possibly migrating veiled cells) then phagocytose this conjugate and transport it via the afferent lymphatics to peripheral draining lymph nodes, where it is presented to antigen-receptive T lymphocytes (Braathen, 1980; Braathen and Thorsby, 1980; Res et al., 1987). This interaction promotes T-cell expansion and the generation of a nickel-specific T-cell clone. As outlined above, the functional activity of these cells may be demonstrated by incubating peripheral blood lymphocytes from nickel-sensitive individuals with Ni(II) in vitro and monitoring cellular proliferation in terms of tritiated thymidine incorporation (Braathen, 1980; Botham et al., 1988; Res et al., 1987; Sinigalia et al., 1985). The nickel-reactive cells have been partially characterized both from biopsy studies and within the circulating pool as $CD4^+8^-$ T lymphocytes (inducer T cells), and recently Kapsenberg et al. (1987) obtained and cloned inflammatory T cells from the lesions of volunteers with experimentally induced nickel-contact dermatitis. In a series of experiments, 7–15% of $CD4^+8^-$ T-lymphocyte clones (TLCs) appeared to be specific for nickel in a nickel-dependent proliferation assay. More importantly the nickel-proliferative response was shown to require the presence of antigen-presenting cells and were restricted by HLA class II molecules (see Chapter 18). All TLCs recognized nickel presented by Langerhans cells from epidermal skin and some (three out of eight clones) also recognized nickel presented by peripheral blood monocytes, indicating that antigen presentation is not totally restricted to Langerhans cells.

Once sensitized, reexposure to even minute quantities of nickel will elicit a type IV contact hypersensitivity reaction at the site of contact. The typical reaction is characterized by a localized inflammatory/eczematous response that appears after about 12 hr and is fully developed at 48–72 hr. The threshold of nickel sensitivity after percutaneous exposure was investigated in nickel-allergic patients by Wahlberg and Skog (1971), who found that the mean threshold of sensitivity in 53 patients was 0.43% nickel sulfate in distilled water and 0.51% in petrolatum. Ten patients however reacted to doses as low as 0.039% (w/v). Although initial sensitization is associated with transient lesions, nickel allergy may, in susceptible individuals, result in the development of more persistent hand eczema. This is characterized by recurrent, itching eruptions and deep vesicles expanding into persistent widespreading lesions that may extend from initial sites of contact to produce secondary eruptions localized to the elbow flexures, neck, and eyelids. According to Calnan

(1956), nickel dermatitis presents a uniquely characteristic distribution pattern: phase 1, primary lesions develop at site of contact with the element; phase 2, secondary symmetrical spreading; and phase 3, the appearance of lesions at more distant sites with the affected sites having no relationship with the phase 1 sites. Furthermore, the lesions may persist some time after the removal of obvious sources of contact. Exacerbations may occur at weekly or monthly intervals and can result in disability. The prognosis in such individuals is generally poor.

Whereas initial sensitization is known only to occur following contact and absorption through the skin, provocation and exacerbation of existing lesions may occur following subsequent noncutaneous exposure to nickel. Thus Christensen and Möller (1975a,b) reported that the oral administration of nickel to allergic individuals aggravated hand eczema even though prolonged handling of nickel-containing objects was without effect. Although such observations were initially viewed with some skepticism, Kaaber et al. (1978) reported a marked improvement in a proportion of patients with chronic nickel dermatitis when maintained on a low-nickel diet and a return of their condition on return to their normal diet. More recent studies have confirmed this phenomenon and provide good evidence to link dietary nickel with hand eczema (Jordan and King, 1979; Cronin et al., 1980; Christensen and Lagesson, 1981; Vienen et al., 1983a,b; Chapter 16).

What is unusual about this effect is that in both man and experimental animals the oral administration of most antigens tends to induce a state of anergy (immunological tolerance) to subsequent parenteral/cutaneous challenge with the same antigen. The intestinal uptake of a variety of soluble and particulate antigens is well documented (Nicklin, 1987). The process occurs continuously and appears to be a prerequisite for the initiation of both intestinal antibody production and perhaps, more importantly, the induction of systemic antigen-specific immune tolerance. This state of tolerance involves a number of mechanisms, the most important of which stem from the generation of antigen-specific suppressor/contrasuppressor cells. These lymphocytes block the initiation/elicitation of hypersensitivity reactions and help maintain a status quo within the immune system [see Challacombe and Tomasi (1987) for a review]. Whereas such mechanisms are clearly by-passed in the atopic, orally elicited nickel sensitivity adds an extra degree of complexity to the story. It is accepted that incoming nickel enters the tissues and interacts with a protein carrier thereby rendering it antigenic. However, the ensuing tissue reaction is often localized to the initial sensitizing site, an effect that appears to be unique to nickel dermatitis. This reaction suggests that the antigen-presenting cells must remain relatively static (cf. analogy with "tattoo monocytes") and that the initial sensitization event selects a population of nickel-"specific" presenting cells. In order to understand this phenomenon more clearly, further studies must be designed to elucidate more precisely the immunochemical interactions that occur between ingested nickel and the various immunoregulatory elements controlling the mammalian immune system.

At present, no animal model is fully capable of mimicking the diverse range of symptoms presented by nickel-intolerant individuals. Nilzén and Wikstrom (1955) appear to be the first to report sensitization of guinea pigs by the repeated topical application of nickel sulfate via a detergent carrier. However, follow-up studies by Samitz and Pomerantz (1958) were unable to reproduce this work, and these authors challenged the earlier observations suggesting that the previous results may reflect local irritation rather than true hypersensitivity [see Hunziker (1969) and Anderson (1986) for historical perspectives and current methodologies]. In later studies, the same group (Samitz et al., 1975) similarly failed to induced sensitization using nickel complexed to amino acid carriers or guinea pig skin extracts.

More recent intradermal injection regimens have produced only marginally more encouraging results. Wahlberg (1976) demonstrated "sensitization" to nickel sulfate using a high-responder strain of guinea pigs. Turk and Parker (1977) reported a allergic-type granuloma formation in animals sensitized using a split adjuvant technique consisting of Freunds complete adjuvant (tuberculin, oil, and detergent) followed by weekly injections of 25 μg nickel sulfate after 14 days. A contact dermatitis/delayed type hypersensitivity reaction developed in two out of five animals after five weeks. Further work (by the authors) utilizing the guinea pig maximalization technique of Magnusson and Kligman (1969) elicited a weak but transient reaction in two out of eight guinea pigs. However, the reactions were considered borderline and difficult to interpret. Similarly, all attempts to sensitize rats have proved negative, even though anticarrier responses were regularly elicited (authors' unpublished observations). In mice, sensitivity is also more an exception than a rule. Although it is easy to sensitize mice to antigens such as picryl chloride, sensitization to nickel requires extensive applications to produce even moderate effects (Möller, 1984). Recent studies and improvements in sensitization techniques have been recently reviewed by Wahlberg (1989), and it is clear that more precise animal sensitization and elicitation regimens will be developed in the not too distant future.

3.3. Nickel-associated Respiratory Reactions

Whereas nickel-induced asthma is more the exception than the rule, a gradually increasing number of respiratory illnesses have been reported in relation to occupational exposure to soluble nickel compounds. Nickel-provoked respiratory problems have thus been reported in individuals employed in the nickel-plating industry (McConnell et al., 1973; Malo et al., 1982; Novey et al., 1983; Cirla et al., 1985), in workers involved in the production of nickel catalysts (Davies, 1986), among metal polishers (Block and Yeung, 1982), and occasionally among welders after inhaling welding fumes (Keskinen et al., 1980; Bjornerem et al., 1983). In addition to these relatively well characterized clinical reactions, nickel exposure has also been related to what is loosely refered to as "fume fever," a reaction triggered in susceptible individuals following the inhalation of freshly formed nickel oxides as produced by welding, smelting, and founding practices (Parkes, 1982). Loefflers' syndrome has also been reported following exposure to nickel tetracarbonyl (Sunderman and Sunderman, 1961). This is a pulmonary hypersensitivity pneumonitis reaction of unknown aetiology characterized by transient alveolar monocyte accumulation and eosinophil infiltration of the connective tissue. Whereas the reaction is usually self-limiting, granulomatous reactions and irreversible lung damage have been recorded. Although an immune aetiology remains to be proven, similarities with progressive type IV granulomatous reactions cannot be ignored.

As outlined above, a number of reaction pathways may be involved in nickel-induced allergic respiratory hypersensitivity. Type I (IgE-dependent) immediate reactions develop within minutes of exposure, reach a maximum after 10–20 min, and decline within 2 hr. Late phase (Type III immune complex–dependent) sensitivity usually appears some hours after exposure and peaks 6–12 hr later. In many instances mixed responses are apparent, and dual forms of reactivity involving early, late, and dual asthmatic responses have been reported (Block and Yeung, 1982; Malo et al., 1982, 1985; McConnell et al., 1973: Novey et al., 1983). In some individuals, urticaria (Kaplan and Zeligman, 1963; Malo et al., 1982) and rhinitis (Niordson, 1981) are also associated with occupational exposure, and occasionally both asthma and delayed (presumed type IV) contact sensitivity have been observed within the same individuals (McConnell et al., 1973; Block and Yeung, 1982; Cirla et al., 1985).

It is now generally accepted that these nickel-induced respiratory effects are antibody-mediated reactions (with the possible exception of Loefflers' syndrome). Antibodies with specificity for nickel–protein conjugates have been detected in the serum of immediate-type nickel-reactive individuals (McConnell et al., 1973; Malo et al., 1982; Novey et al., 1983; Cirla et al., 1985). Furthermore, the results of more recent studies by Dolovich et al. (1984) and Nieboer et al. (1984b) strongly suggest that at least in one patient studied the binding of nickel to the primary copper-binding site of albumin induces conformational changes in the molecule sufficient to render it allergenic. Thus antibodies isolated from serum of the individual were shown to recognize Ni(II)–HSA but not Cu(II)–HSA or Co(II)–HSA. Similarly, blocking experiments demonstrated that Ni(II), Cu(II), and Co(II) competitively inhibited the formation of the ^{63}Ni(II)–HSA–antibody complexes in vitro, whereas metals that fail to interact at the Cu(II)–HSA binding site [e.g., Zn(II), Mn(II), and Cr(III)] were without effect.

4. IMMUNOTOXIC EFFECTS OF NICKEL

Whereas traditional pathological approaches have already implicated various lymphoid tissues as possible organs for toxic insult, it is only relatively recently that the immune system per se has been demonstrated to be a sensitive and perhaps early indicator of toxicity. This increasing awareness stems in part from modern advances in immunology and molecular biology that have provided new and precise techniques for the detection and analysis of immune dysfunction. These novel approaches have not only increased our understanding of the regulatory mechanisms that govern the normal functions of the immune system but also have raised a note of caution concerning the immunotoxic potential of certain metal compounds.

Koller (1980) reviewed the literature relating to the immunotoxic potential of heavy metals, including nickel, and noted particularly that nickel exposure can reduce host resistance to both viral and bacterial infectious agents, depress antibody responses to T-phages, inhibit gamma-interferon production, and suppress the phagocytic capacity of macrophages. Similar effects have been reported in vitro. Rae (1983) demonstrated a significant reduction in the ability of human polymorphonuclear leucocytes to phagocytose and kill *Staphylococcus epidermidis* after nickel exposure (0.05 μM for 18 hr). Smialowicz et al. (1984) investigated the effects of parenterally administered nickel chloride on humoral and cell-mediated immunity in mice. In these studies, a single intramuscular injection of 18.3 mg/kg (equivalent to 4.5 mg Ni/kg) was shown to cause rapid thymic involution and a reduction in splenic T cells. The T-cell blastogenic responses to mitogens (phytohaemagglutinin and concanavalin A) were likewise reduced, whereas the B-cell population of the spleen and in vitro responsiveness to the B-cell mitogen lipopolysaccharide were essentially unchanged. In vivo, these effects were reflected in terms of a reduced primary antibody response to sheep erythrocyte, a well-characterized T cell–dependent antigen challenge. Macrophage activity as judged by endotoxin sensitivity and phagocytic capacity was not significantly influenced by the treatment. Taken together these observations highlight T-cell maturation processes as the target for nickel toxicity (within the dose range employed). Recent studies (Knight et al., 1987) in rats confirm the high sensitivity of the thymus to nickel ions. However, both of the studies outlined used relatively high-dose treatment regimens.

Immunologically relevant data have also been furnished by numerous respiratory studies designed to investigate the effect of nickel exposure on the bronchus-associated immune system (see Haley et al., 1988). Bingham et al. (1972) exposed rats to respirable aerosols as soluble chloride or insoluble oxide at levels consistent with acceptable human industrial exposure and demonstrated that this treatment induced hyperplasia of bronchial epithelium and peribronchial lymphocyte infiltration. Port et al. (1975) noted that the intratracheal injection of nickel oxide into Syrian hampsters 48 hr after exposure to influenza A/PR/8 virus led to a significant increase in mortality and morbidity compared to controls. Other studies relating to respiratory exposure report epithelial cell proliferation (Wehner et al., 1981; Johansson et al., 1983), impaired alveolar macrophage activity, increased susceptibility to respiratory tract infections (Graham et al., 1975a; Castranova et al., 1980; Johansson et al., 1980; Casarett-Bruce et al., 1981; Lundborg and Cramner,

1982; Murthy et al., 1983; Wiernik et al., 1983; Haley et al., 1988), and reduced antibody plaque-forming cells in the spleens of mice and rats following the inhalation of nickel chloride and oxide, respectively (Graham et al., 1975b; Spiegelberg et al., 1984). Respiratory tract toxicity has also been demonstrated in vitro by Dubreuil et al. (1984), who demonstrated dose-dependent cytotoxic effects in cultured human respiratory epithelial cells.

The possibility of adverse effects of nickel on immunosurveillance has also attracted increased interest in recent years. Smialowicz et al. (1985) set out to examine the effects of nickel on natural killer (NK) cell activity. These cells potentially play a pivotal role in immunosurveillance, where they appear to act as first-line, "nonimmune" cytotoxic effector cells. Smialowicz's studies revealed that intramuscular injection of 18.5 mg/kg nickel chloride either as a single injection or over a period of 14 days produced a reduction in the NK activity of splenocytes from CBA/J and C57BL/10 mice. The observation that the treated mice also presented decreased tumor cell clearance from the lungs and a concomitant enhanced "take" of tumor cell colonies in the lung clearly lends support to both the immunotoxic potential of nickel and the importance of NK cells in the control of neoplastic progression.

It is clearly tempting to link immune suppression, reduced NK and alveolar macrophage function, and tissue-specific changes within the lungs as pre-disposing factors in the aetiology of respiratory disease. However, as mentioned earlier, many studies are not necessarily compatible with occupational exposure, and thus the health effects and clinical significance of such studies are difficult to evaluate in any satisfactory manner. It is also necessary to raise a note of caution with regard to the interpretation of the immuno-suppression data. True or classical immunosuppression refers to decreased immune competence; it relates only to the immune system or an individual component or components of an immune response and should occur in the absence of overt pathological or toxic changes in other organ systems. In some studies relatively high doses of nickel have been administered. Consequently, all "immunotoxic" effects should be considered in the light of toxicity within other systems. We therefore suggest that future studies be designed so as to include low-level, "nontoxic" exposure regimens and that all immunological assessment studies be performed in parallel with conventional pathological/toxicological evaluation.

5. CONCLUDING REMARKS

Nickel has the potential to interact with the immune system in a number of ways. Low-level exposure in susceptible individuals either by cutaneous contact or following the inhalation of respirable ambient nickel may result in nickel-specific hypersensitization. Both cell-mediated (type IV) contact sensitivity and antibody-mediated (types I and III) immediate, early, and late-phase asthmatic reactions are recognized. The type IV reactions are the more common, especially among young females. In some cases, the reactions may lead to hand eczema and may result in incapacitation. Such individuals remain highly nickel sensitive, and dietary nickel intake may be sufficient to exacerbate symptoms. Both the nature of the sensitizing Ni(II)–(hapten)–protein conjugate and the various possible immune responses to this determinant remain to be proprly elucidated. Experimentally, nickel possesses certain immunotoxic capacities. The thymus and T cell–dependent processes appear to be particularly sensitive and are perhaps carly indicators of nickel-induced immunotoxicity. Alveolar macrophages and NK cells have also been highlighted as sensitive target cells. Systemically these effects are reflected in terms of reduced host resistance to infection and possibly reduced immunosurveillance. Although it is easy to link immunological dysfunction with an increased risk of tumor development, many studies on immune function have been performed with relatively high levels of compounds. Although local immunotoxic effects may augment the development of cancer and respiratory problems, more detailed studies are required in order to evaluate more fully the biological significance of low-level occupational exposure.

REFERENCES

Adorni, L. (1987). The cells of the immune system: an up-date on their interactions and signals. In A. Berlin, J. Dean, M. H. Draper, E.M.B. Smith, and F. Spreafico (Eds.), *Immunotoxicology*. Martinus Nijhoff, The Hague, Netherlands, pp. 12–23.

Anderson, K. E. (1986). Contact allergy to chlorocresol, formaldehyde and other biocides. Guinea-pig tests and clinical studies. *Acta Dermato-Venereol* **125**, 5–21.

Andrews, R. K., Blakeley, R. L., and Zerner, B. (1988). Nickel in proteins and enzymes. In H. Sigel and A. Sigel (Eds.), *Metal Ions in Biological Systems. Nickel and Its Role in Biology*. Marcel Dekker, New York, pp. 165–263.

Bingham, E., Barkley, W., Zerwas, M., Stemmer, K., and Taylor, P. (1972). Responses of alveolar macrophages to metals. I. Inhalation of lead and nickel. *Arch. Environ. Health* **25**, 406–414.

Bjornerem, H., Thomassen, L. M., and Wergeland, E. (1983). Sveiserastma. *Tidsskr Nor Laegeforen.* **103**, 1286–1288.

Blaese, R. M. (1975). Macrophages and the development of immune competence. In J. A. Bellanti and D. H. Dayton (Eds.), *The Phagocytic Cell in Host Resistance*. Raven, pp. 6–63.

Block, G. T., and Yeung, M. (1982). Asthma induced by nickel. *J. Am. Med. Assoc.* **247**, 1600–1602.

Botham, P., Everness, K. M., Gawkrodger, D. J., and Hunter, J.A.A. (1988). *In vitro* antigen presentation in nickel-induced contact sensitivity. *Proc. Eur. Symp. Contact Dermatitis*, Heidelberg, pp. 81–82.

Braathen, L. R. (1980). Studies on human epidermal Langerhans cells: III. Induction of T lymphocyte response to nickel sulphate in sensitized individuals. *Br. J. Dermatol.* **103**, 517–526.

Braathen, L. R., and Thorsby, E. (1980). Studies on human epidermal Langerhans cells. Allo-activating and antigen-presenting capacity. *Scand J. Immunol.* **11**, 401–419.

Calnan, C. D. (1956). Nickel dermatitis. *Br. J. Dermatol.* **68**, 229–236.

Casarett-Bruce, M., Camner, P., and Curstedt, T. (1981). Changes in pulmonary lipid composition of rabbits exposed to nickel dust. *Environ. Res.* **26**, 353–362.

Castranova, V., Bowman, L., Reasor, M. J., and Miles, P. R. (1980). Effect of heavy metal ions on selected oxidative metabolic processes in rat alveolar macrophages. *Toxicol. Appl. Pharmacol.* **53**, 14–23.

CEC (1987). *Proceedings of the International Symposium on the Immunological System as a Target for Toxic Damage*, A. Berlin, J. Dean, M. H. Draper, E.M.B. Smith, and F. Spreafico (Eds.). Martinus Nijhoff, for the Commission of the European Communities and UNEP, ILO, WHO International Programme on Chemical Safety, Luxembourg, p. 495.

Challacombe, S. J., and Tomasi, T. B. (1987). Oral tolerance. In J. Brostoff and S. J. Challacombe (Eds.), *Food Allergy and Intolerance*. Bailliere Tindall, London, pp. 255–269.

Christensen, O. B., and Lagesson, V. (1981). Nickel concentration of blood and urine after oral administration. *Ann. Clin. Lab. Sci.* **11**, 119–125.

Christensen, O. B., and Möller, H. (1975a). Nickel allergy and hand eczema. *Contact Dermatitis* **1**, 129–141.

Christensen, O. B., and Möller, H. (1975b). External and internal exposure to the antigen in the hand eczema of nickel allergy. *Contact Dermatitis* **1**, 136–141.

Cirla, A. M., Bernabeo, F., Ottoboni, F., and Ratti, R. (1985). Nickel-induced occupational asthma: immunological and clinical aspects. In S. S. Brown and F. W. Sunderman, Jr. (Eds.), *Progress in Nickel Toxicology*. Academic Press, New York, pp. 215–219.

Cronin, E., DiMichiel, A. D., and Brown, S. S. (1980). Oral challenge in nickel-sensitive women with hand eczema. In S. S. Brown and F. W. Sunderman, Jr. (Eds.), *Progress in Nickel Toxicology*. New York, Academic, pp. 149–152.

Davies, J. E. (1986). Occupational asthma caused by nickel salts. *J. Soc. Occup. Med.* **36**, 29–31.

Dolovich, J., Evans, S. L., and Nieboer, E. (1984). Occupational asthma for nickel sensitivity. I. Human serum albumin in the antigenic determinant. *Br. J. Ind. Med.* **41**, 51–55.

Dubreuil, A., Bouley, G., Duret, S., Mestre, J.-C., and Boudene, C. (1984). *In vitro* cytotoxicity of nickel chloride on a human pulmonary epithelial cell line. *Arch. Toxicol. Suppl.* **7**, 391–393.

Environmental Protection Agency (EPA) (1986). Health assessment document for nickel and nickel compounds. EPA/600/8-83/012FF, Final Report. U.S. EPA, Research Triangle Park, NC.

Epstein, S. (1956). Contact dermatitis due to nickel and chromate. *Arch. Dermatol.* **73**, 236–255.

Fabris, N., Garaci, E., Hadden, J., and Mitchison, N. A. (1983). *Immunoregulation.* Plenum, New York, p. 259.

Graham, J. A., Gardner, D. E., Waters, M. D., and Coffin, D. L. (1975a). Effect of trace metals on phagocytosis by alveolar macrophages. *Infect. Immunol.* **11**, 1278–1283.

Graham, J. A., Gardner, D. E., Miller, F. J., Daniels, M. J., and Coffin, D. L. (1975b). Effect of experimental respiratory infection following nickel inhalation. *Environ. Res.* **20**, 33–42.

Hahn, G. S. (1982). Antibody structure, function and active sites. In S. E. Ritzman (Ed.), *Physiology of Immunoglobulins: Diagnostic and Clinical Aspects*. Alan Liss, New York, pp. 1–49.

Haley, P. J., Bice, D. E., Muggernberg, B. A., Hann, F. F., and Benjamin, S. A. (1988). Immunopathologic effects of nickel subsulfide on the primate pulmonary immune system. *Toxicol. Appl. Pharmacol.* **88**, 1–12.

Herberman, R. B. (1985). Immunological mechanisms of host resistance to tumours. In J. H. Dean, M. I. Luster, A. E. Munson, and H. Amos (Eds.), *Immunotoxicology and Immunopharmacology*. Raven, New York, pp. 69–78.

Hunziker, N. (1969). *Experimental Studies on Guinea Pig's Eczema*. Springer-Verlag, New York, p. 98.

Johansson, A., Lundborg, M., Hellstrom, P. A., Camner, P., Keyser, T. R., Kirton, S. E., and Natusch, D.F.S. (1980). Effect of iron, cobalt and chromium dust on rabbit alveolar macrophages: a comparison with the effects of nickel dust. *Environ. Res.* **21**, 165–176.

Johansson, A., Curstedt, T., Robertson, B., and Camner, P. (1983). Rabbit lung after inhalation of soluble nickel. II. Effects on lung tissue and phospholipids. *Environ. Res.* **31**, 399–412.

Jordan, W. P., Jr., and King, S. E. (1979). Nickel feeding in nickel-sensitive patients with hand eczema. *J. Am. Acad. Dermatol.* **1**, 506–508.

Kaaber, K., Veien, N. K., and Tjell, J. C. (1978). Low nickel diet in the treatment of patients with chronic nickel dermatitis. *Br. J. Dermatol.* **98**, 197–201.

Kaplan, I., and Zeligman, I. (1963). Urticaria and asthma from acetylene welding. *Arch. Dermatol.* **88**, 188–189.

Kapsenberg, M. L., Res. P., Bos, T. D., Schootemijer, A., Teunissen, M.B.M., and Schooten, W. V. (1987). Nickel-specific T lymphocyte clones derived from allergic nickel-contact dermatitis lesions in man: heterogeneity based on requirement of dendritic antigen-presenting cell subsets. *Eur. J. Immunol.* **17**, 861–865.

Karvonen, J., Silvennoninen-Kassinen, S., Ilonen, J., Jakkula, H., and Tiilikainen, A. (1984). HLA antigen in nickel allergy. *Ann. Clin. Res.* **16**, 211–212.

Katz, D. H. (1977). *Lymphocyte Differentiation, Recognition and Regulation*. Academic, New York, pp. 26–49.

Keskinen, H., Kalliomaki, P. I.., and Alanko, K. (1980). Occupational asthma due to stainless steel welding fumes. *Clin. Allergy* **10**, 151–159.

Knight, J. A., Rezuke, W. ., Wong, S.H-Y, Hopfer, S. M., Zaharia, O., and Sunderman, F. W., Jr. (1987). Acute thymic involution and increased lipoperoxides in the thymus of nickel chloride-treated rats. *Res. Commun. Chem. Pathol. Pharmacol.* **55**, 291–302.

Koller, L. D. (1980). Immunotoxicology of heavy metals. *Int. J. Immunopharmacol.* **2**, 269–279.

Lundborg, M., and Camner, P. (1982). Decreased level of lysozyme in rabbit lung lavage fluid after inhalation of low nickel concentrations. *Toxicology* **22**, 353–358.

McConnell, L. H., Fink, J. N., Schlueter, D. P., and Schmidt, M. G. (1973). Asthma caused by nickel sensitivity. *Ann. Intern. Med.* **78**, 888–890.

Magnusson, B., and Kligman, M. A. (1969). The identification of contact allergens by animal assay. The guinea pig maximization test. *J. Invest. Dermatol.* **52**, 268–276.

Malo, J.-L., Cartier, A., Doepner, M., Nieboer, E., Evans, S., and Dolovich, J. (1982). Occupational asthma caused by nickel sulfate. *J. Allergy Clin. Immunol.* **69**, 55–59.

Malo, J.-L., Cartier, A., Gagnon, G., Evans, S., and Dolovich, J. (1985). Isolated late asthmatic reaction due to nickel sulphate without antibodies to nickel. *Clin. Allergy* **15**, 95–99.

Martin, R. B. (1988). Nickel ion binding to amino acids and peptides. In H. Sigel and A. Sigel (Eds.), *Metal Ions in Biological Systems. Nickel and Its Role in Biology*. Marcel Dekker, New York, pp. 123–159.

Menné, T., and Holm, N. V. (1983). Hand eczema in nickel sensitive female twins. Genetic predisposition and environmental factors. *Contact Dermatitis* **9**, 289–296.

Miller, J.F.A.P. (1978). The cellular basis of immune responses. In M. Samter (Ed.), *Immunological Diseases*. Little, Brown, New York, pp. 35–48.

Miller, K., and Nicklin, S. (1986). Immunological aspects. In A. N. Worden, D. V. Parke, and J. Marks (Eds.), *The Future of Predictive Safety Evaluation*. MTP Press, Lancaster, England, pp. 181–194.

Möller, G. (1982). Interleukins and lymphocyte activation. *Immunol. Rev.* **63**, 1–205.

Möller, H. (1984). Attempts to induce contact allergy to nickel in the mouse. *Contact Dermatitis* **10**, 65–68.

Murthy, R. C., Barkely, W., Hollingworth, L., and Bingham, E. (1983). Enzymatic changes in alveolar macrophages of rats exposed to lead and nickel by inhalation. *J. Am. Coll. Toxicol.* **2**, 193–199.

National Academy of Sciences (NAS) (1975). *Nickel*. NAS, Washington, DC, p. 210.

Nicklin, S. (1987). Intestinal uptake of antigen:immunological consequences. In K. Miller and S. Nicklin (Eds.), *Immunology of the Gastrointestinal Tract*. CRC Press, Boca Raton, FL, pp. 87–110.

Nieboer, E., Yassi, A., Jusys, A. A., and Muir, D.C.F. (1984a). The technical feasibility and usefulness of biological monitoring in the nickel producing industry. Special document, McMaster University, Hamilton, Canada. (Available from the Nickel Producers Environmental Research Association, Chicago, IL.

Nieboer, E., Evans, S. L., Dolovich, J. (1984b). Occupational asthma from nickel sensitivity: II Factors influencing the interaction of Ni^{2+}, HSA and serum antibodies with nickel related specificity. *Br. J. Ind. Med.* **41**, 56–63.

Nieboer, E., Rickey, T. T., and Sandford, W. E. (1988a). Nickel metabolism in man and animals. In H. Sigel and A. Sigel (Eds.), *Metal Ions in Biological Systems. Nickel and Its Role in Biology*. Marcel Dekker, New York, pp. 91–116.

Nieboer, E. R., Rossetto, F. E., and Menon, C. R. (1988b). Toxicology of nickel compounds. In H. Sigel and A. Sigel (Eds.), *Metal Ions in Biological Systems. Nickel and Its Role in Biology*. Marcel Dekker, New York, pp. 359–392.

Nilzén, A., and Wikstrom, K. (1955). The influence of lauryl sulphate on the sensitization of guinea pigs to chrome and nickel. *Acta Dermatol. Venereol.* **35**, 292–299.

Niordson, A-M. (1981). Nickel sensitivity as a cause of rhinitis. *Contact Dermatitis* **7**, 273–274.

Novey, H. S., Habib, M., and Wells, I. D. (1983). Asthma and IgE antibodies induced by chromium and nickel salts. *J. Allergy Clin. Immunol.* **72**, 407–412.

Owen, J.J.T., and Jenkinson, E. J. (1982). Early events in T lymphocyte genesis in the fetal thymus. *Am. J. Anat.* **170**, 301–321.

Parkes, W. R. (1982). *Occupational Lung Disorders*, 2nd ed. Butterworths, London, p. 210.

Polak, L. (1983). In D. Burrows (Ed.), *Chromium Metabolism and Toxicology*. CRC, Boca Raton, FL, pp. 51–136.

Port, C. D., Fenters, J. D., Ehrlich, R., Coffin, D. L., and Gardner, D. (1975). Interactions of nickel oxide and influenza infection in the hamster. *EHP Environ. Health Perspect.* **10**, 268 (abstract).

Rae, T. (1983). The action of cobalt, nickel and chromium on phagocytosis and bacterial killing by human polymorphonuclear leucocytes; its relevance to infection after total joint arthroplasty. *Biomaterials* **4**, 175–180.

Res, P., Kapsenberg, M. L., Bos, J. D., and Stiekema, F. (1987). The crucial role of human dendritic antigen presenting cell subsets in nickel-specific T-cell proliferation. *J. Invest. Dermatol.* **88**, 550–554.

Roitt, I., Brostoff, J., and Male, D. (1986). *Immunology*. Gower Medical Publishing, London, p. 257.

Samitz, M. H., and Pomerantz, H. (1958). Studies of the effects on the skin of nickel and chromium salts. *Arch. Ind. Health* **18**, 473–479.

Samitz, M. H., Katz, S. A., Schneiner, D. M., and Lewis, J. E. (1975). Attempts to induce sensitization in guinea pigs with nickel complexes. *Acta Derm. Venereol. (Stockh.)* **55**, 475–480.

Sinigalia, F., Scheidegger, D., Garrotta, G., Scheper, R., Pletcher, B., and Lanzavecchia, R. (1985). Isolation and characterisation of Ni-specific T cell clones from patients with Ni-contact dermatitis. *J. Immunol.* **135**, 3929–3935.

Smialowicz, R. J. (1985). The effect of nickel and manganese on natural killer cell activity. In S. S. Brown and F. W. Sunderman, Jr. (Eds.), *Progress in Nickel Toxicology*. Blackwell Scientific, Oxford, pp. 161–164.

Smialowicz, R. J., Rogers, R. R., Riddle, M. M., and Stott, G. A. (1984). Immunologic effects of nickel. I. Suppression of cellular and humoral immunity. *Environ. Res.* **33**, 413–427.

Smialowicz, R. J., Rogers, R. E., Riddle, M. M., Garner, R. J., Rowe, D. G., and Luebke, R. W. (1985). Immunologic effects of nickel. II. Suppression of natural killer cell activity. *Environ. Res.* **36**, 56–66.

Spiegelberg, T., Kordel, W., and Hochrainer, D. (1984). Effects of NiO inhalation on alveolar macrophages and the humoral immune systems of rats. *Ecotoxicol. Environ. Safety* **8**, 516–525.

Stanworth, D. R. (1973). *Immediate Hypersensitivity*. Amsterdam, North-Holland, p. 120.

Stanworth, D. R. (1985). Current concepts of hypersensitivity. In J. H. Dean, M. I. Luster, A. E. Munson, and H. Amos (Eds.), *Immunotoxicology and Immunopharmacology*. Raven, New York, pp. 91–98.

Sunderman, F. W., and Sunderman, F. W., Jr. (1961). Loeffler's syndrome associated with nickel sensitivity. *Arch. Intern. Med.* **107**, 405–408.

Svejgaard, E., Morling, N., Svejgaard, A., and Veien, N. K. (1978). Lymphocyte transformation induced by nickel sulphate: an *in vitro* study of subjects with and without a positive nickel patch test. *Acta Dermatol. Venereol.* **58**, 245–250.

Thulin, H. (1976). The leukocyte migration test in nickel contact dermatitis. *Acta Dermatol. Venereol.* **56**, 377–380.

Turk, J. L. (1980). *Delayed Hypersensitivity*, 3rd ed. *Research Monographs in Immunology*, Vol. 1. Elsevier, Holland, p. 125.

Turk, J. L., and Parker, D. (1977). Sensitisation with Cr, Ni, and Zr salts and allergic type granuloma formation in the guinea pig. *J. Invest. Dermatol.* **68**, 341–345.

Vienen, N. K., Hattel, T., Justesen, O., and Nørholm, A. (1983a). Oral challenge with metal salts (I). Vesicular patch-test-negative hand eczema. *Contact Dermatitis* **9**, 402–406.

Vienen, N. K., Hattel, T., Justesen, O., and Nørholm, A. (1983b). Oral challenge with metal salts (II) Various types of eczema. *Contact Dermatitis* **9**, 407–410.

Wahlberg, J. E. (1976). Sensitisation and testing of guinea pigs with nickel sulfate. *Dermatologica* **152**, 321–329.

Wahlberg, J. E. (1989). Nickel: Animal sensitization assays. In T. Ménne and H. I. Maibach (Eds.), *Nickel and The Skin: Immunology and Toxicology*. CRC Press, Boca Raton, FL, 65–73.

Wahlberg, J. E., and Skog, E. (1971). Nickel allergy and atopy. Threshold of nickel sensitivity and immunoglobulin E determination. *Br. J. Dermatol.* **85**, 97–104.

Wall, L.M., and Calnan, C. D. (1980). Occupational nickel dermatitis in the electroforming industry. *Contact Dermatitis* **6**, 414–420.

Wehner, A. P., Dagle, G. E., and Milliman, E. M. (1981). Chronic inhalation exposure of hamsters to nickel-enriched fly ash. *Environ. Res.* **26**, 195–216.

Wiernik, A., Johansson, A., Jarstrand, C., and Camner, P. (1983). Rabbit lung after inhalation of soluble nickel. I. Effects on alveolar macrophages. *Environ. Res.* **30**, 129–141.

Williams, R. C. (1980). *Immune Complexes in Clinical and Experimental Medicine*. Harvard University Press, Cambridge, MA, p. 145.

World Health Organisation (WHO) (1977). The role of immune complexes in disease. Technical report 606. WHO, Geneva, p. 501.

20

PROBLEMS IN THE TOXICOLOGY, DIAGNOSIS, AND TREATMENT OF NICKEL CARBONYL POISONING

Lindsay G. Morgan

The Medical Department, INCO Europe Limited, Swansea, United Kingdom SA6 5QR

Nickel and Human Health: Current Perspectives, Edited by Evert Nieboer and Jerome O. Nriagu.
ISBN 0-471-50076-3   © 1992 John Wiley & Sons, Inc.

1. INTRODUCTION

This chapter outlines some of the problems that have been experienced by occupational health physicians practicing at a refinery where the unique process of using nickel carbonyl gas to refine nickel to 99.98% purity has been in use for nearly a century. Future avenues for research are also considered.

Mond et al. (Mond, 1890) discovered nickel carbonyl gas [Ni(CO)₄] when it was noticed that the flame of burning carbon monoxide (CO) turned white after the latter was passed over finely divided nickel. Very pure nickel was deposited if the gas was heated to above 250°C. From this discovery they developed a commercial process for the refining of nickel. Mond also instituted research into the toxicity of this gaseous nickel compound, and by the following year it was shown to be highly toxic (McKendrick and Snodgrass, 1891). Despite the knowledge of the toxicity of nickel carbonyl, the first accident came as a surprise. It had not been realized that this gas could remain adsorbed on powder only to be released when the powder was disturbed (Armit, 1907, 1908). Famous individuals such as J.B.S. Haldane of Oxford were involved in the early experiments, and a variety of treatments were considered that may now seem bizarre, such as the administration of potassium ferrocyanide.

The clinical course of a case of accidental exposure to nickel carbonyl gas has been well documented (Amor, 1932). It involves two stages. The first is characterized by headache, pains in the chest, and a metallic taste in the mouth and weakness. Many cases do not go beyond this stage. There is then generally a remission of 8–24 hr followed by a second phase characterized by chemical pneumonitis but with evidence in severe cases of cerebral poisoning. This stage reaches its greatest severity in about four days, but convalescence is often protracted.

At Clydach there have been six fatal cases: four in the first year (1901), one when the refinery was flooded in 1932, and the last in 1937 when a man fell into a decomposer unit and could not be rescued. The scientific literature has not necessarily reported all cases that have occurred, but nine published reports have been traced; three other incidents are known. From this it is believed that the total number of fatalities from nickel carbonyl gas since it was discovered is 20 (Table 20.1). Apart from Clydach (Amor, 1932), other reports of exposure come from the United States (Brandes, 1934; Strong et al., 1973; Jones, 1973), Italy (Sorinson, 1957), Finland (Vuopala et al., 1970), Germany (Bayer, 1939; Ludewigs and Thiess, 1970), Canada (Sutherland, 1958), China (Shi et al., 1986), and Russia (Revnova, 1978).

A review of the experience at Clydach indicates that in the period 1925–1950 there were approximately four lost-time cases per annum and (with the exception of a serious incident in 1958 involving 25 men who lost time) 2.5 cases per annum between 1950 and 1980. However, with the introduction of plant modernization in 1977 and 1978 and resultant improvement in en-

Table 20.1 Fatal Cases of Nickel Carbonyl Poisoning: 1890–1988

Country	Number of Fatalities	Date of Incident	Reference
Wales	4	1904	
	1	1932	Amor, 1932
	1	1937	
Canada			
(Port Colborne)	4	1958	Sutherland, 1958
United States	2	1934	Brandes, 1934
	2	1952	Strong et al., 1954
	1	—	Jones, 1973
Germany	4	—	Bayer, 1939
	1	—	Ludewigs and Thiess, 1970

Nonfatal exposures are also known to have occurred in Canada (Copper Cliff, Ontario), Italy, China, Finland, and Russia (see text).

vironmental conditions, there have been only four lost-time cases in the 1990s. These figures are important considering that at present, at Clydach, over 6000 kg of nickel oxide per hour are converted into nickel carbonyl gas. At INCO's Copper Cliff Refinery in Canada, there have only been two cases of lost-time accidents since commencement of operations in 1974 (see Section 2).

2. ATMOSPHERIC MONITORING*

While insisting that good engineering is the answer to safe processing, no further mention will be made of it. In the plants, atmospheric monitoring is important, particularly for those who are involved in occupational hygiene at the shop floor level. A number of approaches to the measurement or continuous monitoring of nickel carbonyl are or have been employed:

(i) Discoloration of a flame (lamp testing): The original discovery of carbonyl occurred because Langer noticed that a flame was white when it should have been blue. Under ideal conditions a trained observer can detect this discoloration when the level of $Ni(CO)_4$ in the air is as low as 250 ppb. This discoloration phenomenon has been used to monitor the refinery atmosphere for many years and is still employed for leak detection either in plant apparatus or in a particular working area.

(ii) The Draeger nickel carbonyl detector tube requires 20 pumps of air, which takes about 45 sec and gives no indication until the test is complete.

* A number of units are employed for ambient nickel carbonyl: 1 ppb (μL/L) = 0.007 mg $Ni(CO)_4/m^3$ = 0.0024 mg Ni/m^3.

It is therefore not very satisfactory as a monitoring tool but can be used to confirm whether Ni(CO)$_4$ is present or not.

(iii) At INCO's Clydach refinery a "Miran" multipoint sampler based on the absorption by Ni(CO)$_4$ of infrared light is used. This instrument is sensitive to as low as 10 ppb and has provided remarkably good protection, although cross sensitivity can occur with exhaust gases and others. Each instrument can sample 12 points, including one control point at the rate of approximately one point per minute. The sampling heads are situated at strategic points in the carbonyl plant.

(iv) At INCO's Copper Cliff Nickel Refinery in Canada, a chemilluminescence detection system has been installed. It has a sensitivity of 0.1 ppb. The sampling head is placed in the roof at a point where all refinery air is exhausted, and this also has given a very high degree of protection. This plant processes on the order of 7500 kg of nickel each hour and was designed with the hazards of nickel carbonyl exposure in mind. As already indicated, the number of accidental gassings have been few.

(v) The Billionaire, which relies on the effect of Ni(CO)$_4$ pyrolysis products on gaseous conduction in an ionization chamber, has a detection limit of 1 ppb. The system is now obsolete but was reliable and gave a high degree of protection at Rohm and Haas and also at INCO's research unit at Port Colborne and its refinery at Copper Cliff. Because of problems with the presence of particulates, the instrument was not found to be suitable for use at the Clydach refinery.

Continuous monitoring systems play a vital part in warning the worker of the presence of Ni(CO)$_4$ gas and thus help to avoid exposure. However, the sense of smell, eternal vigilance, and a very high standard of operator awareness continue to be essential to avoid accidents. The Health and Safety Executive (HSE, 1987) in the United Kingdom are proposing a short-term occupational exposure standard of 100 ppb on the basis that it is acute exposure that should be avoided, and in a modern plant the ambient level is too low to be measured. Environmental measurements carried out in 1957 indicate that the average day-to-day levels were on the order of 70 ppb, reaching as high as 200 ppb. It is probable that they were even higher in earlier years, but there are no firm data.

3.   BIOLOGICAL MONITORING

3.1.   Current Practices

Biological monitoring of nickel carbonyl poisoning is generally by measurement of urinary nickel (Sunderman et al., 1986). Strong et al. (1954) and Morgan (1960) have both reported grossly elevated levels of urinary nickel excretion in persons exposed to the gas. Strong et al. (1954) recommended routine daily measurement of urinary nickel to identify men who might have been exposed to the gas and devised a protocol of treatment based on the levels observed (Sunderman, 1971).

At the Clydach (Wales) refinery it is normal practice for men to report if an exposure is suspected and for all subsequent urine specimens over the next few days to be analyzed for nickel content. Chelation therapy is started before any urinary nickel results are available, but subsequent treatment is dependent upon the clinical history and urine results. At the Copper Cliff (Ontario) refinery the administration of chelation therapy is dependent upon the first urine result which can be made available within an hour of receipt in the laboratory. It is started if the urinary nickel exceeds 150 μg/L.

3.2. What Does It Tell?

Nickel powder and nongaseous compounds may enter the body by inhalation or ingestion. Measurements of nickel in body fluids cannot detect the source of the material, and therefore, if there is nickel in the body for any reason other than carbonyl exposure, it can confuse the diagnosis. This is particularly true for men working with nickel powder in the respirable range. Men working with these powders may have elevated urinary nickel levels without having been exposed to $Ni(CO)_4$. In the nickel powder packing area, dust exposures are on the order of 0.2 mg Ni/m^3, which is within the proposed control limit of 0.5 mg Ni/m^3 set by the HSE (1987). In the Clydach packing area, workers frequently have urinary nickel levels near 50 μg Ni/L or higher, whereas the normal for nonexposed persons is less than 10 μg Ni/L Sunderman et al. (1986). When such workers take chelating agents, they have a nickeluresis/ irrespective of whether or not they have been gassed; if they have, it is not clear whether the increase in urinary nickel is due to the gas or due to the previous nickel burden. One other confounding factor is that some chelating agents increase the absorption of nickel from the gut and can cause nickeluresis/ even in the absence of any occupational nickel exposure.

3.3. Is It Important?

In the first serious episode in five years, four men had significantly elevated urinary nickel levels. Two of these lost time because of the severity of the illness; a third man felt unwell for a few days but continued to work, and the fourth never felt ill and did a double shift immediately after the incident. The data in Figure 20.1 indicate that all four men had grossly elevated urinary nickel levels when examined only an hour or so after the incident. In two of the cases, the concentration rose to 800 and 1300 μg Ni/L, respectively; in the two who did not lose any time it rose to 600 and 1100 μg Ni/L. Although one of the four men complained of a headache when first seen, there was little in their histories or signs or symptoms to suggest that the exposures for these men had been markedly different. Clearly, urinary nickel can only

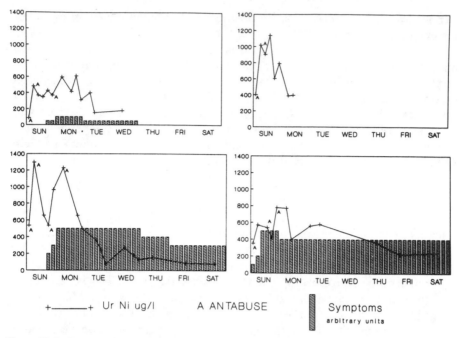

Figure 20.1. Four examples of urinary nickel and symptoms following nickel carbonyl exposure.

provide complementary information and cannot be used as the sole index of the severity of an exposure.

What is needed is a nickel carbonyl breathalyzer. Gas chromatography can detect $Ni(CO)_4$ at levels as low as 0.001 ppb (Sunderman et al., 1968), but instrument operation is complex and frequent calibration is required. At present, the levels of nickel carbonyl in expired air after exposure or their relevance to clinical symptoms are not known.

4.   TREATMENT

4.1.   Dithiocarbamate–Antabuse Regime

After World War II, the idea of removing nickel inhaled as nickel carbonyl was researched by Barnes and Denz (1955), Chiningelli (1957), and West and Sunderman (1958). These studies showed experimentally that, in rats, sodium diethyldithiocarbamate (DDC) was the best drug out of a number tested. However, there are certain drawbacks to its administration, such as that it be coated and that sodium bicarbonate be coadministered to help in passing the acid medium of the stomach without decomposition. Especially prepared sugar-coated capsules were very effective, but problems arose when it became uneconomical for the pharmaceutical company to fill small orders and new legislation was implemented regulating the registration of drugs. New enteric-coated capsules were subsequently made available. Unfortunately, the new preparation did not work well, possibly because some gastric absorption was necessary and did not take place. To avoid any problems with registration in the United Kingdom, the proprietary drug Antabuse (disulfiram) is currently used. It is given as soon as the employee reports to the medical department after a suspected exposure at a dose of 500 mg immediately and a further 500-mg dose 2, 8, and 24 hr later depending upon circumstances. Disulfiram is metabolized to DDC and is nearly as effective in treating exposed rats (Horak et al., 1976). However, it must be stressed that joint disulfiram administration and alcohol intake are contraindicated, and persons given this drug must be warned against the use of alcohol for at least a week.

Radiotracer studies by Oskarsson and Tjalve (1979) have shown that after rats have been exposed to ^{63}Ni-labeled nickel carbonyl gas, nickel is widely distributed in the tissues, particularly in the lung, brain, and adrenal glands. Similar work by Nicholls and Luscombe (1974) using orally administered ^{35}S-labeled DDC in rats showed that this drug is widely but more unevenly distributed in the body. The lowest concentrations occur in the brain, myocardium, and eye, with moderate levels in the blood, lung, muscle, and bone and high amounts in the liver and kidney (related to metabolism and excretion). It would thus appear that the early administration of chelating agent is important to prevent cerebral damage.

4.2. Other Chelating Agents

Using a computer-based model, Jones (1983) has theorized that DDC removes nickel from the serum but is not effective in removing intracellular nickel. He suggested that triethylpentamine could be an efficient adjunct in nickel carbonyl therapy since it could mobilize the intracellular nickel into the extracellular compartment, from which it would subsequently be removed by DDC. This has never been tested in practice.

4.3. Steroids

The pathological changes in the injured lungs include a thickening of the alveolar wall and a subsequent production of a thick alveolar exudate similar to that seen from other causes of chemical pneumonitis. Vuopala et al. (1970), reporting an accident involving 25 men in Finland, detailed studies of lung changes. They observed that the more serious cases had a lowering of diffusing and vital capacities, while in the mild cases these remained normal. It seemed appropriate, therefore, to try and reduce the lung reaction by the use of steroids. In a study reported by Clark et al. (1980), rats were given median lethal doses (LD_{50}) of $Ni(CO)_4$ (350 mg/m^3) together with prednisolone with no observed reduction in mortality. Another group of animals given DDC intraperitoneally were remarkable in that none even looked ill. At first it

Table 20.2   Effect of Prophylactic Treatment on Microscopic Changes in Pulmonary Tissue in 20 Mice Treated with LD$_{50}$ of Nickel Carbonyl[a]

Change	DDC Only	DDC and Prednisolone
Interstitial thickening	10	1
Macrophage aggregation	8	0
Perivascular oedema	5	0
Inflammatory cell infiltration	5	3

[a]Based on the study by Clark et al. (1980).

appeared that steroids could not substitute for DDC in the early treatment of nickel carbonyl poisoning. However, further review of the pathology (Table 20.2) showed that joint treatment by DDC and prednisolone did reduce the microscopic tissue changes. Consequently, their combined use appears to have a place in treatment based upon the animal work.

5. PSYCHOLOGICAL ILLNESS

During the 1970s at the Clydach refinery there were cases with psychological problems of sufficient severity to merit psychiatric help some months after an accidental incident.

In severe debilitating illness in which there are overtones of compensation, it is difficult to sort out organic disease from an anxiety state. The psychiatrist considered that the illness could not be attributed to the toxicity of Ni(CO)$_4$ and was more probably due to the psychological trauma of the accident. In the recent incident mentioned in Section 3.3 the two men who lost time complained of difficulty in concentrating on conversation or on simple manual tasks; one had a very bad headache. The symptoms lasted two weeks in one case and much longer in the other. The respiratory symptoms of one of the men also lasted a long time. It is not clear whether these symptoms were caused by or precipitated the psychological symptoms. The recent paper by Shi et al. (1986) seems particularly relevant, as these authors showed that people chronically exposed to low doses of nickel carbonyl for 10–20 years had a reduction in serum monoamine oxidase and abnormal electroencephalograms, but the relevance to psychological illness was not reported. It is possible that Antabuse can also play a part in the etiology of this depression, as it is known to influence catecholamine metabolism (Eneanya et al., 1986). However, no such psychological reaction was reported by Kaaber et al. (1979), who used disulfiram in a trial treatment of nickel dermatitis.

6. X-RAY CHANGES

Morgan (1962) noted that men severely affected by nickel carbonyl gas had characteristic X-ray appearances. These included fluffy exudates and patchy infiltration and there was elevation of the right hemidiaphragm. X-ray screening has confirmed this to be precipitated by partial diaphragmatic paralysis. The etiology of this condition has never been established. Vuopala et al. (1970) reported that the changes were typical of an interstitial pneumonitis, and although he did not comment on the right hemidiaphragm, similar right diaphragmatic elevation is visible in the published radiographs.

7. TERATOGENICITY

Sunderman et al. (1979, 1980) exposed pregnant rats and hamsters to nickel carbonyl and observed that their offspring were born without any eyes, among other ophthalmic defects. Although it should be considered anecdotal evidence, extensive enquiry revealed that no such deformities could be re-membered by pediatricians or ophthalmic surgeons in the Swansea Valley, the area in which the Clydach refinery is situated. This was reassuring as women had been employed in the refinery during two world wars when ambient $Ni(CO)_4$ levels were believed to have been quite high and women were known to have suffered lost-time gassing accidents.

   INCO in Canada felt that because of potential risks to the fetus from both nickel carbonyl and DDC, it would be unwise to employ women in their nickel refinery. Recently INCO (Canada) lost a human rights hearing instigated by a female employee who considered she was being unjustly denied a job.

8. CARCINOGENICITY

Sunderman et al. (1959) exposed rats to nickel carbonyl gas and observed neoplastic changes. The subject is important in the context of the carcino-genicity of nickel and its compounds. The epidemiological evidence reported by Morgan (1958) is that the cancer cases that occurred at Clydach were related to the dusty parts of the refining process and not to the plants where carbonyl was found. In order to assess whether nickel carbonyl could have caused pulmonary cancer in persons exposed occupationally to the gas, the causes of death of 69 men who lost time as a result of accidental exposure between 1933 and 1966 have been investigated (unreported study). The results listed in Table 20.3 indicate that the pulmonary cancer deaths in this small group did not significantly exceed those expected on age-specific rates.

Table 20.3   Mortality Experience of 69 Employees Exposed to Nickel Carbonyl Gas: 1933–1964[a]

Disease Group	Expected	Observed	SMR[b]
All causes	35.8	38	106
Pulmonary cancer	3.9	6	152

[a] Unreported study.
[b] Standardized Mortality not statistically significant ($p > 0.05$).

REFERENCES

Amor, A. J. (1932). The toxicology of the carbonyls. *J. Ind. Hyg.* **14**(6), 216–221.

Armit, H. W. (1907). The toxicology of nickel carbonyl, Part 1. *J. Hyg.* **7**, 525–551.

Armit, H. W. (1908). The toxicity of nickel carbonyl Part 2. *J. Hyg.* **8**, 565–600.

Barnes, J. W., and Denz, F. W. (1951). The effect of 2,3 dimercaptopropanol (BAC) on experimental nickel carbonyl poisoning. *Br. J. Ind. Med.* **8**, 117–126.

Bayer, O. (1939). Nickel carbonyl poisoning. *Arch. Gewerbepathol.* **9**, 592–608.

Brandes, W. W. (1934). Nickel carbonyl poisoning. *J. Amer. Med. Assoc.* **102**, 1204–1206.

Chiningelli, L. (1957). Utilisation of B.A.L. and thioctic acid in the therapy of nickel carbonyl poisoning. *Soc. Lombarde Sci. Med. E. Biol.* **12**, 1–5.

Clark, G. C., Jackson, G. C., Lewis, D., and Morgan, L. G. (1980). The effect of administrations of prednisone on nickel carbonyl induced lung lesions in the rat in nickel toxicology. In S. S. Brown and F. W. Sunderman, Jr. (Eds.), *Nickel Toxicology*. Academic, London, pp. 117–120.

Eneaya, D. I., Bianchine, J. R., Duran, D. O., and Andresen, B. D. (1981). The actions and metabolic fate of disulfiram. *Ann. Rev. Pharmacol. Toxicol.* **21**, 575–596.

Health and Safety Executive (HSE) (1987). Nickel and nickel compounds. Review paper for ACTS 23/87. HSE, United Kingdom, unpublished document.

Horak, F., Sunderman, F. W., Jr., and Sarkar, B. (1976). Comparison of antidotal efficacy of chelating drugs upon acute toxicity of Ni (II) in rats. *Res. Commun. Chem. Pathol. Pharmacol.* **14**, 153–165.

Jones, C. C. (1973). Nickel carbonyl poisoning. *Arch. Environ. Health* **26**, 245–248.

Jones, D. C. (1983). Chelation in the treatment of nickel toxicity. Ph.D. Thesis, University of Wales.

Kaaber, K., Menné, T., Tjell, J. C., and Veien, N. (1979). Antabuse treatment of nickel dermatitis. Chelation a new principle in the treatment of nickel dermatitis. *Contact Dermatitis* **5**, 221–228.

Ludewigs, V.H.J., and Thiess, A. M. (1970). Arbeitsmedizinische erkenntnisse bei der nickel carbonyl vergiftung. *Zentrablatt Asbeitsmed* **11**, 329–339.

McKendrick, J. G., and Snodgrass, W. (1891). On the physiological action of carbon monoxide of nickel. *J. Phil. Soc. Glasgow* **22**, 204–208.

Mond, L., Langer, C., and Quinke, F. (1890). Action of carbon monoxide on nickel. *J. Chem. Soc. (Lond.)* **57**, 749–753.

Morgan, J. G. (1958). Some observations on the incidence of respiratory cancer in nickel workers. *Br. J. Ind. Med.* **15**, 224–234.

Morgan, J. G. (1960). A simplified method of estimation of nickel in urine. *Br. J. Ind. Med.* **17**, 209–212.

Morgan, J. G. (1962). An enquiry into the medical hazards of nickel refining. M.D. Thesis, London University, pp. 39–40.

Nicholls, D., and Luscombe, J. (1974). Experimental studies with sodium diethyldithiocarbamate in rats. Unpublished report to INCO Europe Limited.

Oskarsson, A., and Tjalve, H. (1979). The distribution and metabolism of nickel carbonyl in mice. *Br. J. Ind. Med.* **36**, 326–335.

Revnova, N. V., (1978). Peculiarities of the blood system reaction in persons having sustained acute poisoning with nickel carbonyl against the background of a systematic action of low concentrations of this substance in conditions of modern industry. Gig Tr Prof Zabol **2**.20–22.

Shi, S., Lata, A. and Yuhua, H. (1986). A study of serum monoamine oxidase (MAO) activity and the E.E.G. in nickel carbonyl workers. *Br. J. Ind. Med.* **43**, 425–426.

Sorinson, G. N. (1957). Acute poisoning with nickel carbonyl. *Gigiena Sanitariya* **22**(11), 30–35 (in Italian).

Strong, J. S., Sunderman, F. W., and Kincaid, J. K. (1954). Nickel poisoning, 2 studies on patients suffering acute exposure to vapours of nickel carbonyl. *J. Am. Med. Assoc.* **155**, 889–894.

Sunderman, F. W. (1971). The treatment of acute nickel carbonyl poisoning by sodium diethyl-dithiocarbamate. *Ann. Clin. Res.* **3**, 182–185.

Sunderman, F. W., Donnelly, A. J., West, B., and Kincaid, J. F. (1959). Nickel poisoning. 9 Carcinogenicity in rats exposed to nickel carbonyl. *AMA Arch. Ind. Health* **20**, 36–41.

Sunderman, F. W., Jr., Roszel, N. O., and Clark, R. J. (1968). Gas chromatography of nickel-carbonyl in blood and breath. *Arch. Environ. Health* **16**, 836–843.

Sunderman, F. W., Jr., Allpass, P. R., Mitchell, J. M., and Baselt, R. C. (1979). Eye malfunction in rats: induction by pre-natal exposure to nickel carbonyl. *Science* **203**, 550–552.

Sunderman, F. W., Jr., Shen, S. K., Reid, M. R., and Allpass, P. R. (1980). Teratogenicity and embryotoxicity of nickel carbonyl in Syrian hamsters. *Terat. Carcin. Mutag.* **1**, 223–233.

Sunderman, F. W., Jr., Aitio, A., Morgan, L. G., and Norseth, T. (1986). Biological monitoring of nickel. *Toxicol. Ind. Health* **2**(1), 17–77.

Sutherland, S. (1958). Personal communication to J. G. Morgan of INCO Ltd Clydach Wales UK.

Vuopala, U., Huhti, E., Takkumen, J., and Huikko, M. (1970). Nickel carbonyl poisoning. Report of 25 cases. *Ann. Clin. Res.* **Z**, 214–222.

West, B., and Sunderman, F. W. (1958). Nickel poisoning 7. The therapeutic effectiveness of ethyldithiocarbamate in experimental animals exposed to nickel carbonyl. *Am. J. Med. Sci.* **236**, 15–25.

21

LONG-TERM EFFECTS OF EXPOSURE TO LOW CONCENTRATIONS OF NICKEL CARBONYL ON WORKERS' HEALTH

Zhi-Cheng Shi

Department of Occupational Medicine and Occupational Diseases Research Center, Third Hospital, Beijing Medical University, Beijing, China

Nickel and Human Health: Current Perspectives, Edited by Evert Nieboer and Jerome O. Nriagu.
ISBN 0-471-50076-3 © 1992 John Wiley & Sons, Inc.

1. INTRODUCTION

Since 1980, in collaboration with other institutions, we have made a periodic survey of workers' health in a factory where nickel carbonyl is produced for the preparation of nickel of high purity. Our objectives are to determine the harmful chronic effects of this gaseous nickel compound on workers who are exposed to relatively low concentrations. About 200 workers have been exposed, consisting of two-thirds males and one-third females. Their ages ranged from 19 to 48 years with years of employment between 2 and 20 years.

2. ENVIRONMENTAL SAMPLING

Air concentrations of nickel carbonyl in the workplace are summarized in Table 21.1 (by gas chromatographic technique). It is clear that ambient levels of $NiCO_4$ exceeded the maximum acceptable concentration (MAC) by 3–88 times (MAC in China is 0.001 mg/m³). Note the systematic decrease in the means and ranges over time.

3. MEDICAL EXAMINATION

3.1. Worker Complaints

By medical examinations, we found that most workers had some complaints that most frequently referred to the nervous and respiratory systems. The prevalence of complaints increased with the duration of exposure. The results of a worker questionnaire are presented in Table 21.2

Table 21.1   Air Concentrations of Ni(CO)₄ in Nickel Carbonyl Refinery

Year	Ambient Levels (Mean ± SD)	Range[a] (mg Ni(CO)₄/m³)	Time-weighted Average Concentration
1975	0.088 ± 0.046	0.05–0.818	0.026
1976	0.074 ± 0.015	0.03–0.468	0.022
1977	0.035 ± 0.026	0.01–0.240	0.016
1978	0.032 ± 0.047	0.008–0.320	0.017
1979	0.012 ± 0.038	0.008–0.300	0.018
1980	0.0075 ± 0.0014	0.008–0.450	0.014
1981	0.0070 ± 0.0062	0.004–0.320	0.012
1982	0.0068 ± 0.0078	0.004–0.282	0.0086
1983	0.0053 ± 0.0045	0.004–0.200	0.0058
1984	0.0032 ± 0.0027	0.001–0.097	0.0043

[a]The average air concentration of CO was 1.91 mg/m³ (0.13–5.20 mg/m³), and the average air dust of Ni was 8.0 mg/m³ (0.059–17.6 mg/m³).

Table 21.2   Summary of Complaints by Ni(CO)$_4$ Workers

Complaint	Controls[a] ($n - 68$)		Subjects ($n - 104$)		P Value[b]
	n	%	n	%	
Headache	22	32.4	58	55.8	<0.01
Dizziness	21	30.9	70	67.3	<0.001
Dreams	16	23.5	62	59.6	<0.001
Weakness	6	8.8	66	63.5	<0.001
Sleeplessness	18	26.5	45	43.2	<0.001
Poor memory	17	25.0	43	41.3	<0.001
Sexual frigidity	0	0	14	13.5	<0.001
Dryness in throat	13	19.1	33	31.7	<0.01
Sore throat	3	4.4	20	19.2	<0.001
Cough	4	5.9	23	22.1	<0.001
Difficulty breathing	3	4.4	36	34.6	<0.001
Chest tightness and pain	8	11.8	49	47.1	<0.001
Palpitation	3	4.4	16	15.4	<0.001
Loss of appetite	3	4.4	15	14.4	<0.001

[a] Machinists with no Ni(CO)$_4$ exposure in the same refinery.
[b] Based on t-test.

3.2.   Lung Function Tests

The lung functions of 112 nickel carbonyl workers were measured; 75 normal persons of similar ages who had no exposure served as controls. The following lung function tests were performed: forced vital capacity (FVC), forced expiratory volume exhaled in 1 sec (FEV$_1$), maximum midexpiratory flow (MMF), peak expiratory flow rate over height (PEFR/H), flow volume at 50 and 25% of FVC over height ($\dot{V}_{50}/H$, $\dot{V}_{25}/H$).

The results showed that $\dot{V}_{50}/H$ and MMF in males employed for more than 14 years and $\dot{V}_{50}/H$ in females employed for more than 10 years were lower than those in the controls ($P < 0.01$). The FVC, FEV$_1$, and $\dot{V}_{25}/H$ in these male workers and FEV$_1$, MMF, $\dot{V}_{25}/H$, and PEFR/H in the females were also lower than those in the controls ($P < 0.05$). Except for $\dot{V}_{50}/H$ and MMF, the other metrics in male workers employed for less than eight years were not different from the controls. The abnormality of lung function in smokers was higher than that in nonsmokers: $\dot{V}_{50}/H$, $\dot{V}_{25}/H$, MMF, and FEV$_1$ all were significantly different between smokers and nonsmokers ($P < 0.01$) (Shi and Chang, 1986).

3.3.   Chest X-ray Findings

The X-ray findings among 153 workers included the following observations: (1) fine irregular linear opacities in 26 cases (16.99%); (2) small miliary opacities on small patches in 12 cases (7.84%); (3) expanding and deformity of hilus

Table 21.3   Chest X-ray Findings

Finding	Number of Workers	Number of cases	%
Fine irregular linear opacities	153	26	16.99[a]
Small miliary opacities or small patches	153	12	7.84[a]
Expanding and deformity of hilus of lungs	153	9	5.88[a]
Emphysema	153	7	4.57
Tuberculosis	153	6	3.92

[a] In excess over controls.

of lung in 9 cases (5.88%); (4) emphysema findings in 7 cases (4.57%); and (5) tuberculosis in 6 cases (3.92%) (Table 21.3).

4.   BIOLOGICAL MONITORING

4.1.   Serum Monoamineoxidase and Electroencephalogram Measurements

The serum monoamineoxidase (SMAO) level was determined and an encephalogram (EEG) performed in 78 nickel carbonyl workers. Workers were divided by years of employment into two groups: group A, 42 cases employed between 10 and 20 years; group B, 36 cases employed for 2–8 years. In addition, 40 healthy persons who had had no exposure to nickel carbonyl were designated as group C (the controls). The mean values of SMAO found in groups A, B, and C were 3.23 ± 2.37, 4.66 ± 2.81 and 5.24 ± 4.07 µg/mL, respectively (Table 21.4). The difference between groups A and C was significant ($P < 0.05$). The result of the EEG examinations also showed an obvious difference between groups A (abnormality rate 42.9%) and C (abnormality rate 10.0%) (Table 21.5). The paired test showed there was a close correlation between SMAO levels and an EEG. Both SMAO level and EEG can be used as objective indices for assessing the chronic effect of nickel carbonyl upon workers (Shi et al., 1986).

4.2.   HbCO Measurements

The data obtained from subjects and controls are given in Table 21.6. The percentage of HbCO both in nonsmokers and smokers of NiCO₄ workers was higher than in controls.

Table 21.4   Comparison of SMAO Results (Mean ± SD)

Group	Working Duration	Cases	SMAO (µg/mL)	P Value
A	Long	42	3.23 ± 2.37 (0–12.8)	<0.05
B	Short	36	4.66 ± 2.81 (1.7–15.9)	>0.05
C	Controls	40	5.24 ± 4.07 (2–14.3)	—

Table 21.5 Comparison of EEG Results

Group	Working Duration	Number of Cases	Abnormal EEG Cases	%	P Value
A	Long	42	18	42.9	<0.01
B	Short	36	5	13.9	>0.05
C	Controls	40	4	10.0	—

4.3. Measurement of Urinary Nickel

The urinary nickel was determined by Zeeman atomic absorption spectrometry (EAAS), and the results indicated that the values in workers were elevated (Table 21.7). There was a significant difference between workers exposed to $NiCO_4$ and controls ($P < 0.05$) (Lin, 1985).

4.4. Measurement of Nickel and Other Microelements in Hair

The hair nickel and other microelements were measured by emission spectrometry (ICPES). The results are shown in Table 21.8. Hair nickel and zinc levels in male nickel carbonyl workers were higher than those in controls ($P < 0.05$). By contrast, the hair manganese content in male and female workers of the nickel carbonyl refinery was lower than that in controls ($P < 0.05$) (Tian and Zheng, 1981).

4.5. Cytogenetic Measurements

The incidence of cancer, especially respiratory cancer, among $NiCO_4$ workers has to our knowledge not increased until now, although the average duration of employment is still short (Lu, 1985). However, some chromosomal anomalies in peripheral lymphocytes have occurred (Table 21.9), and an increase of dyskaryosed cells in sputum (Table 21.10) has been observed (Wang and Sun, 1984).

Table 21.6 Concentration of Carboxyhemoglobin in Blood (%)

	Number of Cases	Mean ± SD	Range	P Value[a]
Subjects				
A. Nonsmokers	31	2.84 ± 0.68	0.80–8.60	<0.05
B. Smokers	29	4.96 ± 0.82	1.80–9.20	<0.05
Controls				
C. Nonsmokers	36	1.72 ± 0.62	0.60–2.50	—
D. Smokers	11	3.22 ± 0.78	1.36–5.30	—

[a] Based on t-test.

Table 21.7   Urinary Nickel Levels

Group	Number of Cases	Mean ± S.D. (µg/L)	Range	P Value[a]
Controls	105	1.06 ± 1.18	0.12–5.60	—
Ni(CO)₄ workers	129	2.45 ± 4.54	0.12–16.00	<0.05
Workers exposed to Ni dust	68	4.10 ± 5.72	2.40–60.00	<0.01

[a] Based on t-test.

Table 21.8   Mean Value of Hair Microelements (µg/g)

Group	Sex	Number of Cases	Ni	Mg	Cu	Fe	Zn	Mn
Controls	M	16	2.45	28.5	19.25	10.25	63.0	2.96
Subjects	M	24	3.90[a]	31.0	22.50	10.00	240.0[a]	1.60[a]
Controls	F	28	2.22	23.5	16.70	5.60	23.5	2.00
Subjects	F	30	1.90	26.5	15.50	7.50	21.4	1.30[a]

[a] $P < 0.05$, employing t-test.

Table 21.9   Chromosomal Anomalies in Peripheral Lymphocytes

	Controls	Subjects	P Value[a]
Number of cases	21	64	
Number of observed cells	2010	6270	
Teratogenized cells (%)	1.25 (0.96–2.07)	2.06 (1.13–2.71)	<0.01
Chromatic-type aberration (%)	1.00 (0.57–1.48)	1.34 (0.96–1.81)	>0.05
Chromosome-type aberration (%)	0.41 (0.14–0.61)	0.73 (0.20–0.87)	<0.05
Breakage and deletions (%)	0.47 (0.20–0.90)	0.20 (0.10–0.35)	<0.05
Sister chromatid exchange	5.60 ± 1.49	9.44 ± 1.77	<0.01
Micronuclei (%)	0.15 ± 0.32	1.02 ± 0.46	<0.001

[a] Based on t-test.

Table 21.10   Dyskaryotic Cells[a] in Sputum (%)

	Mild	P-value[b]	Severe	P Value[b]
Controls ($n = 47$)	10.0	—	0	—
(NiCO)₄ workers ($n = 67$)	26.1	< 0.01	3.4	< 0.01
Workers exposed to Ni dust ($n = 30$)	12.0	< 0.05	1.3	< 0.05

[a] Mild dyskaryosis means DNA content in the nucleus exceeds $6n$ (normal DNA content is $2n$); severe dyskaryosis means DNA content in nucleus exceeds $10n$.
[b] Based on t-test.

5. DISCUSSION

Nickel carbonyl [NiCO$_4$] is a liquid metal compound of high toxicity (boiling point 43°C). Exposure to high concentrations of nickel carbonyl vapor may result in acute poisoning, even death. The number of acute poisonings have been reduced significantly since the early 1970s. Recently, more attention is paid to the long-term effects of exposure to low concentrations of nickel carbonyl on workers' health. The current study showed that long-term exposure to relatively low concentrations of nickel carbonyl definitely affects workers' health. The most frequent complaints pertained to the nervous and respiratory systems; pulmonary function tests such as FEV$_1$, MMF, and $\dot{V}_{50}/H$ were reduced, and the decrease correlated with the duration of employment. Reduced SMAO activity and abnormal changes in EEG patterns are evident, especially among the older workers. Together, these findings show that some degree of functional, biochemical, and electrophysiological impairment of the nervous and respiratory systems develops in workers with long-term exposure to low levels of nickel carbonyl. Although the incidence of cancer, especially respiratory cancer, appears not to have increased, some chromosomal abnormalities in peripheral blood lymphocytes and an increase of dyskaryosed cells in sputum have been documented. Further, elevation of urinary and hair nickel contents have been observed. We consider it necessary to have a follow-up study over a rather prolonged period to clarify the cancer and other issues.

REFERENCES

Lin, B. (1985). Measurement of urinary nickel by EAAS. *Chinese J. Hyg. Occupat. Dis.* **3**, 101–103.

Lu, J. (1985). A report of medical surveillance of NiCO$_4$ workers' health. Unpublished data

Shi, Z. and Chang, Li. (1986). Study on lung function of nickel carbonyl workers. *Chinese J. Prev. Med.* **20**(6), 351–354.

Shi, A., Lata, A., and Har, Y. (1986). A study of serum monoaminoxidase (MAO) activity and EEG in nickel carbonyl workers. *Br. J. Ind. Med.* **43**, 425–426

Tian, Z. and Zheng, X. (1981). Study on hair microelements on nickel workers. *J. Ind. Hyg. Metall.* **7**, 280–283

Wang, C. and Sun, S. (1984). Study on carcinogenicity and teratogenecity of NiCO$_4$ among workers exposed to low air concentration of NiCO$_4$. Unpublished data.

22

USE OF SODIUM DIETHYLDITHIOCARBAMATE IN THE TREATMENT OF NICKEL CARBONYL POISONING

F. William Sunderman, Sr.

Institute for Clinical Science, Pennsylvania Hospital, Philadelphia, Pennsylvania 19107

Nickel and Human Health: Current Perspectives, Edited by Evert Nieboer and Jerome O. Nriagu.
ISBN 0-471-50076-3   © 1992 John Wiley & Sons, Inc.

1. INTRODUCTION

This chapter is primarily concerned with the use of sodium diethyldithio-carbamate (Dithiocarb) in the treatment of acute nickel carbonyl poisoning. Reference will be made to earlier studies pertaining to diagnosis in relation to treatment (Sunderman and Sunderman, 1958; West and Sunderman, 1958; Sunderman, 1959, 1964, 1971, 1979–1981; Sunderman et al., 1963, 1967, 1983, 1984). In addition, recent studies will be briefly considered on the administration of Dithiocarb to rats developing tumors following the muscular implantations of nickel subsulfide (Ni_3S_2).

In 1943, during World War II, it became apparent that exposure to nickel carbonyl presented a serious health hazard and a deterrent to research in atomic energy. As a consequence, studies were initiated at that time (and have been continued to the present) to develop methods for the early detection of nickel carbonyl poisoning in persons who might have been unknowingly exposed to it and to establish therapeutic measures for its treatment.

Obviously, the simplest method of determining the presence of a noxious substance in air is its detection by odor. However, nickel carbonyl, even in hazardous concentrations, has only a mild, nonpenetrating odor, often de-scribed as "sooty" or "musty." In our laboratory, an attempt was made to establish the limits in which nickel carbonyl could be detected by smell. Six laboratory workers were asked to indicate the presence of nickel carbonyl by smelling whiffs of samples of nickel carbonyl in air ranging in concentrations from 0 to 5 ppm. The results were totally erratic, and it was concluded that the presence of nickel carbonyl is likely to be undetected by those unfamiliar with it. For example, in an accident that occurred in an oil refinery 35 years ago in which more than 100 persons were exposed to nickel carbonyl, there was no suspicion of exposure until the workmen became acutely ill.

The first monitoring system (Kincaid et al., 1956), which was developed in our laboratories, for the detection of nickel carbonyl in air was a simple, manually operated rotameter and suction pump that permitted air to be drawn into an absorber in which nickel carbonyl vapor was converted to a nickel halide by its reaction with bromine or chlorine. The smoke formed was capable of scattering light, and the intensity of the scattered light could be related to the concentration of nickel carbonyl in the contaminated air. This device, called "the snifter," had a sensitivity of less than 1 ppm and was reasonably precise. The system was later modified and adapted for the con-tinuous, automatic monitoring of working areas. For the past 20 or more years, the procedure was replaced by sophisticated conductimetric and chemiluminescent instrumentation.

2. DIAGNOSIS

In our initial efforts to establish criteria for the early diagnosis of acute nickel carbonyl poisoning, it was realized that in addition to monitoring the working areas, it became essential to develop a reliable chemical method for the analysis of nickel that would not only detect trace amounts in biological fluids but also provide an estimate of the severity of exposure (Kincaid et al., 1956). Furthermore, it was desirable that such a method be rapid, economical of material, of easy manipulation, and adaptable for routine purposes. To meet these criteria, the method of Alexander et al. (1946) was modified and became well adapted for the routine analysis of nickel in urine. It should be noted that this method afforded the clue that led to the investigations and discovery of sodium diethyldithiocarbamate as a specific antidote for nickel carbonyl poisoning. This chemical is the chelating color reagent used in the procedure for measuring nickel in urine. In recent years, measurements by atomic absorption spectrometry have replaced the early colorimetric measurements.

In studies of nickel metabolism in dogs (Tedeschi and Sunderman, 1958), it will be seen in Figure 22.1 that after exposure to nickel carbonyl more

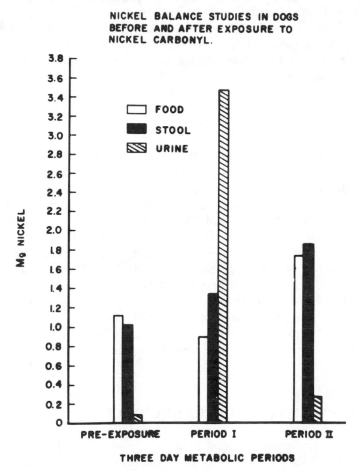

Figure 22.1. Nickel balance studies in dogs exposed to Ni(CO)$_4$.

Table 22.1   Nickel Concentration[a] in Urine (μg/100 mL)

Normal population	
Mean concentration (N = 107)	2.0 ± 1.1
One analysis in 50 (calculated)	> 5.3
18,815 Routine analyses	< 6.0

[a] *Editor's note*: The nickel concentrations reported are about 10-fold higher than current reference intervals. It needs to be mentioned that these nickel measurements were made during the period 1955–1965 employing a colorimetric procedure (Kincaid et al., 1956) well before the advent of modern methods and exhaustive quality assurance practices for nickel.

than twice as much nickel was excreted in the urine as in the feces. The observation that there is a sharp increase in nickel excretion in urine immediately after exposure to nickel carbonyl proved to be of major practical importance. It enabled the detection of exposure in workers to minimal amounts of nickel carbonyl frequently before the onset of symptoms. Furthermore, measurement of the nickel concentration in urine collected after exposure proved to be a valuable aid in classifying patients as a guide to Dithiocarb therapy.

3.   NICKEL IN URINE

The mean concentration of nickel in the urine of 107 normal subjects measured in our laboratory was found to be 2.0 μg/100 mL with a standard deviation of +1.1 over an 8-hr working period (Table 22.1). Statistical analysis of the data have led to the conclusion that only one specimen in 50 selected from a normal population will be found to exceed a value of 5.3 μg/100 mL of urine. This value was therefore selected as the upper limit of normal. In 10 years, 18,815 routine analyses were undertaken on urine specimens collected over 8-hr periods. All of these specimens had a nickel concentration below 6 μg/100 mL of urine. From a practical standpoint, measurements of nickel concentration in urine have proved to be more satisfactory than estimations of total nickel excretion because of the difficulty of obtaining from industrial workers reliable estimations of the volume of urine excreted within stipulated periods of time. Furthermore, the time-saving factor has proved important in critical cases.

4.   THE SEARCH FOR A CHELATING DRUG

When we first undertook to treat patients exposed to nickel carbonyl, the only available chelating drugs were BAL (dimercaprol), *d*-penicillamine, and

EDTA (calcium disodium ethylenediaminetetraacetic acid). Our studies on experimental animals (Kincaid et al., 1953; West and Sunderman, 1958) showed that administration of *d*-penicillamine had doubtful antidotal effectiveness and produced severe toxic side reactions; EDTA provided no antidotal effects (Sunderman, 1958); and BAL was only partially effective. With BAL, the median lethal dose (LD_{50}) in rats exposed to nickel carbonyl is increased by a factor of approximately 2.

In the early 1950s, attention was attracted to a number of metabolic studies on dithiocarbamates that were appearing in the literature at that time. These studies were of special interest since sodium diethyldithiocarbamate (Dithiocarb) is the chemical used as the nickel-binding reagent in routine method for measuring nickel in urine.

The metal-binding property of the dialkyldithiocarbamates was first reported by Delepine in 1908 (Anonymous, 1908; Delepine, 1908; Alexander et al., 1946; Domar et al., 1949; Vaciago and Fasana, 1958). It was not, however, until 25 years later that this property found application in analytical chemistry and led to the development of a method for the measurement of trace amounts of nickel (Alexander et al., 1946). The structures of sodium diethyldithiocarbamate and its nickel chelate were studied by Vaciago and Fasana (1958) and are portrayed in Figure 22.2. The nickel in the complex is described as having square-coplanar (dsp^2) hybrid orbitals.

Domar and associates (1949) found that Dithiocarb was involved in the metabolism of disulfiram (Antabuse). After the administration of Antabuse to man and experimental animals, Dithiocarb was found to be present in blood, tissues, urine, bile, and feces. The metabolic pathway of Antabuse, and presumably Dithiocarb, is shown in Figure 22.3. A portion undergoes oxidation to form free and ethereal sulfates as well as metal complexes.

The chemical and biological properties of Dithiocarb are given in Table 22.2 (Sunderman and Sunderman, 1958). Recognition of the nickel-binding and biological properties of the dithiocarbamates as well as their low toxicity prompted the initiation of studies to determine their possible chemotherapeutic properties as an antidote to acute nickel carbonyl poisoning.

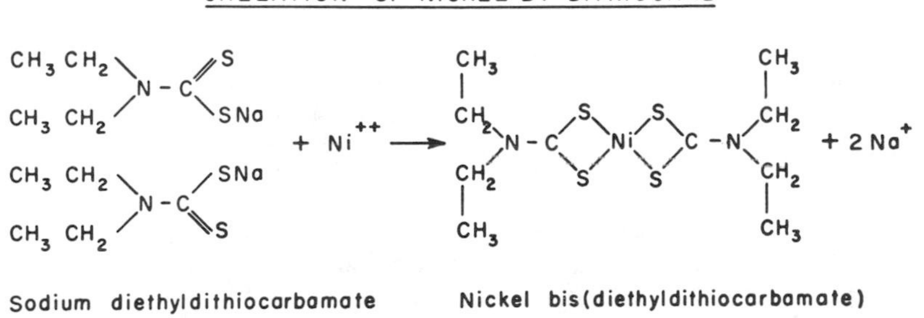

CHELATION OF NICKEL BY DITHIOCARB

Sodium diethyldithiocarbamate   Nickel bis(diethyldithiocarbamate)

Figure 22.2. Chelation of nickel by Dithiocarb.

METABOLISM OF DISULFIRAM AND SODIUM DIETHYLDITHIOCARBAMATE

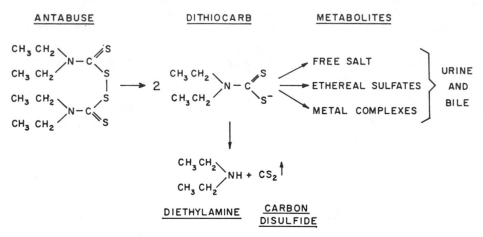

Figure 22.3. Metabolism of disulfiram and Dithiocarb.

The therapeutic effectiveness of 13 alkyl dithiocarbamates was studied in experimental animals receiving lethal inhalations of nickel carbonyl (West and Sunderman, 1958). Of the various derivatives tested, Dithiocarb proved to be the least toxic and one of the most effective. The LD_{50} value for the sodium salt administered to mice and rats was 1.5 g/kg of body weight (West and Sunderman, 1958).

Table 22.2 Properties of Sodium Diethyldithiocarbamate Trihydrate (Dithiocarb)

$$(C_2H_5)_2N\overset{\displaystyle S}{\overset{\|}{-C}}-S\ Na \cdot 3H_2O$$

	Molecular weight 225.3
Appearance	White crystaline solid
Melting point	90–92°C
Solubility	Soluble in water, methanol, ethanol, and acetone; insoluble in ether and benzene
Stability	Stable at room temperature; unstable in acid solutions
LD_{50}	Mice and rats (intraperitoneal and oral) 1.5 g/kg body weight

A 10% aqueous solution of Dithiocarb yields a pH value of 11.6 at room temperature. This solution may be buffered with monosodium phosphate to 7.4. At pH concentrations lower than 7.4, the mixture becomes turbid and decomposes, developing an odor of H_2S.

5. ANTIDOTAL ACTIVITY OF DITHIOCARB

The antidotal activity of Dithiocarb in mice and rats exposed to nickel carbonyl are given in Table 22.3. Of 30 mice exposed to nickel carbonyl vapors in a concentration of 6 ppm for 30 min, only 6 survived a period of five days following exposure. In concentrations of 8 ppm and above, practically all of the exposed mice died. It will be seen in the table that of 30 mice exposed to nickel carbonyl in a concentration amounting to several times the LD_{100} dose and receiving Dithiocarb parenterally in dosages of 50 and 100 mg/kg of body weight immediately after exposure, all of the animals survived. It will also be seen that of 390 mice exposed to nickel carbonyl at 10 ppm, all but 2 died within five days. On the other hand, of mice exposed to this same concentration of nickel carbonyl and given Dithiocarb parenterally, all survived for five days and were in good health.

It will also be seen in Table 22.3 that of 30 rats exposed to nickel carbonyl in a concentration of 67 ppm for 30 min, only 11 survived for five days. In concentrations of 168 ppm and above, none survived. However, rats exposed to lethal concentrations of nickel carbonyl and given Dithiocarb parenterally in doses of 50 and 100 mg/kg of body weight, all survived.

The dramatic effectiveness of Dithiocarb in counteracting the lethal effects of nickel carbonyl in experimental animals led us to employ this chemical in humans who were accidentally exposed to nickel carbonyl.

Table 22.3   Antidotal Activity of Dithiocarb

$Ni(CO)_4$[a] Concentration (ppm)	Number Surviving 5 Days		
		Treated with Dithiocarb Intraperitoneal	
	Untreated	50 mg/kg	100 mg/kg
Mice			
6	6/30	30/30	30/30
8	0/30	30/30	30/30
10	2/390	390/390	30/30
16	0/30	30/30	30/30
24	0/30	30/30	30/30
Rats			
67	11/30	30/30	30/30
105	6/30	30/30	30/30
168	0/30	30/30	30/30
266	0/30	30/30	30/30

[a] Nickel carbonyl was administered by inhalation for 30 min.

In 1957, our first patient, who was severely exposed to nickel carbonyl, was treated with Dithiocarb (Sunderman and Sunderman, 1958). After I served as the first control subject by taking a test dose of Dithiocarb without ill effects, Dithiocarb was administered to a workman who had been accidentally sprayed with nickel carbonyl. This man (patient D) had to be resuscitated by oxygen inhalation. Patient D received 1 g of Dithiocarb twice daily for 10 days after exposure. He became asymptomatic after the second day of hospitalization and developed no delayed reactions. The nickel concentrations in Patient D's urine during the 16 days after exposure are shown in Figure 22.4. It will be noted that the initial concentration of nickel in the patient's urine was 200 µg/100 mL. The highest concentration we have observed in an exposed person was 247 µg/100 mL.

In Figure 22.5 are plotted the concentrations of nickel in urine from 13 patients with nickel carbonyl poisoning treated with Dithiocarb. The total amounts of Dithiocarb administered to each of these men ranged from 6 to 28 g during periods from 2 to 14 days. The administration of Dithiocarb was attended by nickel uresis. Dithiocarb therapy was maintained until the concentrations of nickel in the urine reached normal levels. No adverse effects have been observed from the administration of Dithiocarb to these subjects or to normal volunteers (Sunderman, 1964).

During the past 30 years, more than 375 persons exposed to the inhalation of nickel carbonyl vapor have been treated under our supervision with Di-

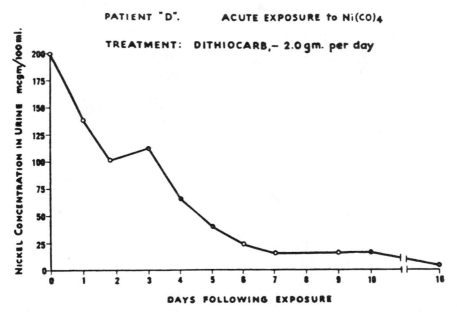

Figure 22.4. Nickel concentrations in 8-hr urine samples of a patient after acute accidental exposure to nickel carbonyl. Dithiocarb therapy (2 g oral dose) was initiated on the day of exposure.

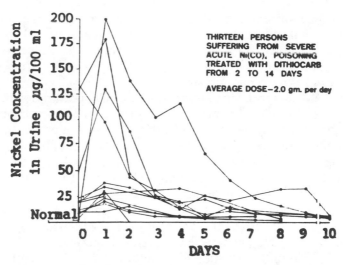

Figure 22.5. Thirteen persons suffering from severe Ni(CO)₄ poisoning treated with Dithiocarb from 2 to 14 days. Average dose: 2.0 g/day.

thiocarb. To our knowledge, no death from acute nickel carbonyl poisoning occurred in any person who received adequate Dithiocarb medication within four days after exposure. It should be noted that patients receiving Dithiocarb who ingest alcoholic beverages may experience symptoms similar to those described for Antabuse (Sunderman, 1964).

Certain manufacturing plants using nickel carbonyl in their operations measure the concentration of nickel in urine of workmen at the end of each working period. In those instances in which increased concentrations of nickel above 10 μg/100 mL of urine are reported, the workmen are given Dithiocarb as a preventive measure even though they may have developed no symptoms of exposure or may not have been aware that they had been exposed to nickel carbonyl. Such preventive measures have proved to be exceptionally effective in reducing the hazards of exposure.

6. GUIDE TO TREATMENT OF ACUTE NICKEL CARBONYL POISONING

6.1. Therapeutic Regimes

The following procedure has been found to be effective for subjects known or suspected of having been exposed acutely to hazardous concentrations of nickel carbonyl (Sunderman, 1970, 1971). If there is any doubt regarding the extent or severity of exposure of a worker to nickel carbonyl, an initial course of 1 g of Dithiocarb is given in divided doses. (Formerly 2 g was given.) When given in one dose, nausea occasionally develops. This may be lessened by administering the Dithiocarb in divided doses as follows: 0.2 g of Dithiocarb with water every 2 min for five doses along with 0.2 g of sodium bicarbonate. If the symptoms of nickel carbonyl poisoning are minimal, decision regarding further therapy may be deferred until the results of the urine analysis for nickel are obtained.

If the initial 8-hr specimen of urine has a nickel concentration of less than 10 μg/100 mL, the exposure may be classified as *mild*. In such cases, it is probable that delayed symptoms will either not develop or will be minimal. Most patients in this group are able to continue work, although a few may complain of fatigue and require rest. If severe delayed symptoms develop unexpectedly, such patients are hospitalized and given Dithiocarb in a dosage schedule outlined for the moderately severe group.

If the concentration of nickel in the first 8-hr collection of urine is above 10 μg but less than 50 μg/100 mL, the exposure may be classified as *moderately severe*. Since delayed symptoms may develop in these patients, they should remain under careful observation for at least a week. Dithiocarb should be administered orally to these patients so that the total daily dosage on the first day of exposure amounts to between 15 and 20 mg/lb of body weight (35–45 mg/kg). For a man weighing 160 lb (72.7 kg), the daily dosage is therefore between 2.5 and 3.3 g. The suggested dosage schedule is as follows:

1.0 g (five 0.2-g capsules), 0 hr

0.8 g (four 0.2-g capsules), 4 hr

0.6 g (three 0.2-g capsules), 8 hr

0.4 g (two 0.2-g capsules), 16 hr

On subsequent days, Dithiocarb therapy should be continued in a dosage of 0.4 g every 8 hr until the patients are free of symptoms and the concentration of nickel in urine has decreased to the normal range.

If the concentration of nickel in the first 8-hr collection of urine is above 50 μg/100 mL, the exposure may be classified as *severe*. These patients are apt to be seriously ill and require hospitalization. Most of these patients can be maintained with oral Dithiocarb therapy as outlined for the moderately severe group. However, if the patient's condition is critical, it is suggested that Dithiocarb be administered *parenterally** in an initial dosage of 12.5 mg/lb of body weight. Additional doses should be given in accordance with the clinical evaluation.

6.2. Comments on Delayed Symptoms

Initial symptoms including frontal headache, nausea, cough, dyspnea, constriction in chest, and so on, frequently pass off rapidly after the patients have received Dithiocarb. In some cases, the patients remain symptom free

* The solution of Dithiocarb for parenteral injection may be prepared by adding 10 mL of a sterile solution of phosphate buffer (0.5 g NaH₂PO₄/100 mL) to 1 g of powdered Dithiocarb contained in a sterile ampule.

for a period of a few hours to a week or more. During this quiescent period, it is essential that the subjects be observed carefully for the appearance of delayed reactions. In cases of severe exposure, the initial symptoms may merge gradually into the more severe delayed type. The delayed symptoms include a return of dyspnea, cough, and sense of constriction over the sternum and epigastrium as well as nausea, vomiting, cyanosis, sleeplessness, and delirium. Seriously ill patients usually have little or no fever. Oral temperatures rarely exceed 101°F (38.3°C). Fatalities usually occur between the days 4 and 11 after exposure.

6.3. Effects of Protracted Administration of Dithiocarb

Studies were undertaken to evaluate the toxicity and metabolic effects that may result from the daily administration of Dithiocarb to albino rats and beagle dogs in dosages of 30, 100, and 300 mg/kg of body weight for a period of 90 days (Sunderman et al., 1967). Throughout this period of observation, the animals receiving Dithiocarb were comparable in appearance, behavior, and appetite to the control group. No significant difference in the mean body weights of the test group were observed during the course of this study.

7. ANTITUMORIGENIC EFFECTIVENESS OF DITHIOCARB

Studies were undertaken to ascertain the possible antitumorigenic effectiveness of Dithiocarb on the development of tumors in rats following the muscular implantations of nickel subsulfide (Ni_3S_2) (Sunderman et al., 1984). These tumors have histologic characteristics that suggest origin from striated muscle and have been classified as rhabdomyosarcoma. Most of the tumors metastasize.

In two separate studies over a period of two years, a total of 100 four-month-old Fischer rats received muscular implantations of Ni_3S_2. Of this total, 50 rats (25 males and 25 females) were treated with Dithiocarb for a period of four to six weeks beginning one week after implantation. Seventy-eight percent of the untreated rats developed sarcomas as compared to 32% of the treated rats (Table 22.4).

Table 22.4 Incidence of Tumors

	Males	Females	Combination
Untreated rats, %	84	72	78
Treated rats, %	52	12	32
Effect of treatment	$P = 0.03$	$P = 0.000003$	
Difference between sexes			
Untreated: $P = 0.5$ (n.s.)			
Treated: $P = 0.005$			

Twenty-five rats per group were used.

Analyses of the data revealed a striking difference in the sex responsiveness to Dithiocarb. Of the 25 female rats with Ni_3S_2 implantations and treated with Dithiocarb, only 12% developed sarcomas; of the male rats similarly treated, 52% developed sarcomas. The difference in sex response is statistically significant ($p = 0.005$).

The mechanism by which Dithiocarb causes an increased inhibition of tumorigenic activity in female rats as compared to males is speculative.

It is recognized that patients with implanted prostheses may on occasion develop malignant tumors. It seems probable that their development is related to the presence of nickel. It is suggested that in such circumstances, chelation therapy with Dithiocarb might be considered.

8. SUMMARY

A brief review and summary has been presented of our studies that led to the therapeutic use of Dithiocarb in the treatment of nickel carbonyl poisoning.

REFERENCES

Anonymous (1908). Properties des thiosulfocarbamates metalliques. *Comptes Rend. Acad. Sci.* **146,** 981.

Alexander, O. R., Godar, E. M., and Linde, N. J. (1946). Spectrophotometric determination of traces of nickel. *Ind. Eng. Chem. Anal. Ed.* **18,** 206–208.

Delepine, M. M. (1908). Composes sulfures et azotes derives du sulfure de carbone (XII). Thiosulfocarbamates metalliques. *Bull. Soc. Chim. Paris* **3,** 643.

Domar, G., Fredga, A., and Linderholm, H. (1949). The determination of tetraethylthiuramdisulfide (Antabuse, Abstinyl) and its reduced form, diethyldithiocarbamic acid, as found in excreta. *Acta Chem. Scand.* **3,** 1441.

Hald, J., Jacobsen, E., and Larsen, V. (1948). The sensitizing effect of tetraethylthiuramdisulphide (Antabuse) to ethylalcohol. *Acta Pharmacol.* **4,** 285–296.

Hald, J., Jacobsen, E., and Larsen, V. (1952). The Antabuse effect of some compound. Related to Antabuse and cyanamide. *Acta Pharmacol. Toxicol.* **8,** 329–337.

Kincaid, J. F., Strong, J. S., and Sunderman, F. W. (1953). Nickel poisoning. I. Experimental study of the effects of acute and subacute exposure to nickel carbonyl. *AMA Arch. Ind. Hyg. Occup. Med.* **8,** 48–60.

Kincaid, J. F., Stanley, E. L., Beckworth, C. H., and Sunderman, F. W. (1956). Nickel poisoning. III. Procedures for detection, prevention, and treatment of nickel carbonyl exposure including a method for the determination of nickel in biologic materials. *Am. J. Clin. Pathol.* **26,** 107–119.

Sunderman, F. W. (1958). Nickel poisoning. VI. A note concerning the ineffectiveness of edathamil calcium disodium (calcium disodium ethylene-diamine-tetraacetic acid). *AMA Arch. Ind. Health* **18,** 480–482.

Sunderman, F. W. (1959). Dithiocarbamates for treatment of nickel poisoning. U.S. Patent 2,876,159.

Sunderman, F. W. (1964). Nickel and copper mobilization by sodium diethyldithiocarbamate. *J. New Drugs* **4,** 154–161.

Sunderman, F. W. (1970). Nickel poisoning. In F. W. Sunderman (Ed.), *The Laboratory Diagnosis of Diseases Caused by Toxic Agents*. Warren H. Green, St. Louis, pp. 387–396.

Sunderman, F. W. (1971). The treatment of acute nickel carbonyl poisoning with sodium diethyldithiocarbamate. *Ann. Clin. Res.* **3**, 182–185.

Sunderman, F. W. (1979). Efficacy of sodium diethyldithiocarbamate (Dithiocarb) in acute nickel carbonyl poisoning. *Ann. Clin. Lab. Sci.* **9**, 1–10.

Sunderman, F. W. (1980). Chelation therapy in nickel poisoning. *Ni. 3. Spurenelement-Symposium 1980*. Karl Marx-Universitat, Leipzig and Friedrich Schiller-Universitat, Jena, Germany, pp. 359–368.

Sunderman, F. W. (1981). Chelation therapy in nickel poisoning. *Ann. Clin. Lab. Sci.* **11**, 1–8.

Sunderman, F. W., and Sunderman, F. W., Jr. (1958). Nickel poisoning. VIII. Dithiocarb: a new therapeutic agent for persons exposed to nickel carbonyl. *Am. J. Med. Sci.* **236**, 26–31.

Sunderman, F. W., Jr., White, J. C., Sunderman, F. W., and Lucyszyn, G. W. (1963). Metabolic balance studies in hepatolenticular degeneration treated with diethyldithiocarbamate. *Am. J. Med.* **34**, 875–888.

Sunderman, F. W., Paynter, O. E., and George, R. B. (1967). The effects of the protracted administration of the chelating agent sodium diethyldithiocarbamate (Dithiocarb). *Am. J. Med. Sci.* **254**, 46–56.

Sunderman, F. W., Schneider, H. P., and Lumb, G. D. (1983). Sodium diethyldithiocarbamate administration in nickel-induced rhabdomyosarcoma. In S. S. Brown and J. Savory (Eds.), *Chemical Toxicology and Clinical Chemistry of Metals*. Academic, London, pp. 399–400.

Sunderman, F. W., Schneider, H. P., and Lumb, G. D. (1984). Sodium diethyldithiocarbamate administration in nickel-induced malignant tumors. *Ann. Clin. Lab. Sci.* **14**, 1–9.

Tedeschi, R. E., and Sunderman, F. W. (1958). Nickel poisoning. V. The metabolism of nickel under normal conditions and after exposure to nickel carbonyl. *AMA Arch. Ind. Health* **18**, 480–482.

Vaciago, A., and Fasana, A. (1958). Crystal structure and polymorphism of nickel bis(diethyldithiocarbamate). *Atti Acad. Nazl. Lincei Rend. (Classe Sci. Fis. Mat. Nat.)* **25**, 528.

West, B., and Sunderman, F. W. (1958). Nickel poisoning. VII. The therapeutic effectiveness of alkyldithiocarbamates in experimental animals exposed to nickel carbonyl. *Am. J. Med. Sci.* **236**, 15–25.

23

INVOLVEMENT OF HETEROCHROMATIN DAMAGE IN NICKEL-INDUCED TRANSFORMATION AND RESISTANCE

Max Costa, Kathleen Conway, Richard Imbra, and Xin Wei Wang

Institute of Environmental Medicine,
New York University Medical Center,
New York, New York 10016

Nickel and Human Health: Current Perspectives, Edited by Evert Nieboer and
Jerome O. Nriagu.
ISBN 0-471-50076-3   © 1992 John Wiley & Sons, Inc.

1. OVERVIEW OF GENOTOXIC EFFECTS OF NICKEL COMPOUNDS

In intact mammalian cells carcinogenic nickel compounds produce single-strand breaks and DNA–protein crosslinks (Robison et al., 1982; Patierno and Costa, 1985; Patierno et al., 1985). The single-strand breaks are easily repaired while the DNA–protein crosslinks are relatively persistent lesions. The mechanism by which nickel ions induce these DNA lesions is not known. Nickel ions are thought to bind weakly to DNA. There is some evidence that at high concentrations carcinogenic nickel compounds induce the formation of oxygen radicals (for review see Coogan et al., 1988; however, because of the high concentrations of nickel required for this effect, it is unlikely to be biologically relevant. The ultimate carcinogen of most nickel compounds is thought to be hexacoordinate [$Ni(H_2O)_6^{2+}$] since this product is produced following dissolution of a number of nickel compounds (for review see Coogan et al., 1988). However, there is also evidence that Ni^{3+} can be formed under certain conditions (Thomson, 1982), and the formation of this species may have an involvement in the nickel carcinogenesis process, particularly in the oxidation of DNA caused by Ni^{2+}/Ni^{3+} redox cycling.

Specific particulate nickel compounds display striking differences in their carcinogenic potency in experimental animals (Sunderman, 1984). The reasons for these differences appear to be due to the bioavailability of nickel ions to the DNA. Certain particulate nickel compounds that are potent carcinogens are actively phagocytized by cancer target cells whereas others that lack carcinogenic activity are not phagocytized (Costa and Mollenhauer, 1980). The phagocytized particulate nickel compounds result in a high and sustained delivery of nickel ions to the DNA.

Despite the potent carcinogenic activity displayed by nickel compounds, Ni^{2+} is not mutagenic in most gene mutation assays (Biggart and Costa, 1986; Coogan et al., 1988). However, many of these assays measure a limited endpoint, and therefore all of the genotoxic effects of nickel compounds cannot be assessed. Recent evidence has shown that carcinogenic nickel compounds are potently mutagenic provided the appropriate assay is utilized (Chapter 24). Although it is too early to state with certainty, it is likely that this assay is capable of measuring deletions that may be a major mutagenic response of nickel. Nickel compounds are potent inducers of sister chromatid exchanges and chromosomal aberrations (Sen and Costa, 1985, 1986; Sen et al., 1987).

2. SELECTIVE DAMAGE PRODUCED BY NICKEL IN HETEROCHROMATIN

Numerous studies indicate that carcinogenic nickel compounds induce non-random damage in the genome (Patierno et al., 1985; Sen and Costa, 1985,

1986; Sen et al., 1987). At the chromosomal level, a large fraction of the damage induced by nickel compounds occurs in heterochromatic regions (Sen and Costa, 1985, 1986; Sen et al., 1987). Studies with alkaline elution have shown that a major lesion induced by nickel compounds is the DNA–protein crosslink (Patierno et al., 1985). However, alkaline elution curves of DNA–protein crosslinks induced by nickel compounds demonstrate a biphasic elution pattern indicating that nickel induces DNA–protein complexes nonrandomly. Chromatin fractionation studies have also shown that nickel induces DNA–protein complexes selectively in a $MgCl_2$ insoluble fraction consisting of heterochromatin (Patierno et al., 1987). The most striking example of the selectivity displayed by nickel compounds toward heterochromatin is found in chromosome studies (Sen and Costa, 1985, 1986; Sen et al., 1987). Nickel induces a high incidence of chromosome damage in heterochromatic regions of mouse and Chinese hamster genomes. A large fraction of the DNA in the heterochromatic regions of mouse is made up of highly repetitive satellite DNA sequences whereas there is less repetitive DNA in Chinese hamster heterochromatin. The reason nickel ions interact with heterochromatin selectively is not known. In the Chinese hamster there is a highly selective damage in the heterochromatic region of the long arm of the X chromosome that is higher than the damage seen in other heterochromatic regions of this species. The damage produced by nickel in heterochromatin is reversed by increasing the extracellular levels of magnesium ions (Conway et al., 1987). Interestingly, the euchromatic damage induced by nickel is not diminished by increasing the Mg^{2+} concentration. However, elevation of the Mg^{2+} tends to inhibit nickel-induced cell transformation. These studies were the first indication that nickel-induced damage in heterochromatic DNA may be important in nickel carcinogenesis.

3. NICKEL RESISTANCE INVOLVES CHANGES IN HETEROCHROMATIN

Recently, our laboratory has been successful in producing nickel-resistant mouse cells (Wang et al., 1988). The level of nickel resistance is dependent upon the initial $NiCl_2$ exposure concentration (Wang et al., 1988). Interestingly, $NiCl_2$-resistant cells exhibit a high incidence of centromeric fusions. The level of centromeric fusion is directly correlated to the degree of $NiCl_2$ resistance, as shown in Table 23.1. Figure 23.1 shows a C-band staining pattern of normal and $NiCl_2$-resistant Balb-3T3 cells. There is a high degree of centromeric fusions present in the $NiCl_2$-resistant cells. In other studies, we have shown that the steady-state levels of $^{63}NiCl_2$ in wild-type or resistant cells is quite similar at short as well as long time intervals of exposure to $^{63}NiCl_2$ (Wang et al., 1988). Thus, the resistance does not seem to be due to the alterations in the levels of this metal ion. Prolonged incubation of cells in the absence of $NiCl_2$ results in the loss of nickel resistance. The loss of

Table 23.1   Chromosome Analysis of Wild-Type Balb/c-3T3 Cells and Nickel-resistant Sublines[a]

Cell Lines	Number of Centric Fusion per Metaphase	Total Chromosomes per Metaphase
Balb/c-3T3	1.1 ± 1.1	69.7 ± 5.2
B50[b]	2.3 ± 1.5	63.3 ± 3.5
B100[b]	7.2 ± 3.2	55.0 ± 4.7
B200[b]	9.6 ± 2.8	54.6 ± 4.8

[a]Chromosome numbers are based on the counts of at least 30 metaphase cells per line.
[b]Nickel-resistant cells were adapted for at least 2 months to medium containing 50, 100, or 200 μM NiCl$_2$ and expressed as B50, B100, B200, respectively.

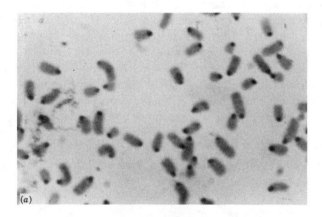

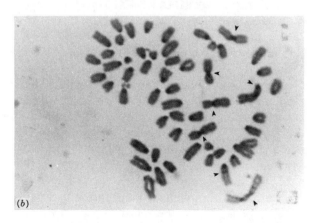

Figure 23.1. The C banding of (a) normal or (b) NiCl$_2$-resistant Balb-3T3 cells. Arrow shows centromeric fusions.

NiCl$_2$ resistance correlates directly with a similar loss of centromeric fusion. Thus, the degree of heterochromatic centromeric fusions appears to correlate directly with the degree of NiCl$_2$ resistance. Recent evidence has demonstrated that nickel resistance can be transfected, suggesting that nickel resistance is inherited and the genes involved in this phenomenon can be isolated. We are currently attempting to isolate genes coding for proteins involved in nickel resistance.

4.   NICKEL-INDUCED CELL TRANSFORMATION INVOLVES CHANGES IN HETEROCHROMATIN

Since nickel selectively damages heterochromatin and alterations involving heterochromatin may be genetically significant in nickel resistance (Wang et al., 1988), we have begun to investigate if alterations in heterochromatin were involved in the nickel-induced cell transformation process. For these studies, we used Chinese hamster embryo (CHE) cells since the entire long arm of the X chromosome is heterochromatic. This fraction represents a substantial portion of the heterochromatin in the genome of this species. If nickel-induced heterochromatic damage is involved in carcinogenesis, then there may be a difference in the incidence of transformation observed in male and female cells. Table 23.2 summarizes the data obtained thus far showing that only the male cells transform to anchorage-independent growth. Previous studies have shown that both male and female rats and mice exhibit an equal incidence of cancer induction by nickel compounds (Sunderman, 1983). Thus, the sex difference in the incidence of transformation was not evident in other species. Table 23.2 also shows that in four out of five of the male nickel-transformed cells there was a deletion of the heterochromatic long arm of the X chromosome. This was the major chromosome aberration seen in these four cell lines. Figure 23.2 shows an example of a normal G-banded karyotype as well as one from a nickel-transformed cell line. Note the deletion of the long arm of the X chromosome in the nickel-transformed cell line. This is the primary chromosome aberration observed in this cell line.

We have recently obtained data showing that nickel-treated normal cells do not have a high incidence of deletions of heterochromatin. This was an important control because the deletions may have been associated with nickel treatment and not necessarily with the transformation process. Additionally, all of the male nickel-transformed lines having deletions in the long arm of the X chromosome were also able to form tumors in athymic nude mice. Thus, the deletion of the long arm of the X chromosome is associated with the neoplastic transformation of cells.

Figure 23.3 summarizes potential steps that may be involved in the nickel-induced transformation process. It should be noted that with a metal ion, such as nickel, there are probably many mechanisms involved in its carcino-

Table 23.2 Chinese Hamster Embryo Cell Transformation

Number of Cultures Tested	Transforming Agent	Sex	Number of Cultures Exhibiting Initial Anchorage-independent Growth	Number of Cultures That Continued to Proliferate in Culture and Exhibit Anchorage Independence	Number of Cultures Having Deletions of Heterochromatic Long Arm of X Chromosome
13	NiS or $NiCl_2$	Male	8	5/8	4/5
10	NiS or $NiCl_2$	Female	2	0/2	0
2	MCA	Male	2	2	0/1
2	MCA	Female	2	2	0/1
1	Chromate	Female	1	1	0

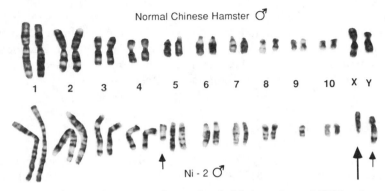

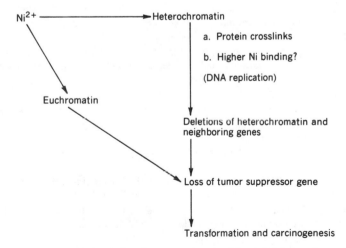

Figure 23.2. The G banding karyotype of normal and nickel-transformed CHE embryo cells.

genesis, and the model presented is only an example of one such pathway. Nickel, by virtue of its ability to form DNA–protein complexes or by its association with protein in chromatin, may cause deletions to occur during DNA replication. These deletions may result from an inhibitory effect of Ni^{2+} on the DNA polymerase or interference with template replication through formation of DNA–protein crosslinks. These effects occur to a greater extent in heterochromatin compared with euchromatin leading to deletions predominantly in heterochromatic regions. In the Chinese hamster cells, the long arm of the X chromosome is entirely heterochromatic, and we expect that the deletions of this region explain the high incidence of neoplastic transformation in the males. Our hypothesis is that the deletion of the heterochromatic long arm of the X chromosome results in the loss of a tumor

Figure 23.3. Illustration of the molecular events involved in nickel-induced transformation.

suppressor gene. We hypothesize that the tumor suppressor gene is located either in the heterochromatic region that is deleted or in neighboring euchromatin that is affected by the deletion of heterochromatin. Recent studies have demonstrated that nickel causes a deletion or inactivation of a senescence gene on the X chromosome (Klein et al., 1991).

ACKNOWLEDGMENTS

This work is supported by Grant No. CA43070 from the National Cancer Institute and Grant No. R813140-010 from the U.S. Environmental Protection Agency.

REFERENCES

Biggart, N. W., and Costa, M. (1986). Assessment of the uptake and mutagenicity of nickel chloride in Salmonella tester strains. *Mutation Res.* **175**, 209–215.

Conway, K., Wang, X.-W., Xu, L.-S, and Costa, M. (1987). Effect of magnesium on nickel-induced genotoxicity and cell transformation. *Carcinogenesis* **8**, 1115–1121.

Coogan, T. P., Latta, D. M., Snow, E. T., and Costa, M. (1988). Toxicity and carcinogenicity of nickel compounds. In R. O. McClellan (Ed.), *CRC Critical Reviews in Toxicology*, Vol. 19. CRC Press, Boca Raton, FL, pp. 341–394.

Costa, M., and Mollenhauer, H. H. (1980). Carcinogenic activity of particulate nickel compounds is proportional to their cellular uptake. *Science* **209**, 515–517.

Klein, C. B., Conway, K., Wang, X. W., Bhamra, R. K., Lin, X., Cohen, M. D., Annab, L., Barret, J. C., and Costa, M. (1991). Senescence of nickel-transformed cells by an X chromosome: possible epigenetic control. Science **251**, 796–799.

Patierno, S. R., and Costa, M. (1985). DNA–protein cross-links induced by nickel compounds in intact cultured mammalian cells. *Chem. Biol. Interact.* **55**, 75–91.

Patierno, S. R., Sugiyama, M., Basilion, J. P., and Costa, M. (1985). Preferential DNA-protein cross-linking by NiCl$_2$ in magnesium-insoluble regions of fractionated Chinese hamster ovary cell chromatin. *Cancer Res.* **45**, 5787–5794.

Patierno, S. R., Sugiyama, M., and Costa, M. (1987). Effect of nickel(II) on DNA-protein binding, thymidine incorporation, and sedimentation pattern of chromatin fractions from intact mammalian cells. *J. Biochem. Toxicol.* **2**, 13–23.

Robison, S. H., Cantoni, O., and Costa, M. (1982). Strand breakage and decreased molecular weight of DNA induced by specific metal compounds. *Carcinogenesis* **3**, 657–662.

Sen, P., and Costa, M. (1985). Induction of chromosomal damage in Chinese hamster ovary cells by soluble and particulate nickel compounds: preferential fragmentation of the heterochromatic long arm of the X-chromosome by carcinogenic crystalline NiS particles. *Cancer Res.* **45**, 2320–2325.

Sen, P., and Costa, M. (1986). Incidence and localization of sister chromatid exchanges induced by nickel and chromium compounds. *Carcinogenesis* **7**, 1527–1533.

Sen, P., Conway, K., and Costa, M. (1987). Comparison of the localization of chromosome damage induced by calcium chromate and nickel compounds. *Cancer Res.* **47**, 2142–2147.

Sunderman, F. W., Jr. (1983). In R. Langenbach, S. Nesnow, and J. M. Rice (Eds.), *Organ and Species Specificity in Chemical Carcinogenesis*. Plenum, New York, p. 107.

Sunderman, F. W., Jr. (1984). Carcinogenicity of nickel compounds in the animals. In F. W. Sunderman, Jr. (Ed.), *Nickel in the Human Environment*. IARC Scientific, Lyon, France, p. 127.

Thomson, A. J. (1982). Proteins containing nickel. *Nature* **298,** 602–603.

Wang, X. W., Imbra, R. J., and Costa, M. (1988). Characterization of mouse cell lines resistant to nickel(II) ions. *Cancer Res.*, **48,** 6850–6854.

24

ROLE OF Ni(II) IN MUTATION

Nelwyn T. Christie, Donna M. Tummolo,
Catherine B. Klein, and Toby G. Rossman

Institute of Environmental Medicine, New York University Medical
Center, New York, New York 10016

1. INTRODUCTION

While there is clear documentation for nickel-induced carcinogenicity as
well as other types of nickel genotoxicity, studies of the mutagenicity of
nickel have demonstrated only weak or null responses in both bacterial and
mammalian assays. Several reviews of metal mutagenesis have noted the

Nickel and Human Health: Current Perspectives, Edited by Evert Nieboer and
Jerome O. Nriagu.
ISBN 0-471-50076-3   © 1992 John Wiley & Sons, Inc.

anomalous mutagenic behavior of this metal ion although it has been clearly established as a carcinogen (Flessel, 1978; Leonard et al., 1981; Heck and Costa, 1982; Gebhart and Rossman, 1991). Nickel compounds have been previously associated with increased risk of cancer of the nasal sinuses, lung, larynx, and bladder in exposed refinery workers (USEPA, 1985). Nickel in the form of alpha nickel subsulfide (Ni_3S_2) is one of most potent known metal carcinogens (Sunderman, 1981, 1984). Phagocytic uptake of this poorly soluble, particulate nickel compound targets the nucleus by aggregation of particles at the nuclear membrane (Costa, 1983). The directed release of the metal ions in the vicinity of the deoxyribonucleic acid (DNA) causes preferential decondensation of a heterochromatic region of the X chromosome (Sen and Costa, 1985). Nickel ions also decrease the fidelity of DNA replication in vitro (Sirover and Loeb, 1976) and can substitute for magnesium as the activating metal cation for DNA polymerases (Sirover et al., 1979; Christie and Tummolo, 1989; Christie et al., 1991). An effect of nickel on RNA synthesis has also been observed with an initial stimulation of in vitro transcription by *Escherichia coli* ribonucleic acid (RNA) polymerase followed by a progressive decrease in synthesis (Niyogi et al., 1981).

Epidemiological studies support a cocarcinogenic role of Ni(II) in the enhancement of respiratory carcinogenesis by benzo(*a*)pyrene (BaP), a component of cigarette smoke (Sunderman, 1973), and experimental exposure of animals has substantiated the results of human exposures (Lau et al., 1972). A similar enhancement phenomenon was observed for the induction of rat respiratory tumors by *N*-ethyl, *N*-hydroxyethyl nitrosamine (Kurokawa et al., 1985). While there is evidence supporting a cocarcinogenic role for other metal ions, the strength of the nickel-induced response typically exceeds that of other metals such as Cr(VI), Cd(II), or As(III). The extremely low toxicity of Ni(II) in comparison to other metal ions may be partly responsible for the higher levels of synergistic action observed for nickel compounds (Williams et al., 1982; Christie et al., 1983, 1984; Christie and Costa, 1983, 1984).

The failure to demonstrate a clear mutagenic response with compounds of Ni(II) in spite of prevailing evidence of other forms of genotoxicity has caused many investigators to propose nonmutational pathways to explain the tumorigenicity for this ion compared to organic carcinogens or mutagenic metals such as Cr(VI). In contrast, there have been several demonstrations of a comutagenic effect of Ni(II) in different test systems. Synergistic enhancement of mutagenesis by Ni(II) in mammalian cells has been observed in combination with benzo(*a*)pyrene for induction of ouabain resistance (Rivedal and Sanner, 1980) and in combination with ultraviolet light for induction of 6-thioguanine resistance (Hartwig and Beyersmann, 1987). The Ni(II) enhancement of the mutagenicity of alkylating agents has also been demonstrated in bacteria (Dubins and LaVelle, 1986). In addition to synergistic responses in mutational assays, we have demonstrated enhancement of benzo(*a*)pyrene cell transformation using Syrian hamster embryo cells (Christie, 1989).

There are at least two plausible explanations for the low-level mutational response previously reported for Ni(II) compounds. First, many of the earlier studies used soluble compounds of Ni(II) rather than the insoluble compounds that have been associated with carcinogenesis in humans and animals. A second consideration is that some of the studies have used an X-linked genetic locus that is less sensitive to the detection of deletions or rearrangements than hemizygous autosomal loci. Several reports have shown that Ni(II) produces chromosomal gaps, deletions, and rearrrangements in different cell types (Nishimura and Umeda, 1979; Larramendy et al., 1981; Sen and Costa, 1985; Christie et al., 1988a). Based on these previous studies of the genotoxicity of Ni(II), we wish to test for the induction of deletions as a mechanism of direct mutagenesis by Ni(II) or in combination with DNA-damaging agents. In this chapter we use a mutational system that consists of the *gpt* bacterial gene stably integrated into one of the V79 chromosomes. This cell line is termed G12 and is apparently able to detect deletions in that it has an enhanced mutagenic response to X-rays compared to the *hrpt* locus in the parental V79 cells (Klein and Rossman, 1990). Using the G12 mutational assay, we show that soluble nickel sulfate is weakly mutagenic while insoluble nickel sulfide is highly mutagenic.

2. MATERIALS AND METHODS

2.1. Cell Culture

The parental Chinese hamster V79 cells (strain 743-3-6) and all derived cell lines were grown in Ham's F12 medium (Gibco, Grand Island, NY) supplemented with 5% fetal bovine serum and with additions as indicated below for specific cell lines at 37°C in a humid atmosphere of 5% CO_2.

2.2. Cell Lines

The parental V79 cell line was mutagenized with ultraviolet light, to obtain the UV1-6.1 cell line that was *hprt*[-] and consistently nonrevertible at this locus either spontaneously or by treatment with several mutagens (Klein and Rossman, 1990). The UV1-6.1 cell line was selected for transfection of the pSV2*gpt* plasmid. This plasmid carries the bacterial gene for xanthine phosphoribosyl transferase (XPRT), and its molecular organization has been described (Mulligan and Berg, 1980).

2.3. Mutational Assay

The assay we have used for the detection of mutations at the inserted *gpt* locus in the G12 cell line and at the *hprt* gene in the parental V79 cells is essentially as described for the detection of 6-TG[R] mutations at the endogenous *hprt* locus in Chinese hamster cells (O'Neill et al., 1977). The G12 cell line was maintained in HAT medium (Ham's F12 containing 10^{-4} M hypoxanthine, 10^{-6} M aminopterin and 10^{-4} M thymidine) supplemented with 5% fetal bovine serum. For each treatment condition two 10-cm plates were seeded with 2×10^5 cells at least 4 hr before mutagen treatment. Cytotoxicity was determined separately by plating 400 cells on three 6-cm dishes for each mutagen treatment and comparing the clonal survival in treated dishes to untreated controls. Seven days were allowed for the expression of mutagenic alterations. Cells were reseeded and treated with selection medium containing 10 μg/mL 6-thioguanine (6-TG). The 6-TG was prepared fresh for each experiment. A minimum of 10 days was allowed for the selection period before colonies were stained with 0.5% crystal violet and counted.

3. RESULTS

3.1. Development of the Mutational Assay

A cell line has been isolated following transfection of the bacterial *gpt* gene into V79 Chinese hamster cells (Klein, 1988). A diagrammatic representation of the isolation process is shown in Figure 24.1*a*. The endogenous *hprt* gene of the parental V79 cells was inactivated by ultraviolet light, and a nonrevertible cell line was selected for transfection with the pSV2*gpt* plasmid containing the *gpt* gene. The 57 positive transfectants identified by selection in HAT medium were subjected to two additional challenges of HAT medium after which only 22 clones remained. The G12 clone was selected for this study because of its low spontaneous mutation frequency at the *gpt* locus of approximately 20 mutants per 10^6 clonable cells. All other clones exhibited frequencies at least a hundred times higher. Southern blot analyses of the organization of the *gpt* gene in the V79 genome indicate that there is only one copy of the gene located at a single site (Klein and Rossman, 1990). The procedure used for the assessment of mutation at the *gpt* locus is diagrammed in Figure 24.1*b* and is essentially identical to assays for mutation at the endogenous *hprt* locus (see Section 2).

3.2. Mutagenicity of the G12 Cell Line

The mutagenic response to several mutagens was monitored at the *gpt* locus in the G12 cell line and compared to the response at the *hprt* locus in the parental cell line (Table 24.1). There was little detectable difference in the number of mutants observed for the G12 cells compared to the V79 cell line for each of the two concentrations used for MNNG and BPL. In the case of the two fluences of ultraviolet light used for mutagenesis, there was an approximately 50% increase in the response of G12 cells over that of the V79 cells. The most striking difference between the two cell lines, however, was observed following irradiation with X-rays.

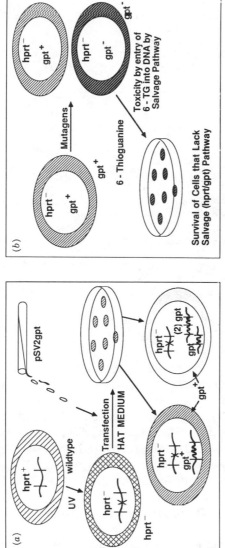

Figure 24.1. Development of mutational assay. (*a*) Inactivation of the endogenous *hprt* locus, transfection of the pSV2*gpt* plasmid, and isolation of the positive transfectants is diagrammatically presented (*b*). Selection of the *hprt⁻* cells was performed in 6-thioguanine (10 μg/ml), and positive transfectants were selected in F12 containing 10^{-4} *M* hypoxanthine, 10^{-6} *M* aminopterin, and 10^{-4} *M* thymidine (HAT medium). Mutation of the *gpt* gene in positive transfectants is detected by resistance to 6-thioguanine.

Table 24.1   Mutational Response of the G12 Cell Line[a]

Treatment	Mutation Frequency[b]	
	G12	V79
Untreated	30	5
MNNG		
1 µg/ml	211	192
2 µg/ml	505	395
BPL		
75 µg/ml	52	66
150µg/ml	109	127
Ultraviolet		
2 J/m^2	81	58
4 J/m^2	202	138
X-rays		
5 Gy	299	75
7.5 Gy	446	198

[a] A more extensive presentation of these results is given by Klein (1988).
[b] Given in mutants/10^6 clonable cells

3.3. Mutational Analysis with X-rays

Mutagenesis with X-rays is known to produce DNA deletions in mammalian cell lines (Vrieling et al., 1985; Urlaub et al., 1986). The induction of multilocus deletions by X-rays has been presented to explain the occurrence of mutant colonies of smaller than average size in which the extent of the deletion is beyond the target locus and results in inactivation of a gene required for a normal growth rate (Evans et al., 1986). The mutational response to X-rays of the endogenous X-linked *hprt* gene in our parental V79 cell line is weaker than for the *gpt* gene in G12. A deletion of sufficient size to extend into a neighboring essential gene would not be recovered at the *hprt* locus, which is present on the X chromosome found in one copy in the V79 cell line. This supposition is supported, but not proven, by the increase in the number of small colonies only in the G12 cell line and not in the V79 parental cell line (Fig. 24.2). The characterization of small and large colonies is given in the legend to Figure 24.2.

The increase in small colonies with increasing dose of X-rays provides suggestive evidence that multilocus deletions can be detected in the mutational assay that we have constructed. This is shown by the comparison in mutagenic response between the *gpt* gene in G12 cells and the response in the *hprt* gene in V79 cells (Fig. 24.2). The frequency of large mutant colonies increases in a similar fashion in both the G12 and V79 cell lines while the presence of small colonies is observed only in the G12 cell line. These results suggest

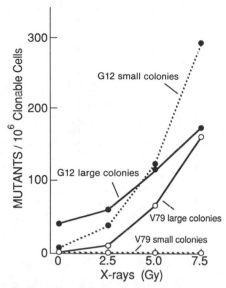

Figure 24.2. Mutational response to X-rays in G12 and V79 cells. The mutation frequency following X irradiation was determined as described in Section 2, and the results given are the average of two experiments. The discrimination of small colonies was made for colonies less than 0.5 mm.

that the G12 cell line is capable of detecting deletions of large segments of the mammalian genome and therefore will be a good test system for compounds expected to produce such deletions. We thus tested the mutagenic response of soluble and crystalline nickel compounds in these cells.

3.4. Mutational Response with Ni(II)

The mutational response of G12 cells to $NiSO_4$ is shown in Figure 24.3. The strongest mutagenic response was observed following a 16-hr treatment of cells with 0.1 mM $NiSO_4$; however, the extent of this response was only

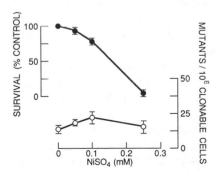

Figure 24.3. Mutational response to $NiSO_4$ in G12 cells. The mutation frequency (○) was determined as in Fig. 24.2. The cell survival (●) was determined immediately after the treatment with Ni(II) as described in Section 2.

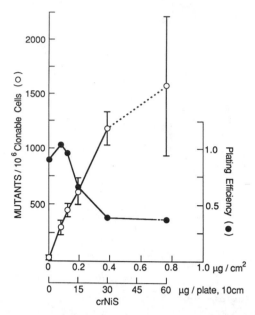

Figure 24.4. Mutational response to NiS in G12 cells. The mutation frequency (○) was determined as in Fig. 24.2. The measure of cell survival for particulate compounds such as NiS cannot be determined in a similar manner to soluble compounds as discussed in the text. The cytotoxicity given (●) represents the plating efficiency relative to untreated cells determined at the time of trypsinization and reseeding into selection medium.

twofold over the background level of mutation. The mutagenic response following 0.25 mM NiSO$_4$ was only slightly higher than the background level. Due to the extent of cellular toxicity observed at 0.25 mM, it is possible that the mutagenic response for concentrations of NiSO$_4$ between 0.1 and 0.25 mM would give a significantly higher response. An increase to 5- or 6-fold over the background rate would be consistent with prior studies, but a strong mutagenic response of 30- to 40-fold over background is not expected.

The magnitude of the mutational response of G12 cells to NiS was distinctly different (Fig. 24.4). There was a linear increase in the gpt^- mutants observed for 24-hr treatments of 10–30 μg/plate (10 cm). Beyond 30 μg/plate the toxicity of the NiS resulted in considerable variability in the mutational response. The number of observed mutants at 30 μg/plate (1235 mutants) represents a very strong mutational response compared to the other mutagens studied (see Table 24.1). If an average spontaneous mutation frequency of 20 is utilized, the increase in mutation rate over the spontaneous background is greater than 60-fold.

The determination of cytotoxicity to particulate NiS cannot be performed in the same manner as for soluble NiSO$_4$. Cytotoxicity is normally determined as clonal survival by treating cells seeded at greatly lower densities in the same manner as the cells plated at higher densities for the mutational assay. Since the probability of cellular engulfment of a particle will be considerably higher in the cells seeded for the mutation assay compared to the cytotoxicity assay, this approach does not provide an accurate measure of toxicity in the mutation assay. We therefore used the plating efficiency of the cells trypsinized from the more densely seeded treatment plates of the mutational assay at the time of transfer to the selection media as our measure of cytotoxicity.

4. DISCUSSION

The mutagenic response that we observed for NiS in this study was considerably higher than previously reported for any nickel compound in mammalian mutational assays. A review of earlier studies reveals that most investigators have used soluble salts rather than the insoluble, crystalline compounds that are phagocytized by cells in culture (Costa and Mollenhauer, 1980; Costa et al., 1982a). Weakly positive mutational responses of two- to threefold over the spontaneous rate was observed with $NiCl_2$ for the induction of mutations at the hypoxanthine–guanine phosphoribosyl transferase locus in Chinese hamster V79 cells (Miyaki et al., 1979). A slightly higher rate of three- to sevenfold over the spontaneous rate was observed for the induction of tri-fluorothymidine-resistant mutants in mouse $L5178Y/TK^{+/-}$ cells (Amacher and Paillet, 1980). These mutation rates are considered weak in comparison to responses observed for more potent mutagens, for example, MNNG, that produced a 300-fold increase in $hprt^-$ mutants (Miyaki et al., 1979).

Variations in the treatment times or the concentrations of metal salts are potential factors in the differences observed for the magnitudes of the mutagenic responses by different investigators. Since nickel salts are known to inhibit DNA replication (Costa et al., 1982b; Conway et al., 1986; Christie and Tummolo, 1989), prolonged treatments with nickel may actually reduce the yield of mutants recovered. Conversely, shorter treatments using relatively high concentrations of nickel salts may be more effective at mutant induction if the inhibitory effect of Ni(II) on replication is reversible. In support of this idea Hartwig and Beyersmann (1987) report a 15-fold increase over the spontaneous yield of $hprt^-$ mutants in V79 cells following a 5-hr treatment with 2.0 mM $NiCl_2$. Similarly, we have found that the use of shorter treatment periods are more effective than longer ones in the induction of cell trans-formation of Syrian hamster cells by $NiSO_4$ (Christie, 1989). A 15-fold increase in mutation rate at the $hprt$ locus may represent an even higher rate of mutagenic potential if the compound being tested is capable of producing deletions since multilocus deletions cannot be detected at this X-linked gene.

Chromosomal studies with compounds of Ni(II) support the hypothesis that induction of deletions or rearrangements is a possible source of mutant induction (Nishimura and Umeda, 1979; Larramendy et al., 1981; Sen and Costa, 1985; Christie et al., 1988a). The hypothesis has been presented that the ability of Ni(II) to alter the efficiency of DNA replication may lead to the induction of deletions or rearrangements (Christie, 1989). This hypothesis predicts that Ni(II) may also cause an increase in other types of mutational events. The inhibitory effects of Ni(II) on the replication complex responsible for closure of single-strand breaks may produce many types of mutagenic events including base changes, deletions, rearrangements, and gene conversion. Similar to the cytogenetic analyses mentioned above, studies of mutational assays using *Drosophila* and yeast have also supported a deletion-type mechanism of mutagenesis. The test for sex-linked recessive lethals (SLRLs) in *Drosophila melanogaster* was used to evaluate the mutagenicity of nickel sulfate (Rodriguez-Arnaiz and Ramos, 1986). The SLRL assay is capable of detecting deletion mutations as a modified sex ratio in the progeny of a female fly carrying the sex-linked mutation. All the concentrations of nickel sulfate tested were shown to induce SLRLs. The same investigators measured sex chromosome loss as a monitor of aneuploidy induction by following nondisjunction of sex chromosomes in female flies. Only a weak response for sex chromosome loss was observed for treatments with nickel sulfate. A study in *Saccharomyces cerevisiae* monitored the induction of gene conversion and reverse mutation by nickel sulfate (Singh, 1984). A weakly positive response with Ni(II) was observed only for gene conversion. This process is thought to involve inappropriate strand displacement and new DNA synthesis to convert one strand of one allele into its heterozygous counterpart. Interference by Ni(II) with the replication events at a strand break in the testor allele could explain this result. While these results coupled with our study provide support for the idea that Ni(II) may increase deletion events, they do not eliminate the possible involvement of other mechanisms including induction of base substitutions.

The number of mutational studies using particulate compounds has been very few. Costa and co-workers (1980) treated cultured CHO cells with insoluble crystalline Ni_3S_2 and monitored mutants at the *hprt* locus as 6-thioguanine-resistant colonies. Only a weak response was noted, but the highest concentration used was only 1.5 μg/ml. A comparable treatment with NiS in our assay system also produced a relatively weak response. Another study used Ni_3S_2 to treat human foreskin fibroblasts to monitor ouabain resistance and 6-thioguanine resistance, but none was detected (Biedermann and Landolph, 1987).

In conclusion, it appears that the major reason for our finding of a strong mutagenic response with Ni(II) is that we have used a crystalline, particulate compound with potent transforming properties. Several explanations are possible to explain the absence of a strong mutagenic response in the two other studies that used similarly potent compounds (Costa et al., 1980; Biedermann and Landolph, 1987). Differences in the storage of the particulate compounds have been shown to alter the extent of cellular uptake by phagocytosis (Abbracchio et al., 1982). A second factor may be the difficulty in the removal of the particles after the treatment period, thus leading to extended toxicity. Another possible explanation is that the high mutagenic response that we observe may be related to some feature of our mutational assay that monitors mutation at an exogenous bacterial gene inserted into a mammalian genome. Experiments are in progress to distinguish between these possibilities.

ACKNOWLEDGMENTS

This work was supported by grant ES-00260 and ES-04895 from the National Institute of Environmental Health Sciences, by grant CA-292589 from the National Cancer Institute, and by grant R184751 from the U.S. Environmental Protection Agency. The authors also acknowledge the excellent secretarial assistance of Maureen Kohler.

REFERENCES

Abbracchio, M. P., Heck, J. D., and Costa, M. (1982). The phagocytosis and transforming activity of crystalline metal sulfide particles are related to their negative surface charge. *Carcinogenesis* **3**, 175–180.

Amacher, D. E., and Paillet, S. C. (1980). Induction of trifluorothymidine-resistant mutants by metal ions in L5178Y/TK$^{+/-}$ cells. *Mutat. Res.* **78**, 279–288.

Biedermann, K. A., and Landolph, J. R. (1987). Induction of anchorage independence in human diploid foreskin fibroblasts by carcinogenic metal salts. *Cancer Res.* **47**, 3815–3823.

Christie, N. T. (1989). The synergistic interaction of Ni(II) and DNA damaging agents. *Toxicol. Environ. Chem.* **22**, 51–59.

Christie, N. T., and Costa, M. (1983). In vitro assessment of the toxicity of metal compounds. III. Effects of metals on DNA structure and function in intact cells. *Biol. Trace Elements Res.* **5**, 55–71.

Christie, N. T., and Costa, M. (1984). In vitro assessment of the toxicity of metal compounds. IV. Disposition of metals in cells: interactions with membranes, glutathione, metallothionein and DNA. *Biol. Trace Elements Res.* **6**, 139–158.

Christie, N. T., and Tummolo, D. M. (1989). The effect of Ni(II) on DNA replication. *Biol. Trace Element Res.*, **21**, 3–12.

Christie, N. T., Gosslee, D. G., Bate, L. C., and Jacobson, K. B. (1983). Quantitative estimates of genetic variability in metal ion toxicity in *Drosophila melanogaster*. *Toxicology* **26**, 295–312.

Christie, N. T., Cantoni, O., Evans, R. M., Meyn, R. E., and Costa, M. (1984). Use of mammalian DNA repair deficient mutants to assess the effects of toxic metal compounds on DNA. *Biochem. Pharmacol.* **33**, 1661–1670.

Christie, N. T., Sen, P., and Costa, M. (1988a). Chromosomal alterations in cell lines derived from mouse rhabdomyosarcomas induced by crystalline nickel sulfide. *Biol. Metals* **1**, 43–50.

Christie, N. T., Tummolo, D. M., Biggart, N. W., and Murphy, E. C., Jr. (1988b). Chromosomal alterations in cell lines derived from mouse rhabdomyosarcomas induced by crystalline nickel sulfide. *Cell Biol. Toxicol.* **4**(4), 427–445.

Christie, N. T., Chin, Y. E., Snow, E. T., and Cohen, M. (1991). Kinetic analysis of Ni(II) effects on DNA replication by polymerase α. *J. Cell Biochem.* **15D**, 114.

Conway, K., Sen, P., and Costa, M. (1986). Antagonistic effect of magnesium chloride on the nickel chloride-induced inhibition of DNA replication in Chinese hamster ovary cells. *J. Biochem. Toxicol.* **1**, 11–26.

Costa, M. (1983). Sequential events in the induction of transformation in cell culture by specific nickel compounds. *Biol. Trace Elements Res.* **5**, 285–295.

Costa, M., and Mollenhauer, H. H. (1980). Carcinogenic activity of particulate nickel compounds is proportional to their cellular uptake. *Science* **209**, 515–517.

Costa, M., Jones, M. K., and Lindberg, O. (1980). Metal carcinogenesis in tissue culture systems. In A. E. Martell (Ed.), ACS Symposium Series, No. 140, *Inorganic Chemistry in Biology and Medicine*, pp. 45–73.

Costa, M., Heck, J. D., and Robison, S. H. (1982a). Selective phagocytosis of crystalline metal sulfide particles and DNA strand breaks as a mechanism for induction of cellular transformation. *Cancer Res.* **42**, 2757–2763.

Costa, M., Cantoni, O., De Mars, M., and Swartzendruber, D. E. (1982b). Toxic metals produce an S phase-specific cell cycle block. *Res. Commun. Chem. Pathol. Pharmacol.* **38**, 405–419.

Dubins, J. S., and LaVelle, J. M. (1986). Nickel(II) genotoxicity: potentiation of mutagenesis of simple alkylating agents. *Mutat. Res.* **162**, 187–199.

Evans, H. H., Mencl, J., Horng, M. F., Ricanati, M., Sanchez, C., and Hozier, J. (1986). Locus specificity in the mutability of L5178Y mouse lymphoma cells: the role of multilocus lesions. *Proc. Natl. Acad. Sci. USA* **83**, 4379–4383.

Flessel, C. P. (1978). Metals as mutagenic initiators of cancer. In G. N. Kharasch (Ed.), *Trace Metals in Health and Disease*. Raven, New York, pp. 117–128.

Gebhart, E., and Rossman, T. G. (1991). Mutagenicity, carcinogenicity, and teratogenicity. In E. Merian (Ed.), *Metals and Their Compounds in the Environment*. VCH Verlagsgesellschaft, Weinheim, Germany, 617–640.

Hartwig, A., and Beyersmann, D. (1987). Enhancement of UV and chromate mutagenesis by nickel ions in the Chinese hamster *hgprt* assay. *Toxicol. Environ. Chem.* **14**, 33–42.

Heck, J. D., and Costa, M. (1982). In vitro assessment of the toxicity of metal compounds. I. Mammalian cell transformation. *Biol. Trace Element Res.* **4**, 319–330.

Klein, C. B. (1988). The genotoxicity and mutagenic mechanisms of beta-propiolactone in Chinese hamster and bacterial cells. Ph.D Thesis, New York University, New York.

Klein, C. B. and Rossman, T. G. (1990). Transgenic Chinese hamster V79 cell lines exhibit variable levels of *gpt* mutagenesis. *Env. Mol. Mutagen.* **16**, 1–12.

Kurokawa, Y., Matsushima, M., Imazawa, T., Takamura, N., and Hayashi, Y. (1985). Promoting effect of metal compounds on rat renal tumorigenesis. *J. Am. Coll. Toxicol.* **4**, 321–330.

Larramendy, M. L., Popescu, N. C., and DiPaolo, J. A. (1981). Induction by inorganic metal salts of sister chromatid exchanges and chromosome aberrations in human and Syrian hamster cell strains. *Environ. Mutagen.* **3**, 597–606.

Lau, T. J., Hackett, R. L., and Sunderman, F. W., Jr. (1972). The carcinogenicity of intravenous nickel carbonyl in rats. *Cancer Res.* **32**, 2253–2258.

Leonard, A., Gerber, G. B., and Jacquet, P. (1981). Carcinogenicity, mutagenicity and teratogenicity of nickel. *Mutat. Res.* **87**, 1–15.

Miyaki, M., Akamatsu, N., Ono, T., and Koyama, H. (1979). Mutagenicity of metal cations in cultured cells from Chinese hamster. *Mutat. Res.* **68**, 259–263.

Mulligan, R. C., and Berg, P. (1980). Expression of a bacterial gene in mammalian cells. *Science* **209**, 1422–1427.

Nishimura, M., and Umeda, M. (1979). Induction of chromosomal aberrations in cultured mammalian cells by nickel compounds. *Mutat. Res.* **68**, 337–349.

Niyogi, S. K., Feldman, R. P., and Hoffman, D. J. (1981). Selective effects of metal ions on RNA synthesis rates. *Toxicology* **22**, 9–21.

O'Neill, J. P., Couch, D. D., Machanoff, R., San Sebastian, J. R., Brimer, D. A., and Wilshire, A. (1977). A quantitative assay of mutation induction at the hgprt locus in Chinese hamster ovary cells: utilization with a variety of mutagenic agents. *Mutat. Res.* **45**, 103–109.

Rodriquez-Arnaiz, R., and Ramos, P. (1986). Mutagenicity of nickel sulphate in *Drosophila melanogaster*. *Mutat. Res.* **170**, 115–117.

Rivedal, E., and Sanner, T. (1980). Synergistic effect on morphological transformation of hamster embryo cells by nickel sulfate and benzo(*a*)pyrene. *Cancer Lett.* **8**, 203–208.

Sen, P., and Costa, M. (1985). Induction of chromosomal damage in Chinese hamster ovary cells by soluble and particulate nickel compounds: preferential fragmentation of the heterochromatic long arm of the X-chromosome by carcinogenic crystalline NiS particles. *Cancer Res.* **45**, 2320–2325.

Singh, I. (1984). Induction of gene conversion and reverse mutation by manganese sulphate and nickel sulphate in *Saccharomyces cerevisiae*. *Mutat. Res.* **137**, 47–49.

Sirover, M. A., and Loeb, L. A. (1976). Infidelity of DNA synthesis in vitro: screening for potential metal mutagens or carcinogens. *Science* **194**, 1434–1436.

Sirover, M. A., Dube, D. K., and Loeb, L. A. (1979). On the fidelity of DNA replication: Metal activation of *Escherichia coli* DNA polymerase I. *J. Biol. Chem.* **254**, 107–111.

Sunderman, F. W., Jr. (1973). The current status of nickel carcinogenesis. *Ann. Clin. Lab. Sci.* **3**, 156–180.

Sunderman, F. W., Jr. (1981). Recent research on nickel carcinogenesis. *Environ. Health Perspect.* **40**, 131–141.

Sunderman, F. W., Jr. (1984). Recent advances in metal carcinogenesis. *Ann. Clin. Lab. Sci.* **14**, 93–122.

Urlaub, G., Mitchell, P. J., Kas, E., Chasin, L. A., Funanage, V. L., Myoda, T. T., and Hamlin, J. (1986). Effect of gamma rays at the dihydrofolate reductase locus. *Somatic Cell Molec. Gen.* **12**, 555–566.

U.S. Environmental Protection Agency (USEPA) (1985). Health assessment document for nickel. USEPA, U.S. Department of Commerce, National Technical Information Service.

Vrieling, H., Simons, J.W.I.M., Arwert, F., Natarajan, A. T., and van Zeeland, A. A. (1985). Mutations induced by x-rays at the HPRT locus in cultured Chinese hamster cells are mostly large deletions. *Mutat. Res.* **141**, 281–286.

Williams, M. W., Hoeschele, J. D., Turner, J. E., Jacobson, K. B., Christie, N. T., Paton, C. L., Smith, L. H., Witschi, H. R., and Lee, E. H. (1982). Chemical softness and acute metal toxicity in mice and drosophilia. *Toxicol. Appl. Pharmacol.* **63**, 461–469.

25

COMPARATIVE IN VITRO CYTOTOXICITY OF NICKEL COMPOUNDS TO PULMONARY ALVEOLAR MACROPHAGES AND RAT LUNG EPITHELIAL CELLS

J. M. Benson, A. L. Brooks, and R. F. Henderson

Inhalation Toxicology Research Institute,
Lovelace Biomedical and Environmental Research Institute,
Albuquerque, New Mexico 87185

1. **Introduction**
2. **Methods**
    2.1. Test Compounds
    2.2. Alveolar Macrophage Studies
    2.3. Lung Epithelial Cell Studies
3. **Results**
    3.1. Alveolar Macrophage Studies
    3.2. LEC Studies
4. **Discussion**
    References

Nickel and Human Health: Current Perspectives, Edited by Evert Nieboer and Jerome O. Nriagu.
ISBN 0-471-50076-3 © 1992 John Wiley & Sons, Inc.

1. INTRODUCTION

Pulmonary alveolar macrophages (AMs) and lung epithelial cells (LECs) are targets for nickel toxicity in the lung. Prolonged insult to these cells could cause inflammation, fibrosis, emphysema, alveolar proteinosis, and cancer. The nature of response to nickel inhalation appears to depend on the animal species and the chemical form of the inhaled nickel compound (Bingham et al., 1972; Mastromatteo, 1967; Wehner et al., 1975; Wehner and Craig, 1972; Ottolenghi et al., 1974; Takenaka et al., 1985; Benson et al., 1987, 1988a; Dunnick et al., 1988). Species differences in the response of AMs and LECs to nickel compounds may qualitatively and quantitatively affect the toxic response of the lung upon nickel compound exposure. The objectives of this study were to evaluate the cytotoxicity of various nickel compounds to AMs and rat LECs in vitro, to begin characterization of species differences in AM sensitivity to nickel compounds, and to compare the genotoxicity of several nickel compounds in LECs in vitro.

2. METHODS

2.1. Test Compounds

Nickel subsulfide (α Ni_3S_2; CAS No. 12035-72-2) and nickel oxide (green oxide, calcined at 1200°C; CAS No. 1313-99-1) were supplied by the International Nickel Company (INCO) (Toronto, Ontario, Canada). Nickel sulfate hexahydrate (referred to as nickel sulfate, CAS No. 10101-97-0) was from Aldrich Chemical Company (Milwaukee, WI). These compounds were obtained through the National Institutes of Environmental Health Sciences, National Toxicology Program. Elemental analyses conducted on these compounds by the Midwest Research Institute (Kansas City, MO) indicated overall purities of 97.0, 99.5, and 99.5% for Ni_3S_2, $NiSO_4$, and NiO, respectively (Dunnick et al., 1988).

The nickel oxides (compounds A to F) and nickel-copper oxides (compounds G to J) were produced and characterized by collaboration between the Metallurgical Technology Centre of Falconbridge, Sudbury, Ontario, and the J. Roy Gordon Research Laboratory of INCO, Sheridan Park, Mississauga, Ontario. They were obtained through the courtesy of the Nickel Producers Environmental Research Association (NiPERA). The physical and chemical characteristics of the compounds used in this study are given in Table 25.1. More detail regarding the physicochemical properties of the compounds may be found in Sunderman et al. (1987). The test compounds were used as received.

2.2. Alveolar Macrophage Studies

Adult F344/N rats and Beagle dogs used in these studies were from the Lovelace Inhalation Toxicology Research Institute colonies. The B6C3F₁

Table 25.1 Nickel Compounds and Their Relevant Physical Characteristics[a]

Compound	Calcination Temperature (°C)	Nickel Content (wt %)	Cu Content (wt %)	Dissolution[b] Half-time (Rat Serum)
Ni_3S_2[c]	NA	73	ND	34 days
$NiSO_4 \cdot 6H_2O$[c]	NA	22	ND	Soluble
NiO (Green, INCO)[c]	1200	78	<0.01	310 days
(A) NiO[c]	<650	78	<0.01	>11 yr
(B) NiO	735	78	<0.01	>11 yr
(C) NiO	800	78	<0.01	>11 yr
(D) NiO	850	78	<0.01	>11 yr
(E) NiO	918	78	<0.01	>11 yr
(F) NiO	1045	79	<0.01	>11 yr
(G) NiCuO[c]	850	44	27.9	>11 yr
(H) NiCuO[c]	850	52	20.6	3.2 yr
(I) NiCuO	850	63	12.6	7.2 yr
(J) NiCuO	850	69	6.9	4.5 yr

[a] NA, not applicable; ND, not determined. From Sunderman et al. (1987).
[b] Defined as the time required for one-half the material to dissolve in rat serum in vitro. Methods used are described in Sunderman et al. (1987).
[c] Compounds tested in both AMs and LECs.

mice were from Simonsen Laboratories (Gilroy, CA). Rodent AMs were obtained from excised lungs of rats and mice, while canine AMs were isolated by pulmonary endotracheal lavage according to previously described procedures (Benson et al., 1988b). Monkey AMs were obtained from excised lungs (left) of cynomologous monkeys sacrificed for another study. Macrophages from all species were prepared and treated with nickel compounds according to previously described procedures (Benson et al., 1988b).

Nickel oxides (A to F) and nickel copper oxides (G to J) were initially screened for toxicity to rat AM. The most toxic NiO and two most toxic NiCuO compounds were evaluated for toxicity in rat, mouse, and dog AMs. The cytotoxicities of these compounds were then compared to those of Ni_3S_2, $NiSO_4$, and NiO (INCO, calcined at 1200°C). Only Ni_3S_2 was evaluated for cytotoxicity to monkey AMs due to the limited availability of these cells.

Median lethal concentrations for the compounds (LC_{50}) were calculated using the probit procedure of the SAS Software Statistical Package (SAS Institute, 1982) installed on a VAX 11/780 computer.

2.3. Lung Epithelial Cell Studies

Rat LECs used in these studies have been previously described (Li et al., 1983). Cells were grown in Ham's F12 medium (Difco) containing 10% fetal calf serum and penicillin–streptomycin. The LECs (10^6) were seeded in 75-

cm^2 flasks (Falcon) and were incubated for 24 hr before addition of nickel compounds. Graded concentrations of nickel compounds were then added, and cells were incubated for an additional 20 hr. After incubation of LECs for 20 hr with test compounds, the extent of cell killing was determined by removing the cells, washing with Ham's F12 medium, and reseeding 500 LECs into flasks. Relative colony-forming ability of the reseeded LEC was determined by counting colonies formed after 10–11 days in culture. Median lethal concentrations for each test compound were determined using probit analysis as described above for AMs.

The mitotic index, fraction of cells containing nuclear fragments, and extent of chromosome damage were determined in cells exposed to Ni_3S_2 (0.075 mM Ni), $NiSO_4$ (0.5 mM Ni), NiCuO-G (1.0 mM Ni), and NiO (INCO, 5.0 mM Ni). Concentrations used were the median lethal concentration for each compound. Cells were exposed for 20 hr. During the final 2 hr of incubation, colcemid was added at a final concentration of 0.1 µg/mL. Cells were harvested using standard techniques (Bender et al., 1988), and the frequency of chromosome aberrations was determined using coded slides. At least 100 cells were scored per experimental treatment. The mitotic index and fraction of cells that contained nuclear fragments were determined using a total of 1000 cells per experimental treatment.

3.   RESULTS

3.1.   Alveolar Macrophage Studies

The LC_{50} values of NiO compounds A–F in rat AMs were greater than 1000 µg compound/mL (> 13 µmol Ni/mL) (Table 25.2). The cytotoxicities of the NiCuO compounds in rat AMs were generally greater than those of the pure NiO (Table 25.2). The cytotoxicities of the NiCuO increased with increasing copper content and was inversely related to their nickel content. Expressed on the basis of copper concentration, the LC_{50} values of compounds NiCuO-G, NiCuO-H, and NiCuO-I were approximately 2 µmol Cu/mL in all cases, strongly suggesting that copper contributed to the toxicities of these compounds.

Median lethal concentrations of NiO-A, NiCuO-G, and NiCuO-H in F344/ N rat and $B6C3F_1$ mouse AMs were similar, with overlapping 95% confidence intervals, indicating similar sensitivity of AMs from the two rodent species (Table 25.3). On the other hand, the LC_{50} values for these compounds in dog AMs were three to five times lower than those for these compounds in rodent AMs (Table 25.3). The sensitivity of monkey AMs to Ni_3S_2 was between those of dog and rat AMs (Table 25.3).

3.2.   LEC Studies

Median lethal concentrations of the nickel compounds in LECs are given in Table 25.4. Cells exposed to the particulate forms of nickel concentrated

Table 25.2 Median Lethal Concentrations of Nickel Oxide and Nickel–Copper Oxide Compounds to Rat Alveolar Macrophages in Vitro[a]

Compound	Nickel Content[b] (wt %)	Copper Content[b] (wt %)	Median Lethal Concentration (95% Confidence Interval)		
			μg Compound/mL	mM Ni	mM Cu
NiO					
A–F	78.2	0.1	>1000[c]	>13	—
NiCuO					
G	43.8	27.9	530 (320–730)	4.0 (2.4–5.5)	2.0 (1.8–2.3)
H	51.7	20.6	570 (470–730)	5.1 (4.2–6.5)	1.9 (1.5–2.4)
I	62.6	12.6	1000 (850–1270)	11 (9.4–14)	2.0 (1.7–2.5)
J	68.8	6.9	>1000	>12	>1.1
Ni$_3$S$_2$[d]	72.2	<0.1	110[d]	1.4[d]	—
NiO (INCO)[d]	78.8	<0.1	>1000	>13	—

[a]Results represent the median lethal concentration (95% confidence level). Experiments were repeated at least 3 times when the LC$_{50}$ <1000 μg compound/mL and twice when the LC$_{50}$ >1000 μg compound/mL.
[b]From Sunderman et al. (1987).
[c]Median lethal concentration was greater than the highest concentration of compound tested (1000 μg/mL).
[d]From Benson et al. (1988b).

Table 25.3   Differences in Sensitivity of Alveolar Macrophages from Rat, Mouse, Dog, and Monkey to Nickel Compounds

	Median Lethal Concentration			
Compound	Rat	Mouse	Dog	Monkey
Ni_3S_2				
μg compound/mL	110	ND	6.0	26
mM Ni	1.4	—	0.08	0.41
$NiSO_4 \cdot 6H_2O$				
μg compound/mL	820	ND	80	ND
mM Ni	3.1	—	0.3	
NiO (INCO)				
μg compound/mL	>1000	ND	290	ND
mM Ni	>13	—	3.9	
NiO (A)				
μg compound/mL	>1000	>1000	900	ND
mM Ni	>13	>13	12	
NiCuO (G)				
μg compound/mL	530	630	100	ND
mM Ni	4.0	4.7	0.72	
mM Cu	2.0	2.9	0.42	
NiCuO (H)				
μg compound/mL	520	460	150	ND
mM Ni	5.1	4.1	1.3	
mM Cu	1.9	2.3	0.78	

ND, experiments not performed.

the particles in both dividing and nondividing cells. The particles remained with the cells throughout the 10-day culture period. The presence of these particles might have influenced plating efficiency and cell division, thus causing an overestimation of the cytotoxicity of the particulate nickel compounds.

The effect of nickel compound exposure on the mitotic index of LECs is illustrated in Figure 25.1. Exposure of LECs to the median lethal concentration of $NiSO_4$, NiCuO-G, or NiO (INCO) depressed cell division, while administration of Ni_3S_2 at the median lethal concentration elevated cell division.

Only $NiSO_4$ exposure (0.5 mM Ni as $NiSO_4$) caused nuclear fragmentation in LECs (Fig. 25.2). The frequency of occurrence of all types of chromosome aberration occurring in LECs exposed to any of the nickel compounds was low and not significantly different from the frequency of aberrations in control cells (Fig. 25.3). If data from all nickel compound–exposed LECs are pooled, the frequency of exchange-type aberrations in LECs exposed to high concentrations of the nickel compounds was increased over the frequency of these aberrations in control LECs. The background frequencies of exchange aberrations in control and nickel-exposed LECs were 0.003 (0.006) and 0.04 (0.01) exchange per cell [mean (95% confidence interval)].

Table 25.4 Median Lethal Concentrations of Nickel Compounds in Rat Lung Epithelial Cells[a]

Compound	Nickel Content (wt %)	Copper Content (wt %)	Median Lethal Concentration (95% Confidence Interval)		
			μg Compound/mL	mM Ni	mM Cu
NiO (A)	78.2	0.1	900	12[b]	—
NiCuO (G)	43.8	27.9	2100	1.2[b]	9.0
NiCuO (H)	51.7	20.6	1400	1.2[b] (0.98–1.32)	4.4
Ni₃S₂	72.2	<0.1	7.3 (5.7–8.1)	0.09 (0.07–0.10)	—
NiSO₄·6H₂O	22.2	ND	140 (110–170)	0.54 (0.43–0.65)	—
NiO (INCO)	78.8	<0.1	210 (130–370)	2.8 (1.7–5.0)	—

[a]Results obtained in epithelial cells are those from three separate experiments; 95% confidence intervals are given in parentheses. ND, not determined.
[b]The 95% confidence interval was not available for NiO-A and NiCuO-G.

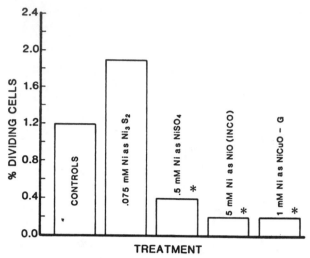

Figure 25.1. Effect of nickel compound exposure on mitotic index of LECs. Cells were exposed to each compound at its respective median lethal concentration. The only source of error is in counting. The same relative percentage of error exists for each data set. There was a significant depression in the mitotic index in LECs, induced by 0.5 mM NiSO$_4$, 1 mM NiCuO, and 5 mM NiO (Z-test; $p < 0.05$).

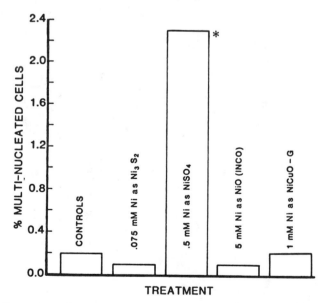

Figure 25.2. Percentage of occurrence of multinucleated LECs following nickel compound exposure. Cells were exposed to each compound at its respective median lethal concentration. The only source of error is in counting. The same relative percentage of error exists for each set. Only the frequency of multinucleated cells induced by NiSO$_4$ exposure was significantly elevated relative to the controls (Z-test; $p < 0.001$).

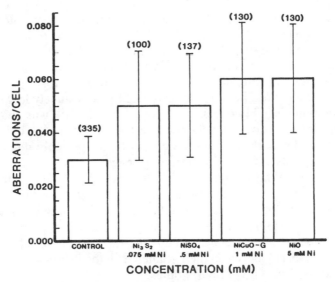

Figure 25.3. Induction of chromosome aberrations in LECs. Error bars represent the 95% confidence interval. Number of cells scored are listed in parentheses. No significant increase in aberration frequency was observed for any treatment group relative to the controls (Z-test; $p > 0.05$).

4. DISCUSSION

The NiO compounds examined in this study displayed low cytotoxicity in AMs. As expected, NiO compound A, having the lowest calcination temperature and greatest solubility in rat serum, was the most toxic of the six NiO compounds (A–F) tested. Calcination of the NiO at 735°C and above greatly reduced dissolution half-time and toxicity of NiO compounds B–F. The cytotoxicities of NiCuO compounds G–J did not correlate with their nickel content, surface area, or dissolution half-times in rat serum but did correlate with their copper content. These results indicated that NiO compounds formed at high temperatures may be less toxic to AMs following inhalation than NiO compounds formed at low temperatures [compare NiO (INCO) with other NiO]. In addition, NiCuO compounds may be more cytoxic to AMs upon inhalation than are NiO compounds.

Ranking of nickel compounds, in decreasing order of cytotoxicity to AMs, was $Ni_3S_2 > NiSO_4 > NiCuO$-G, NiCuO-H > NiO (all). Ranking of species sensitivity of AM to nickel compounds was dog > rat ≥ mouse, with some evidence suggesting that the sensitivity of monkey AMs lies between those of the dog and rat.

The cytoxicity ranking of the nickel compounds tested in LECs was $Ni_3S_2 > NiSO_4 > NiCuO$-G, NiCuO-H > NiO (INCO) > NiO-A. This is similar to the cytotoxicities reported for alveolar macrophages and suggests that for this group of compounds, the amount of bioavailable nickel may influence their cytotoxicity. It has been postulated that the cytotoxicities of some forms of nickel are related to the ability of the cells to phagocytize or internalize the material, thereby increasing the actual dose to cell and allowing solubilization of the compound to Ni^{2+} within the cell. In the present study, NiO compounds accumulated in the cells by phagocytosis but were not cytotoxic or genotoxic. This was probably due to the relatively low solubilities of these compounds in biological fluids. In contrast, Ni_3S_2 was phagocytized by LECs and was toxic, probably because the Ni_3S_2 was solubilized within the cells. Nickel sulfate was less toxic than Ni_3S_2, possibly due to the fact that entry into cells was by diffusion, not phagocytosis.

Decreases in the mitotic index of LECs exposed to $NiSO_4$, NiCuO-G, or NiO (INCO) were consistent with earlier findings of Sen and Costa (1986a). These investigators found a decrease in the rate of division of Chinese hamster ovary cells and Syrian hamster embryo cells exposed to NiS or $NiCl_2$. Without a complete time course to define the cell kinetics, it is not possible to resolve the apparent disparity between their studies with the increased mitotic index in LECs exposed to Ni_3S_2 in our studies.

The very low level of cytogenetic damage reported in this chapter relative to that in other reports (Sen and Costa, 1986a,b) may be related to differences in the treatment time, time between nickel exposure and harvest, or sensitivity of the lung epithelial cells to the action of the nickel. In the studies of Sen and Costa (1986a), CHO cells were treated for 2, 6, 24, and 48 hr, and the cells were evaluated for damage 24 hr after the end of the treatment. In the current study, lung epithelial cells were evaluated at the end of a 20-hr exposure period without the 24-hr recovery time. During the recovery time, the cells would have passed through at least one cell division, which may be necessary to express the chromosome damage produced by the nickel.

The nuclear fragmentation observed following exposure to $NiSO_4$ supports the observation by Anderson (1985) that one of the modes of action of nickel is disruption of spindle fiber function. Anderson's study demonstrated that exposure to $NiSO_4$ for 4–20 hr at concentrations similar to those used in the present study caused both chromosome contraction and inhibition of chromatid segregation. These are early indicators of alterations in spindle fiber formation. Problems in spindle fiber formation and function could be reflected in improper segregation of the chromosomes into the nucleus during the next cell division after treatment. This may be one of the mechanisms by which exposure to nickel results in cell killing.

Since disruption of spindle fiber formation can lead to aneuploidy, and since many tumors have aneuploid karyotypes, this action of nickel may be related to its ability to produce tumors. Even though this was observed only at very high concentrations of nickel, it may still be relevant to nickel inhalation exposure, since there can be high concentrations of nickel ions around the nickel particle.

This study defined the relative toxicities of different chemical forms of nickel compounds to AMs and LECs. The results indicated that the nickel compounds tested produce little chromosome damage and that one form of nickel (NiSO$_4$) can produce nuclear fragmentation. These data support the concept that most of the nickel toxicity is related to the solubility of the nickel. The mode of action for cytotoxicity in LECs may be related to spindle fiber function but is not related to the clastogenic activity of the nickel compounds.

ACKNOWLEDGMENT

This research was performed under Funds in Agreement, No. DE-FI04-87L44742, between the U.S. DOE/OHER under Contract No. DE-AC04-76EV01013 and the Nickel Producers Environmental Research Association in facilities fully accredited by the American Association for Laboratory Animal Care. The excellent technical assistance of J. Holmes, K. McCloud, V. White, and M. Waltman is gratefully acknowledged.

REFERENCES

Anderson, O. (1985). Evaluation of the spindle-inhibiting effect of Ni^{++} by quantitation of chromosomal super-contraction. *Res. Commun. Chem. Pathol. Pharmacol.* **50**, 379–386.

Bender, M. A., Awa, A. A., Brooks, A. L., Evans, H. J. Groer, P. G., Littlfield, L. G., Pereira, C., Preston, R. J., and Wachholz, B. W. (1988). Current status of cytogenic procedures to detect and quantify previous exposure to radiation. *Mutat. Res.* **196**, 103–159.

Benson, J. M., Carpenter, R. L., Hahn, F. F., Haley, P. J., Hanson, R. L., Hobbs, C. H., Pickrell, J. A., and Dunnick, J. K. (1987). Comparative inhalation toxicity of nickel subsulfide to F344/N rats and B6C3F$_1$ mice exposed for 12 days. *Fundam. Appl. Toxicol.* **9**, 251–265.

Benson, J. M., Burt, D. G., Carpenter, R. L., Eidson, A. F., Hahn, F. F., Haley, P. J., Hanson, R. L., Hobbs, C. H., Pickrell, J. A., and Dunnick, J. K. (1988a). Comparative inhalation toxicity of nickel sulfate to F344/N rats and B6C3F$_1$ mice exposed for 12 days. *Fundam. Appl. Toxicol.* **10**, 164–178.

Benson, J. M., Henderson, R. F., and Pickrell, J. A. (1988b). Comparative *in vitro* cytotoxicity of nickel oxide and nickel copper oxide to rate, mouse, and dog pulmonary alveolar macrophages. *J. Toxicol. Environ. Health.* **24**, 373–383.

Bingham, E., Barkley, W., Zerwas, M., Stemmer, K., and Taylor, P. (1972). Response of alveolar macrophages to metals. I. Inhalation of lead and nickel. *Arch. Environ. Health.* **25**, 406–414.

Dunnick, J. K., Benson, J. M., Hobbs, C. H., Hahn, F. F., Cheng, Y. S., and Eidson, A. F. (1988). Comparative toxicity of nickel oxide, nickel sulfate hexahydrate, and nickel subsulfide after 12 days of inhalation exposure to F344/N rats and B6C3F$_1$ mice. *Toxicology* **50**, 145–156.

Li, A. P., Hahn, F. F., Zamora, P. O., Shimizu, R. W., Henderson, R. F., Brooks, A. L., and Richards, R. (1983). Characterization of a lung epithelial cell strain with potential applications in toxicological studies. *Toxicology* **27**, 257–272.

Mastromatteo, E. (1967). Nickel: a review of its occupational health aspects. *Occup. Med.* **9**, 127–136.

Ottolenghi, A. D., Haserman, J. K., Payne, W. W., Falk, H. L., and MacFarland, H. N. (1974). Inhalation studies of nickel sulfide in pulmonary carcinogenesis in rats. *J. Natl. Cancer Inst.* **54**, 1165–1172.

Sen, P., and Costa, M. (1986a). Incidence and localization of sister chromatid exchanges induced by nickel and chromium compounds. *Carcingenesis* **7**, 1527–1533.

Sen, P., and Costa, M. (1986b). Pathway of nickel uptake influences its interaction with hetero-chromatic DNA. *Toxicol. Appl. Pharmacol.* **84**, 278–285.

Sunderman, W. F., Jr., Hopfer, S. M., Knight, J. A., McCully, K. S., Cecutti, A. G., Thornhill, P. G., Conway, K., Miller, C., Patierno, S. R., and Costa, M. (1987). Physicochemical characteristics and biological effects of nickel oxides. *Carcinogenesis* **8**, 305–313.

Takenaka, S., Hochrainer, D., and Oldiges, H. (1985). Alveolar proteinosis induced in rats by long-term inhalation of nickel oxide. In S. S. Brown and F. W. Sunderman, Jr. (Eds.), *Progress in Nickel Toxicology*. Blackwell Scientific Publications, Oxford, pp. 89–92.

Wehner, A. P., and Craig, D. K. (1972). Toxicology of inhaled NiO and CoO in golden hamsters. *Am. Ind. Hyg. Assoc. J.* **33**, 146–155.

Wehner, A. P., Busch, R. H., Olson, R. J., and Craig, D. K. (1975). Chronic inhalation of nickel oxide and cigarette smoke by hamsters. *Am. Ind. Hyg. Assoc. J.* **36**, 801–810.

26

PHAGOCYTOSIS AND METABOLIZATION OF WELDING FUMES BY GUINEA PIG ALVEOLAR MACROPHAGES IN CULTURE

A. Ben Otmane, A. M. Decaestecker, H. F. Hildebrand, and D. Furon

Institut de Médecine du Travail, F 59045 Lille Cedex, France

R. Martinez

CNRS UA 234, Université des Sciences et Techniques de Lille, F 59655 Villeneuve D'Asco Cedex, France

Nickel and Human Health: Current Perspectives, Edited by Evert Nieboer and Jerome O. Nriagu.
ISBN 0-471-50076-3   © 1992 John Wiley & Sons, Inc.

1. INTRODUCTION

Welders are exposed to a variety of substances contained in the welding fumes that arise during the welding process. The most current technologies are manual metal arc (MMA) welding using stick electrodes and metal inert gas (MIG) welding employing coutinuous wire on mild steel, stainless steel (SS), or aluminum. Certain welding fumes contain significant amounts of Cr, Mn, Ni, and traces of other metals, which have been shown to be mutagenic in one or more in vitro bioassays (Norseth, 1981; Costa et al., 1980; Hansen and Stern, 1983, 1985). Some of these metals are strongly suspected human carcinogens (Norseth, 1981; Sunderman, 1981), and it might therefore be anticipated that welders experience an excess risk of respiratory tract cancer because of their occupation (Beaumont and Weiss, 1981; Stern, 1983, 1987). This risk may be increased by the alveolar retention of welding fume particles as characterized by in vivo studies on shipyard arc welders (Kalliomaki et al., 1978) and on experimental animals (Kalliomaki et al., 1983).

In order to assess the retention, metabolization, and cytotoxic effects in lung cells, we investigated the in vitro incorporation of two MIG and one MMA welding fumes on human embryonic epithelial pulmonary cells (L132) in culture. This work has shown that the cytotoxic effect of the MMA/SS welding fume is about 100 times higher than that of MIG welding fumes (Hildebrand et al., 1986). Since L132 is a cell line adapted to cell culture, these cells may be considered from the physiological point of view as "ghost cells." Guinea pig alveolar macrophages (GPAMs) in primary culture are physiologically nearer to reality. So the present investigation deals with the metabolism and incorporation of welding fumes by this cell type.

2. MATERIALS AND METHODS

2.1. Welding Fume Preparation

Welding fumes were obtained from the Danish Welding Institute. They were produced in a welding robot, collected on large paper filters from which they are immediately scraped, placed in glass vials, and stored at room temperature (Stern and Pigott, 1983).

The MMA/SS fume contains Cr(VI), mostly as a sodium or potassium chromate. The main constituents are F 14.7%, Si 7.3%, K 22%, Ti 1.4%, Cr(VI) 3.6%, Mn 2.7%, Fe 36%, and traces of Ni and Cu. The MIG/SS fume has an iron oxide spinel structure with various amounts of chromium as a substitutional atom. The main constituents are Si 3.2%, Cr(III) 9.8%, Cr(VI) 1%, Mn 10.4%, Ni 4.1%, Fe 37%, and traces of Cu and Z. The MIG/Ni fumes occur in welding on cast iron with pure nickel wire, which produces a particulate that is completely crystalline with a Ni–NiO mixture. The main constituents are Ni 60.8%, Fe 3%, and traces of Cr, Cu, As, and Z.

These chemical compositions have been assessed by Thomsen at the Technological Institute in Tastrup (DK) (Thomsen and Stern, 1979).

2.2. Cell Culture

Alveolar macrophages were obtained after ablation of guinea pig (Hartley albinos) lungs by bronchoalveolar lavage with Hanks saline solution. The cell suspension was centrifuged at 500g, and the pellet was resuspended in MEM 2011 medium and cultured in plastic bottles after adding 5% fetal calf serum.

The GPAMs were incubated with MMA/SS at the following concentrations: 2.5, 5, and 10 $\mu g/mL$ of culture medium; both MIG fumes were added into the cell culture at concentrations of 50, 100, and 200 $\mu g/mL$. Prior to the addition of cells (10^6), the fumes were sonicated in the culture medium for 1 hr. The incubation time was three days.

2.3. Electron Microscopy

Ultrastructural investigation was carried out for all fumes at each concentration. The material was prepared for electron microscopy employing routine glutaraldehyde–OsO_4 fixation and Epoxy resin embedding methods (Hildebrand and Biserte, 1979). X-ray microprobe analyses by energy-dispersive spectrometry (EDS) was performed on unstained sections on carbon-coated nylon grids (Jeol 200 CX electron microscope connected to a Link analytical system QX 200). In order to avoid impurities and secondary precipitates, all electron micrographs were taken on unstained sections.

2.4. β-Glucuronidase Activity Measurements

Determination of β-glucuronidase activity in GPAM homogenate was performed on cells after 48 hr incubation with the welding fumes at two different concentration: 2.5 and 5 $\mu g/mL$ for the MMA/SS fume and 50 and 100 $\mu g/mL$ for the MIG fumes. We applied the method described by Szasz (1967) using p-nitrophenyl glucuronide as substrate in these measurements.

3. RESULTS

3.1. Ultrastructural Studies

As previously observed in L132 cells (Hildebrand et al., 1986), the welding fumes were phagocytized by GPAMs in primary culture. The relatively high cytotoxic effect of MMA/SS welding was confirmed also for these cells. Light microscopy observations permitted estimation of the survival rate as 10% in cultures with 10 μg MMA/SS per milliliter. This rough estimate must, however, be confirmed by appropriate enzymatic investigations.

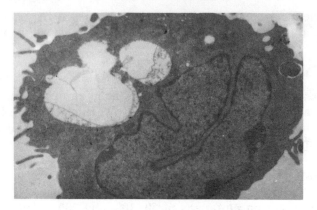

Figure 26.1. Guinea pig alveolar macrophage after three days in control culture (×5,700).

The primary ultrastructural modification of GPAMs is an increase in the number of microvilli with respect to the control cells (compare Figs. 26.1 and 26.2). As seen in Figure 26.2, areas with a high density of microvilli are often observed for all the welding fumes tested. In GPAM incubated with MMA/SS, we generally observe cytoplasmic vacuolization with two types of particles (Fig. 26.3): big round particles with a size ranging from 0.1 to 0.3 μm and aggregates of fine particles. In addition, we observe very tiny particles on cell and cytoplasmic membranes and in the nucleus, in particular on euchromatin. Heterochromatin is generally devoid of these particles. One also observes the association of these microparticles to surfactant (Fig. 26.3). The GPAMs incubated with MIG/Ni fume (Fig. 26.2) and MIG/SS fume (Fig. 26.4) show similar ultrastructural changes, allthough the fumes occur in lower amounts in phagocytic vacuoles.

Figure 26.2. Guinea pig alveolar macrophage after three days incubation with 200 μg of MIG/Ni/mL culture medium. Many phagocytic vacuoles contain this fume. Also notice the high density of microvilli at the left of the micrograph (×6,300).

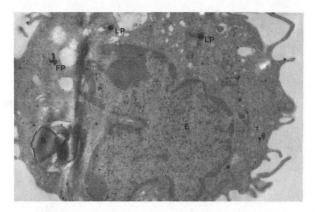

Figure 26.3. Guinea pig alveolar macrophage after three days incubation with 5 μg of MMA/
SS/mL. Note the presence of large round particles (LP); aggregates of fine particles
(FP); and microparticles on surfactant (S), euchromatin (E), and other cytoplasmic
structures (×7,200).

3.2. X-ray Microprobe Analysis

As already described in a previous investigation of L132 cells (Hildebrand
et al., 1986), the MIG/Ni fume does not show any change in its elemental
composition before and after phagocytosis during three days incubation.
Nickel is the only major element that can be detected (Fig. 26.5). Concerning
the MIG/SS fume, there is evidence of a decrease in the amount of manganese
in phagocytized particles when compared to extracellular areas.

The essential difference in the MMA/SS fume between the big round
particles (Fig. 26.6) and the aggregates of fine particles (Fig. 26.7) is the
presence of chromium in the latter. The other principal elements are present

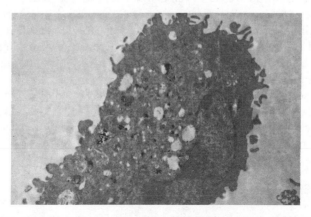

Figure 26.4. Guinea pig alveolar macrophage after three days incubation with 100 μg of MIG/
SS/mL (× 4,600).

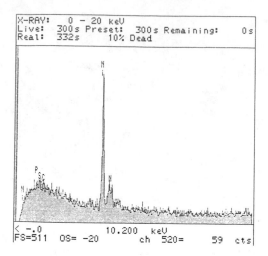

Figure 26.5. An EDS diagram of intracellular particles of MIG/Ni fume.

in both forms. After performing a more precise analysis of EDS diagrams in the window from 0.7 to 3.3 keV, we observed a phosphorus peak (Figs. 26.6*b* and 26.7*b*), which is absent in the native fume. The low density and the small size of microparticles observed in euchromatin and surfactant do not allow an accurate assessment of their composition.

3.3. β-Glucuronidase Activity

Since alveolar macrophages are differentiated cells that do not divide in culture, the only precise method to determine cytotoxic effects is the study of activity modification of some appropriate enzymes. We have started our investigation in this field by the measurements of β-glucoronidase activity. After 48 hr incubation we observed a significant increase of β-glucuronidase activity as a function of concentration of fumes. With the MMA/SS fume, a very significant increase was observed at very low concentrations (2.5 and 5 μg/mL) (Fig. 26.8). A somewhat lower increase is also seen with MIG fumes. This activation, however, is obtained with a 20-fold higher concentration of the fumes.

These results clearly demonstrate an activation of GPAMs. A similar increase in β-glucuronidase activity was also observed in the culture medium, confirming cellular damage that was much more prominent for the MMA/SS fume than for the two MIG fumes.

4. DISCUSSION

All welding fumes are phagocytized by GPAMs and induce significant cell death (measured by cell counting before and after incubation with the fumes).

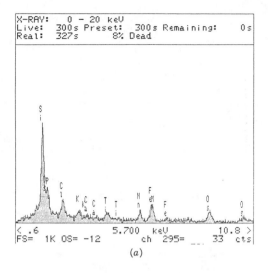

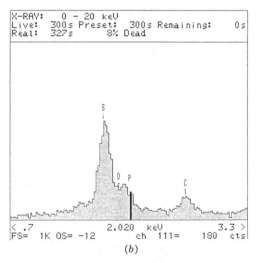

Figure 26.6. An EDS diagram of MMA/SS fume; large round particles: (a) whole diagram; (b) enlargement of phosphorus area.

The cytotoxic effect of the fumes, and especially that of MMA/SS particulates, was confirmed by an increase in the culture medium β-glucuronidase activity. The welding fumes examined induced approximatively the same cytotoxic effect in GPAMs than in L132 cells (Hildebrand et al., 1986), although it is generally accepted that alveolar macrophages are more refractory to metallic compounds than to established cell lines (Frazier and Andrews, 1979). The high cytotoxic effect of the MMA/SS fume may be due to the presence of a soluble Cr(VI) compound: estimated 3.6% by content. Indeed, growth

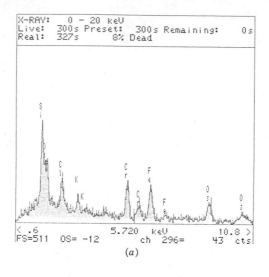

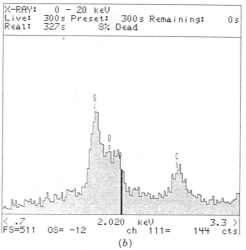

Figure 26.7. An EDS diagram of MMA/SS fume; aggregate of fine particles: (a) whole diagram; (b) enlargement of phosphorus area.

curves of L132 cells are similar when cells are grown at Cr(VI) concentrations comparable to that contained in the MMA/SS fume (Hildebrand et al., 1986). Additional effects can also be assigned to the high concentrations of water-soluble fluorine (14.7%) and potassium (22%).

Welding fume particles can be localized in nearly all GPAMs. The EDS analyses revealed some differences in the elementary composition between extracellular and intracellular areas of MIG/SS, especially with respect to the amount of manganese. Its decrease in intracellular areas indicates a dissolution and consequently a metabolization of this compound, which probably is bound to cell structures and/or proteins. No remarkable difference can be found for the MIG/Ni fume between extracellular and intracellular areas. This fume contains a high amount of nickel in the form of NiO or other nickel oxides of the spinel type, which are known to be highly insoluble. This may explain that no breakdown of this fume occurred during at least the first three days of incubation.

The MMA/SS welding fumes have a rapid and important cytotoxic effect on GPAMs. Electron microscopy followed by EDS analyses reveals the existence of different types of particles with characteristic elemental composition. These results are rather similar to the observations described in previous investigations analyzing welding fume particles contained in alveolar

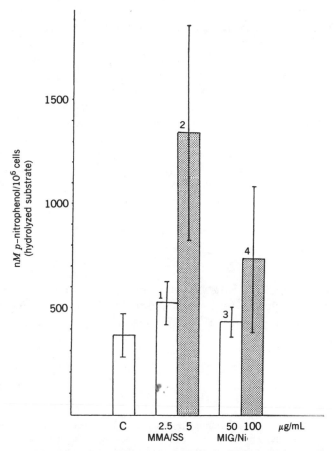

Figure 26.8. Measurements of β-glucuronidase activity in GPAMs after 48 hr incubation with MMA/SS and MIG/Ni; note the strong increase of activity with respect to the welding fume concentration: (1) $p < 0.3$, (2) $p < 0.1$, (3) $p < 0.9$, (4) $p < 0.3$, with respect to the control (C).

and interstitial macrophages in lung tumors (Anttila et al., 1984) and in rat lungs after inhalation experiments (Anttila, 1986).

For MMA/SS welding fumes, some other remarkable events occur in alveolar macrophages indicating an obvious metabolization of at least some of its components. Most importantly, a phosphorus peak appears in the EDS spectra. The phosphorus could be either of exogenous origin (deriving from stick electrodes, wires, or buffer) or from an endogenous source from the cell itself. Phosphorus has been revealed neither by quantitative chemical analysis of the native fumes nor in EDS spectra of medium-generated precipitates (see Chapter 27). So the exogenous origin may be excluded. In our previous in vitro experiments and in vivo observations, we frequently have observed the presence of phosphorus in other cell assays with nickel compounds where no exogenous phosphorus was present (Hildebrand et al., 1985, 1986, 1987; Chapter 27) and in precipitates next to orthopedic implants (Hildebrand et al., 1988). These observations lead us to suggest that there may exist one or more organic or inorganic metal–phosphorus complexes that may be considered as metabolites.

The second observation is the presence of tiny particles bound preferentially to membranes, surfactant, and euchromatin. Since similar precipitates deriving from crystalline Ni_3S_2 have been observed in in vitro assays and have been shown to contain phosphorus (Hildebrand et al., 1985, 1987; Chapter 27), it is possible that also the microparticles in the in vitro assay with MMA/SS fumes represent metabolites interacting with the phosphate groups of phospholipids of DNA or RNA and/or of phosphorus-transferring proteins.

It can be concluded that the cytotoxic effect of welding fumes depends primarily on its solubilization rate and in this way on the bioavailibility of its constituents.

ACKNOWLEDGMENTS

We thank Anne-Sophie de Saint Mahieu for her skillful technical assistance. This work was supported by a grant from the European Communities of Steel and Coal (Convention No. 7248/22/048 and by regional funds (G. I. S. "Travail-Santé").

REFERENCES

Anttila, S. (1986). Transmission electron microscopy and X-ray microanalysis in environmental lung pathology. Studies on inhaled inorganic particles and related tissue reactions. Acta Univ. Oul. D 137. *Anat. Pathol. Microbiol.* **21,** 1–65.

Anttila, S., Sutinen, S., Paakko, P., Alapieti, T., Peura, R., and Sivonen, S. J. (1984. Quantitative X-ray microanalysis of pulmonary mineral particles in a patient with pneumoconiosis and two primary lung tumours. *Br. J. Ind. Med.* **41,** 468–473.

Beaumont, J. J., and Weiss, N. S. (1981). Lung cancer among welders. *J. Occup. Med.* **23**, 839–844.

Costa, M., Mollenhauer, H. M., and Jones, M. K. (1980). Carcinogenic activity of nickel compounds may be related to their cellular uptake. *Fed. Proc.* **39**, 395.

Frazier, M. E., and Andrews, T. K. (1979). *In vitro* clonal growth assay for evaluating toxicity of metal salts. In N. Kharash (Ed.), *Trace Metals in Health and Disease*. Raven Press, New York, pp. 71–81.

Hansen, K., and Stern, R. (1983). *In vitro* toxicity and transformation potency of nickel compounds. *Environ. Health Perspect.* **51**, 223–226.

Hansen, K., and Stern, R. M. (1985). Welding fumes and chromium compounds in cell transformation assay. *J. Appl. Tox.* **5**, 306–314.

Hildebrand, F. F., and Biserte, G. (1979). Nickel-subsulfide-induced leiomyosarcoma in rabbit white skeletal muscle. *Cancer*, **43**, 1358–1374.

Hildebrand, F. F., Collyn-D'Hooge, M., and Herlant-Peers, M. C. (1985). Incorporation of — Ni3S2 and NiS into human embryonic pulmonary cells in culture. In S. S. Brown and F. W. Sundermann, Jr. (Eds.), *Progress in Nickel Toxicology*. Blackwells Ltd., London, pp. 61–66.

Hildebrand, H. F., Collyn-D-Hooge, M., and Stern, R. M. (1986). Cytotoxic effects of welding fumes on human embryonic epithetlial pulmonary cells in culture. In *Proceedings of the Intern. Conf. on Health Hazards and Biol. Effects of Welding Fumes*. Elsevier, Amsterdam, pp. 319–324.

Hildebrand, H. F., Decaestecker, A. M., and Hetuin, D. (1987). Binding of nickel sulfides to lymphocyte subcellular structural. In *Trace Elements in Human Health and Disease*. WHO–Environmental Health Series, Copenhagen, Vol. 20, 82–85.

Hildebrand, H. F., Ostapczuk, P., Mercier, J. F., Stoeppler, M., Roumazeille, B., and Decoulx, J. (1988). Orthopaedic implants and corrosion products. Ultrastructural and analytical studies of 65 patients. In H. F. Hildebrand and M. Champy (Eds.), *Biocompatibility of Co-Cr-Ni Alloys*. NATO-ASI Series A, Plenum, New York, Vol. 158, pp. 133–153.

Kalliomaki, P. L., Korhonen, O., Vaaranen, V., Kalliomaki, K., and Koponen, M. (1978). Lung retention and clearance of shipyard welders. *Ind. Arch. Occup. Environ. Health* **42**, 83–90.

Kalliomaki, P. L., Tuomisaari, M., Lakomaa, E. L., Kalliomaki, K., and Kivela, R. (1983). Retention and clearance of a stainless steel shieldgas welding fumes in rat lungs. *Am. Ind. Hyg. Assoc. J.* **44**, 649–654.

Norseth, T. (1981). The carcinogenicity of chromium. *Environ. Health Perspect.* **40**, 121–130.

Stern, R. M. (1983). Assessment of risk of lung cancer for welders. *Arch. Environ. Health* **38**, 148–155.

Stern, R. M. (1987). Cancer incidence among welders: possible effects of exposure to extremely low frequency electromagnetic radiation (ELF) and to welding fumes. *Environ. Health Perspect.* **76**, 221–229.

Stern, R. M., and Pigott, G. H. (1983). *In vitro* RPM fibrogenic potential assay of welding fumes. *Environ. Health Perspect.* **51**, 231–236.

Sunderman, F. W., Jr. (1981). Recent research on nickel carcinogenesis. *Environ. Health Perspect.* **40**, 131–141.

Szasz, G. (1967). Die Bestimmung derB -Glucuronidase Aktivität im Serum mit *p*-Nitrophenyl Glucuronid. *Clin. Chim. Acta* **15**, 275–282.

Thomsen, E., and Stern, R. M. (1979). A simple anlytical technique for the determination of hexavalent chromium in welding fumes and other complex matrices. *Scand. J. Work Health* **5**, 386–403.

27

INTERACTION OF Ni₃S₂ WITH PLASMA MEMBRANE OF LUNG CELLS

P. Shirali, H. F. Hildebrand, and A. M. Decaestecker

Institut de Médecine du Travail, 59045 Lille Cedex, France

C. Bailly and J. P. Henichart

INSERM Unité 16, 59045 Lille Cedex, France

R. Martinez

CNRS UA 234, Université des Sciences et Techniques de Lille, F 59655 Villeneuve d'Ascq Cedex, France

Nickel and Human Health: Current Perspectives, Edited by Evert Nieboer and Jerome O. Nriagu.
ISBN 0-471-50076-3 © 1992 John Wiley & Sons, Inc.

1. INTRODUCTION

Clinical, pathological and experimental investigations have shown that lung is the major target organ of nickel incorporation (Sunderman and Sunderman, 1963) and that certain nickel-compounds may induce pulmonary tumors in humans and animals (see Chapter 30, 46–50). In experimental animals, one of the most carcinogenic nickel compound is αNi_3S_2 (Suderman, 1984). Although studies on the metabolism, toxicity, and possible carcinogenicity of nickel compounds in humans and animals have generated a great deal of interest in metal-induced diseases, little information exists on the cell structures and proteins implicated in the transport, binding, and/or storage of nickel in vivo and even less of its metabolism.

Sunderman et al. (1981) have demonstrated nickel-binding proteins in renal cytosol, and in our laboratory we have obtained evidence for the existence of several nickel-binding proteins in lung and liver subcellular fractions (Herlant-Peers et al., 1983). In both investigations $^{63}NiCl_2$ incorporation was carried out in vitro and in vivo. We have also studied the incorporation of αNi_3S_2 and βNiS into human embryonic pulmonary epithelial cells (L132) (Hildebrand et al., 1986) and into lymphocytes from rat and human peripheral blood (Hildebrand et al., 1987). We concluded that the uptake of two related compounds was different: βNiS crystals are preferentially phagocytized in their original form; αNi_3S_2 crystals, however, are metabolized into very tiny particles (10–30 nm) that are observed bound to the cell membranes, euchromatin, mitochondria, and Golgi system in rat lymphocytes. Interestingly, the binding to cellular structures of human lymphocytes depends on the allergic sensitization state of the donors; that is, lymphocytes of sensitized persons bind these tiny particles to a greater extent than lymphocytes of unsensitized persons. One of the most important results was obtained by X-ray microprobe analysis (EDS). The EDS diagram of extracellular αNi_3S_2 crystals showed strong sulfur and nickel peaks, whereas the tiny particles bound to cell membranes and euchromatin no longer contained sulfur but only phosphorus and nickel. This observation strongly suggests the formation of a nickel–phosphorus complex with the phosphate groups of membranous phospholipids and of DNA in euchromatin, at least in the cell types studied.

The L132 cells are excellent for cytotoxic tests but should be considered from the physiological point of view as "ghost cells." For this reason, we have decided to perform our metabolism and incorporation experiments on cells physiologically more realistic, namely, alveolar macrophages of guinea pig (GPAMs).

2. MATERIALS AND METHODS

The L132 cells derived from normal human embryonic lung epithelium were cultured in plastic bottles in MEM 0011 medium (Eurobio) supplemented with 5% fetal calf serum. Alveolar macrophages were obtained after ablation of guinea pig (Hartley albinos) lungs by bronchoalveolar lavage with Hanks saline solution. The cell suspension was centrifuged at $500g$ and the pellet was resuspended in MEM 2011 medium and cultured in plastic bottles after adding 5% fetal calf serum.

The L132 and GPAM cells were cultured in the presence of 50 μM αNi_3S_2 (INCO, Toronto) for three days. For GPAM cells, the incorporation of αNi_3S_2 was assessed in a time course study with observations at 12, 24, 48, and 72 hr. Ultrastructural investigation was conducted after each incubation time. The material was prepared for electron microscopy employing routine fixation and embedding methods (Hildebrand and Biserte, 1979). The EDS X-ray microprobe analysis was performed on unstained sections on Pioloform F–coated titanium grids or carbon-coated nylon grids (Jeol 200 CX electron microscope connected to a Link analytical system QX 200).

Membrane fluidity measurements involved both L132 and GPAM cells (three days/incubation with αNi_3S_2) by steady-state fluorescence polarization analysis (Van Blitterswijk et al., 1981). It involves the labeling of cells in 0.9% NaCl at 37°C with 1,6-diphenyl-1,3,5-hexatriene (DPH) during a 30-min incubation period. The lipid structural order parameter (S_{DPH}) was calculated according to the formula given by Van Blitterswijk et al.

3. RESULTS

3.1. Electron Microscopy

The ultrastructural study reveals prominent changes in the shape and the size of L132 cells as well as of their nuclei. In contrast to control cells, Ni_3S_2-treated cells are retracted and globular and exhibit a high number of microvilli (Fig. 27.1). The Ni_3S_2 is observed in crystalline form in extracellular

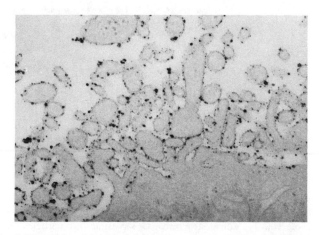

Figure 27.1. An L132 cell incubated for three days with 50 μM Ni_3S_2. Note the high number of microvilli strongly labeled with tiny electron-dense particles ($\times 24,000$).

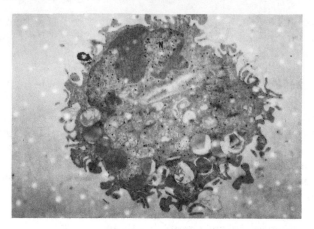

Figure 27.2. Guinea pig alveolar macrophage incubated for three days with 50 μM Ni₃S₂. Tiny particles are present on the cell membrane, intracellular structures, and euchromatin of the nucleus (N) (×5,800).

spaces. In association with cells, however, it is observed as small particles with an average size of 10 nm that are bound to the membrane of intracellular vesicles, to lysosomal structures, and especially to the outer layer of cell membranes (Fig. 27.1). Some of these particles can also be observed in extracellular areas.

Similar ultrastructural changes are observed in GPAMs, which also exhibit a significant increase in microvilli with respect to the control cells (Fig. 27.2). For these cells we also observe the tiny Ni₃S₂-derived particles (10–30 nm). They are bound preferentially, although less marked than in lymphocytes, to mitochondria, Golgi apparatus, and peroxisomes (Fig. 27.3). A remarkably

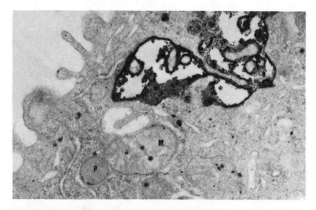

Figure 27.3. A GPAM cell incubated for three days with 50 μM Ni₃S₂ presenting a phagocytic vacuole with lysosomal structures. Cytoplasmic tiny particles are preferentially bound to mitochrondria (M) and peroxisomes (P) (×21,500).

high concentration of the Ni_3S_2-derived particles is observed on the cell membrane and in nuclei, with a striking preference for euchromatin (Fig. 27.2). Only in some GPAM cells did we detect crystalline particles in phagocytic vacuoles; the membranes and the lysosomal structures of the vacuoles are totally covered with tiny particles (Fig. 27.3).

In order to assess the binding of the tiny particles to the GPAM cell membrane, we performed a time course study. At 12 hr incubation, most particles were bound to the outer layer of the cell membrane; at 24 hr, they occurred on both sides, and after 48 hr and especially after 72 hr, they were localized at the inner layer of the membrane (Fig. 27.4).

3.2.   X-ray Microprobe Analysis

In extracellular Ni_3S_2 crystalline particles, Ni and S are found in a ratio very near to the initial chemical formula. Particles in phagocytic structures such as shown in Figure 27.3 essentially contain Ni, some Cl, and very little S. The main constituents of the tiny particles localized on cell membranes (cf. Fig. 27.1) and on euchromatin in strongly labeled nuclei, however, are Ni and P. They also contain some Cl and Ca, but significant S peaks can no longer be detected (Fig. 27.5).

3.3.   Steady-State Fluorescence Polarization Analysis

Fluorescence polarization analysis of DPH-labeled L132 and GPAM cells reveals a significant increase of membrane fluidity with respect to the control

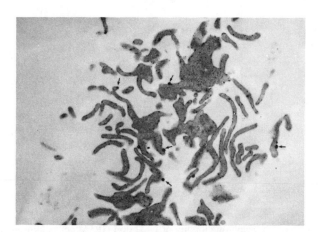

Figure 27.4. Microvilli area of a GPAM incubated for three days with 50 μM Ni_3S_2. Note the presence of tiny particles on the inner layer of cell membrane; some bigger particles (arrows) on the outer layer can still be detected (×9,100).

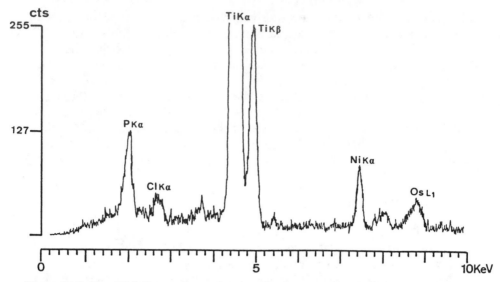

Figure 27.5. The EDS X-ray microprobe analysis of area shown in Figure 27.1 (L132 cell). Note the presence of a phosphorus peak. Sulfur is no longer detectable. Section was analyzed on a titanium grid.

cells (Fig. 27.6). Since the lipid structural order parameter (S_{DPH}) may assume a value between 0 and 1, representing a free fluid and an immobilized environment, respectively, the binding of Ni_3S_2 metabolite particles to cell membrane modifies its lipidic organization, which reduces the membrane rigidity, at least for these two cell types.

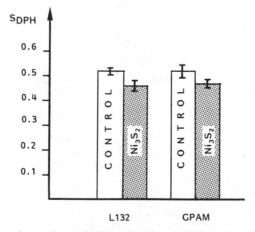

Figure 27.6. Difference in membrane fluidity of L132 and GPAM cells cultured in the presence of 50 μM Ni_3S_2 for three days: S_{DPH}, lipid structural order parameter as revealed by diphenyltrihexene (DPH) incorporation. The effect αNi_3S_2 was statistically significant ($p < 0.1$).

4. DISCUSSION

Our previous studies on L132 cells (Hildebrand et al., 1985) and on rat and human lymphocytes (Hildebrand et al., 1987) demonstrate clearly that there exist, at least in these three cell types, two different pathways of Ni_3S_2 incorporation and metabolism. The first mechanism involves the phagocytosis of the Ni_3S_2 crystalline particles, with subsequent lysis after fusion with lysosomes. The EDS analysis of these vacuoles reveals a decrease of sulfur depending on the degree of lysosomal action. The second pathway probably begins with an initial extracellular degradation of crystalline structures into tiny particles, which rapidly bind to the outer layer of the cell membrane (12 hr). In L132 cells, they rest primarily on the outer layer, while in GPAMs and lymphocytes, they pass through the membrane and can consequently be found on cytoplasmic and nuclear membranes and on euchromatin in nuclei. These membrane-bound, cytoplasmic and intranuclear particles are shown by EDS to contain phosphorus but not sulfur.

Similar results for the binding of Ni_3S_2 to phagocytic membranes of tumor cells, peritoneal macrophages, and rat spleen cells are described by Berry et al. (1984) and in CHO cells by Costa and Mollenhauer (1980). None of these authors, however, mentioned the sulfur–phosphorus shift, which we now have clearly demonstrated for three different cell types. All these cells can be considered as target cells either in lung disease or in immunological reactions. These new observations of Ni_3S_2 metabolization—both structural and chemical changes—and its preferential binding to cell membranes and euchromatin reinforce our hypothesis of the formation of a complex with the phosphate groups of membranous phospholipids and of nuclear DNA and/or RNA.

There is also evidence that MEM medium itself generates small particles, and it may be suggested that Ni_3S_2 dissolved in the medium precipitate with the NaH_2PO_4 contained in it. The EDS spectra of these medium-generated particles, however, have never revealed a significant increase in the phosphorus peak but have shown a progressive decrease in the sulfur peak. So we think that these particles represent organic nickel metabolites and that the cells are needed as a cofactor that transforms the primary medium–generated particles into phosphorus or phosphate-containing metabolites during the binding process to cellular structures. In addition, the phosphorus cannot derive from the phosphate buffer used for electron microscope fixation, since identical particles are also observed after fixation with the Na–cacodylate buffer system.

It is well established that one of the principal binding sites for metals on the nucleic acid molecules are the phosphate groups and the 2-OH groups in the case of RNA (Eichhorn et al., 1979). Thus the observed sulfur–phosphorus shift for intranuclear particles can be explained by the binding of a Ni_3S_2 metabolite to phosphate groups of the euchromatinic part of DNA or of RNA, the latter likely resulting in the weakening of the internucleotide linkages requiring the presence of the 2'-hydroxyl group. The resulting oligomers and mononucleotides may precipitate with nickel as the tiny aggregates observed by electron microscopy. Such a suggestion is in agreement with Lee et al. (1982), who demonstrated the interaction of Ni_3S_2 with DNA. Furthermore, Beach and Sunderman (1969, 1970) and Sunderman and Esfahani (1968) have shown evidence for the inhibition of RNA polymerase activity in hepatic nuclei. It is generally known that RNA polymerase is located in nucleoli and euchromatin and that euchromatin is the active part of nuclei in the s phase. The cellular pathway proposed by Costa and Heck (1982) and Sen and Costa (1986) for crystalline nickel sulfide, namely, uptake by phagocytosis and interaction with heterochromatin in studies with Chinese hamster ovary cells, is not confirmed by our observations. However, this difference may be due to the different cell types studied and the length of the incubation time employed.

The interaction of Ni_3S_2 with cell membranes is a rather complex process, and several phenomena should be considered. The tiny phosphorus-containing particles may appear on membranes after cleavage of choline from phosphatidylcholine and consequent binding of a primary Ni_3S_2 metabolite to the phosphate group. This possibility does not explain the increase of membrane fluidity, which is induced either by desaturation of aliphatic chains (i.e., formation of double bonds) or directly or consequently by cleavage of the fatty acid chains. On the other hand, the degradation of unsaturated fatty acids is also induced by lipid peroxidation, and Sunderman et al. (1985) have shown evidence for a significant increase in lipid peroxidation in liver, kidney, and lung of rats after injection of $NiCl_2$. Furthermore, the transmembranous passage in GPAM cells may be facilitated by the binding of a Ni_3S_2 metabolite to certain proteins and enzymes participating in the transport of phospholipids and the transfer of phosphate groups, respectively.

In conclusion, our investigation demonstrates clearly that there exist different pathways for the breakdown of αNi_3S_2 and of its incorporation into the cell and its compartments. Consequently, phagocytosis should not be considered as the only criterion determining cytotoxic effects. At least this is so for the cell types studied. In addition, our investigation reinforces our hypothesis of simultaneous epigenetic and genotoxic effects of αNi_3S_2. One of the most important aspects is the interaction with and the passage through the cellular membrane.

ACKNOWLEDGMENT

We thank Anne-Sophie de Saint Mahieu for skillful technical assistance. This work was supported by a grant from the European Communities of Steel and Coal (Convention No. 7248/22/048) and by regional funds (G. I. S. Travail-Santé).

REFERENCES

Beach, D. J., and Sunderman, F. W., Jr. (1969). Nickel carbonyl inhibition of ^{14}C-orotic acid incorporation into rat liver RNA. *Proc. Soc. Exptl. Biol. Med.* **131,** 321–322.

Beach, D. J., and Sunderman, F. W., Jr. (1970). Nickel carbonyl inhibition of RNA synthesis by a chromatin-RNA polymerase complex from hepatic nuclei. *Cancer Res.* **30,** 48–50.

Berry, J. P., Galle, P., Poupon, M. F., Pot-Deprun, J., Chouroulinkov, I., Judde J. G., and Dewailly, D. (1984). Electron microprobe *in vitro* study of interaction of carcinogenic nickel compounds with tumor cells. In *Nickel in the Human Environment.* IARC Scientific Publication, Lyon, Vol. 53, pp. 153–164.

Costa, M., and Mollenhauer, H. H. (1980). Phagocytosis of nickel subsulfide particles during the early stages of neoplastic transformation in tissue culture. *Cancer Res.* **40,** 2688–2694.

Costa, M., and Heck, J. D. (1982). Specific nickel compounds as carcinogens. *Trends Pharm. Sci.* **3,** 408–420.

Eichhorn, G. L., Shin, Y. A., Clark, P., Rifkind, J., Pitha, J., Tarien, E., Rao, G., Karlik, S. J., and Crapper, D. R. (1979). Essential and deleterious effects in the interaction of metalions with nuclei acids. In N. Kharasch (Ed.), *Trace Metals in Health and Disease*, Raven Press, New York, pp. 123–133.

Herlant-Peers, M. C., Hildebrand, H. F., and Kerckaert, J. P. (1983). *In vitro* and *in vivo* incorporation of [63]Ni (II) into lung and liver subcellular fractions of Balb/C mice. *Carcinogenesis* **4,** 387–392.

Hildebrand, H. F., and Biserte, G. (1979). Nickel-subsulfide-induced leiomyosarcoma in rabbit white skeletal muscle. *Cancer* **43,** 1358–1374.

Hildebrand, H. F., Collyn-d'Hooghe, M., and Herlant-Peers, M. C. (1985). Incorporation of Ni_3S_2 and NiS into human embryonic pulmonary cells in culture. In *Progress in Nickel Toxicology*, S. S. Brown and F. W Sunderman, Jr. (Eds.), Blackwells Ltd., London, pp. 61–66.

Hildebrand, H. F., Collyn-d'Hooghe, M., and Herlant-Peers, M. C. (1986). Cytotoxicité de dérivés du Nickel et leur incorporation dans les cellules épithéliales pulmonaires embryonaires humaines. *LARC Méd.* **6,** 249–251.

Hildebrand, H. F., Decaestecker, A. M., and Hetuin, D. (1987). Binding of Nickel sulfides to lymphocyte subcellular structures. In *Trace Elements in Human Health and Disease.* WHO-CEC-EPA Environmental Health Series, Vol. 20, pp. 82–85.

Lee, J. E., Cicarelli, R. B., and Jennette, K. W. (1982). Solubilization of the carcinogen nickel subsulfide and its interaction with deoxyribonucleic acid and protein. *Biochemistry* **21,** 771–778.

Sen, P., and Costa, M. (1986). Pathway of nickel uptake influences its interaction with heterochromatic. DNA. *Toxicol. Appl. Pharm.* **84,** 278–285.

Sunderman, F. W., Jr. (1984). Carcinogenicity of nickel compounds in animals. IARC Scientific Publication, in *Nickel in the Human Environment.* Lyon, Vol. 53, pp. 127–142.

Sunderman, F. W., Jr., and Esfahani, M. (1968). Nickel carbonyl inhibition of RNA polymerase activity in hepatic nuclei. *Cancer Res.* **28,** 2565–2567.

Sunderman, F. W., Jr., and Sunderman, F. W. (1963). Studies of nickel carcinogenesis. The subcellular partition of nickel in lung and liver following inhalation of nickel carbonyl. *Am. J. Clin. Pathol.* **40,** 563–575.

Sunderman, F. W., Jr., Costa, E. R., Fraser, C., Hui, G., Levin, J. J., and Tse T.P.H. (1981). [63]Nickel-constituents in renal cytosol of rats after injection of [63]$NiCl_2$. *Ann. Clin. Lab. Sci.* **11,** 448–496.

Sunderman, F. W., Jr., Zaharia, O., Marzouk, A., and Reid, M. C. (1985). Increased lipid peroxidation in liver, kidney, and lung of NiCl$_2$ treated rats. In S. S. Brown and F. W. Sunderman, Jr. (Eds.), *Progress in Nickel Toxicology*, Blackwells Ltd., London, pp. 88-93.

Van Blitterswijk, W. J., Van Hoeven, R. P., and Van Der Meer, B. W. (1981). Lipid structural order parameters (reciprocal of fluidity) in biomembranes derived from steady-state fluorescence polarisation measurements. *Biochem. Biophys. Acta* **644,** 323–332.

28

PHAGOCYTOSIS AND TRANSFORMATION OF PRIMARY EPITHELIAL RESPIRATORY CELLS BY αNi_3S_2

Kari Feren, George Farrants, and Albrecht Reith

Laboratory for Experimental Pathology and Image Analysis, Institute for Cancer Research, The Norwegian Radium Hospital, Montebello, N-0310 Oslo 3, Norway

1. INTRODUCTION

In humans, the respiratory tract is the predominant target organ for nickel and its compounds. Several studies have shown an excess risk of respiratory cancers in nickel refinery workers (Bridge, 1953; Doll, 1958; Pedersen et al., 1973; see Chapters 46 to 50). This has been attributed to the inhalation of

Nickel and Human Health: Current Perspectives, Edited by Evert Nieboer and Jerome O. Nriagu.
ISBN 0-471-50076-3 © 1992 John Wiley & Sons, Inc.

nickel compounds (Sunderman, 1977). In experimental animals the situation has not been as clear-cut (Reith and Brøgger, 1984), although particulate nickel compounds have been found to induce respiratory tumors (Yarita and Nettesheim, 1978; Sunderman, 1984; Pott et al., 1987). In vitro studies suggest that both the cytotoxic and carcinogenic effects of particulate compounds are dependent on the phagocytotic uptake of the particles (Costa et al., 1981).

Studies of phagocytosis and transformation in vitro commonly employ cells of fibroblastic origin, which, in addition to being nonrespiratory, often have a low capacity to phagocytize particles. Moreover, these cells are usually established cell lines and thus are already partially transformed. To avoid these disadvantages, we have chosen a bioassay using normal primary epithelial cells from the rat trachea (RTE cells). These cells have a better phagocytotic capacity than fibroblasts and have a limited growth potential (Pai et al., 1983). Fortunately, this limitation can be overcome by carcinogens, which give rise to transformed cells in a dose-dependent manner. Some of these transformed colonies produce cell lines that in turn may lose their anchorage dependence and/or give rise to tumors in nude mice (Nettesheim and Barrett, 1984).

Our studies show that RTE cells are capable of phagocytosis and are transformed by particulate nickel compounds such as αNi_3S_2 and welding fumes. We also report the use of scanning electron microscopy in the backscattered electron mode to detect phagocytosed αNi_3S_2 particles. This method gives more precise characterization of particles than light microscopy.

2. MATERIALS AND METHODS

Cells were isolated from tracheas of specific pathogen-free Fisher 344 male rats using 1% pronase in minimal essential medium (MEM) as described by Pai et al. (1983). Isolated cells were grown on collagen films in a medium consisting of a 1 : 1 mixture of Ham's F12 and 3T3 conditioned Dulbecco's modified Eagle medium (DMEM) with hormones and growth factors added (Wu et al., 1982).

The αNi_3S_2 was obtained from Falconbridge (Ontario, Canada) and was ground in an agar mortar and filtered through a 5-μm Millipore filter to obtain particles <5 μm. Fumes from manual metal arc welding of stainless steel (MMA/SS) with Nichroma 160 electrodes (Norweld, Norway) were collected on filters by welding in a Swedish fume box. All particles were <5 μm as confirmed by transmission electron microscopy.

For the phagocytosis studies, $3 \times 10^5 - 4 \times 10^5$ cells were seeded on 60-mm dishes and allowed 24 hr for attachment. They were then exposed to 20 μg/mL αNi_3S_2 or MMA/SS fumes for different time intervals. After exposure the cells were fixed in 2% glutaraldehyde in 0.1 M cacodylate/0.1 M sucrose buffer (pH 7.3) at room temperature, dehydrated in ethanol, and subjected to critical-point drying. They were examined in a Phillips scanning electron

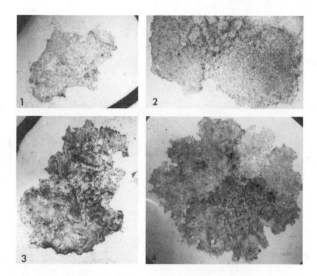

Figure 28.1. Transformation of RTE cells gives rise to four different types of colonies. Of these, types 1 and 2, composed of relatively small colonies with large pale cells, are considered nontransformed. The type 3 and 4 colonies, which are larger with small densely stained cells often occurring in several layers, are both considered transformed.

microscope (PSEM 500) using both secondary (SEI) and backscattered (BEI) electron imaging.

In the cytotoxicity and transformation tests 10^4 and 5×10^4 cells, respectively, were seeded on 60-mm dishes. One day later, the cells were exposed to different concentrations of αNi_3S_2 for 24 hr. Cytotoxicity was measured by counting Giemsa stained colonies (over eight cells) four to five days after exposure. In the transformation studies, cells were allowed to grow for five weeks before fixation and staining (Fig. 28.1).

3. RESULTS AND CONCLUSIONS

3.1. Phagocytosis

Phagocytosis was studied by scanning electron microscopy and image analysis. Particles inside the cells were detected using BEI, while surface particles were excluded from counting in the secondary electron image. Particle size and distribution were determined using image analysis of the combined BEI–SEI pictures. As indicated in Figure 28.2, welding fumes and αNi_3S_2 were phagocytized to the same extent after 6 hr; at earlier stages welding fumes were more actively phagocytized. This phenomenon is probably due to the existence of very small particles in the welding fumes that are not found in αNi_3S_2 suspensions. At 24 hr, 75% of the αNi_3S_2-exposed cells

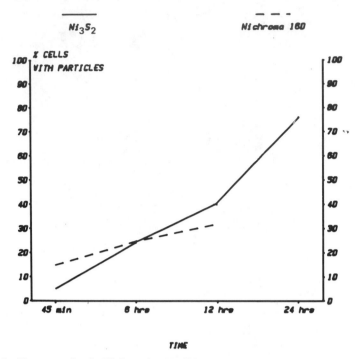

Figure 28.2. Phagocytosis of αNi_3S_2 and MMA/SS welding fumes expressed as percentage of cells with one or more intracellular particles.

contained one or more particles. The cytotocity of welding fumes was, however, so acute that all cells were dead at this stage (Fig. 28.3a). This severe cytotoxic effect is probably a consequence of the release of soluble chromates. The chromates are also thought to cause the blackening of the nuclear area seen in Figure 28.3b. Combined SEI–BEI showed no preferential localizations of the particles within the cells (Figs. 28.4a,b). Image analysis of the cells revealed large variations in the size of the phagocytized particles (Fig. 28.4c).

3.2. Cytotoxicity and Transformation

The cytotoxicity in the exposed cultures is measured by the percentage of colonies surviving relative to the control cultures (100%). Transformation frequencies are expressed as the total number of type 3 and 4 colonies (Fig. 28.1) per 1000 surviving cells found in the Giemsa stained dishes. The number of surviving cells were calculated as follows: number of cells seeded × CFE × total number of dishes, where CFE is the number of colonies divided by the number of cells seeded.

As is often found in biological studies, the cytotoxicity and transformation frequencies varied from one experiment to another. A 50% variation in

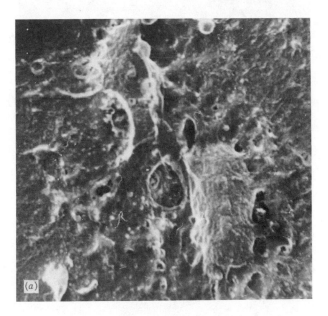

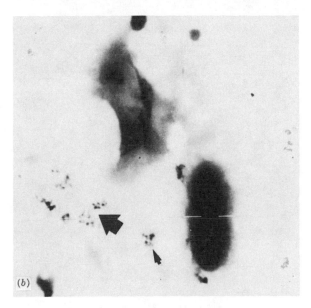

Figure 28.3. (*a*) Secondary electron image of cells exposed to welding fumes for 24 hr. Note that most cells are dead due to acute toxicity. (*b*) Backscattered electron image of the cells in (*a*). Phagocytized particles are marked with arrows. The blackening in the nuclear areas is suspected as originating from soluble Cr(VI), likely localized as Cr(III).

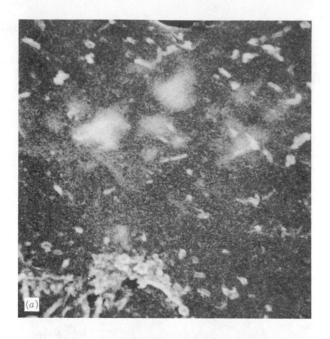

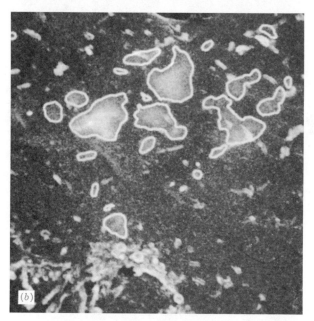

Figure 28.4. (*a*) Secondary electron image of a cell that has been exposed to αNi₃S₂ for 24 hr. By contrast to the cells in Fig. 28.3, this cell is still healthy. (*b*) Computer-combined SEI–BEI picture of the cell in (*a*) with phagocytized particles encircled for size measurement. (*c*) Particle diameter distribution of the particles inside the cell. The distribution of number of particles (*Y* axis) is plotted as a function of maximum diameter (arbitrary units) along the *X* axis. A total of 28 particles are present in the cell, with minimum diameter 0.1 μm, maximum 2.3 μm, mean 0.8 μm, and standard deviation 0.6 μm.

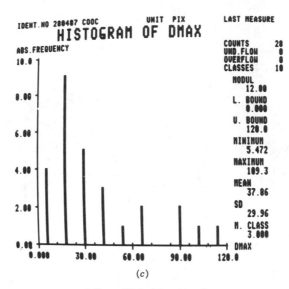

IDENT.NO 280487 COOC   UNIT PIX   LAST MEASURE
            HISTOGRAM OF DMAX
ABS.FREQUENCY   COUNTS   28
10.0   UND.FLOW   0
                                                  OVERFLOW   0
                                                  CLASSES   10
                                                  MODUL
8.00   12.00
                                                  L. BOUND
                                                     0.000
                                                  U. BOUND
6.00   120.0
                                                  MINIMUM
                                                     5.472
                                                  MAXIMUM
4.00   109.3
                                                  MEAN
                                                    37.86
                                                  SD
2.00   29.96
                                                  M. CLASS
                                                     3.000
0.00   DMAX
    0.000   30.00   60.00   90.00   120.0

                        (c)

Figure 28.4. (*Continued*)

relative survival of RTE cells after exposure to the same concentration of a substance may be expected (Paul Nettesheim, personal communication). This is also found with other primary cells, and the variation is often greater when the cells are grown on plastic or artificial "membranes" than on feeders or in collagen gels. This variation may reflect differences from cell batch to batch in adaptation to artificial growth substrates, in the ability to utilize growth factors, and in the bioavailability and biotransformation of toxic substances. Figure 28.5 shows the results from two separate transformation experiments with corresponding cytotoxicities. Note that despite differences in the response to αNi_3S_2 in the two experiments, the median lethal concentration (LC_{50}) doses produced the same number of transformants. The direct-acting carcinogen N-methyl-N-nitro-n-nitrosoguanidin (MNNG) was included as a positive control in each run. A similar variation in transformation frequencies was found with MNNG as for αNi_3S_2. Nevertheless, the ratio between the transformation frequencies for MNNG and αNi_3S_2 in different experiments remained constant. A similar invariant ratio was found when transformation frequencies of other substances were compared to MNNG, as can be seen from the comparison of transformation potentials at LC_{50} doses listed in Table 28.1. The most interesting observation in Table 28.1 is that αNi_3S_2 was almost three times as potent as MNNG as a transforming agent at the LC_{50} dose.

In conclusion, we have shown that the RTE cells actively phagocytize particulate matter and are suitable for testing the transformation potential of particulate compounds, even if the sensitivity varies from run to run. We

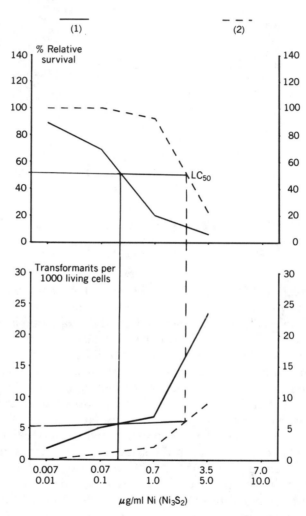

Figure 28.5. Relative survival and number of transformed colonies per 1000 survivors (transformation frequency) as function of dose (log scale) in two experiments. The concentrations are given both as micrograms of nickel per milliliter (upper scale) and of αNi_3S_2 (lower scale). Note that although both survival and transformation frequencies differ, the number of transformants at the LC_{50} dose is the same in both experiments.

Table 28.1   Comparisons of Transformation Potentials at LC$_{50}$ Doses of Transforming Agents in RTE Cell System[a]

Substance[b]	LC$_{50}$ ± SE (μg/ml)	Relative Transformation Potential ± SE
MNNG[c]	0.3 (12)	100% (8)
αNi$_3$S$_2$	2.5 ± 1.3 Ni (8)	270 ± 8% (3)
K$_2$Cr$_2$O$_7$	0.3 ± 0.07 Cr (8)	75 ± 15% (2)
Welding fumes	0.3 ± 0.09 Cr (10)	75 ± 14% (3)
	0.05 ± 0.01 Ni	

[a] Numbers in parentheses denote the number of experiments.
[b] For MNNG the exposure time was 1 hr; for the other substances it was 24 hr.
[c] For MNNG no standard error (SE) could be detected since the same dose was used in all experiments. The relative survival at this concentration varied from 40 to 80% with a mean of 53 ± 18%.

have not yet been able to establish cell lines and test their tumorigenicity in nude mice, but this work is in progress.

ACKNOWLEDGMENTS

This study is part of a joint German–Norwegian project supported by Bundesministerium für Forschung und Technologie and the Royal Norwegian Council for Scientific and Industrial Research.

REFERENCES

Bridge, J. C. (1933). *Annual Report of the Chief Inspector of Factories and Workshops for Year 1932*. HMSO, London, pp. 103–109.

Costa, M., Abbracchio, M. P., and Simmons-Hansen, J. (1981). Factors influencing phagocytosis, neoplastic transformation, and cytotoxicity of particulate nickel compounds in tissue culture systems. *Toxicol. Appl. Pharma.* **60**, 313–323.

Doll, R. (958). Cancer of the lung and nose in nickel workers. *Br. J. Ind. Med.* **15**, 217–223.

Nettesheim, P., and Barrett, J. C. (1984). Tracheal epithelial cell transformation: A model system for studies on neoplastic progression. *Crit. Rev. Toxicol.* **12**, 215–239.

Pai, S. B., Steele, V. E., and Nettesheim, P. (1983). Neoplastic transformation of primary epithelial cell cultures. *Carcinogenesis* **4**, 369–374.

Pedersen, E., Hoegetveit, A. C., and Andersen, A. (1973). Cancers of respiratory organs among workers at a nickel refinary in Norway. *Int. J. Cancer* **12**, 32–41.

Pott, F., Ziem, U., Reiffer, F.-J., Ernst, H., and Mohr, U. (1987). Carcinogenicity studies on fibres, metal compounds and some other dust in rats. *Exp. Pathol.* **32**, 129–132.

Reith, A., and Brøgger, A. (1984). Carcinogenicity and mutagenicity of nickel and nickel compounds. In F. W. Sunderman, Jr. (Ed.), *Nickel in the Human Environment*. IARC Scientific Publication, Lyon, No. 53, pp. 175–192.

Sunderman, F. W., Jr. (1977). A review of the metabolism and toxicology of nickel. *Ann. Clin. Lab. Sci.* **7**, 377–398.

Sunderman, F. W., Jr. (1984). Carcinogenicity of nickel compounds in animals. In F. W. Sunderman, Jr. (Ed.), *Nickel in the Human Environment*. IARC Scientific Publication, Lyon, No. 53, pp. 127–142.

Wu, R., Groelke, J. W., Chang, L. Y., Porter, M. E., Smith, D. and Nettesheim, P. (1982). Effects of hormones on the multiplication and differentiation of tracheal epithelial cells in culture. In D. Sirbasku, G. H. Sato, and A. Pardu (Eds.), *Growth of Cells in Hormonally Defined Medium*. Cold Spring Harbor Press, New York, pp. 641–656.

Yarita, T., and Nettesheim, P. (1978). Carcinogenicity of nickel subsulfide for respiratory tract mucosa. *Cancer Res.* **38**, 3140–3145.

29

CHEMILUMINESCENT DETECTION OF FREE RADICAL GENERATION BY STIMULATED POLYMORPHONUCLEAR LEUKOCYTES: IN VITRO EFFECT OF NICKEL COMPOUNDS

Peter H. Evans

MRC Dunn Nutrition Unit, Milton Road, Cambridge, England CB4 1XJ

Lindsay G. Morgan

Medical Department, INCO Europe Ltd., Clydach, Swansea, SA6 5QR, United Kingdom

Eiji Yano and Naoko Urano

Department of Public Health, Teikyo University School of Medicine, Itabashi Ku, Tokyo 173, Japan

Nickel and Human Health: Current Perspectives, Edited by Evert Nieboer and Jerome O. Nriagu.
ISBN 0-471-50076-3   © 1992 John Wiley & Sons, Inc.

1. INTRODUCTION

The present investigation examines the generation of oxygen free radicals by activated phagocytic cells that have been exposed to a variety of nickel compounds. The purpose is twofold: first, to see whether it is possible to develop a rapid, economical in vitro carcinogen screening assay for nickel compounds based upon the property of certain known human carcinogens to stimulate phagocyte-derived reactive oxygen metabolities (ROMs) and second, to investigate mechanisms of nickel-induced carcinogenesis.

The ROMs, namely superoxide and hydroxyl free radicals and hydrogen peroxide, may be produced in the process of normal cell oxidative metabolism or by activation of the respiratory burst in phagocytic cells such as macrophages and polymorphonuclear (PMN) cells. As well as having beneficial bactericidal activities, ROMs are also potentially damaging to a variety of tissue components, for example, membranes, enzymes, and DNA (Babior, 1984; Badwey and Karnovsky, 1980).

The possible pathogenic role of ROMs in nickel-induced cancer has been raised by a number of authors (Furst, 1984; Schlatter, 1985). Other authors have considered specific toxic effects that are of particular relevance to obtaining a greater understanding in the study of ROM-related carcinogenicity. These include tumor promotion (Goldstein et al., 1981; Cerutti, 1985), mutation (Weitzman and Stossel, 1981), DNA modification (Frenkel and Chrzan, 1987), and malignant cell transformation of cocultured cells (Weitzman et al., 1985).

Furthermore, inhalation of mineral dusts (Rola-Pleszczynski et al., 1984) and metallic nickel particles (Hueper, 1958) has been shown to stimulate the subsequent infiltration of phagocytes, alveolar macrophages, and PMN into the lung. It has been proposed (Costa and Mollenhauer, 1980) that endocytosis of particulate nickel compounds is correlated to their carcinogenic capacity. Evidence that the generation of ROMs by macrophages (Hatch et al., 1980) and PMNs (Doll et al., 1982; Evans et al., 1987a) stimulated by phagocytosed mineral dusts may be of direct etiological significance to their pathogenic activity prompted the present investigation into the ability of various nickel compounds to likewise stimulate the PMN respiratory burst, the cellular production of phagocyte-derived ROMs being monitored by means of luminol-dependent chemiluminescence using a dedicated luminometer. Luminol chemiluminescence is a complex chemical reaction of uncertain mechanism, whereby the interaction of luminol with oxidants (e.g., hydrogen peroxide and superoxide free radical) and catalysis by various transitional metal cations result in the emission of light (Campbell, 1988).

The utility of the PMN chemiluminescent technique to detect and evaluate putative tumorigenic nickeliferous particulate materials and hence provide a potentially valuable in vitro assay system in the toxicological assessment of nickel and possibly other metallic compounds is also examined.

2. METHODOLOGY AND MATERIALS

2.1. *In Vitro* PMN Chemiluminescent Analysis

The PMN cells were purified from samples of heparinized human blood by sedimentation in dextran to remove the majority of unwanted red blood cells and subsequently centrifuged with Ficoll-Paque to remove the remaining contaminating monocyte and lymphocyte white cells (Ferrante and Thong, 1980). Residual red cells were lysed in NH_4Cl–EDTA solution (Andersen and Amirault, 1979). The resultant purified (>95%) PMN cells were washed in calcium-free Krebs-Ringer HEPES buffer, pH 7.35, before final resuspension in Krebs-Ringer HEPES buffer in the presence of the chemiluminescent amplifier reagent ($10\mu M$) luminol (5-amino-2-dihydro-1,4-phthalazinedione). The chemiluminescent response of PMN (Campbell et al., 1985) to added suspensions of the various nickel compounds was measured at 37°C for 20 min on a Biolumat six-channel luminometer equipped with automatic printout of peak-height (cpm) and integral peak-area counts.

2.2. Sample Nickel Compounds

Samples supplied by the Nickel Producers Environmental Research Association (NiPERA) for examination included nickel subsulfide and nickel oxide samples letter coded as follows: A, B, F, G, H, I, and J, of determined physicochemical characteristics with median particle diameters occurring in the range $2–5\mu M$ (Sunderman et al., 1987). Replicate chemiluminescent assays were performed together with a zymosan test particulate (the positive standard sample) and a control blank buffer solution. Selection of an appropriate biologically "inert" particulate as a suitable negative control sample is problematical. Hence in order to standardize the presentation of the results obtained with the various nickel samples, they have been expressed relative to the chemiluminescent response levels obtained by using buffer alone and termed the PMN chemiluminscent index.

3. PMN CHEMILUMINESCENT RESPONSE TO NICKEL COMPOUNDS

3.1. PMN Chemiluminescent Dose–Response Relationship

The dose–response relationship of luminol-dependent PMN chemiluminescent activity to varying concentrations of nickel subsulfide is illustrated (Fig. 29.1). Evidence of a diminution of activity at the highest dose (800 μg/mL) due to a toxic effect on PMN viability is apparent; otherwise a direct quantitive, positive dose–response relationship is evident.

3.2. PMN Chemiluminescent Rank Response to Nickel Compounds

The rank order of the PMN chemiluminescent index for peak height of the tested nickel compounds is illustrated (Fig. 29.2). Comparison with the closely related index of integrated peak area showed that in each case nickel subsulfide exhibited the greatest activity, though the ranking of some of the nickel oxide samples (I and J and G and B) was dependent, due to interchange of near equivalent levels, on the particular parameter selected (i.e., peak height or peak area). The zymosan positive standard particulate chemiluminescent peak height index was 7.70 (SE = 2.20). The rank order of the PMN chemiluminescent index was as follows:

Peak height: nickel subsulfide $\gg$ I $\geq$ J $>$ F $>$ A $\gg$ G $\geq$ B $>$ H.
Peak area: nickel subsulfide $\gg$ J $\geq$ I $>$ F $>$ A $\gg$ B $\geq$ G $>$ H.

Whether the peak height (indicative of the maximal rate of free radical generation) or the peak area (indicative of the sum total of free radicals

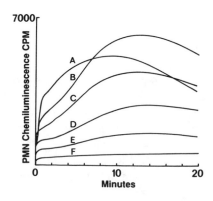

Figure 29.1. The PMN luminol-dependent chemiluminescent time reaction and dose–response relationship to stimulation by nickel subsulfide. Nickel subsulfide concentrations (μg/mL): *A*, 800; *B*, 400; *C*, 200; *D*, 100; *E*, 50; *F*, 0.

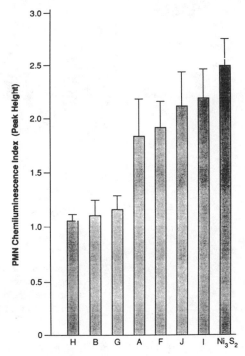

Figure 29.2. Rank order of luminol-dependent PMN chemiluminescent response to nickel subsulfide and letter-coded nickel oxide samples (400 µg/mL) (mean and standard
error). Nickel subsulfide and samples I, J, F, and A exhibit significantly enhanced
chemiluminescent activity ($p < 0.05$). The index value for this zymosan standard
was 7.70 ± 2.20.

generated, represents the more pertinent pathogenic process remains a matter
of conjecture, but this may be related to the level and regenerative capacity
of endogenous tissue antioxidants (e.g., superoxide dismutase, catalase, vi-
tamins C and E, selenium, and zinc).

3.3. Correlative Relationships of PMN Chemiluminescent Response

The relationship between the in vitro PMN chemiluminescent index (peak
height) and the incidence of tumors induced in rats in vivo by the nickel
compounds (data supplied by Dr. F. W. Sunderman to NiPERA) is shown
in the scattergram in Figure 29.3. It is of interest and perhaps of pathogenic
significance that both nickel compounds that produce 100% of rat tumors,
namely nickel subsulfide and nickel oxide code I, also exhibit the highest
peak chemiluminescent activity. A general positive, though weak, correlation
between the chemiluminescent index and tumorigenicity is apparent, with
samples H and F possibly representing false-negative and false-positive results,
respectively, in the PMN chemiluminescent assay.

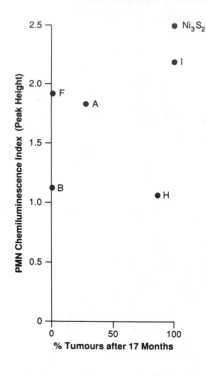

Figure 29.3. Relationship of luminol-dependent PMN chemiluminescent index (peak activity) to in vivo incidence of tumors at 17 months induced by nickel compounds in rats. Kendall's correlation coefficient is 0.35 ($p < 0.33$).

The relationship between the in vitro PMN chemiluminescent index (peak height) and the mean peak haematocrit (Sunderman et al., 1987), a proposed in vivo experimental index of nickel-related carcinogenicity, is shown in Figure 29.4. Again a general positive, though weak, correlation is evident.

4. PMN CHEMILUMINESCENCE AS *IN VITRO* PREDICTIVE ASSAY FOR CARCINOGENICITY

Several criteria for optimizing short-term, in vitro tests for toxicity and carcinogenicity may be considered to be desirable. Of these, a high predictive validity and reproducibility, an objective endpoint, a quantitative dose–response relationship, economy of scientific and technical expertise, time, effort, facilities and materials, general applicability over a range of toxic agents, and a scientific rationale that supports any empirical test correlation may be recognized as of particular pertinence and value.

Necessary caution in the use of the PMN luminol-dependent chemiluminescent technique as an in vitro assay system should be exercised for a variety of reasons.

(a) The specificity and reactivity of particular ROMs with luminol are uncertain with superoxide and hydrogen peroxide as reactants.

(b) Direct chemical interference of the test sample with the generated ROM may possibly either enhance or reduce the associated cell-mediated, luminol-dependent chemiluminescence.

(c) Differential contributions of intra- and extracellular derived luminol-dependent chemiluminescence may modify the overall light emission. One may assume that extracellular ROMs are of more direct pathogenic relevance to the potential carcinogenic effect.

(d) Absorption of emitted chemiluminescence by colored or opaque agents may limit detection of the emitted light.

(e) Purified human blood PMN should be of high purity, viability, and consistency.

However, despite these caveats, the PMN luminol-dependent chemiluminescent technique possesses several useful scientific and operational attributes. Scientific considerations include: the use of human, not animal-derived, cells eliminates species-specific differences; the use of primary cells eliminates cell line–acquired modifications; the technique permits the direct living cellular response to be monitored in real time without necessitating subsequent destructive biochemical assays; and the target PMN cell is of known pathogenic relevance to particle-mediated pathological damage. Likewise, several operational attributes are of special usefulness: the target PMN

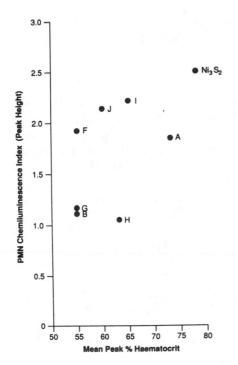

Figure 29.4. Relationship of luminol-dependent PMN chemiluminescent index (peak activity) to hematocrit index in rats exposed to nickel compounds. Kendall's correlation coefficient is 0.42 ($p < 0.16$).

cells are conveniently freshly available in large numbers and are readily purified; the luminometer detection system is sensitive, accurate, reproducible, and rapid, and the reagents required are stable, inexpensive, and readily available.

Despite the listed useful characteristics, additional studies are required to determine the complex cellular mechanism involved and to evaluate and validate properly the PMN chemiluminescent technique before the system can be confidently promoted as an appropriate in vitro predictive assay for metal- or particulate-induced carcinogens.

However, the proposed use of a similar spectrometric in vitro test system utilizing peritoneal macrophages to detect phorbol ester and related soluble tumor promoters (Ohkama et al., 1984) indicates that the described PMN chemiluminescence assay could well provide an alternative simple and worthwhile method for the assessment of tumor-promoting activity. Nevertheless, the complex methodological and interpretive problems in achieving a clear understanding of the relevant mechanisms involved in the phagocytic and ROM interactions with the multiple and interdependent particulate parameters (i.e., composition, crystal structure, size, shape, and charge) are formidable.

5. RATIONALE OF FREE RADICAL INVOLVEMENT IN NICKEL CARCINOGENESIS

Evidence for the possible pathogenic involvement of oxygen-derived free radicals in nickel-related toxicological reactions has been provided by studies showing lipid peroxidative alterations occurring in acute nickel toxicity (Sunderman, 1987), and similar oxidative changes have been implicated in nickel carcinogenesis (Athar et al., 1987). The finding that soluble nickel salts may depress the oxidative metabolism of phagocytosing macrophages (Castranova et al., 1980) while nickeliferous dusts stimulate PMN chemiluminescence (Evans et al., 1987b), a process evidently related to particulate surface charge (Abbracchio et al., 1982), indicates the complexity of the cellular interactions involved. The capacity of both the polypeptide GlyGlyHis (Nieboer et al., 1986) and human serum albumin protein (Nieboer et al., 1984) nickel complexes to participate in Ni(II)/(III) redox interactions provides additional evidence of the potential toxicological role of nickel-mediated oxidative interactions. Recent studies have demonstrated the capacity of nickel ion (Kawanishi et al., 1989 and nickel subsulfide (Kasprzak et al., 1989), in the presence of hydrogen peroxide, to produce hydroxyl-mediated DNA damage. Subsequent confirmation of the capacity of nickel sulfide compounds to generate PMN oxyradicals has also recently been demonstrated by Zhong et al. (1990).

In addition, it is also interesting to consider whether the ability of the tumor promoter phorbol myristate acetate (PMA) to stimulate ROM generation by PMNs is of relevance to the enhancement of iron uptake by erythroid cells (Hebbert and Morgan, 1985). This may be of significance to the biochemical mechanism of nickel-mediated erythropoiesis and hematocrit changes utilized to assess the carcinogenicity of nickel-related compounds (Sunderman and Hopfer, 1983). Carcinogenic bioassays involving intramuscular injection of nickel oxides into rats have also been related to their catalytic ROM reactivity (Sunderman et al., 1990).

Confirmation of the pathogenic role of ROM in the etiology of nickel-related cancers could have significant and important implications in evaluating possible antagonists to oxyradical stress. Epidemiological (Sasco et al., 1985) and experimental (Chatterjee et al., 1979) studies into the dietary and nutritional modulation of nickel toxicity together with consideration of the general role of antioxidant agents and micronutrients in countering tissue oxidative stress (Pryor, 1987; Evans et al., 1989) suggest avenues of research endeavor worthy of future study.

In the present study, the general positive correlation between PMN chemiluminescence activity and the putative carcinogenicity of nickel subsulfide and the other nickel oxide particulates may be considered as offering some supportive evidence for the proposed hypothetical pathogenic role of free radicals and related ROM in the etiology of nickel-related cancers. In the light of these findings, further detailed investigations to clarify the relevant specific physicochemical particulate parameters and cellular mechanisms involved, would seem to be of particular value.

ACKNOWLEDGMENTS

The authors are grateful to the Nickel Producers Environmental Research Association (NiPERA) for funding the project.

REFERENCES

Abbracchio, M. P., Heck, J. D., and Costa, M. (1982). The phagocytosis and transforming activity of crystalline metal sulfide particles are related to their negative surface charge. *Carcinogenesis* **3**, 175–180.

Andersen, B. R., and Amirault, H. J. (1979). Important variables in granulocyte chemiluminescence. *Proc. Soc. Exp. Biol. Med.* **162**, 139–145.

Athar, M., Hasan, S. K., and Srivastava, R. C. (1987). Evidence for the involvement of hydroxyl radicals in nickel mediated enhancement of lipid peroxidation: implications for nickel carcinogenesis. *Biochem. Biophys. Res. Commun.* **147**, 1276–1281.

Babior, B. M. (1984). Oxidants from phagocytes: agents of defence and destruction. *Blood* **64**, 959–966.

Badwey, J. A., and Karnovsky, M. L. (1980). Active oxygen species and the functions of phagocytic leukocytes. In E. E. Snell (Ed.), *Ann. Rev. Biochem.* **49**, 695–726.

Campbell, A. K. (1988). *Chemiluminescence: Principles and Applications in Biology and Medicine.* Ellis Horwood, Chichester, England.

Campbell, A. K., Hallett, M. B., and Weeks, I. (1985). Chemiluminescence as a tool in cell biology and medicine. In D. Glick (Ed.), *Methods in Biochemical Analysis.* **31**, 317–416.

Castranova, V., Bowman, L., Reasor, M. J., and Miles, P. R. (1980). Effects of heavy metal ions on selected oxidative metabolic processes in rat alveolar macrophages. *Toxicol. Appl. Pharm.* **53,** 14–23.

Cerutti, P. A. (1985). Proxidant stress and tumor promotion. *Science* **227,** 375–381.

Chatterjee, K., Charkarborty, D., Majumdar, K., Bhattacharyya, A., and Chatterjee, G. C. (1979). Biochemical studies of nickel toxicity in weanling rats—influence of vitamin C supplementation. *Intern. J. Vit. Nutr. Res.* **49,** 264–275.

Costa, M., and Mollenhauer, H. H. (1980). Carcinogenic activity of particulate nickel compounds is proportional to their cellular uptake. *Science* **209,** 515–517.

Doll, N. J., Stankus, R. P., Goldbach, S., and Salvaggio, J. E. (1982). In vitro effects of asbestos fibres on polymorphonuclear leukocyte function. *Int. Arch. Allergy Appl. Immun.* **68,** 17–21.

Evans, P. H., Campbell, A. K., Yano, E., and Goodman, B. (1987a). Phagocytic oxidant stress and antioxidant interactions in the pneumoconioses and dust-induced tumorigenic lung disease. In C. Rice-Evans (Ed.), *Free Radicals, Oxidant Stress and Drug Action.* Richelieu Press, London, pp. 213–235.

Evans, P. H., Morgan, L. G., Yano, E., and Campbell, A. K. (1987b). Chemiluminescent detection of free radical oxidant production by PMN phagocytosis of nickeliferous dusts. In S. Brown and Y. Kodama (Eds.), *Toxicology of Metals: Clinical and Experimental Research.* Ellis Horwood, Chichester, pp. 375–376.

Evans, P. H., Campbell, A. K., Yano, E., and Morgan, L. G. (1989). Environmental cancer, phagocytic oxidant stress and nutritional interactions. In J. C. Somogyi and H. R. Müller (Eds.), *Nutritional Impact of Food Processing.* Karger, Basle, pp. 313–326.

Ferrante, A., and Thong, T. H. (1980). Optimal conditions for simultaneous purification of mononuclear and polymorphonuclear leukocytes from human blood by the Hypaque-Ficoll method. *J. Immunol. Methods* **36,** 107–117.

Frenkel, K., and Chrzan, K. (1987). Hydrogen peroxide formation and DNA base modification by tumor promoter-activated polymorphonuclear leukocytes. *Carcinogenesis* **8,** 455–460.

Furst, A. (1984). Mechanism of action of nickel as a carcinogen: needed information. In F. W. Sunderman (Ed.), *Nickel in the Human Environment*, IARC Scientific Publication, Lyon, No. 53, 245–252.

Goldstein, B. D., Witz, G., Amoruso, M., Stone, D. S., and Troll, W. (1981). Stimulation of polymorphonuclear leukocyte superoxide anion radical production by tumor promoters in an epithelial cell line. *Cancer Lett.* **11,** 257–262.

Hatch, G. E., Gardner, D. E., and Menzel, D. B. (1980). Stimulation of oxidant production in alveolar macrophages by pollutant and latex particles. *Environ. Res.* **23,** 121–136.

Hebbert, B., and Morgan, E. H. (1985). Calmodulin antagonists inhibit and phorbol esters enhance transferrin endocytosis and iron uptake by immature erythroid cells. *Blood* **65,** 758–765.

Hueper, W. C. (1958). Experimental studies in metal cancerigenesis. *Arch. Pathol.* **65,** 600–607.

Kasprzak, K. S., and Hermandez, L. (1989). Enhancement of hydroxylation and deglycosylation of 2′-deoxyguanosine by carcinogenic nickel compounds. *Cancer Res.* **49,** 5964–5968.

Kawanishi, S., Inoue, S., and Yamamoto, K. (1989). Hydroxyl radicals and singlet oxygen production and DNA damage induced by carcinogenic metal compounds and hydrogen peroxide. *Biol. Trace Element Res.* **21,** 367–372.

Nieboer, E., Stetsko, P. I., and Hin, P. Y. (1984). Characterization of the Ni(III/II) redox couple for nickel (II) complex of human serum albumin. *Ann. Clin. Lab. Sci.* **14,** 409.

Nieboer, E., Maxwell, R. I., Rossetto, F. E., Stafford, A. R., and Stetsko, P. I. (1986). Concepts in nickel carcinogenesis. In A. V. Xavier (Ed.), *Frontiers of Bioinorganic Chemistry.* VCH, Weinheim, pp. 142–151.

Ohkawa, Y., Iwata, K., Shibuya, H., Fujiki, H., and Inui, N. (1984). A rapid, simple screening method for skin-tumor promoters using peritoneal macrophages in vitro. *Cancer Lett.* **21,** 253–260.

Pryor, W. A. (1987). Views on the wisdom of using antioxidant vitamin supplements. *Free Rad. Biol. Med.* **3,** 189–191.

Rola-Pleszczynski, M., Gouin, S., and Begin, R. (1984). Asbestos-induced lung inflammation. Role of local macrophage-derived chemotactic factors in accumulation of neutrophils in the lung. *Inflammation* **8,** 53–62.

Sasco, A. J., Hubert, A., and De The, G. (1985). Diet and nasopharygeal carcinoma: epidemiological approach to comparative dietary assessment in different populations. In J. V. Joosens, M. J. Hill, and J. Gevoers (Eds.), *Diet and Human Carcinogenesis.* Excerpta Medica, Amsterdam, pp. 63–69.

Schlatter, C. (1985). Speculation on mechanisms of metal carcinogenesis. In E. Merian, R. W. Frei, W. Hardi, and C. Schlatter (Eds.), *Carcinogenic and Mutagenic Metal Compounds.* Gordon & Breach, New York, pp. 529–539.

Sunderman, F. W. (1987). Lipid peroxidation as a mechanism of acute nickel toxicity. *Toxicol. Environ. Chem.* **15,** 59–69.

Sunderman, F. W., and Hopfer, S. M. (1983). Correlation between carcinogenic activities of nickel compounds and their potencies for stimulating erythropoiesis in rats. In B. Sarker (Ed.), *Biological Aspects of Metal Related Diseases.* Raven Press, New York, pp. 171–181.

Sunderman, F. W., Hopfer, S. M., Knight, J. A., McCully, K. S., Cecutti, A. G., Thornhill, P. G., Conway, K., Miller, C., Patierno, S. R., and Costa, M. (1987). Physicochemical characteristics and biological effects of nickel oxides. *Carcinogenesis* **8,** 305–313.

Sunderman, F. W., Hopfer, S. M., Plowman, M. C., and Knight, J. A. (1990). Carcinogenesis bioassays of nickel oxides and nickel-copper oxides by intramuscular administration to Fischer-344 rats. *Res. Commun. Chem. Pathol. Pharmacol.* **70,** 103–113.

Weitzman, S. A., and Stossel, T. P. (1981). Mutation caused by human phagocytes. *Science* **212,** 546–547.

Weitzman, S. A., Weitberg, A. B., Clarke, E. P., and Stossell, T. P. (1985). Phagocytes as carcinogens: malignant transformation produced by human neutrophils. *Science* **227,** 1231–1233.

Zhong, Z., Troll, W., Koenig, K. L., and Frenkel, K. (1990). Carcinogenic sulfide salts of nickel and cadmium induce H_2O_2 formation by human polymorphonuclear leukocytes. *Cancer Res.* **50,** 7564–7570.

30

EFFECTS OF NICKEL ON CATALASE AND THE GLUTATHIONE PEROXIDASE–REDUCTASE SYSTEM IN VITRO

Ricardo E. Rodriguez and Kazimierz S. Kasprzak

Laboratory of Comparative Carcinogenesis, National Cancer Institute, FCRDC, Frederick, Maryland 21702

Nickel and Human Health: Current Perspectives, Edited by Evert Nieboer and Jerome O. Nriagu.
ISBN 0-471-50076-3   © 1992 John Wiley & Sons, Inc.

1. INTRODUCTION

Recent studies have shown that nickel (Sunderman et al., 1985; Athar et al., 1987, Knight et al., 1986; Rungby, 1987), like other transition metals [cobalt (Hasan and Ali, 1981; Sasame and Boyd, 1978), cadmium (Eaton et al., 1980; Stacey et al., 1980), and copper (Fujimoto et al., 1984; Ding and Chan, 1984)] causes lipid peroxidation, which may contribute to its acute toxicity. Similar oxidative reactions resulting in modifications of nuclear chromatin may also contribute to the carcinogenic activity of nickel (Kasprzak and Bare, 1989). The mechanisms of these effects appear complex and are poorly understood. They may involve redox reactions with a direct participation of ambient oxygen and the Ni(II)/(III) redox couple (Kasprzak and Bare, 1989) as well as indirect effects of nickel mediated through inflammatory cells, neutrophils, and phagocytes attracted to tissues by this metal-caused cell damage. The latter is the usual situation in all carcinogenic bioassays with insoluble nickel compounds, for example, nickel subsulfide (Ni_3S_2). The neutrophils and phagocytes release relatively large amouns of hydrogen peroxide (Iyer et al., 1961; Paul and Sbarra, 1968), which may cause lipid peroxidation (Comporti, 1985; Recknagle et al., 1982; Ames, 1982), DNA damage (Szmigiero and Studzian, 1988; Imlay et al., 1988, and cell death (Hirschelmann and Bekemeier, 1982; Becker, 1988). Among others, these cytotoxic effects are controlled by several enzymes that prevent hydrogen peroxide buildup in the tissue. One of these is catalase (CAT; EC 1.11.1.6), which decomposes hydrogen peroxide very rapidly ($k_1 \sim 10^7 M^{-1} sec^{-1}$) to water and oxygen by either a two-step catalatic reaction (Kremer, 1975) or a peroxidatic reaction (Brill, 1966; Kremer, 1970). The other is glutathione peroxidase (GSH-Px; EC 1.11.1.9), which is thought to be one of the primary defense enzymes against oxidative damage (Cohen and Hochstein, 1963; Little and O'Brien, 1968) and decomposes hydrogen peroxide ($k_1 \sim 10^8 M^{-1} sec^{-1}$) in a peroxidatic reaction in which two molecules of the reduced form of glutathione (GSH) are converted to the oxidized form (GSSG) (Wendel, 1980). In order to sustain GSH-Px action, the GSH concentration has to be restored by reduction of GSSG by glutathione reductase (GSSG-R; EC 1.6.4.2). It is easy to comprehend that lack of or inhibition of these enzymes would greatly enhance the cell-damaging effects of H_2O_2.

Nickel has been found to inhibit GSH-Px in vivo (Athar, 1987; Athar et al., 1987). However, its effects on the entire GSH-Px/GSSG-R system and on CAT have not been investigated. Our recent works have shown that essential metals, magnesium [Mg(II)] and iron [FeO and Fe(III)] substantially reduce the carcinogenic activity of Ni_3S_2 (Kasprzak, 1988; Kasprzak et al., 1985, 1987; see Chapter 42) and that this reduction is preceded by a marked modification by Mg(II) and Fe(III) or FeO of the early local inflammatory response to the carcinogen. This finding focused our attention on possible cause–effect relationship between Ni_3S_2-induced local inflammatory response (with all its peroxidative consequences) and carcinogenesis. Hence, the present experiment was carried out to determine if nickel [Ni(II)] had any influence on the activity of CAT and the GSH-Px/GSSG-R system and whether Mg(II) and/or Fe(III) could prevent those effects.

2. MATERIALS AND METHODS

The catalytic activity of CAT from bovine liver was determined at 25°C in terms of effective rate constants according to Beers and Sizer (1952). The enzyme solution, 16.3 units/mL upon final dilution, was injected into 12 mM H_2O_2 in a 10 mM Tris-HCl buffer, pH 7.2, in the presence or absence of Ni(II), Mg(II), and/or Fe(III). A unit of CAT is defined as the amount of enzyme that will decompose one micromole of H_2O_2 per minute at 25°C and pH 7.0. The rate of decrease of the ultraviolet absorbance of H_2O_2 at 240 nm was measured in 1-cm cells in a Gilford model 250 spectrophotometer in conjunction with a Gilford model 6051 recorder, a Gilford 2527-C thermostatic cell holder, and a Gilford Thermoset temperature controller. Hydrogen peroxide concentration of the stock solutions was determined iodometrically and spectrophotometrically by measuring the absorbance at 240 mM. Tris and H_2O_2 were from Sigma Chemical Co. (St. Louis, MO) and CAT was purchased from Worthington Biochemicals (Freehold, NJ).

Stock solutions of Ni(II), Mg(II), and Fe(III) were made by dissolving appropriate amounts of $NiCl_2 \cdot 6H_2O$, $FeCl_3 \cdot 6H_2O$ (Baker Analyzed Reagents; J. T. Baker Chemical Co., Phillipsburg, NJ) in 10 mM Tris-HCl buffer, pH 7.2, and the concentration of the metals was checked by atomic absorption spectroscopy.

The catalytic activity of GSH-Px from bovine erythrocytes was assayed by the method of Wendel (1981) in the presence of GSSG-R Type IV derived from yeast. The latter reduces GSSG back to GSH. The assay was carried out at 25°C in 10 mM Tris-HCl buffer, pH 7.0, in the absence or presence of Ni(II), Mg(II), and/or Fe(III) by measuring the decrease in absorbance of reduced nicotinamide-adenine dinucleotide phosphate (NADPH) at a wavelength of 340 nm. The final concentrations of the reagents in this system were GSH-Px, 1.05 units/mL; GSSG-R, 6.0 units/ml; GSH, 2 mM; NADPH, 0.25 mM. A unit of GSH-Px or GSSG-R is defined as the amount of enzyme that will decompose one micromole of NADPH per minute. The concentrations of the metal salts studied are specified in Section 3. The enzymatic activity of the system was calculated by determining the amount of NADPH consumed per minute by the given concentration of the enzymes.

The catalytic activity of GSSG-R was determined at 25°C, separately, according to Carlberg and Mannervik (1985) by following the loss of optical density of NADPH at a wavelength of 340 nm in the presence and absence of the Ni(II) or Mg(II). The final concentration of the reagents were GSSG-R, 0.6 units/mL; GSSG, 20 mM; and NADPH, 2 mM. The catalytic activity of GSSG-R was expressed in the same manner as the activity of the GSH-

Px/GSSG-R system. All chemicals used in the GSH-Px/GSSG-R experiments, unless otherwise specified, were obtained from Sigma Chemical Company (St. Louis, MO). Four to six measurements were made to obtain an experimental point with the standard deviation of the final results not exceeding $\pm 10\%$ of the given mean values. Significance of the differences between the mean values was tested using Student's t-test.

3.   RESULTS

3.1.   Catalase

The CAT activity, when measured by the rate of H_2O_2 consumed, was found to be $(11.6 \pm 0.41) \times 10^{-3}$ sec^{-1}. In the presence of Ni(II) this rate of H_2O_2 decomposition decreased linearly with increasing concentration of the metal (Fig. 30.1). At a Ni(II) concentration of 24 mM, the rate of H_2O_2 decomposition decreased by 41.4% of the original value. Enzymatic activity of CAT in the presence of Mg(II) at concentrations up to 21.6 mM remained unchanged (Fig. 30.1). Owing to its limited solubility at pH 7.2, Fe(III) could not be studied at concentrations exceeding 0.2 mM. In the presence of 0.2 mM Fe(III), the rate of H_2O_2 decomposition was slightly increased by 6%. Combination of 0.2 mM Fe(III) and 10.6 mM Ni(II) resulted in a 23.0% activity reduction, as opposed to a 26.6% loss when only 10.6 mM Ni(II) was added. However, this difference was not statistically significant (Fig. 30.2). At maximum concentrations, neither of the tested metals caused any detectable decomposition of H_2O_2 in the absence of CAT.

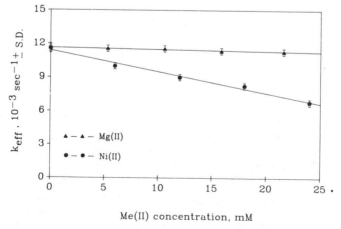

Figure 30.1. Effect of Ni(II) and Mg(II) concentration on the catalytic activity of catalase. The vertical bars indicate standard deviation.

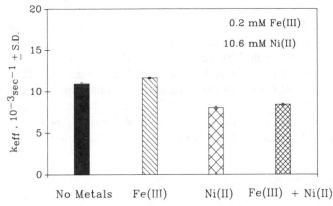

Figure 30.2. Catalase activity in the presence of Ni(II) and Fe(III) alone or combined at the given concentrations. Standard deviations are shown on top of the bars.

3.2. Glutathione Peroxidase–Reductase System

Under the present experimental conditions, the average value of the GSH-Px/GSSG-R catalytic activity in the absence of metals was found to be 15.32 ± 0.26 μmol NADPH consumed/min. The addition of Ni(II) resulted in a gradual activity decrease with increasing concentrations of Ni(II) (Fig. 30.3). The catalytic activity of this system decreased to only 12% of the original value at a Ni(II) concentration of 5.3 mM. The activity measured in the presence of Mg(II) (up to 5.3 mM) remained essentially constant (Fig. 30.3). In the presence of 0.3 mM Fe(III) there appeared to be a slight (7.3%)

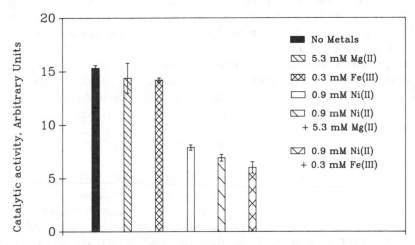

Figure 30.3. Effect of Ni(II), Mg(II), and Fe(III) concentration on the catalytic activity of the glutathione peroxidase–glutathione reductase system. See Section 2 for definition of arbitrary units. The vertical bars indicate standard deviation.

decrease in activity. Higher concentrations of Fe(III) were not investigated because of the limited solubility of this metal under the present experimental conditions. To determine whether the inhibition of the enzyme system by Ni(II) could be prevented by Fe(III) or Mg(II), the enzymatic activity was tested with 0.9 mM Ni(II) plus 0.3 mM Fe(III) or 5.3 mM Mg(II). The effects shown in Figure 30.4 were not statistically different from each other; neither Fe(III) nor Mg(II) could prevent Ni(II) action. None of the investigated metals at maximum concentration in the presence of H_2O_2 had any influence on the NADPH optical density in the absence of the enzymes.

3.3.   Glutathione Reductase

The enzymatic activity of GSSG-R was investigated in the presence of 0–0.9 mM Ni(II). The activity decreased nonlinearly with increasing concentration of Ni(II) (Fig. 30.5). At 0.09 mM Ni(II), the activity of the enzyme decreased by 19% and at 0.54 mM Ni(II) by 82%. At 0.9 mM Ni(II), the catalytic activity of GSSG-R dropped to 10% of the original value. Up to 0.9 mM, Mg(II) had no significant effect on the activity of GSSG-R.

4.   DISCUSSION

As shown in the present study, the catalytic activity of CAT in vitro can be substantially reduced by Ni(II). This effect seems to be specific and not simply dependent on the ionic strength of the solution since neither Mg(II) at the same concentration range as Ni(II) (5–24 mM) nor Fe(III) up to 0.2 mM (solubility limit) had any significant effect on the catalytic performance of CAT. Also, these two metals could not prevent the inhibition of CAT by Ni(II). The mechanisms of the Ni(II)-caused inhibition are not clear. Most likely, they involve a direct binding of the metal to CAT, causing a distortion in the protein structure of the enzyme. This may hinder binding of the substrate, H_2O_2, to one or more of the four active sites of the CAT molecule (Dounce and Deisseroth, 1970). An interaction of nickel with the CAT prosthetic group by the direct replacement of the heme iron can be excluded on kinetic grounds since such a substitution is very slow. Also, no direct interaction between Ni(II) or the organic buffer constituents and H_2O_2 were observed in our experiments.

The GSH-Px activity as measured in the presence of Ni(II) in the in vitro system recommended by Wendel (1981) showed a great decrease with increasing Ni(II) concentration. The Mg(II) or Fe(III) had no significant influence on this enzyme and they did not prevent the inhibition by Ni(II). All these effects, however, may be somewhat deceiving since a second enzyme, GSSG-R, was also present in the assay, not to mention the two forms of glutathione [the reduced (GSH) and oxidized form (GSSG)], as well as NADPH and

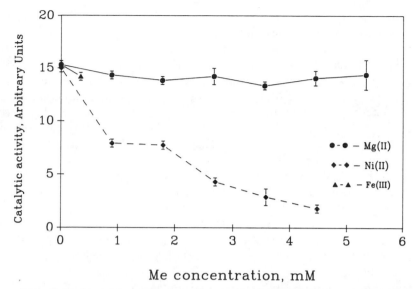

Figure 30.4. Catalytic activity of the glutathione peroxidase–glutathione reductase system in the presence of Ni(II), Mg(II), and Fe(III) alone or combined at given concentrations. See Section 2 for definition of arbitrary units. Standard deviations are shown on top of the bars.

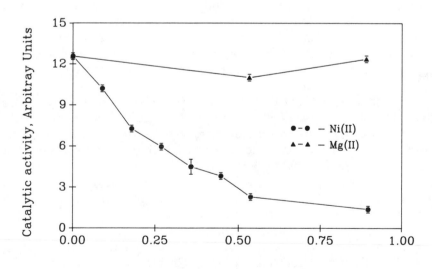

Figure 30.5. Effect of Ni(II) and Fe(III) concentration on the catalytic activity of glutathione reductase. See Section 2 for definition of arbitrary units. The vertical bars indicate standard deviation.

H_2O_2, most of which could interact with Ni(II) and thus disturb the efficiency of this system. Other studies on GSH-Px inhibition by various metals have also used this method (Splittgerber and Tappel, 1979; Jamal and Smith, 1985a,b), and therefore the results must be interpreted with caution. Hence, the best way to view our results is to relate them to the whole system. Contribution of possible Ni(II) interactions with the individual constituents of this system to the final outcome cannot be easily determined. However, some information related to this subject may be furnished by experiments in which one of the enzymes, GSSG-R, was assayed separately.

In the present study, GSSG-R appeared to be more sensitive to inhibition by Ni(II) than CAT or the GSH-Px/GSSG-R system. This conclusion is, however, weakened by the differences in the conditions under which the enzymatic activities were measured and requires further investigation. Also, the effects of possible binding of Ni(II) to GSH, GSSG, and/or NADPH on the activity of GSSG-R and GSH-Px constitute a separate set of problems for further study. Attempts to solve these problems and to make the enzymatic assay conditions more comparable and suitable for testing higher Fe(III) concentrations are currently underway in our laboratory.

The inhibition by Ni(II) of the activity of two key enzymes that decompose H_2O_2 may greatly contribute to this metal's toxicity and carcinogenicity. For example, a marked increase in cellular H_2O_2 can be expected following Ni(II) administration. It may result in greater production of highly toxic and/or mutagenic hydroxyl radical (Becker, 1988; Loeb et al., 1988). Hydrogen peroxide and hydroxyl radical, in turn, are members of a greater family of biologically active oxygen species including organic peroxides, superoxide anion, and singlet oxygen that are produced and handled by a variety of enzymes (Babior, 1978a,b). Determination of Ni(II) effects on the enzymes in question (e.g., xanthine oxidase, myeloperoxidase, superoxide dismutases) would provide more complete insight into possible mechanisms of nickel toxicity and carcinogenicity.

So far, neither Mg(II) nor Fe(III) show any signs of counteraction against Ni(II)-produced inhibition of CAT and the GSH-Px/GSSG-R enzymatic couple. Hence, the inhibition of nickel carcinogenesis by these metals does not involve any direct interaction at these enzymes. Perhaps, iron could substitute for CAT and/or GSH-Px by decomposing H_2O_2 directly through the Fenton and Haber–Weiss reactions. This would, however, lead to even a more reactive specie, the hydroxyl radical, which should increase rather than suppress carcinogenesis (Loeb et al., 1988). The anticarcinogenic effect of Mg(II) cannot involve any direct interaction with H_2O_2, although this metal is capable of indirectly influencing active oxygen generation and handling through the enhancement of phagocytes (see Chapter 42). Obviously, a better understanding of the Ni(II), Fe(III), and other metal's influence on the biological production and detoxification of active oxygen species deserves further experimental studies.

5. CONCLUSIONS

Nickel inhibits in a dose-dependent manner the in vitro catalytic activity of CAT, GSSG-R, and the GSH-Px/GSSG-R system. This inhibition may contribute to the toxic and carcinogenic influence of nickel observed in animals. Mg(II) and Fe(III) exert no effect on these enzymes and do not prevent the nickel-caused inhibition.

REFERENCES

Ames, B. N. (1982). Lipid peroxidation and oxidative damage to DNA. In K. Yagi (Ed.), *Lipid Peroxides in Biology and Medicine*. Academic Press, New York, pp. 339–351.

Athar, M. (1987). Role of glutathione metabolizing enzymes in nickel mediated induction of hepatic glutathione. *Res. Commun. Chem. Pathol. Pharm.* **57**, 421–424.

Athar, M., Hasan, S. K., and Srivastava, R. C. (1987). Evidence for the involvement of hydroxyl radicals in nickel mediated enhancement of lipid peroxidation: implications for nickel carcinogenesis. *Biochem. Biophys. Res. Commun.* **147**, 1276–1281.

Babior, B. M. (1978a). Oxygen-dependent microbial killing by phagocytes: Part I. *New Engl. J. Med.* **1978**, 659–668.

Babior, B. M. (1978b). Oxygen-dependent microbial killing by phagocytes: Part II. *New Engl. J. Med.* **1978**, 721–725.

Becker, L. (1988). The cytotoxic action of neutrophils on mammalian cells in vitro. *Prog. Allergy* **44**, 183–208.

Beers, R. F., and Sizer, I. W. (1952). A spectrophotometric method for measuring the breakdown of hydrogen peroxide by catalase. *J. Biol. Chem.* **195**, 133–140.

Brill, A. S. (1966). Peroxidases and catalases. In M. Florkin and E. H. Stotz (Eds.), *Comprehensive Biochemistry*, Vol. 14, Elsevier, New York, pp. 471–477.

Carlberg, I., and Mannervik, B. (1985). Glutathione reductase. *Methods Enzymol.* **113**, 484–490.

Cohen, G., and Hochstein, P. (1963). Glutathione peroxidase: The primary agent for the elimination of hydrogen peroxide in erythrocytes. *Biochemistry* **2**, 1420–1428.

Comporti, M. (1985). Biology of disease: Lipid peroxidation and cellular damage in toxic liver injury. *Lab. Invest.* **53**, 599–623.

Ding, A-H., and Chan, P. C. (1984). Singlet oxygen in copper-catalyzed lipid peroxidation in erythrocyte membranes. *Lipids* **19**, 278–284.

Dounce, A. L., and Deisseroth, A. (1970). Catalase: Physical and chemical properties, mechanism of catalysis and physiological role. *Physiol. Rev.* **50**, 319–375.

Eaton, D. L., Stacey, N. H., Wong, K-L., and Klaassen, C. D. (1980). Dose-response of various metal ions on rat liver metallothionein, glutathione, heme oxygenase and cytochrome P-450. *Toxicol. Appl. Pharm.* **55**, 393–402.

Fujimoto, Y., Maruta, S., and Yosida, A. (1984). Effect of transition metal ions on lipid peroxidation of rabbit renal cortical mitochondria. *Res. Commun. Chem. Pathol. Pharm.* **44**, 495–498.

Hasan, M., and Ali, S. F. (1981). Effects of thallium, nickel, and cobalt administration on the lipid peroxidation in different regions of the rat brain. *Toxicol. Appl. Pharm.* **57**, 8–13.

Hirschelmann, R., and Bekemeier, H. (1982). Anti-inflammatory activity of catalase in carrageenin edema, manganese chloride edema and adjuvant arthritis of rats. *Agents Actions Suppl. AAS* **10,** 157–172.

Imlay, J. A., Chin, S. M., and Linn, S. (1988). Toxic DNA damage by hydrogen peroxide through the Fenton reaction *in vivo* and *in vitro*. *Science* **240,** 640–642.

Iyer, G. Y., Islam, M. F., and Quastel, J. H. (1961). Biochemical aspects of phagocytosis. *Nature* **192,** 535–541.

Jamall, I. S., and Smith, J. C. (1985a). Effects of cadmium treatment on selenium-dependent and selenium-independent peroxidase activities and lipid peroxidation in the kidney and liver of rats maintained on various levels of dietary selenium. *Arch. Toxicol.* **58,** 102–105.

Jamall, I. S., and Smith, J. C. (1985b). Effects of cadmium on glutathione peroxidase, superoxide dismutase, and lipid peroxidation in the rat heart: A possible mechanism of cadmium cardio-toxicity. *Toxicol. Appl. Pharm.* **80,** 33–42.

Kasprzak, K. S. (1988). Inhibitory effect of iron on the carcinogenicity of nickel subsulfide in F344/NCr rats. *Toxicologist* **8,** 193.

Kasprzak, K. S., and Bare, R. M. (1989). *In vitro* polymerization of histones by carcinogenic nickel compounds. *Carcinogenesis* **10,** 621–624.

Kasprzak, K. S., Quander, R. V., and Poirier, L. A. (1985). Effects of calcium and magnesium salts on nickel subsulfide carcinogenicity in Fischer rats. *Carcinogenesis* **6,** 1161–1166.

Kasprzak, K. S., Ward, J. M., Poirier, L. A., Reichardt, D. A., Denn, A. C., III, and Reynolds, C. W. (1987). Nickel–magnesium interaction in carcinogenesis: dose effects and involvement of natural killer cells. *Carcinogenesis* **8,** 1005–1011.

Knight, J.A., Hopfer, S. M., Reid, M. C., Wong, S.H-Y., and Sunderman, F. W., Jr. (1986). Ethene and ethane exhalation in Ni[II]-treated rats, using an improved rebreathing apparatus. *Ann. Clin. Lab. Sci.* **16,** 386–394.

Kremer, M. L. (1970). Peroxidatic activity of catalase. *Biochem. Biophys. Acta* **198,** 199–209.

Kremer, M. L. (1975). The reaction of bacterial catalase with H_2O_2. *Isr. J. Chem.* **13,** 1975.

Little, C., and O'Brien, P. J. (1968). An intracellular GSH-peroxidase with a lipid peroxide substrate. *Biochem. Biophys. Res. Commun.* **31,** 145–150.

Loeb, L. A., James, E. A., Waltersdorph, A. M., and Klebanoff, S. (1988). Mutagenesis by the autooxidation of iron with isolated DNA. *Proc. Natl. Acad. Sci. USA* **85,** 3918–3922.

Paul, B. B., and Sbarra, A. J. (1968). The role of the phagocyte in host-parasite interactions. XII. The direct quantitative estimation of H_2O_2 in phagocytosing cells. *Biochem. Biophys. Acta* **156,** 168–178.

Recknagle, R. O., Glende, E. A., Jr., Waller, R. L., and Lowery, K. (1982). Lipid peroxidation: Biochemistry, measurement, and significance in liver cell injury. In G. Plaa and W. R. Hewitt (Eds.), *Toxicology of the Liver*, Raven Press, New York, pp. 213–241.

Rungby, J. (1987). Silver-induced lipid peroxidation in mice: Interactions with selenium and nickel. *Toxicology* **45,** 135–142.

Sasame, H. A., and Boyd, M. R. (1978). Paradoxical effects of cobaltous chloride and salts of divalent metals on tissue levels of reduced glutathione and microsomal mixed-function oxidase components. *J. Pharm. Exp. Ther.* **205,** 718–724.

Splittgerber, A. G., and Tappel, A. L. (1979). Inhibition of glutathione peroxidase by cadmium and other metal ions. *Arch. Biochem. Biophys.* **197,** 534–542.

Stacey, N. H., Cantilena, L. R., Jr., and Klaassen, C. D. (1980). Cadmium toxicity and lipid peroxidation in isolated rat hepatocytes. *Toxicol. Appl. Pharm.* **53,** 470–480.

Sunderman, F. W., Jr., Marzouk, A., Hopfer, S. M., Zaharia, O., and Reid, M. C. (1985).

Increased lipid peroxidation in tissues of nickel chloride-treated rats. *Ann. Clin. Lab. Sci.* **15,** 229–236.

Szmigiero, L., and Studzian, S. (1988). H_2O_2 as a DNA fragmenting agent in the alkaline elution interstrand crosslinking and DNA-protein crosslinking assays. *Anal. Biochem.* **168,** 88–93.

Wendel, A. (1980). Glutathione Peroxidase. In W. B. Jakoby (Ed.), *Enzymatic Basis of Detoxification*, Vol. I, Academic Press, New York, pp. 333–353.

Wendel, A. (1981). Glutathione peroxidase. *Methods Enzymol.* **77,** 325–333.

31

ANIMAL STUDIES: AN OVERVIEW

Kazimierz S. Kasprzak

Laboratory of Comparative Carcinogenesis, National Cancer Institute, FCRDC, Frederick, Maryland 21702

Nickel and Human Health: Current Perspectives, Edited by Evert Nieboer and Jerome O. Nriagu.
ISBN 0-471-50076-3 © 1992 John Wiley & Sons, Inc.

---

1. INTRODUCTION

Experimental evidence relative to nickel carcinogenesis in animals has been reviewed and summarized in several monographs and articles. A detailed summary of the results collected over the pioneering period of investigations, up to 1975–1976, can be found in the U.S. National Academy of Sciences monograph on nickel (Sunderman et al., 1975) and in the subsequent IARC monograph (IARC, 1976). The results gathered in the ensuing years have been reviewed by Sunderman (1977, 1979, 1981, 1984a,b, 1986a,b), Cecutti and Nieboer (1981), Leonard et al. (1981), Rigaut (1983), Grandjean (1986), Sky-Peck (1986), Kasprzak (1987), and Norseth (1988). Also, the evidence for carcinogenicity of nickel to animals has been regularly updated by IARC (IARC, 1982, 1987, 1990). The present overview will attempt to update, summarize, and reevaluate the existing data, including relevant papers published in this book, with special emphasis on carcinogenic activity of certain classes of nickel derivatives to find possible structure –effect relationships and draw some conclusions as to possible molecular mechanisms of nickel carcinogenesis. The needs for further experimental work in animals will also be defined in order to clear some discrepancies and/or test certain hypotheses.

2. A SURVEY OF CARCINOGENESIS BIOASSAYS

Nickel derivatives that were tested for carcinogenicity in animals and the corresponding main results are listed in Table 31.1 (updated from Kasprzak, 1987). These compounds were administered to various experimental animal species and strains by different routes, alone or combined with other treatments, and observed for different time periods. Of the routes directly relevant to human exposures, only the inhalation or intratracheal instillation has been tested more widely; oral, cutaneous, and subcutaneous administration of nickel compounds did not draw much attention. Although carcinogenic results following intravenous, intraperitoneal, intrarenal, intratesticular, intraocular, and intraosseous routes have been published, the most popular are studies using intramuscular injections.

Table 31.1 Nickel Derivatives Tested for Carcinogenesis in Animals

Compound	Result[a]	Reference
A. Water-insoluble compounds		
1. Metal Ni dust, Ni_3S_2, crystalline NiS, amorphous NiS, NiS_2, Ni_4FeS_4, Ni–Fe–S matte, Ni_3Se_2, NiSe, crystalline $Ni(OH)_2$, NiO, Ni_2O_3, $NiCO_3$, $Ni(CO)_4$, $Ni(C_5H_5)_2$,[b] $Ni_{11}AsS_8$, Ni_5As_2, NiAsS, NiSb, NiTe; Ni–Ga, Ni–Al, Ni–Fe–Cr alloys	Local tumors after i.m. injection to rats; local tumors by some compounds after s.c., i.p., i.h., i.r., i.pl., i.t., i.th., i.ts., i.oc., i.cr., i.os. injection to rats, mice, or hamsters	Sunderman et al., 1975; IARC, 1976, 1982, 1987, 1990; Grandjean, 1986; Kasprzak, 1987
2. Colloidal $Ni(OH)_2$, NiAs, $NiTiO_3$, Ni–Fe alloy	No tumors after a single i.m. injection to rats	Sunderman et al., 1975; IARC, 1976, 1982, 1987, 1990; Grandjean, 1986
B. Water-soluble compounds		
1. NiF_2, $Ni(CH_3COO)_2$	Tumors after i.m. or i.p., especially by multiple injections to rats and/or mice	Sunderman et al., 1975; Poirier et al., 1984; Kasprzak, 1987; Kasprzak et al., 1990; Pott et al., 1987, Chapter 37
2. $NiCl_2$, $NiSO_4$, $Ni(NO_3)_2$	No tumors following single i.m., i.p. injections; local tumors after multiple i.t., or i.p. injections to rats and/or mice	Sunderman et al., 1975; Kasprzak et al., 1983; Pott et al., 1987, Chapter 37

[a] Abbreviations: i.cr., intracranial; i.h., intrahepatic; i.m., intramuscular; i.oc., intraocular; i.p., intraperitoneal; i.pl., intrapleural; i.r., intrarenal; i.t., intratracheal; i.ts., intratesticular; s.c., subcutaneous; i.os., intraosseous.
[b] Bis(cyclopentadienyl)nickel, nickelocene.

2.1. Inhalation and Intratracheal Instillation

2.1.1. Insoluble Nickel Derivatives

Metallic Nickel. Hueper (1958) exposed C57BL mice, Wistar rats, NIH Black rats, and guinea pigs to 15 mg/m^3 metallic nickel powder (<4-μm particles) in air for 6 hr/day, 4–5 days/week, for the duration of the experiment. In 15 months, lymphosarcomas developed in 2/20 mice while benign lung tumors developed in 15/50 rats of both strains and in 42/42 guinea pigs; one of the guinea pigs also bore a malignant intra-alveolar carcinoma. However, no tumors were observed by Hueper and Payne (1962) in lungs of 120 rats and 100 male hamsters following inhalation of nickel powder mixed with powdered limestone in the presence of 20–35 ppm of sulfur dioxide, 5–6 hr/day for over 18 months. No lung tumors were seen in Wistar rats inhaling 0.06–0.2 mg/m^3 of nickel powder for 18 months (Chapter 33). Pott et al. (1987) administered nickel powder intratracheally to Wistar rats in 20 weekly doses of 0.3 mg Ni powder each, or 10 weekly doses of 0.9 mg Ni each, and observed the rats for over 2 years; lung tumors developed in 10/39 rats given the lower dose and in 8/32 rats given the higher dose of nickel versus 0/40 rats given 20 × 0.3 mL of normal saline. Recently, Muhle et al. (Chapter 35) administered 12 intratracheal biweekly doses of nickel powder to Syrian golden hamsters, to a total of 9.6 mg per hamster, and observed the animals for over 21 months. Only one lung tumor and sporadic metaplasia were noted in 56 hamsters; under the same experimental conditions, no tumors were produced by three brands of nickel–chromium stainless steel.

Nickel Oxides. The exposure of hamster or rat lungs to nickel monoxide (NiO) did not produce any significant carcinogenic response. Only one tumor developed in over 1 year in 50 hamsters given weekly intratracheal doses of 4 mg NiO (0.5–1-μm particles) each (IARC, 1976; Farrel and Davis, 1974). Likewise, inhalation of nickel-enriched fly ash (70 mg/m^3, 9% Ni) by Syrian golden hamsters resulted in the development of only two lung tumors in 102 hamsters compared to no tumors in the control groups (Wehner et al., 1984). No lung tumors were observed in another study, in which 51 Syrian golden hamsters were exposed for their life span to pure NiO dust (53 mg/m^3, 0.3-μm particles; 7 hr/day, 5 days/week). Additional exposure to cigarette smoke did not change that result (IARC, 1976; Wehner et al., 1975).

Intratracheal administration of black NiO to rats (20–40 mg per rat) produced squamous cell lung carcinoma in 1 of 26 treated animals in 15 months, with no lung tumors in the control rats (Saknyn and Blokhin, 1978). More recently, Pott et al. (1987) found NiO to be carcinogenic in the lungs of rats that had received 10 × 5 mg or 10 × 15 mg intratracheal weekly doses of NiO; in 2 years, pulmonary tumors developed in 10/37 rats or 12/38 rats given the higher or the lower dose, respectively. Horie et al. (Chapter 39) observed single tumors of the lung, pancreas, and other tissues in male Wistar rats over a 1-year period after exposure by inhalation to green NiO (0.96 mg/m^3;

4.1-μm particles; 6 hr/day, 5 days/week) either alone for 6 months or in combination with tobacco smoke. For example, pulmonary adenomas were found in 1/47 rats treated with NiO alone, 2/23 rats treated with NiO plus tobacco smoke, and 3/22 rats treated with tobacco smoke alone, versus 3/38 untreated rats; the incidences of pancreatic adenomas were 2/47, 4/23, and 1/22, versus 1/38, respectively. One renal carcinoma, one osteosarcoma, and two fibrosarcomas were also found that had no relation to the treatments. The authors concluded that NiO was not a highly carcinogenic substance. Takenaka et al. (1985) exposed 60 male Wistar rats by inhalation to 60 and 200 μg/m^3 of nickel oxide, 23 hr/day, 7 days/week, for up to 31 weeks; lung tumors were not observed in any rat. Heinrich et al. (1987) studied carcinogenic effects in rats, hamsters, and mice of intratracheal NiO combined with inhalation of polycyclic aromatic hydrocarbons (PAHs) from diesel engine and coal oven exhaust, 16–19 hr/day, 5 days/week, for a maximum of 2.5 years. The results published only in abstract form do not include NiO among agents potentiating or having at least additive effects on PAH. A large-scale experiment aimed at long-term effects of NiO inhalation on rats and mice is currently under way at the U.S. National Toxicology Program (Chapter 33).

Nickel Subsulfide (Ni$_3$S$_2$). Ottolenghi et al. (1975) exposed Fischer rats to Ni$_3$S$_2$ by inhalation (1 mg/m^3; 1–1.5-μm particles; 6 hr/day, 5 days/week, for 78–80 weeks) and observed them up to 110 weeks. During that time, 32 of the 226 Ni$_3$S$_2$-treated rats developed lung tumors (14 of them malignant), compared with only 2 of 241 untreated rats (one malignant tumor). Coexisting lung infarction did not increase the incidence of lung tumors in the Ni$_3$S$_2$-exposed rats. Single intratracheal instillation of 5 mg Ni$_3$S$_2$ per rat to 13 Wistar rats produced no tumors over 15 months in the study of Kasprzak et al. (1973); however, one carcinoma developed in 13 rats treated with Ni$_3$S$_2$ plus benzo[a]pyrene. Sunderman (1983) tested Ni$_3$S$_2$ (<2-μm particles) by multiple intratracheal instillations to Syrian hamsters of either four monthly doses of 2.5 mg or 18 weekly doses of 2 mg per animal and found no tumors during a 7-month observation (all hamsters died by this time). Yarita and Nettesheim (1978) studied the effect of Ni$_3$S$_2$ on the epithelium of rat tracheas transplanted subcutaneously into isogenic Fischer rats; Ni$_3$S$_2$ inserted into the transplants at 1- and 3-mg doses produced 15 and 70% tumor incidence, respectively, with carcinomas prevailing at the lower dose and sarcomas at the higher dose. Pott et al. (1987) introduced Ni$_3$S$_2$ to three groups of female Wistar rats in 15 weekly intratracheal doses of 0.063, 0.125, and 0.25 each for over 2 years and found pulmonary tumors in 7/47, 13/45, and 12/40 rats, respectively; most of the tumors were squamous cell carcinomas and adenocarcinomas. Muhle et al. (Chapter 35) administered intratracheally Ni$_3$S$_2$ (2.2-μm particles) to 62 male and female Syrian golden hamsters at 12 doses of 0.1 mg each every 2 weeks and found no pulmonary tumors during the 114-week observation period. Under the same experimental conditions, *pentlandite* (a mixed Ni–Fe sulfide) produced bronchioalveolar adenoma in 1/

60 rats. A large inhalation experiment aimed at determination of the long-term effects of Ni_3S_2 on rats and mice is currently under way at the U.S. National Toxicology Program (Chapter 33; Dunnick et al., 1988).

Nickel Carbonyl, Ni(CO)$_4$. Exposure of animals to $Ni(CO)_4$ by inhalation would be relatively easy because of its high volatility. However, because of its high acute toxicity, the doses that can be safely administered are only marginally carcinogenic (if at all). The high toxicity of $Ni(CO)_4$ may also be the cause of the low popularity of this important industrial compound among cancer research scientists. In fact, the data available thus far originate from the pioneering studies of Sunderman et al. (1957, 1959) and Sunderman and Donelly (1965). They exposed Wistar rats to $Ni(CO)_4$ by repeated inhalation (0.03–0.06 mg/L; 30 min, three times per week for 12 months or for the life span of the rats) or single 30-min inhalations of 0.25 or 0.6 mg/L. The rats were observed for up to 30 months during which time 2/96 rats given the divided dose for 12 months developed pulmonary carcinomas, 1/64 rats exposed to the divided dose for their life span (2 years) developed adenocarcinomas, 2/80 rats exposed to 0.25 mg/L developed pulmonary carcinoma and adenomas, and 1/285 rats exposed to 0.6 mg/L developed adenocarcinomas of the lung. Therefore, the incidence of tumors was very low. Nevertheless, some specific tumorigenic effect of $Ni(CO)_4$ could not be excluded since no pulmonary tumors were observed in control rats. It is also worth noting that the high mortality caused by the treatment allowed only a few rats to survive longer than 2 years, that is, the time necessary for pulmonary tumors to occur in rats.

2.1.2. Water-soluble Nickel Compounds

Although some epidemiological observations indicate that soluble nickel compounds constitute a potential carcinogenic hazard to workers of nickel refining and electroplating industries (IARC, 1976, 1982, 1987, 1990; Grandjean, 1986; Grandjean et al., 1988), relevant experiments on animals have not been performed yet. An inhalation study is currently being conducted by the U.S. National Toxicology Program at the Lovelace Inhalation Toxicology Research Institute in which the long-term effects of nickel compounds including soluble nickel sulfate will be tested in rats and mice (Chapter 33; Dunnick et al., 1988).

2.2. Oral Administration

Oral administration of potential nickel carcinogens to laboratory animals has not been adequately studied despite its importance to human occupational exposures and certain signals that pharyngeal and stomach cancer incidences may be increased in nickel refinery workers (Saknyn and Shabynina, 1973; IARC, 1987). Only two relevant experiments have been fully described in the literature thus far. In the one by Schroeder et al. (1964), 50 mice were given 5 mg/L of a nickel salt in drinking water for up to 3 years; the incidence of tumors in those mice did not differ significantly from the tumor incidence in control animals. In the second experiment by Sunderman (1983), Ni_3S_2 was applied in multiple doses (54 doses of 1 mg and 108 of 10 mg per hamster, three times per week) into the buccal pouch of Syrian golden hamsters; none of the animals developed tumors in the pouch, oral cavity, or gastrointestinal tract over the 19 months of observation. However, since the survival of animals in both experiments was poor, the negative results obtained must be viewed with caution. An experiment is under way in this author's laboratory in which Fischer rats received 1000–1200 mg nickel acetate/L in drinking water; in a preliminary assessment of this bioassay after 1 year of treatment, no nickel-related increase in tumor incidence has been noticed.

2.3. Intramuscular and Subcutaneous Administration

2.3.1. Insoluble Nickel Derivatives

Powdered *metallic nickel* is much more tumorigenic by the intramuscular route than by inhalation. Thus, Heath and Daniel (1964) induced local sarcomas in all 10 hooded rats given a single intramuscular injection of 28 mg of nickel suspended in fowl serum versus none in control rats. Friedman and Bird (1969) injected Sprague-Dawley rats with 20 mg of nickel sponge; local sarcomas developed in 6/25 rats. Following 5 monthly intramuscular injections of 5 mg of nickel powder in trioctanoin to 50 Fischer rats and 50 hamsters, Furst and Schlauder (1971) observed the development of local sarcomas in 38 rats and 2 hamsters by 11 months after the first injection; no sarcomas occurred in control animals given trioctanoin only. Sunderman and Maenza (1976) administered nickel powder in a single intramuscular injection to two groups of 10 Fischer rats, 3.6 mg Ni per rat in one group and 14.4 mg Ni per rat in the second (<2-µm particles) and in 2 years observed no tumors in the first group and only two local sarcomas in the second group. Sunderman (1984b) tested carcinogenicity of 18 different nickel derivatives in Fischer rats by a single intramuscular injection at a dose equivalent to 14 mg Ni per rat; within 2 years postinjection, nickel powder induced local sarcomas in 13/20 injected rats compared to 0/40 controls and 0/20 rats given powdered ferronickel alloy. Judde et al. (1987) injected 20 mg of nickel powder suspended in paraffin oil intramuscularly to WAG rats and observed local tumors in 14/30 rats in 1 year.

Some carcinogenic response in animals was also obtained following intramuscular administration of *nickel-containing alloys* used for making internal prostheses, bone plates, and dental materials. The response was generally weak (except in dogs) or negligible and could not be linked solely to nickel; relevant data can be found in several publications (Stevenson et al., 1982; Memoli et al., 1986; Vityugov et al., 1986; Waalkes et al., 1987).

Nickel monoxide (NiO) injected intramuscularly to rodents in some experiments produced a higher incidence of local tumors than when given by the intratracheal route or inhalation, but in other experiments the results were likewise negative. Thus, single 5-mg intramuscular doses into both thighs of mice resulted in local sarcomas in 23–35% of animals in approximately 1 year, depending on strain (IARC, 1976). Similar studies in rats are numerous, but the published data are inconsistent. In Sunderman's (1984b) comparative study, intramuscular carcinogenic activity of NiO in male Fischer rats is ranked right after the most potent nickel sulfide carcinogens (85–93% incidence of tumors following unilateral intramuscular injection of 14 mg Ni as NiO to 15 rats, 2-year observation, versus 100% for Ni_3S_2 under the same conditions). Local muscle tumor incidence was also high in hooded, Wistar, and NIH Black rats (Daniel, 1966). Recently, Sunderman et al. (1988) discovered that the intramuscular carcinogenic activity of NiO strongly depended on the manufacturing procedure; in a 2-year bioassay, they found muscle sarcomas in 6 of 15 Fischer rats injected intramuscularly with 20 mg Ni as the black NiO and in none of the 15 rats given the same doses of NiO calcined at either 735 or 1045°C compared with tumors in all 15 rats given equimolar doses of Ni_3S_2. It is not surprising, therefore, that some authors proclaim nickel oxide to be noncarcinogenic (Longstaff et al., 1984; Berry et al., 1985).

From nickel oxides other than NiO, only *Ni(III)oxide* (Ni_2O_3) has been tested thus far by the intramuscular route. The Ni_2O_3 did not induce any tumors during 18 months when implanted in pellets of sheep fat into the muscles of 35 NIH Black rats, 7 mg per site, three sites per rat (IARC, 1976). Likewise, the same oxide injected unilaterally to 10 WAG rats, 20 mg per rat (Judde et al., 1987), or to 20 Wistar rats at a dose of 10 mg per rat (Sosinski, 1975) did not produce muscle tumors during 2 years.

Nickel hydroxide [$Ni(OH)_2$], which may constitute a possible health hazard to workers of alkaline battery plants, was tested for carcinogenicity in only two animal experiments. In the first one, Gilman (1966) reported development of 19 local sarcomas at the intramuscular site of $Ni(OH)_2$ implantation to 40 Wistar rats. In the second experiment, Kasprzak et al. (1983) reported low incidences of local sarcomas following a single intramuscular injection to Wistar rats of three preparations of $Ni(OH)_2$. The preparations differed in carcinogenic activity depending on the completeness of development of their crystal structure; a colloidal preparation did not produce tumors, while two distinctly crystalline preparations induced tumors in 5/20 and 3/20 rats.

Nickel subsulfide (Ni_3S_2) has been tested for carcinogenicity in different species and strains of animals and is used as a prototype substance for nickel carcinogenesis in diverse experimental systems using the intramuscular administration route (Kasprzak, 1987). This compound has been found to be highly carcinogenic to the muscles of rats; the yield of tumors after a single intramuscular dose of 10 mg Ni_3S_2 varied from 60 to over 95% in NIH Black, Wistar, hooded, and Fischer rats, respectively, in 12–14 months. Various physical forms, including powders of different particle size (1 to over 30 μm in diameter), chips, tablets, in different suspending media (e.g., water, saline, water–glycerol, penicillin G), or in Ni_3S_2-containing diffusion chambers, were found to be carcinogenic (IARC, 1976; Kasprzak, 1978, 1987). A clear dose–response relationship was established by Sunderman (1983) for Ni_3S_2 in the skeletal muscles of Fischer rats. Mice have been relatively less susceptible; 2.5–5 mg Ni_3S_2 intramuscularly produced sarcomas in 33–60% of mice in 13–17 months depending on the strain (Kasprzak, 1987). Syrian golden hamsters developed local tumors in 4/15 animals in 10 months after receiving 5 mg of Ni_3S_2 per site, or in 12/17 animals in 11 months after being given 10 mg Ni_3S_2 per site (Sunderman, 1983). Most interestingly, Ni_3S_2 did not induce any tumors within 6 years after intramuscular injection (25 mg per site, two sites per rabbit) to four New Zealand albino rabbits (Sunderman, 1983).

Nickel monosulfide (NiS) appears to be highly carcinogenic to the rat muscle only in its crystalline form. Sunderman (1984b) found this compound to be as active as Ni_3S_2 in the muscles of Fischer rats. However, freshly precipitated amorphous NiS produced local sarcomas in only a few injected rats (3/25 rats in 11 months; all doses were equivalent to 14 mg Ni per rat; same particle size as the crystalline powders). The difference in carcinogenicity between these two forms of the same compound stresses the importance of structural effects in nickel carcinogenesis.

Nickel disulfide (NiS_2) and some *mixed sulfides* as well as various other *insoluble nickel compounds* have been tested for carcinogenicity by the intramuscular route in Fischer rats by Sunderman and McCully (1983a) and Sunderman (1984b, 1986a,b). Their activity under identical experimental conditions (the same dose, 14 mg Ni per rat; same particle size of powdered substances, <2 μm; 2-year observation) has been classified as follows (Sunderman, 1984b); Ni_4FeS_4, like Ni_3S_2 and crystalline NiS, induced local sarcomas in 100% of the injected rats (class A); NiS_2, NiAsS, Ni_3Se_2, Ni_5As_2, like NiO, induced tumors in 85–93% of the rats (class B); Ni metal powder, $Ni_{11}As_8$, NiSe, NiTe, and NiSb induced sarcomas in 50–65% of the rats (class C); $NiCrO_4$ was as poorly carcinogenic as amorphous NiS and induced sarcomas in only 6% of the rats (class D); no tumors were produced by NiAs, $NiTiO_4$, or NiFe alloy (class E).

Nickelocene, a water-insoluble nickel complex with cyclopentadiene, was found by Furst and Schlauder (1971) to be a weak carcinogen in hamsters; single intramuscular injection induced local sarcomas in 4/50 hamsters given 25 mg of this compound, whereas none of the animals responded to eight injections of 5 mg each (trioctanoin solution).

Nickel carbonate ($NiCO_3$) produced local sarcomas in 6/35 NIH Black rats that had been implanted 18 months earlier with 7 mg of this compound mixed with sheep fat by Payne (1964).

Subcutaneous injection of insoluble nickel derivatives has not been used widely in carcinogenicity testing despite its closer relevance to human exposures than the intramuscular route (surgical implants, trapping of occupational nickel dusts in cuts and wounds). The reason for this is most likely related to the high responsiveness of the subcutaneous site in rodents to so-called solid-state carcinogenesis (Grandjean, 1986; Furst, 1984), which may obscure the specificity of response. Nevertheless, fine powders that do not induce solid-state tumors might be tested by this route. As found by Mason et al. (1971) and Mason (1972), Ni_3S_2 injected subcutaneously to Fischer rats at doses of 3.3 or 10 mg per rat produced the same high incidence of local sarcomas (>90%) over the 18-month period as after intramuscular injections. Mitchell et al. (1960), who tested several pelleted (2-mm) dental materials including metallic nickel and a nickel–gallium alloy by subcutaneous implantation into Wistar rats (four pellets per rat), found in 2 years local sarcomas in 5/10 rats given nickel pellets and in 9/10 rats receiving the alloy pellets. Sarcomas were not found at sites of pellets made of other materials including silver, copper, and vitallium.

2.3.2. Water-soluble Nickel Compounds

There were only a few studies in which water-soluble nickel compounds were tested for carcinogenic activity in animals by the intramuscular or subcutaneous route. Most of the experiments gave negative results. Thus, in his pioneering experiment, Payne (1964) found only single sarcomas in a group of 35 NIH Black rats implanted intramuscularly 1.5 years earlier with $NiSO_4$ and nickel acetate tetrahydrate [$Ni(CH_3COO)_2 \cdot 4H_2O$] mixed with sheep fat; $NiCl_2$ and nickel ammonium sulfate [$Ni(NH_4)_2(SO_4)_2$] produced no tumors. The results were, however, confusing since no tumors were found at intramuscular sites of injection of anhydrous nickel acetate; it is unclear how hydration of the same compound could make it carcinogenic. In the experiments of Gilman (1962), no tumors were produced by intramuscular injections of 5 mg $NiSO_4$ to both thighs of 32 Wistar rats during the over 600-day observation period. Kasprzak et al. (1983) administered $NiSO_4$ intramuscularly in 15 doses of 4.4 μmol per rat per dose every second day to male Wistar rats [the dosing was intended to mimic the release of soluble Ni(II) from an intramuscular deposit of 5.5 mg Ni_3S_2] and observed no tumors within 2 years from the beginning of injections. Recently, Teraki and Uchiumi (Chapter 40) injected Fischer rats with five subcutaneous weekly doses of 30 mg $Ni(CH_3COO)_2$ (hydration not specified)/kg body weight (BW) each and found injection site sarcomas in 5/16 rats between weeks 22 and 78 after the injections. This result stresses the importance of prolonged exposure of an animal to nickel compounds for a positive carcinogenic response. Nickel fluoride (NiF_2), another water-soluble nickel salt, was also found to be carcinogenic in Fischer rats (Gilman, 1966).

2.4. Other Routes of Administration

2.4.1. Insoluble Nickel Derivatives

Of administration routes other than intratracheal, subcutaneous, or oral, intraosseous implantation may have the greatest relevance to human exposures because of its resemblance to surgical implants.

Metallic nickel powder tested for carcinogenicity by intraosseous implantation to rats produced inconsistent results. Hueper (1952, 1955) reported three osteogenic local tumors and one carcinoma in 4/17 Osborne–Mendel rats 7 months after a single intraosseous injection of 6.25 mg Ni in lanolin in one experiment, compared with 27 tumors, none of them of osteogenic origin, in 27/100 Wistar rats given intraosseous 5 mg Ni powder in 20% gelatin–saline suspension in a second experiment. The difference in tumor origin was most likely caused by some seepage of nickel from the bone cavity into the adjacent tissues. Nickel powder also induced bone tumors in WAG rats (Berry et al., 1985) and in rabbits (Hueper, 1955) following intraosseous injections . Thus far, no other nickel derivatives have been studied by this route. Carcinogenicity of nickel powder was tested also by intravenous, intraperitoneal, intrathoracic, intrapleural, subcutaneous, and intrarenal administration, with mostly negative or a very weak response. Thus, Hueper (1955) observed no tumors in 25 C57BL mice given two intravenous injections of 2.5 µg Ni each and surviving up to 15 months. Likewise, no tumors were observed by him in 10 rabbits given six intravenous injections of 5 mg/kg BW nickel powder each and surviving for up to 2 years. Only one tumor was observed in 25 Wistar rats 7–8 months after six intravenous injections of 2.5 mg Ni/kg BW Hueper and Payne (1962) administered nickel powder (4 mg per rat) intrathoracically to 34 NIH Black rats by thoracotomy and repeated the treatment 12 months later; only one rat of 14 that survived 18 months developed an injection site sarcoma and no other tumors were observed.

Intrapleural administration of nickel powder (12 µg per mouse; single injection) to 50 C57BL mice that survived up to 18 months was also ineffective. Rats, however, appeared to be a little more susceptible, though to a much higher nickel dose; 5 monthly intrapleural injections of 6.25 mg (Osborne–Mendel rats) (Hueper, 1952) or 5 mg Ni powder each (Fischer rats) (Furst et al., 1973) produced local tumors in 4/12 and 2/10 rats, respectively, in over 1 year. Sunderman et al. (1984) administered 7 mg Ni powder intrarenally into one kidney of 15 Fischer rats and observed no tumors in 2 years. A relatively strong carcinogenic response was achieved in Wistar rats by Pott et al. (1987) using the intraperitoneal route. Multiple intraperitoneal injections of nickel powder (10 weekly doses of 7.5 mg Ni each) resulted in 96% tumor incidence in the abdominal cavities of 48 rats up to 2.5 years postinjection. In a follow-up study, Pott et al. (Chapter 37) tested the carcinogenicity of 10 nickel-containing dusts in female Wistar rats and found a clear dose–response relationship to metallic nickel administered intraperitoneally in multiple injections: 3/34, 5/34, or 25/35 rats developed local tumors (mesotheliomas and sarcomas) after 1×6, 2×6, or 25×1 mg (weekly), respectively. Under the same experimental conditions, a nickel–aluminum alloy (51% Ni) and two types of stainless steel (29% Ni or 70% Ni) produced intraperitoneal tumors in up to 10/35, 1/36, or 24/33 rats, respectively, depending on the multiplicity of doses. Multiple dosing tended to increase the tumor yield.

An unexpected finding was made by Waalkes et al. (1987) in Wistar rats ear-tagged for identification purpose with tags made of a nickel–copper alloy (65% Ni); 14 osteosarcomas developed in a group of 168 rats in 2 years, with no relation to other treatments.

Nickel oxide (NiO) was found by Pott et al. (1987) to be carcinogenic in female Wistar rats following intraperitoneal injection. Two very high weekly doses of 500 mg NiO induced tumors in 46/47 rats over 1.5 years. In the recent experiment of Pott et al. (Chapter 37), a sample of green-gray NiO induced intraperitoneal tumors in 10/34 and 12/36 rats given 1×25 and 1×100 mg Ni as NiO, respectively, during a more than 1.5-year observation period. Eilertsen et al. (1988) tested carcinogenicity of different nickel oxides injected into the pleural cavity in groups of 20 rats per oxide. After 27 months, the most active of the oxides tested (NiO Merck PA) (Joint Committee on Powder Diffraction Standard File No. 4-835) induced 4 tumors per 20 rats, with only single tumors produced by the other oxides. It is worthwhile to notice, however, that in an earlier study (Skaug et al., 1985) under identical experimental conditions the same NiO Merck PA had been much more effective, inducing 31 tumors in 32 rats. That discrepancy might be caused by differences in preparing the injection suspension. Sunderman et al. (1984) injected NiO intrarenally (7 mg Ni per kidney, one kidney per rat) to 15 Fischer rats and found no kidney tumors in over 2 years postinjection. Sosinski (1975) administered Ni(III) oxide (Ni_2O_3), 3 mg per rat, into the cerebral cortex of 20 Wistar rats and found cerebral gliomas in two rats after 14 and 21 months and a meningioma in one rat 21 months postinjection.

Nickel subsulfide (Ni_3S_2) produces local tumors following administration to animals by diverse routes. The results have been recently presented and reevaluated by Sunderman (1983) and reviewed by Kasprzak (1987) and will be only briefly summarized here. Thus, Ni_3S_2 proved carcinogenic following intraperitoneal, intrarenal, intrahepatic, intratesticular, or intraocular injections or injection into the submaxillary gland. Possible transplacental effects in rats were also studied following intramuscular injection (Sunderman, 1983). Administration of Ni_3S_2 into the submaxillary gland and liver (Fischer rats) gave negative results; no tumors were produced in the gland by 2.5 mg or in liver by 5 mg Ni_3S_2, and only one liver tumor occurred in six rats injected intrahepatically with 10 mg of Ni_3S_2. Likewise, the results of tests for transplacental carcinogenicity of Ni_3S_2 in Fischer rats were negative. The intratesticular injection of 10 mg Ni_3S_2 was highly carcinogenic in Fischer rats, inducing testicular tumors in 16/19 animals, but much less in Syrian golden hamsters; only 4/13 animals developed tumors in 1 year. Unilateral intraocular injection of 0.5 mg Ni_3S_2 into Fischer rats resulted in local tumors in 14/15 rats with a median latency of 8 months. Carcinogenic effects of intraperitoneal administration of Ni_3S_2 were studied by Pott et al. (Chapter 37) in Wistar rats; 1×6, 2×6, and 25×1 Ni as Ni_3S_2 (weekly doses) produced local tumors in 20/36, 22/35, and 26/34 rats, respectively, in up to 2 years. Administration of Ni_3S_2 into the kidneys produced local tumors in Fischer, NIH

Black, and Wistar–Lewis rats but not in Long–Evans hooded rats (Sunderman, 1983), mice, or hamsters (Jasmin and Solymoss, 1978).

Sunderman et al. (1984) injected Fischer rats intrarenally with *17 insoluble nickel derivatives*, including Ni_3S_2 and other nickel sulfides, at a dose equivalent to 7 mg Ni. Renal cancers developed within 2 years in rats treated with (in order of decreasing activity) crystalline NiS (8/14 rats) > Ni_3S_2 > NiS_2, NiAsS > Ni_3Se_2 > Ni_4FeS_4, NiSe > NiFe alloy > NiAs (1/20 rats). In the same experiment, amorphous NiS, NiTe, $Ni_{11}As_8$, Ni_5As_2, NiSb, and $NiTiO_3$ did not induce renal tumors (negative results with metallic Ni and NiO were mentioned before).

Nickel(II) carbonate (basic, $NiCO_3 \cdot 2Ni(OH)_2 \cdot 4H_2O$) appeared to be only marginally carcinogenic (if at all) by multiple intraperitoneal injections to Wistar rats. Pott et al. (Chapter 37) found local tumors in 1/35 rats given 25 doses of 1 mg Ni, or in 1/33 rats given 50 doses of 1 mg Ni as the basic carbonate (weekly doses; 88-week observation). Multiple intraperitoneal injections of Fe_2O_3 or Fe_3O_4 did not produce local tumors (except for a single one in 139 rats), which indicates that the experimental conditions established by Pott et al. (Chapter 37) did not favor solid-state carcinogenesis.

Nickel carbonyl [$Ni(CO)_4$] was tested for carcinogenicity by Lau et al. (1972) in Sprague-Dawley rats injected intravenously with this liquid compound. Incidence of malignant tumors of various tissues was found significantly increased in rats given 6 biweekly or monthly doses of 9 mg Ni/kg BW that survived up to 2 years; 19/121 rats developed tumors in the lung, liver, kidney, breast, pancreas, uterus, and other sites, compared to 2/47 rats in the control group. Tumor incidence in rats that survived a single intravenous median lethal dose (LD_{50}) of $Ni(CO)_4$ (22 mg Ni/kg BW) was not significantly increased over the incidence in control animals. It was also discovered that $Ni(CO)_4$ injected intravenously to pregnant rats or hamsters causes malformations of the fetus such as exencephaly, anophthalmia, or microphthalmia (Leonard and Jacquet, 1984).

2.4.2. Water-soluble Nickel Compounds

The carcinogenic potential of $NiCl_2$, $Ni(CH_3COO)_2$, and $NiSO_4$ was recently tested by Pott et al. (Chapter 37) in Wistar rats injected intraperitoneally and observed for up to 2 years. Administered in 50 weekly injections of 1 mg (as Ni), $NiCl_2$ and $NiSO_4$ produced local intraperitoneal tumors in 3/32 and 4/30 rats, respectively. Twenty-five weekly injections of the same dose of nickel as the acetate induced tumors in 2/35 rats, while 50 injections increased the incidence of abdominal tumors to 6/31 rats ($p < 0.01$ vs. control rats given 50 weekly injections of 1 mL normal saline). Administered to strain A mice by multiple intraperitoneal injections, $Ni(CH_3COO)_2$ significantly increased the multiplicity of lung adenomas (Poirier et al., 1984); 24 intraperitoneal injections of 10.7 mg $Ni(CH_3COO)_2 \cdot 4H_2O$/kg BW per injection to 30 mice three times weekly increased the average number of lung tumors per mouse from 0.32 ± 0.12 (controls) to 1.5 ± 0.46.

2.5.   Two-Stage Carcinogenesis

The results of two experiments in which nickel compounds had been tested in two-stage carcinogenesis models were published in 1985. In the first one, Nesnow et al. (1985) attempted to initiate skin papillomas in SENCAR mice by a single intraperitoneal injection of $NiSO_4$ at doses ranging from 0.1 to 400 μg per mouse followed by cutaneous promotion with 12-O-tetradeca-noylphorbol-13-acetate (TPA); no significant effect was observed. In the second experiment, Kurokawa et al. (1985) tested five metal salts, including $NiCl_2$, for potential modifying influence on renal tumorigenesis in Fischer rats treated with dietary N-ethyl-N-hydroxyethylnitrosamine (EHEN). After completion of the EHEN treatment, the rats were given 600 mg $NiCl_2$/L in drinking water for 25 weeks. As a result, a statistically significant increase was found in the incidence of renal cell tumors in rats given EHEN and $NiCl_2$ compared to EHEN alone (8/15 rats vs. 2/15 in the controls given EHEN alone) and dysplastic foci (15/15 rats vs. 10/15 in the control rats). The incidence of renal cell tumors in rats treated with EHEN–$ZnCl_2$, $CrCl_3$, $HgCl_2$, or $CdCl_2$ was not different from that in the control rats. More recently, two other initiation/promotion experiments have been completed in this author's laboratory. In the first one (unpublished), groups of 29–39 strain A mice were given a single intraperitoneal injection of $NiCl_2$ at doses ranging from 50 to 200 μmol/kg BW followed by treatment with a standard diet or diet containing 0.75% of butylated hydroxytoluene (BHT) from day 7 postinjection until termination of the experiment; control mice received intraperitoneal injections of normal saline. The mice were sacrificed 24 weeks after the nickel injection. The results indicated that $NiCl_2$ alone had no statistically significant effect on the incidence of pulmonary adenomas inducing a maximum of 0.30 adenoma per mouse in animals given 200 μmol Ni/kg BW versus 0.23 adenoma per mouse in the control mice. The BHT alone increased the incidence to 0.41 adenoma per mouse while the combination of $NiCl_2$ and BHT produced from 0.37 (with 25 μmol $NiCl_2$/kg BW) to 0.53 adenoma per mouse (with 200 μmol $NiCl_2$/mouse). In the second experiment (Kaspizak et al., 1990), the effects of sodium barbital, a multitissue tumor promoter (Diwan et al., 1985) were studied in rats given a single intraperitoneal injection of $Ni(CH_3COO)_2$. Male Fischer rats were divided into four groups of 24 rats each. One group of rats received a single intraperitoneal injection of 90 μmol Ni/kg BW followed 1 week later by 500 ppm of sodium barbital in drinking water for the remaining experimental period. The next two groups of rats received either the same dose of $Ni(CH_3COO)_2$ alone followed by tap water or the sodium barbital alone preceded by intraperitoneal saline; control rats were given intraperitoneal saline and tap water. The $Ni(CH_3COO)_2$ alone was not carcinogenic at any site. By 92 weeks, renal tubular (RTC) tumors were found in 13/19 rats in the nickel acetate–sodium barbital group while 5/13 rats exposed to sodium barbital alone developed RTC tumors. Four rats in the $Ni(CH_3COO)_2$–sodium barbital group had RTC carcinomas, two of which metastasized to the lung. None of the rats exposed to sodium barbital alone developed RTC carcinomas. No RTC tumors were found in control rats or rats exposed to $Ni(CH_3COO)_2$ alone. Also, three rats in the $Ni(CH_3COO)_2$–sodium barbital group developed thyroid follicular cell tumors; no such tumors were found in other groups. The results of this study provide the first evidence for tumor-initiating activity of intraperitoneal $Ni(CH_3COO)_2$ in the rat kidney. Renal cortical epithelium appeared to be the only target cells conclusively initiated by nickel, although the thyroid data suggest a possible weak initiating effect of $Ni(CH_3COO)_2$ also in thyroid follicular epithelium.

2.6. Histogenesis of Nickel-induced Tumors

Nickel compounds are capable of inducing a wide variety of malignant tumors that are predominantly diagnosed as sarcomas; carcinomas were found less frequently. The histological identification of various types of nickel-induced sarcomas is usually difficult because of the pleomorphic character of these tumors. Tumors induced in the muscle, which is the most commonly used experimental target tissue for Ni_3S_2, have been most often diagnosed as rhabdomyosarcomas followed by fibrosarcomas, undifferentiated sarcomas, and histiocytic sarcomas. Internal injection of Ni_3S_2 produced poorly differentiated tumors classified as spindle cell tumors, mixed cell tumors, undifferentiated cell tumors, or rhabdomyosarcomas. Melanomas and retinoblastomas predominated following intraocular injection, while fibrosarcomas and rhabdomyosarcomas were the most frequent Ni_3S_2-induced testicular tumors (Sunderman, 1983; Bruni, 1986; Sano et al., 1988). Some pathologists have serious doubts about the correctness of most of these diagnoses, especially when rhabdomyosarcomas are concerned (Lumb et al., 1987). A sequential histological analysis of the intramuscular site from day 14 after the injection of Ni_3S_2 to NIH Black, hooded, and Fischer rats was performed by Daniel (1966) with special attention focused on the necrotizing/inflammatory and regenerative responses. Histogenesis of tumors arising at the subcutaneous and intramuscular sites of Ni_3S_2 implantation was recently studied by Lumb et al. (1987) and Lumb and Sunderman (Chapter 38). According to these authors, the Ni_3S_2-induced tumors originate from primitive mesenchymal cells that are either altered or regenerating myocytes when the tumor arises in a muscle or undifferentiated mesenchymal cells when the tumor arises subcutaneously. The tumors are easily transplantable and preserve their histological characteristics until the F3 generation (Lumb et al., 1987).

3. ROLE OF PHYSIOCOCHEMICAL AND BIOLOGICAL SPECIATION IN NICKEL CARCINOGENESIS

3.1. Effects of Physicochemical Properties of Compounds Tested

The relative carcinogenic activity of different nickel compounds had been established quite early in the 1960s (Payne, 1964) when it became clear that the most potent carcinogens originated from the water-insoluble nickel derivatives. This general rule is still valid with, however, many exceptions and conditions that greatly limit its applicability. It appears, for example, that the acute pathogenic and carcinogenic potential of different samples of the same chemical compound, NiO, Ni(OH)$_2$, or NiS, under identical conditions of a bioassay may strongly depend on the preparation mode of a particular sample (compare Section 2.3). On the other hand, a presumed noncarcinogenic soluble nickel salt (e.g., NiSO$_4$) may be made carcinogenic if its interaction with tissue is prolonged by multiple injections (Chapter 37) or when applied in a vehicle that slows down its mobilization and excretion (e.g., fat pellet; Payne, 1964). Another example of "tricks" aimed at prolonging the retention of nickel at the desired site and thus increasing the biological effect is use of tracheal transplants from which the carcinogen cannot be removed (Yarita and Nettesheim, 1978). The effectiveness of a tested compound may as well be influenced by the mode by which its suspension is prepared prior to use (Eilertsen et al., 1988). Hence, ranking of carcinogenic potency of different nickel derivatives is difficult, and the results of corresponding experiments can be generalized only with caution. Such rankings were done by Sunderman (1984b) relative to a single injection into the skeletal muscle and kidney of Fischer rats and most recently by Pott et al. (Chapter 37) relative to multiple injections into the intraperitoneal cavity of Wistar rats. The crystalline sulfides Ni$_3$S$_2$ and NiS were the most active of 17 compounds tested in both the muscle and the kidney, whereas NiO and nickel metal were moderately carcinogenic in the muscle but noncarcinogenic in the kidney; soluble nickel compounds were not tested (Sunderman, 1984b; Sunderman et al., 1984). In the peritoneal cavity, the carcinogenicity of Ni$_3$S$_2$ and nickel metal was ranked as *high*, that of NiO and of two nickel alloys containing over 50% Ni was ranked as *medium*; carcinogenic activity of Ni(CH$_3$COO)$_2$, NiSO$_4$ was *low*, that of NiCl$_2$ was *questionable*, and that of NiCO$_3$ and low-nickel alloy was *not proven*. Thus, the only general conclusions that can be drawn from the results of these and other experiments (IARC, 1976, 1982, 1987, 1990) are (i) that the crystalline nickel sulfides Ni$_3$S$_2$ and NiS are consistently the most active nickel carcinogens irrespective of different physical forms administered to various tissues and (ii) that carcinogenic potential of other nickel compounds strongly depends on many factors either intrinsic, such as the stoichiometry, crystal structure, dispersion, and solubility, or external (e.g., application regime and/or tissue susceptibility).

3.2. Species and Tissue Specificity

Wide variations in experimental conditions under which the carcinogenic effects of nickel and its derivatives were studied in animals make it impossible to draw firm general conclusions relative to the differences in response to every single nickel compound among species, strains, and tissues. Such a comparative study (Sunderman, 1983) has been thus far completed for only

Ni_3S_2. Sunderman's experiments were performed with similar protocols using powdered Ni_3S_2, <2-μm median particle size, from the same producer (INCO, Canada). The carcinogenic effects of Ni_3S_2 were tested in four strains of rats (Fischer, Wistar–Lewis, Long–Evans hooded, and NIH Black), two strains of mice (DBA/2 and C57BL/6), in Syrian golden hamsters, and in albino New Zealand rabbits. The target tissues included skeletal muscle, lung, liver, kidney, testis, eye, submaxillary gland, and oral mucosa (hamster cheek pouch). Based upon this study as well as on some observations of other laboratories, Sunderman (1983) concludes that no absolute species specificity for Ni_3S_2 carcinogenicity seems to exist, "albeit rats are more susceptible than mice, hamsters, or rabbits." The responsiveness of different rat strains to induction of kidney tumors following intrarenal Ni_3S_2 injections may be ranked as follows: Wistar–Lewis > NIH Black > Fischer > Long–Evans hooded (Sunderman, 1983). The relative responsiveness of skeletal muscles to Ni_3S_2 was Long–Evans hooded > NIH Black (Daniel, 1966) and Fischer > Long–Evans hooded (Yamashiro et al., 1980); direct comparison of the results of the latter two experiments should not be made because of different experimental conditions. Regarding the susceptibility of different organs of Fischer rats to local Ni_3S_2 carcinogenesis, Sunderman (1983) suggests the following: eye > skeletal muscle > testis = kidney > liver; lack of tumors in the oral cavity and gastrointestinal tract in hamsters given Ni_3S_2 to the cheek pouches and in rats following intratracheal exposure to Ni_3S_2 (upward clearing) may suggest that the digestive tract of these species is not sensitive to Ni_3S_2 carcinogenesis; however, this speculation awaits confirmation by feeding experiments (Sunderman, 1983). Some effects of age, sex, and castration on carcinogenicity of nickel subsulfide in the muscles of Sprague-Dawley rats were studied by Jasmin (1965); females appeared to be a little more susceptible than males, castrated rats developed fewer tumors than normal animals, and maximum susceptibility to Ni_3S_2 in both sexes occurred at 2 months of age.

4. EFFECTS OF INORGANIC AND ORGANIC COMPOUNDS ON NICKEL CARCINOGENESIS

4.1. Effects of Metals

Carcinogenic activity of nickel can be significantly inhibited by the concurrent administration of some metals such as Mn, Mg, Fe, or Zn, whereas certain other metals have no effect whatsoever. Historically, the first investigation on metal interactions in nickel carcinogenesis was completed by Sunderman et al. (1974). Equimolar doses of metal dusts of Al, Cu, Cr, or Mn were injected together with 2.5 mg Ni_3S_2 (31 μmol Ni) per rat into the muscles of Fischer rats. During the 2-year observation that followed the injection, local sarcomas developed in over 95% of the rats (10–40 rats per group) given

Ni_3S_2 alone or Ni_3S_2 combined with Al, Cu, or Cr, compared with only 63% (15/24 rats) in rats injected with Ni_3S_2–Mn. In the two follow-up studies on the effect of manganese upon nickel carcinogenesis, Sunderman et al. (1976) and Sunderman and McCully (1983b) established the following: (i) The degree of inhibition depended on the mole ratio of the metals injected to nickel. (ii) The effect was strictly local (i.e., injection of Mn at a site distant from that of Ni_3S_2 had no influence on tumor yield). (iii) The effect was produced only by metallic manganese whereas MnS, Mn_2O_3, MnO_2, or $Mn_2(CO)_{10}$ in dosages equivalent to that of manganese metal were ineffective; only MnS slightly prolonged the latent period of tumors. (iv) The inhibition by manganese was accompanied by manganese-caused changes in the distribution of nickel in the subcellular constituents of the injected muscle.

None of the other metals tested with nickel produced any tumors by itself. Metallic manganese was also found to inhibit induction of muscle sarcomas by metallic nickel powder: a mixture of 20 mg Ni and 20 mg Mn produced tumors in 2/10 WAG rats vs. 7/10 rats given 20 mg Ni alone in approximately 1 year (Judde et al., 1987).

The effects of two other physiological metals, calcium and magnesium, on nickel carcinogenesis were checked in mice and rats. In the experiment mentioned above, Poirier et al. (1984) administered 24 intraperitoneal injections to strain A mice of nickel acetate three times weekly, 40 μmol Ni/kg BW per injection alone or with different proportions of calcium acetate or magnesium acetate, the calcium–nickel or magnesium–nickel molar ratio varying from 0 to 30, and observed that both calcium and magnesium even at low doses completely prevented the enhancement of pulmonary adenomas caused by nickel. The inhibitory action of calcium and magnesium was of the same magnitude despite striking differences between these two metals in their effects on nickel uptake and subcellular distribution or the nickel-caused disturbances in pulmonary DNA synthesis in strain A mice (Kasprzak and Poirier, 1985). The effects of calcium and magnesium in rats were different (Kasprzak et al., 1985). Fischer rats injected intramuscularly with Ni_3S_2 plus calcium carbonate developed local sarcomas at the same rate as with Ni_3S_2 alone while admixture of magnesium basic carbonate had a significant inhibitory effect. In a 1.5-year study only 25% of rats treated with Ni_3S_2–Mg developed local sarcomas versus 85% incidence in the Ni_3S_2-only rats. The effects was dose dependent and strictly local since magnesium salts had no effect on intramuscular Ni_3S_2 carcinogenesis when administered subcutaneously or by the dietary route. The mechanism of the magnesium inhibitory action remains to be unveiled. Most likely, this effect is due to the inhibition by magnesium of nickel transport and binding phenomena at cells giving rise to tumors and the immunocompetent cells (Kasprzak et al., 1987a,b). The prevention by magnesium of nickel-induced inhibition of intercellular communication may also play a role (Miki et al., 1987). More detail on related experiments can be found in this book (Chapter 42).

A profound inhibition of Ni_3S_2 carcinogenesis in the muscles of Fischer rats can also be achieved by treatment with either metallic iron powder or Fe(III) sulfate. As in the cases of manganese and magnesium, the inhibition by iron is strictly local and cannot be produced by subcutaneous injection at a site distant from that of Ni_3S_2 (Chapter 42). The observed inhibition is consistent with negative results of carcinogenicity testing of an iron–nickel alloy in the rat muscle by Sunderman (1984b).

The inhibitory effect of zinc on Ni_3S_2 carcinogenesis in Fischer rats differs from those of manganese, magnesium and iron by its weak and systemic character. Unlike those of magnesium and iron, the zinc derivatives inhibited Ni_3S_2 carcinogenesis following both local and distant administration; however, in either case the effect was limited to some retardation of tumor development with no influence on the 2-year incidence of tumors. It is possible that multiple injections of zinc could exert a much stronger effect. For practical reasons (prevention of nickel cancer in humans), such a possibility is worth testing in animals. The mechanisms of zinc's action remains unknown (Chapter 42).

4.2. Effect of Organic Compounds

Treatment of rats with ethylenediamine tetraacetic acid (EDTA) decreases the tumorigenic activity of nickel (Noble and Capstick, 1963), most likely due to chelation and faster excretion of the metal. Gilman (1966) observed inhibition of nickel carcinogenesis following treatment of Ni_3S_2-injected rats with cortisol. However, this finding was not confirmed by Kasprzak and Ward (1991), who found no effect of locally injected cortisol and another anti-inflammatory agent, indomethacin, on Ni_3S_2 carcinogenesis in the muscles of Fischer rats. Moreover, indomethacin, if administered subcutaneously away from the intramuscular Ni_3S_2, tended to increase the occurrence rate of the local muscle tumors. The results of use of the *Mycobacterium bovis* antigen to study the effects of inflammatory response on muscle tumor induction by Ni_3S_2 in Fischer rats may serve as another example of a strong dependence of the final outcome on the administration protocol (Kasprzak and Ward, 1991). When the antigen was injected intramuscularly in a common mixture with Ni_3S_2, only one local tumor developed in 20 rats over the 1.5-year observation period. However, when the antigen was injected subcutaneously separately from Ni_3S_2, all the rats (20/20) developed tumors at the Ni_3S_2 site much sooner than rats treated with Ni_3S_2 alone. Waalkes et al. (1987) observed that microbial toxins and/or antigens might as well be involved in the induction of ear pinna sarcomas in Wistar rats at sites of ear tags made of a high-nickel alloy. Interferon injected intramuscularly to WAG rats together with nickel metal had no effect on the occurrence of tumors (Judde et al., 1987). The mechanisms involved in the inhibitory or enhancing effects on nickel carcinogenesis of substances altering immune responses to nickel are poorly understood. It is worth noting that the inhibition by manganese and magnesium mentioned in Section 4.1 is likely also to be mediated through the cellular immune responses to nickel, especially from the natural killer cells and macrophages (Smialowicz, 1985; Judde et al., 1987; Kasprzak et al., 1987b).

Other conditions that alter the growth and/or metabolism of skeletal muscles, such as treatment with anabolic steroids (Jasmin, 1963) or infection with the *Trichinella spiralis* larvae (the so-called basophilic transformation of some striated muscle fibers occurs even before the larvae invade the muscle), exert a stimulatory effect on Ni_3S_2 carcinogenesis in rat muscles (Kasprzak et al., 1971).

Mutually potentiating carcinogenic effects were also observed for Ni_3S_2 and benzo[*a*]pyrene (BP). Thus, Maenza et al. (1971) noticed a dramatic acceleration of the occurrence rate of local sarcomas in muscles of Fischer rats injected with Ni_3S_2–BP compared to rats given the carcinogens separately. Likewise, Kasprzak et al. (1973) observed more severe pathological changes, including one bronchogenic carcinoma, in 13 rats given intratracheally Ni_3S_2–BP than in rats treated with either of these two compounds alone. These findings may indicate a possible enhancement of tobacco smoke carcinogens upon nickel activity, suggested by human epidemiology (Doll et al., 1970) but not confirmed in animal experiments with NiO (Wehner et al., 1975; Chapter 39).

5. RECENT ANIMAL STUDIES RELEVANT TO NICKEL CARCINOGENESIS

Since the discovery of nickel carcinogenicity in animals, experiments attempting to unveil the sequence and mechanisms of events leading to tumors have been numerous in both in vivo and in vitro systems. The results were already reviewed, certain conclusions drawn, and hypotheses formulated by many authors (IARC, 1976, 1990; Sunderman, 1977, 1979, 1981, 1984a,b, 1986a,b; Sunderman et al., 1975; Furst, 1984; Sky-Peck, 1986; Kasprzak, 1987). Therefore, this chapter is limited to a brief discussion of the more recent papers on the uptake, retention, metabolism, nickel–DNA interactions, and pathogenic effects of nickel compounds in animals that may shed some new light on possible mechanisms of nickel carcinogenesis. Primary attention will be paid to organs that have been targets for nickel carcinogenesis in humans, that is, the respiratory tract and kidney.

The results of exposue to nickel compounds by routes most relevant to human occupational exposures, that is, by inhalation and intratracheal insufflation, have been recently studied at the U.S. Inhalation Toxicology Research Institute. Thus, lung clearance and disposition of nickel in rats after intratracheal administration of nickel sulfate solutions were investigated by Medinsky et al. (1987). Fischer rats were given 17, 190, or 1800 nmol Ni (as $^{63}NiSO_4$). The retention of nickel in the lungs depended on the dose; the long-term half-life for clearance of the pulmonary nickel varied from 21 hr at the highest dose to 36 hr for the lowest dose instilled. The main route for clearance of nickel from the body was urinary excretion (50–80% of the dose with $T_{1/2} = 4.6$ hr for the highest dose and 23 hr for the lowest dose), which points to the possibility of renal and urinary bladder damage. High concentrations of nickel were also found in the trachea and larynx. Fecal excretion reached a maximum of 30%, indicating that some of the nickel had been cleared from the lung into the gastrointestinal tract. This distribution concurs with the target sites for nickel carcinogenesis in workers of electrolytic nickel refineries (Medinsky et al., 1987). The results of 13-week toxicity studies of $NiSO_4$, NiO, and Ni_3S_2 in F344/N rats and B6C3F$_1$ mice are described by Dunnick et al. (Chapter 33); the main aim of that study was to select exposure levels for a 2-year comparative toxicity and carcinogenicity study. The experimental detail and results can be found in the original paper. The main findings of this study may be summarized as follows: (i) F344/N rats are more sensitive than B6C3F$_1$ mice to nickel lung toxicity, (ii) the order of toxicity to the lung of the compounds studied is $NiSO_4 \gg Ni_3S_2 >$ NiO, and (iii) lung toxicity of the three compounds tested reflects their relative solubility in water, the most soluble $NiSO_4$ being the most toxic even though it is cleared from the lung faster than the oxide and the subsulfide. The retention of nickel in the lung of male mice after 13-week exposure to the same air concentration of nickel in different compounds, 0.4 mg Ni/m^3, was 0.8, 13.4, and 80.1 μg Ni/g sample for $NiSO_4$, Ni_3S_2, and NiO, respectively. The very low retention of nickel from $NiSO_4$ and very high retention from NiO are consistent with the toxicokinetics of intratracheal $NiSO_4$ and NiO in rats (English et al., 1981). The retention of Ni_3S_2 is low because of a relatively fast clearance of this compound from the mouse lung (Valentine and Fisher, 1984; Finch et al., 1987); for example, lung clearance of intratracheal Ni_3S_2 (1.7-μm particles) was biphasic with the first half-life of 1.2 days (the first 5 days) and the second half-life of 12.4 days (days 5–35). Benson et al. (1986) compared the acute toxicity of $NiCl_2$, $NiSO_4$, Ni_3S_2, and NiO given intratracheally to Fischer rats at single doses ranging from 0.01 to 1.0 μmol per rat; the rats were sacrificed 1 or 7 days after the compound administration. Retention of nickel in the lung on day 7 was ranked NiO $> Ni_3S_2 > NiSO_4 > NiCl_2$. The relative toxicity of these compounds, assessed by the severity of pulmonary inflammation, was ranked $NiCl_2$, $NiSO_4$, $Ni_3S_2 \gg$ NiO. The long-term carcinogenicity bioassay, which is to follow the study of Dunnick et al. (Chapter 33), should provide more information about the relationships (if any) among the solubility, acute toxicity, and carcinogenicity of nickel compounds.

The lung appears to be the primary target for nickel also after parenteral injections of water-soluble nickel salts as discovered by Herlant-Peers et al. (1982) in Balb/c mice. Following a single intraperitoneal injection of $NiCl_2$, the clearance of nickel from the lung was slowest of 14 tissues studied; therefore, despite lower immediate uptake of nickel, the pulmonary retention of this metal after 6 days predominated over that in other tissues, including kidney, skin, liver, muscle, and others. Also, much more nickel remained bound to subcellular fractions of the lung than liver 1 day after seven consecutive daily intraperitoneal injections of $NiCl_2$ (Herlant-Peers et al., 1983). For example, pulmonary nuclei and mitochondria retained approximately 12 times, microsomes 22 times, and cytosol over 6 times more nickel than the corresponding liver fractions; nickel was preferentially bound to mitochondrial and microsomal proteins. These results concur with the most recent observations made in rats by Knight et al. (1988). Fischer rats were given multiple subcutaneous doses of 62.5 or 125 μmol $NiCl_2$/kg BW, or single subcutaneous doses of 250–750 μmol $NiCl_2$/kg BW (LD_5–LD_{95}) and sacrificed 1, 2, or 3 days after completion of the injections. The retention of nickel was determined in lung, kidney, and eight other tissues; the lungs were examined histologically. Following 42 daily injections of $NiCl_2$, retention of nickel in lung and kidney was much greater than in any other tissue, with the kidney slightly predominating after the 62.5-μmol doses and the lung predominating after the 125-μmol doses (the concentrations were 77, 22, 3, 3, and 2 μg/g dry wt. in the lung, kidney, testis, brain, and liver, respectively). Histopathological changes in the lung of rats given the multiple intraperitoneal doses of $NiCl_2$ were similar to those previously reported in rats or mice exposed by inhalation to $NiCl_2$, $NiSO_4$, or NiO (Benson et al., 1986, 1987; Bingham et al., 1972; Horie et al., 1985; Takenaka et al., 1985), including hyperplasia of alveolar cells (especially of type II pneumocytes), sometimes in association with hyperplasia of bronchial epithelial cells and alveolar proteinosis. Vascular endothelium and pulmonary mononuclear cells were principal targets following the single lethal doses of nickel. Whether or not these changes can lead to lung tumors remains to be determined. Thus far, carcinogenic potential toward the lung of a soluble nickel salt [$Ni(CH_3COO)_2$] by multiple intraperitoneal injections was tested only in strain A mice in which the positive response observed (Poirier et al., 1984) might not be specific; such a response was also produced by $Ca(CH_3COO)_2$. As described earlier, the increase of the lung adenoma incidence by nickel could be prevented by simultaneous treatment with $Mg(CH_3COO)_2$. The anticarcinogenic action of magnesium was accompanied by a significant decrease of total lung nickel and its contents in the pulmonary cell nuclei and cytosol and by inhibition of nickel-caused disturbances in DNA synthesis (Kasprzak and Poirier, 1985). Inhibition by magnesium of nickel uptake in the lung, kidney, and liver was also observed in rats (Kasprzak et al., 1986). However, the action of magnesium could not be solely attributed to prevention of nickel binding since the same antitumorigenic effect was obtained with $Ca(CH_3COO)_2$ which acted differently on nickel uptake; calcium disturbed pulmonary DNA synthesis by itself and did not inhibit but enhanced nickel uptake by the whole lung, pulmonary mitochondria, microsomes, and cytosol. Also, calcium had no effect on nickel uptake by the nuclei and assisted nickel-caused unscheduled DNA synthesis in the mouse lung (Kasprzak and Poirier, 1985). Hence, different mechanisms might have been involved in the inhibitory effects by magnesium and calcium, or the inhibition by both calcium and magnesium was due to phenomena other than nickel uptake and/or nickel-produced disturbances in DNA synthesis.

Toxic effects of nickel in the second main target organ, the kidney, were studied quite extensively following the discovery by Jasmin and co-workers (Jasmin and Riopelle, 1976; Jasmin and Solymoss, 1978) of erythrocytosis and local carcinogenesis by Ni_3S_2 injected into this organ.

Striking rank correlation was established by Sunderman et al. (1984) between the incidence of renal sarcomas and the capacity of 18 nickel derivatives to induce erythrocytosis after intrarenal injection to Fischer rats. This firm correlation suggests that stimulation by nickel of erythropoietin production and oncogenesis may be related mechanistically; the prompt erythrocytic response (1–4 months) can serve as a test for carcinogenic activity of nickel compounds in rats (Sunderman, 1986b). Most interestingly, erythrocytosis induction by Ni_3S_2 may be inhibited by local administration of metallic manganese (Sunderman, 1986b).

Uptake of nickel by renal parenchyma and nuclei following parenteral administration to rats was studied by Ciccarelli and Wetterhahn (1984) and Liber and Sunderman (1985). The nuclear uptake depended on the injected dose and comprised 2–5% of the total renal nickel. However, some methodological pitfalls make these numbers uncertain, and the real uptake in vivo may be much lower. Distribution of nickel bound to various subcellular protein fractions in the rat kidney was extensively invesigated by Sunderman and co-workers (Sunderman et al., 1983). Renal cytosol and microsomes were found to contain six nickel-binding protein fractions of which one major fraction comprised over 80% of the cytosolic nickel. Nickel in this fraction was most likely bound to a low-molecular-weight protein (MW < 7,000) that differed from metallothionein by amino acid and carbohydrate composition and to some lower molecular weight constituents that were not identified. Structures of some of the nickel-sequestering renal proteins were established by Abdulwajid and Sarkar (1983) and Templeton and Sarkar (1985). Unlike the cytosolic proteins, the microsomal proteins contained one high-molecular-weight nickel-binding fraction (MW > 700,000), which probably represented aggregates of microsomal proteins and RNA. This finding suggests formation of relatively stable RNA–protein–nickel complexes in which the protein–RNA moieties are crosslinked to each other with bonds withstanding the isolation and separation procedures. Such a complex may be similar to the ternary complex(es) of microsomal protein(s), nickel and DNA described by Ciccarelli and Wetterhahn (1982) and Lee et al. (1982). Formation of ternary complexes of this type may be damaging to the expression of cellular genetic information and may lead to cell transformation and cancer. Nothing is known about the exact nature of the proteins bound in these complexes. They may be the same DNA-binding proteins that normally regulate gene expression (Sunderman and Barber, 1988). The physiological interaction of these proteins with DNA depends on zinc. Substitution of nickel (or other carcinogenic metal) for zinc would change the DNA–protein bonding. According to Sunderman and Barber (1988), such a substitution may constitute a hypothetical mechanism for metal carcinogenesis. It is worthwhile to notice that nickel binding to the DNA molecule without participation of a protein (and most likely, also some microsomal enzymes) is weak and the ligands dissociate easily [e.g., under the conditions of sodium dodecyl sulfate–polyacrylamide gel electrophoresis (SDS–PAGE)] in the presence of salt or polyethylene glycol (Heerlant-Peers et al., 1983).

Consistent with the strong nickel effects on the DNA–protein complex (chromatin) rather than on DNA itself are recent findings from Costa's laboratory (Sen and Costa, 1985; Sen et al., 1987) regarding the nickel-caused damage to heterochromatin in rodent cells. Interestingly, this damage may be prevented by magnesium (Conway et al., 1987), an antagonist of nickel carcinogenesis (Kasprzak et al., 1987a,b).

A toxic effect that may be of great importance to the mechanisms of nickel carcinogenesis is the lipid peroxidation observed in many tissues of animals injected with this metal. Thus, Sunderman et al. (1985) and Donskoy et al. (1986) reported evidence of increased lipid peroxidation in the lung, kidney, and liver of rats injected with high parenteral doses of $NiCl_2$. Knight et al. (1986) reported exhalation of ethane, a secondary product of lipid peroxidation, by rats after $NiCl_2$ injection. The same authors observed increased lipid peroxidation and acute involution of the thymus in $NiCl_2$-treated rats (Knight et al., 1987); this finding is relevant to the immunotoxic effects of nickel discussed below. Kasprzak et al. (1986) noticed increased contents of peroxidized lipids in the kidneys and livers of rats injected intraperitoneally with $Ni(CH_3COO)_2$; treatment with magnesium had no effect on that increase. Wong et al. (1987) observed elevated levels of lipoperoxides in blood plasma of rats injected subcutaneously with $NiCl_2$. According to Marnett (1987), Klein et al. (1991), Kasprzak (1991), and Standeven and Wetterhahn (1991) there is growing evidence of a direct involvement of oxygen radicals derived from hydrogen peroxide and lipid hydroperoxides with participation of transition metals in both initiation and promotion of tumors.

Another pathogenic property of nickel that may assist in the induction of tumors by this metal is its immunotoxicity (Adkins et al., 1979; Wiernik et al., 1983; Johansson et al., 1983; Smialowicz et al., 1984, 1985, 1987). A single intramuscular injection of $NiCl_2$ to mice caused a significant involution of the thymus and suppression of T-cell-mediated immune responses and of splenic natural killer (NK) cell activity. In the rat, a single intramuscular injection of 10–20 mg $NiCl_2$/kg BW had no effect on the responsiveness of the T and B cells but significantly (though transiently) suppressed cytotoxic activity of the NK cells (Smialowicz et al., 1987). Suppression of the NK cells in rats following intramuscular administration of Ni_3S_2 was also observed by Judde et al. (1987) and Kasprzak et al. (1987b). The NK cells are thought to constitute the primary cytotoxic guard against newly arising tumor cells (Smialowicz et al., 1985). The immunotoxic activity of nickel may depend on its dose and administration regime. For example, a single intraperitoneal dose of $NiCl_2$ had either an impairing or an enhancing effect on the resistance of mice against bacterial infection, depending on the dose and time of its application relative to the time of infection (Laschi-Loquerie et al., 1987). A long-term exposure of B6C3F$_1$ mice to $NiSO_4$ in the drinking water damaged the granulocyte–macrophage progenitor cells in the bone marrow, which led to a secondary systemic immune deficiency (Dieter et al., 1988). Inhalation of metallic nickel dust or $NiCl_2$ was found to decrease bactericidal activity of lung alveolar macrophages in mice (Adkins et al., 1979) and rabbits (Wiernik et al., 1983; Johansson et al., 1983), whereas intratracheal instillation of Ni_3S_2 to cynomolgus monkeys caused a significant suppression of alveolar macrophage activity toward sheep red blood cells; this suppression was accompanied by a secondary increase of NK-cell-mediated cytotoxicity (Haley et al., 1987). The suppression of macrophage and NK-cell activity by nickel seems to be important to the process of carcinogenesis since enhancement of either of these cells [e.g., the NK cells by Mn (Smialowicz, 1985; Smialowicz et al., 1987; Judde et al., 1987) or the macrophages by Mg, Fe, or *M. bovis* antigen (Kasprzak and Ward, 1991; Chapter 42)] is accompanied by a suppression of nickel carcinogenesis. A similar pattern was observed by Daniel (1966), who compared susceptibility of different rat strains to intramuscular Ni_3S_2 carcinogenesis; the more vigorous was the macrophage response to Ni_3S_2 particles in a given strain (NIH Black >> hooded, Fischer rats), the fewer tumors developed in that strain.

6. CONCLUDING REMARKS

Despite growing pressure to use alternate in vitro systems in cancer research, animal studies still constitute the only way of testing suspected nickel (and other) carcinogens and possible mechanisms of their action under dynamic conditions similar to those encountered by man. Animal experiments confirmed the results of epidemiological investigations in nickel refineries and helped to identify the most active nickel carcinogens. Animal experiments have revealed species and tissue susceptibility to nickel carcinogenesis, the metabolism of various nickel compounds administered by different routes, interactions with other substances including carcinogens and metals, and some correlations between acute pathogenic and carcinogenic effects as well as helped to discover the importance of physicochemical form and dosing regime of a nickel compound to its carcinogenic activity. Experiments on laboratory animals led to the discovery of immunotoxic properties of nickel and their possible contribution to nickel carcinogenesis. And finally, pathomorphologic investigations of all stages of the development of nickel-induced tumors to establish their histogenesis and malignancy were also performed in animals. None of these results would be obtained from in vitro models.

Nickel derivatives emerge from the animal studies as multipotent local carcinogens with some affinity for the lung and kidney. There is no distinct difference in response to a particular nickel compound (Ni_3S_2) among rodent species. There are, however, marked differences in the carcinogenic activity in one tissue (lung, kidney, muscle) among various nickel derivatives administered by the same way (e.g., a single injection). Most active are the insoluble crystalline nickel sulfides; least active are water-soluble salts. However, the solubility criterion has to be applied with caution because of some exceptions and quite strong dependence of the carcinogenic result on the administration regime; also, carcinogenic activity of a particular compound may depend on the process of its preparation. More experiments are needed to establish a firm structure–effect relationship for nickel carcinogens.

Thus far, animal experiments have not provided much information regarding the molecular mechanisms of nickel carcinogenesis. The subcellular uptake and distribution experiments often yield equivocal results because of redistribution of loosely bound nickel during the isolation procedures; therefore, any conclusions drawn from increased or decreased nickel uptake (e.g., by cell nuclei) or binding to specific biomolecules in vivo must be viewed with caution. Nevertheless, the in vivo formation of relatively stable ternary nickel complexes with DNA or RNA and proteins seems to bear cell-transforming potential and should be studied in more detail. Likewise, high affinity of nickel for mitochondria and microsomes observed in some experiments may be of significance to the mechanisms of nickel carcinogenesis. The formation by nickel of the DNA–protein–nickel complexes and possible substitution for zinc in the "finger loops" may affect gene expression (Sunderman and Barber, 1988). However, surprisingly enough, no papers on possible oncogene activation in nickel-induced tumors could be found.

Lipid peroxidation observed in nickel-treated animals as well as some other redox reactions catalyzed by nickel (Nieboer et al., 1986; Kasprzak and Bare, 1989) and nickel-caused inhibition of catalase and peroxidase (Chapter 30; Kasprzak, 1991) open up a possibility of indirect, nickel-mediated oxidative damage to the cellular genetic material and deserve further experimental studies.

A new, rapidly evolving research field that may be of great importance to understanding the mechanisms of nickel carcinogenesis concerns the immunotoxicity of this metal. High, strain-related or treatment-enhanced activity of macrophages and/or NK cells correlated well with lower susceptibility of the animal to nickel carcinogenesis in many experiments. Investigations of this type deserve utmost attention.

ACKNOWLEDGMENTS

Helpful critical comments of Drs. J. Rice and B. Diwan and invaluable editorial help of Ms. K. Breeze are gratefully acknowledged.

REFERENCES

Abdulwajid, A. W. and Sarkar, B. (1983). Nickel-sequestering renal glycoprotein. *Proc. Natl. Acad. Sci. USA* **80,** 4509–4512.

Adkins, B., Jr., Richards, J. H., and Gardner, D. E. (1979). Enhancement of experimental respiratory infection following nickel inhalation. *Environ. Res.* **20,** 33–42.

Benson, J. M., Henderson, R. F., McClellan, R. O., Hanson, R. L., and Rebar, A. H. (1986). Comparative acute toxicity of four nickel compounds to F344 rat lung. *Fundam. Appl. Toxicol.* **7,** 340–347.

Benson, J. M., Carpenter, R. L., Hahn, F. F., Haley, P. J., Hanson, R. L., Hobbs, C. H., Pickrell, J. A., and Dunnick, J. K. (1987). Comparative inhalation toxicity of nickel sulfate to F344/N rats and B6C3F₁ mice exposed for 12 days. *Fundam. Appl. Toxicol.* **9,** 251–265.

Berry, J. P., Poupon, M. F., Judde, J. C., and Galle, P. (1985). In vitro electron microprobe of carcinogenic nickel compound interaction with tumor cells. *Ann. Clin. Lab. Sci.* **15,** 109–120.

Bingham, E., Barkley, W., Zerwas, M., Stemmer, K., and Taylor, P. (1972). Response of alveolar macrophages to metals. *Arch. Environ. Health* **25,** 406–414.

Bruni, C. (1986). Abnormal cytodifferentiation in leiomyosarcomas induced with injections of nickel subsulfide into skeletal muscles of rats. *Ann. Clin. Lab. Sci.* **16,** 118–129.

Cecutti, A., and Nieboer, E. (1981). Effects of nickel on animals and humans. In *Effects of Nickel in the Canadian Environment.* Publ. No. NRCC 18568 of the Environmental Secretariat NRCC/CNRC, Ottawa, pp. 217–260.

Ciccarelli, R. B., and Wetterhahn, K. E. (1982). Nickel distribution and DNA lesions induced in rat tissues by the carcinogen nickel carbonate. *Cancer Res.* **42,** 3544–3549.

Ciccarelli, R. B., and Wetterhahn, K. E. (1984). Nickel-bound chromatin, nucleic acids, and nuclear proteins from kidney and liver of rats treated with nickel carbonate *in vivo. Cancer Res.* **44,** 3892–3897.

Conway, K., Xu, L., and Costa, M. (1987). Protective effect of magnesium against nickel-induced DNA damage and cell transformation. *Toxicologist,* **7,** 81.

Daniel, M. R. (1966). Strain differences in the response of rats to the injection of nickel sulphide. *Br. J. Cancer* **20,** 886–895.

Dieter, M. P., Jameson, C. W., Tucker, A. N., Luster, M. I., French, J. E., Hong, H. L., and Boorman, G. A. (1988). Evaluation of tissue disposition, myelopoietic, and immunologic responses in mice after long-term exposure to nickel sulfate in the drinking water. *J. Toxicol. Environ. Health* **24,** 357–372.

Diwan, B. A., Rice, J. M., Ohshima, M., Ward, J. M., and Dove, L. F. (1985). Comparative tumor-promoting activities of phenobarbital, amobarbital, barbital sodium, and barbituric acid on livers and other organs of male F344/NCr rats following initiation with *N*-nitroso-diethylamine. *J. Natl. Cancer Inst.* **74,** 509–516.

Doll, R., Morgan, L. G., and Speizel, F. E. (1970). Cancer of the lung and nasal sinuses in nickel workers. *Br. J. Cancer* **24,** 623–632.

Donskoy, E., Donskoy, M., Forouhar, F., Gillies, C. G., Marzouk, A. B., Reid, M. C., Zaharia, O., and Sunderman, F. W., Jr. (1986). Hepatic toxicity of nickel chloride in rats. *Ann. Clin. Lab. Sci.* **16,** 108–117.

Dunnick, J. K. Benson, J. M., Hobbs, C. H., Hahn, F. F., Cheng, Y. S., and Eidson, A. F. (1988). Comparative toxicity of nickel oxide, nickel sulfate hexahydrate, and nickel subsulfide after 12 days of inhalation exposure to F344/N rats and B6C3F₁ mice. *Toxicology* **50,** 145–156.

Eilertsen, E., Skaug, V., and Norseth, T. (1988). Tumor induction in rats after intrapleural instillation of various nickel oxides. *Abstr. 4th Intl. Conf, Nickel Metabol. Toxicol.,* Espoo. p. 43.

English, J. C., Parker, R.D.R., Sharma, R. P., Oberg, S. G. (1981). Toxicokinetics of nickel in rats after intratracheal administration of a soluble and insoluble form. *Am. Ind. Hyg. Assoc. J.* **42,** 486–492.

Farrell, R. L. and Davis, G. W. (1974). The effects of particulates on respiratory carcinogenesis by diethylnitrosoamine. In E. Karbe and J. F. Park (Eds.), *Experimental Lung Cancer: Carcinogenesis and Bioassays,* Springer Verlag, New York. pp. 219–233.

Finch, G. L., Fisher, G. L., and Hayes, T. L. (1987). The pulmonary effects and clearance of intratracheally instilled Ni_3S_2 and TiO_2 in mice. *Environ. Res.* **42,** 83–93.

Friedman, I., and Bird, E. S. (1969). Electron microscope investigation of experimental rhabdomyosarcoma. *J. Pathol.* **97,** 375–382.

Furst, A. (1984). Mechanism of action of nickel as a carcinogen: Needed information. In F. W. Sunderman, Jr. (Ed.), *Nickel in the Human Environment.* IARC Scientific Publications, Lyon, No. 53, pp. 245–252.

Furst, A., and Schlauder, M. C. (1971). The hamster as a model for metal carcinogenesis. *Proc. West. Pharm. Soc.* **14,** 68–71.

Furst, A., Casetta, D. M., and Sasmore, D. P. (1973). Rapid induction of pleural mesotheliomas in the rat. *Proc. West. Pharm. Soc.* **16,** 150–153.

Gilman, J.P.W. (1962). Metal carcinogenesis. II. A study of the carcinogenic activity of cobalt, copper, iron and nickel compounds. *Cancer Res.* **22,** 158–162.

Gilman, J.P.W. (1966). Muscle tumorigenesis. *Canad. Cancer Conf.* **6,** 209–223.

Grandjean, P. (Ed.) (1986). *Health Effects Document on Nickel Submitted to Ontario Ministry of Labour.* Department of Environmental Medicine, Odense University, Odense, pp. 64–74.

Grandjean, P., Andersen, O., and Nielsen, G. D. (1988). Carcinogenicity of occupational nickel exposures: an evaluation of epidemiological evidence. *Am. J. Ind. Med.* **13,** 193–209.

Haley, P. J., Bice, D. E., Muggenburg, D. A., Hahn, F. F., and Benjamin, S. H. (1987). Immunopathologic effects of nickel subsulfide on the primate pulmonary immune system. *Toxicol. Appl. Pharm.* **88,** 1–12.

Heath, J. C., and Daniel, M. R. (1964). The production of malignant tumors by nickel in the rat. *Br. J. Cancer* **18,** 261–264.

Heinrich, U., Pott, F., Rittinghausen, S., and Peters, L. (1987). Carcinogenic effects after combined exposure to PAH-containing emissions and other respiratory tract carcinogens. *Proc. 11th Symp. Polynucl. Arom. Hydrocarb.*, Gaithersburg, MD, p. 2.

Herlant-Peers, M. C., Hildebrand, H. F., and Biserte, G. (1982). ^{63}Ni(II) incorporation into lung and liver cytosol of Balb/C mouse. An *in vitro* and *in vivo* study. *Zbl. Bakt. Hyg., I. Abt. Orig. B* **176,** 368–382.

Herlant-Peers, M. C., Hildebrand, H. F., and Kerckaert, J. P. (1983). *In vitro* and *in vivo* incorporation of ^{63}Ni(II) into lung and liver subcellular fractions of Balb/C mice. *Carcinogenesis* **4,** 387–392.

Horie, A., Haratake, J., Tanaka, I., Kodama, Y., and Tsuchiya, K. (1985). Electron microscopical findings with special reference to cancer in rats caused by inhalation of nickel oxide. *Biol. Trace Elem. Res.* **7,** 223–239.

Hueper, W. C. (1952). Experimental studies in metal carcinogenesis: I. Nickel cancers in rats. *Texas Rep. Biol. Med.* **10,** 167–186.

Hueper, W. C. (1955). Experimental studies in metal carcinogenesis: IV. Cancer produced by parenterally introduced metallic nickel. *J. Natl. Cancer Inst.* **16,** 55–73.

Hueper, W. C. (1958). Experimental studies in metal carcinogenesis. IX. Pulmonary lesions in guinea pigs and rats exposed to prolonged inhalation of powdered metallic nickel. *Arch. Pathol.* **65,** 600–607.

Hueper, W. C., and Payne, W. W. (1962). Experimental studies in metal carcinogenesis: chromium, nickel, iron, and arsenic. *Arch. Environ. Health* **5,** 445–462.

International Agency for Research on Cancer (IARC) (1976). Cadmium, nickel, some epoxides, miscellaneous industrial chemicals and general considerations on volatile anaesthetics. In *IARC Monographs on the Evaluation of Carcinogenic Risk of Chemicals to Man*, Vol. 11. IARC, Lyon, pp. 75–112.

International Agency for Research on Cancer (IARC) (1982). Chemicals, industrial processes and industries associated with cancer in humans. In *IARC Monographs on the Evaluation of Carcinogenic Risk of Chemicals to Humans*, Suppl. 4. IARC, Lyon, pp. 167–170.

International Agency for Research on Cancer (IARC) (1987). Overall evaluation of carcinogenicity: An updating of IARC monographs, Volumes 1 to 42. In *IARC Monographs on the Evaluation of Carcinogenic Risk of Chemicals to Humans*, Suppl. 7. IARC, Lyon, pp. 264–269.

International Agency for Research on Cancer (IARC) (1990). Chromium, nickel and welding. In *IARC Monographs on the Evaluation of Carcinogenic Risks to Humans*, Vol. 49. IARC, Lyon, pp. 257–445.

Jasmin, G. (1963). Effect of methandrostenolone on muscle carcinogenesis in rats by nickel sulphide. *Br. J. Cancer* **17**, 681–687.

Jasmin, G. (1965). Influence of age, sex and glandular extirpation on muscle carcinogenesis in rats. *Experientia* **21**, 149–150.

Jasmin, G., and Riopelle, J. L. (1976). Renal carcinomas and erythrocytosis in rats following intrarenal injection of nickel subsulfide. *Lab. Invest.* **35**, 71–78.

Jasmin, G., and Solymoss, B. (1978). The topical effects of nickel sulsulfide in renal parenchyma. In G. N. Schrauzer (Ed.), *Inorganic and Nutritional Aspects of Cancer*. Plenum Press, New York, pp. 69–83.

Johansson, A., Camner, P., Jarstrand, C., and Wiernik, A. (1983). Rabbit alveolar macrophages after inhalation of soluble cadmium, cobalt, and copper: a comparison with the effects of soluble nickel. *Environ. Res.* **31**, 340–354.

Judde, J. G., Breillout, F., Clemenceau, C., Poupon, M. F., and Jasmin, C. (1987). Inhibition of rat natural killer cell function by carcinogenic nickel compounds: preventive action of manganese. *J. Natl. Cancer Inst.* **78**, 1185–1190.

Kasprzak, K. S. (1978). *Nickel Subsulfide, Ni_3S_2. Chemistry, Applications, Carcinogenicity* (Pol.). Technical University of Poznan Press, Poznan, pp. 1–66.

Kasprzak, K. S. (1987). Nickel. In L. Fishbein, A. Furst, and M. A. Mehlman (Eds.), *Advances in Modern Environmental Toxicology. XI. Genotoxic and Carcinogenic Metals—Environmental and Occupational Occurrence and Exposure*. Princeton Sci. Publ., Co., Princeton, pp. 145–183.

Kasprzak, K. S. (1991). The role of oxidative damage in metal carcinogenicity. *Chem. Res. Toxicol.*, **4**, 604–615.

Kasprzak, K. S., and Bare, R. M. (1989). Polymerization of histones by carcinogenic nickel compounds. *Carcinogenesis* **10**, 621–624.

Kasprzak, K. S., and Poirier, L. A. (1985). Effects of calcium(II) and magnesium(II) on nickel(II) uptake and stimulation of thymidine incorporation into DNA in the lungs of strain A mice. *Carcinogenesis* **6**, 1819–1821.

Kasprzak, K. S., and Ward, J. M. (1991). Prevention of nickel subsulfide carcinogenesis by local administration of *M. bovis* antigen in male F344/NCr rats. *Toxicology* **67**, 97–105.

Kasprzak, K. S., Marchow, L., and Breborowicz, J. (1971). Parasites and carcinogenesis. *Lancet* **ii**, 106–107.

Kasprzak, K. S., Marchow, L., and Breborowicz, J. (1973). Pathological reactions in rat lungs following intratracheal injection of nickel subsulphide and 3,4-benzopyrene. *Res. Commun. Chem. Pathol. Pharm.* **6**, 237–245.

Kasprzak, K. S., Gabryel, P., and Jarczewska, K. (1983). Carcinogenicity of nickel(II)hydroxides and nickel(II)sulfate in Wistar rats and its relation to the *in vitro* dissolution rates. *Carcinogenesis* **4**, 275–279.

Kasprzak, K. S., Quander, R. V., and Poirier, L. A. (1985). Effects of calcium and magnesium salts on nickel subsulfide carcinogenicity in Fischer rats. *Carcinogenesis* **6,** 1161–1166.

Kasprzak, K. S., Waalkes, M. P., and Poirier, L. A. (1986). Effects of magnesium acetate on the toxicity of nickelous acetate in rats. *Toxicology* **42,** 57–68.

Kasprzak, K. S., Waalkes, M. P., and Poirier, L. A. (1987a). Effects of essential divalent metals on carcinogenicity and metabolism of nickel and cadmium. *Biol. Trace Elem. Res.* **13,** 253–273.

Kasprzak, K. S., Ward, J. M., Poirier, L. A., Reichardt, D. A., Denn, A. C., III, and Reynolds, C. W. (1987b). Nickel-magnesium interactions in carcinogenesis: dose effects and involvement of natural killer cells. *Carcinogenesis* **7,** 1005–1011.

Kasprzak, K. S., Diwan, B. A., Konishi, N., Misra, M., and Rice, J. M. (1990). Initiation by nickel acetate and promotion by sodium barbital of renal cortical epithelial tumors in male F344 rats. *Carcinogenesis* **11,** 647–652.

Klein, C. B., Frenkel, K., and Costa, M. (1991). The role of oxidative processes in metal carcinogenesis. *Chem. Res. Toxicol.* **4,** 592–604.

Knight, J. A., Hopfer, S. M., Reid, M. C., Wong, S.H.Y., and Sunderman, F. W., Jr. (1986). Ethene (ethylene) and ethane exhalation in Ni(II)-treated rats, using an improved rebreathing apparatus. *Ann. Clin. Lab. Sci.* **16,** 386–394.

Knight, J. A., Rezuke, W. N., Wong, S.H.Y., Hopfer, S. M., Zaharia, O., and Sunderman, F. W., Jr. (1987). Acute thymic involution and increased lipoperoxides in thymus of nickel chloride-treated rats. *Res. Commun. Chem. Pathol. Pharmacol.* **55,** 291–302.

Knight, J. A., Rezuke, W. N., Gillies, C. G., Hopfer, S. M., and Sunderman, F. W., Jr. (1988). Pulmonary histopathology of rats following parenteral injections of nickel chloride. *Toxicol. Path.* **16,** 350–359.

Kurokawa, Y., Matsushima, M., Imazawa, T., Takamura, N., Takahashi, M., and Hayashi, Y. (1985). Promoting effect of metal compounds on rat renal tumorigenesis. *J. Am. Coll. Tox.* **4,** 321–330.

Laschi-Loquerie, A., Eyrand, A., Morisset, D., Sanon, A., Tachon, P., Veysseyre, C., Descotes, J. (1987). Influence of heavy metals on the resistance of mice toward infection. *Immunopharm. Immunotox.* **9,** 235–241.

Lau, T. J., Hackett, R. L., and Sunderman, F. W., Jr. (1972). The carcinogenicity of intravenous nickel carbonyl in rats. *Cancer Res.* **32,** 2253–2258.

Lee, J. E., Ciccarelli, R. B., and Wetterhahn-Jenette, K. (1982). Solubilization of the carcinogen nickel subsulfide and its interaction with deoxyribonucleic acid and protein. *Biochemistry* **21,** 771–778.

Leonard, A., and Jacquet, P. (1984). Embryotoxicity and genotoxicity of nickel. In F. W. Sunderman, Jr. (Ed.), *Nickel in the Human Environment.* International Agency for Research on Cancer, Lyon, No. 53, pp. 277–291.

Leonard, A., Gerber, G. B., and Jacquet, P. (1981). Carcinogenicity, mutagenicity and teratogenicity of nickel. *Mutat. Res.* **87,** 1–15.

Liber, C. M., and Sunderman, F. W., Jr. (1985). [63]Ni content in renal parenchyma and nuclei from [63]NiCl$_2$-treated rats. *Res. Commun. Chem. Pathol. Pharmacol.* **49,** 3–16.

Longstaff, E., Walker, A. I., and Jackh, R. (1984). Nickel oxide: potential carcinogenicity— a review and further evidence. In F. W. Sunderman, Jr. (Ed.), *Nickel in the Human Environment.* International Agency for Research on Cancer, Lyon, No. 53, pp. 235–243.

Lumb, G. D., Sunderman, F. W., Sr., Schneider, H. P., and Chou, R. H. (1987). Histogenesis of subcutaneous malignant tumors resulting from nickel subsulfide implantation. *Ann. Clin. Lab. Sci.* **17,** 286–299.

Maenza, R. M., Pradhan, A. M., and Sunderman, F. W., Jr. (1971). Rapid induction of sarcomas in rats by a combination of nickel sulfide and 3,4-benzpyrene. *Cancer Res.* **31,** 2067–2071.

Marnett, L. J. (1987). Peroxyl free radicals: potential mediators of tumor initiation and promotion. *Carcinogenesis* **8**, 1365–1373.

Mason, M. M. (1972). Nickel sulphide carcinogenesis. *Environ. Physiol. Biochem.* **2**, 137–141.

Mason, M. M., Cate, C. C., and Baker, J. (1971). Toxicology and carcinogenesis of various chemicals used in the preparation of vaccines. *Clin. Toxicol.*, **4**, 185–204.

Medinsky, M. A., Benson, J. M., and Hobbs, C. H. (1987). Lung clearance and disposition of ^{63}Ni in F344/N rats after intratracheal instillation of nickel sulfate solutions. *Environ. Res.* **43**, 168–178.

Memoli, V. A., Urban, R. M., Alroy, J., and Galante, J. O. (1986). Malignant neoplasms associated with orthopedic implant materials in rats. *J. Orthop. Res.* **4**, 346–355.

Miki, H., Kasprzak, K. S., Kenney, S., and Heine, U. (1987). Inhibition of intercellular communication by nickel(II): antagonistic effects of magnesium. *Carcinogenesis* **11**, 1757–1760.

Mitchell, D. F., Shankwalker, G. B., and Shazer, S. (1960). Determining the tumorigenicity of dental materials. *J. Dent. Res.* **39**, 1023–1028.

Nesnow, S., Triplett, L. L., and Slaga, T. J. (1985). Studies on the tumor initiating, tumor promoting, and tumor co-initiating properties of respiratory carcinogens. *Carcinog. Compr. Surv.* **8**, 257–277.

Nieboer, E., Maxwell, R. I., Rosetto, F. E., Stafford, A. R., and Stetsko, P. I. (1986). Concepts in nickel carcinogenesis. In A. V. Xavier (Ed.), *Frontiers in Bioinorganic Chemistry*. VCH Verlag, Weinheim, pp. 142–151.

Noble, R. L., and Capstick, V. (1963). Rhabdomyosarcomas induced by nickel sulfide in rats. *Proc. Am. Assoc. Cancer Res.* **4**, 48.

Norseth, T. (1988). Metal carcinogenesis. *Ann. N.Y. Acad. Sci.* **534**, 377–386.

Ottolenghi, A. D., Haseman, J. K., Payne, W. W., Falk, H. J., and MacFarland, H. N. (1975). Inhalation studies of nickel sulfide in pulmonary carcinogenesis in rats. *J. Natl. Cancer Inst.* **54**, 1165–1172.

Payne, W. W. (1964). Carcinogenicity of nickel compounds in experimental animals. *Proc. Am. Assoc. Cancer Res.* **5**, 50.

Poirier, L. A., Theiss, J. C., Arnold, L. J., and Shimkin, M. B. (1984). Inhibition by magnesium and calcium acetates of lead subacetate- and nickel acetate-induced lung tumors in strain A mice. *Cancer Res.* **44**, 1520–1522.

Pott, F., Ziem, U., Reiffer, F. J., Huth, F., Ernst, H., and Mohr, U. (1987). Carcinogenicity studies on fibres, metal compounds, and some other dusts in rats. *Exp. Pathol.* **32**, 129–152.

Rigaut, J. P. (1983). *Rapport Preparatoire sur les Criteres de Sante pour le Nickel*. Direction Generale de l'Emploi, des Affaires Sociales et de l'Education, Luxembourg, pp. 542–585.

Saknyn, A. V., and Blokhin, V. A. (1978). Development of malignant tumors in rats under the influence of nickel-containing aerosols. *Vopr. Onkol.* **24**(4), 44–48.

Saknyn, A. V., and Shabynina, N. K. (1973). Epidemiology of malignant neoplasms in nickel refineries. *Gig. Truda Prof. Zabol.* **14**(11), 10–13.

Sano, N., Shibata, M., Izumi, K., and Otsuka, H. (1988). Histopathological and immunohistochemical studies on nickel sulfide-induced tumors in F344 rats. *Jpn. J. Cancer Res. (Gann)* **79**, 212–221.

Schroeder, H. A., Balassa, J. J., and Vinton, W. H. (1964). Chromium, lead, cadmium, nickel and titanium in mice: effect on mortality, tumors and tissue levels. *J. Nutr.* **83**, 239–250.

Sen, P., and Costa, M. (1985). Induction of chromosomal damage in Chinese hamster ovary cells by soluble and particulate nickel compounds: preferential fragmentation of the heterochromatic long arm of the X-chromosome by carcinogenic crystalline NiS particles. *Cancer Res.* **45**, 2320–2325.

Sen, P., Conway, K., and Costa, M. (1987). Comparison of the localization of chromosome damage induced by calcium chromate and nickel compounds. *Cancer Res.* **47**, 2142–2147.

Skaug, V., Gylseth, B., Palmer-Reiss, A. L., and Norseth, T. (1985). Tumor induction in rats after intrapleural injection of nickel subsulfide and nickel oxide. In S. S. Brown and F. W. Sunderman, Jr. (Eds.), *Progress in Nickel Toxicology*. Blackwell Scientific Publications, Oxford, pp. 37–40.

Sky-Peck, H. H. (1986). Trace metals and neoplasia. *Clin. Physiol, Biochem.* **4**, 99–111.

Smialowicz, R. J. (1985). The effects of nickel and manganese on natural killer cell activity. In S. S. Brown and F. W. Sunderman, Jr. (Eds.), *Progress in Nickel Toxicology*. Blackwell Scientific Publications, Oxford, pp. 161–164.

Smialowicz, R. J., Rogers, R. R., Riddle, M. M., and Stott, G. A. (1984). Immunologic effects of nickel. I. Suppression of cellular and humoral immunity. *Environ. Res.* **33**, 413–427.

Smialowicz, R. J., Rogers, R. R., Riddle, M. M., Garner, R. J., Rowe, D. G., and Luebke, R. W. (1985). Immunologic effects of nickel. II. Suppression of natural killer cell activity. *Environ. Res.* **36**, 56–66.

Smialowicz, R. J., Rogers, R. R., Riddle, M. M., Rowe, D. G., Luebke, R. W., and Fogelson, L. D. (1987). The effects of manganese, calcium, magnesium, and zinc on nickel-induced suppression of murine natural killer cell activity. *J. Toxicol. Environ. Health* **20**, 67–80.

Sosinski, E. (1975). Morphological changes in rat brain and skeletal muscle in the region of nickel oxide implantation. *Neuropatol. Pol.* **13**, 479–483.

Standeven, A. M., and Wetterhahn, K. E. (1991). Is there a role for reactive oxygen species in the mechanism of chromium(VI) carcinogenesis? *Chem. Res. Toxicol.*, **4**, 616–625.

Stevenson, S., Hohn, R. B., Pohler, O.E.M., Fetter, A. W., Olmstead, M. L., and Wind, A. P. (1982). Fracture associated sarcoma in the dog. *J. Am. Vet. Med. Assoc.* **180**, 1189–1196.

Sunderman, F. W., Jr. (1977). Metal carcinogenesis. In R. A. Goyer and M. A. Mehlman (Eds.), *Advances in Modern Toxicology*, Vol. 1. Hemisphere Publishing Co., Washington, pp. 257–295.

Sunderman, F. W., Jr. (1979). Carcinogenicity and anticarcinogenicity of metal compounds. In P. Emmelot and E. Kriek (Eds.), *Environmental Carcinogenesis*. Elsevier Biomedical Press, Amsterdam, pp. 165–192.

Sunderman, F. W., Jr. (1981). Recent research on nickel carcinogenesis. *Environ. Health Perspect.* **40**, 131–141.

Sunderman, F. W. Jr. (1983). Organ and species specificity in nickel subsulfide carcinogenesis. In R. Langenbach, S. Nesnow, and J. M. Rice, (Eds.), *Organ and Species Specificity in Chemical Carcinogenesis*. Plenum Publishing Co., New York, pp. 107–126.

Sunderman, F. W., Jr. (1984a). Recent advances in metal carcinogenesis. *Ann. Clin. Lab. Sci.* **14**, 93–122.

Sunderman, F. W., Jr. (1984b). Carcinogenicity of nickel compounds in animals. In F. W. Sunderman, Jr. (Ed.), *Nickel in the Human Environment*. International Agency for Research on Cancer, Lyon, No. 53, pp. 127–142.

Sunderman, F. W., Jr. (1986a). Carcinogenicity and mutagenicity of some metals and their compounds. In I. K. O'Neill, P. Sculler, and L. Fishbein (Eds.), *Environmental Carcinogens—Selected Methods of Analysis*. International Agency for Research on Cancer, Lyon, No. 71, pp. 17–43.

Sunderman, F. W., Jr. (1986b). Recent progress in nickel carcinogenesis. *Ann. Inst. Super. Sanita.* **22**, 669–680.

Sunderman, F. W., Jr., and Barber, A. M. (1988). Finger-loops, oncogenes, and metals. *Ann. Clin. Lab. Sci.* **18**, 267–286.

Sunderman, F. W., and Donelly, A. J. (1965). Studies of nickel carcinogenesis. Metastasizing pulmonary tumors in rats induced by the inhalation of nickel carbonyl. *Am. J. Clin. Pathol.* **46**, 1027–1041.

Sunderman, F. W., Jr., and McCully, K. S. (1983a). Carcinogenesis tests of nickel arsenides, nickel antimonide and nickel telluride in rats. *Cancer Invest.* **1,** 469–474.

Sunderman, F. W., Jr., and McCully, K. S. (1983b). Effects of manganese compounds on carcinogenicity of nickel subsulfide in rats. *Carcinogenesis* **4,** 461–465.

Sunderman, F. W., Jr., and Maenza, R. M. (1976). Comparisons of carcinogenicities of nickel compounds in rats. *Res. Commun. Chem. Pathol. Pharm.* **14,** 319–330.

Sunderman, F. W., Kincaid, J. R., Donelly, A. J., and West, B. (1957). Nickel poisoning. IV. Chronic exposure of rats to nickel carbonyl: a report after one year of observation. *Arch. Industr. Health* **16,** 480–485.

Sunderman, F. W., Donelly, A. J., West, B., and Kincaid, J. F. (1959). Nickel poisoning. IX. Carcinogenesis in rats exposed to nickel carbonyl. *Arch. Industr. Health* **20,** 36–41.

Sunderman, F. W., Jr., Lau, T. J., and Cralley, L. J. (1974). Inhibitory effect of manganese upon muscle tumorigenesis by nickel subsulfide. *Cancer Res.* **34,** 92–95.

Sunderman, F. W., Jr., Coulston, F., Eichhorn, G. L., Fellows, J. A., Mastromatteo, E., Reno, H. T., Samitz, M. H., Curtis, B. A., Vallee, B. L., West, P. W., McEwan, J. C., Shibko, S. I., and Boaz, T. D., Jr. (1975). *Medical and Biologic Effects of Environmental Pollutants. Nickel.* Natl. Acad. Sci. USA, Washington, D.C., pp. 1–277.

Sunderman, F. W., Jr., Kasprzak, K. S., Lau, T. J., Minghetti, P. P., Maenza, R. M., Becker, N., Onkelinx, C., and Goldblatt, P. J. (1976). Effects of manganese on carcinogenicity and metabolism of nickel subsulfide. *Cancer Res.* **36,** 1790–1800.

Sunderman, F. W., Jr., Mangold, B.L.K., Wong, S.H.Y., Shen, S. K., Reid, M. C., and Jansson, I. (1983). High-performance size-exclusion chromatography of ^{63}Ni-constituents in renal cytosol and microsomes from ^{63}NiCl$_2$-treated rats. *Res. Commun. Chem. Pathol. Pharm.* **39,** 477–492.

Sunderman, F. W., Jr., McCully, K. S., and Hopfer, S. M. (1984). Association between erythrocytosis and renal cancers in rats following intrarenal injection of nickel compounds. *Carcinogenesis* **5,** 1511–1517.

Sunderman, F. W., Jr., Marzouk, A. B., Hopfer, S. M., Zaharia, O., and Reid, M. C. (1985). Increased lipid peroxidation in tissues of nickel chloride-treated rats. *Ann. Clin. Lab. Sci.* **15,** 229–236.

Sunderman, F. W., Jr., Hopfer, S. M., and Knight, J. A. (1988). Biological reactivity of nickel oxides and related compounds. *Abstr. 4th Intl. Conf. Nickel Metabol. Toxicol.* Espoo. p. 15.

Takenaka, S., Hochreiner, D., and Oldiges, H. (1985). Alveolar proteinosis induced in rats by long-term inhalation of nickel oxide. In S. S. Brown and F. W. Sunderman, Jr. (Eds.), *Progress in Nickel Toxicology.* Blackwell Scientific Publications, Oxford, pp. 89–92.

Templeton, D. M., and Sarkar, B. (1985). Peptide and carbohydrate complexes of nickel in human kidney. *Biochem. J.* **230,** 35–42.

Valentine, R., and Fisher, G. L. (1984). Pulmonary clearance of intratracheally administered ^{63}Ni$_3$S$_2$ in strain A/J mice. *Environ. Res.* **34,** 328–334.

Vityugov, I. A., Kotenko, V. V., Gunter, V. E., Itin, V. I. and Kopisova, V. A. (1986). The organism reaction to implantation of polished and unpolished fixators from Ni-Ti. *Ortop. Travmatol. Protez.* (Rus.), August 8, 18–22.

Waalkes, M. P., Rehm, S., Kasprzak, K. S. and Issaq, H. (1987). Inflammatory, proliferative, and neoplastic lesions at the site of metallic identification ear tags in Wistar [Crl:(WI)BR] rats. *Cancer Res.* **47,** 2445–2450.

Wehner, A. P., Busch, R. H., Olson, R. J., and Craig, D. K. (1975). Chronic inhalation of nickel oxide and cigarette smoke by hamsters. *Am. Indust. Hyg. Assoc. J.* **36,** 801–810.

Wehner, A. P., Dagle, G. E., and Busch, R. H. (1984). Pathogenicity of inhaled nickel compounds in hamsters. In F. W. Sunderman, Jr. (Ed.), *Nickel in the Human Environment.* International Agency for Research on Cancer, Lyon, No. 53, pp. 143–151.

Wiernik, A., Johansson, A., Jarstrand, C., and Camner, P. (1983). Rabbit lung after inhalation of soluble nickel. I. Effects on alveolar macrophages. *Environ. Res.* **30,** 129–141.

Wong, S.H.Y., Knight, J. A., Hopfer, S. M., Zaharia, O., Leach, C. N., Jr., and Sunderman, F. W., Jr. (1987). Lipoperoxides in plasma as measured by liquid chromatographic separation of malondialdehyde-thiobarbituric acid adduct. *Clin. Chem.* **33,** 214–220.

Yamashiro, S., Gilman, J.P.W., Holland, T. J., and Abandowitz, H. M. (1980). Nickel sulphide-induced rhabdomyosarcomata in rats. *Acta Pathol. Jp.* **30,** 9–22.

Yarita, T., and Nettesheim, P. (1978). Carcinogenicity of nickel subsulfide for respiratory tract mucosa. *Cancer Res.* **38,** 3140–3145.

32

LUNG DOSIMETRY: EXTRAPOLATION MODELING FROM ANIMALS TO MAN

G. Oberdörster

Environmental Health Sciences Center and Department of Biophysics, University of Rochester Medical Center, Rochester, New York 14642

1. INTRODUCTION

Inhalation of different nickel compounds in man and experimental animals has been found to cause both direct inflammatory reactions in the respiratory tract and secondary effects resulting from the influence of nickel on the respiratory immune system (for summary see EPA, 1986). However, of the

Nickel and Human Health: Current Perspectives, Edited by Evert Nieboer and Jerome O. Nriagu.
ISBN 0-471-50076-3 © 1992 John Wiley & Sons, Inc.

greatest concern is the evidence that some inhaled nickel compounds caused lung cancer in humans and animals. Cancer sites included the nasal cavities and the lung (human) or the lung only (animals). The question as to whether the nickel ion should be considered as the carcinogenic agent—as is suggested for other pulmonary effects of nickel (Wiernik et al., 1983)—is debated and remains an important area for future research into the mechanism of nickel carcinogenicity. Thus, the EPA (1986) classified only nickel refinery dust and nickel subsulfide as known human carcinogens and nickel carbonyl as a probable human carcinogen, whereas the IARC (1987) classified nickel and nickel compounds as having shown sufficient evidence of carcinogenicity in humans, although this may not apply to all nickel compounds, and sufficient evidence in animals.

Although the Carcinogen Assessment Group (CAG) of the EPA (1986) has calculated a unit risk estimate for nickel refinery dust based on epidemiological data, no such estimates could be made for other nickel compounds. With the exception of an inhalation study in rats with Ni_3S_2 by Ottolenghi et al. (1974), inhalation studies in animals with other nickel compounds have been equivocal with regard to tumor induction in the respiratory tract. Injection studies with different nickel compounds caused mostly injection site tumors in experimental animals whose relevance to human carcinogenic risk from inhalation of these compounds is often questioned. Recent studies with repeated intraperitoneal injections of soluble and insoluble nickel compounds in rats (Chapter 37) showed that in addition to Ni_3S_2, nickel powder was carcinogenic, and at higher doses so were two nickel alloys (containing 50 and 70% nickel) and NiO. Although these studies demonstrate the carcinogenic potential of different nickel compounds, they cannot readily be used to extrapolate to the carcinogenic risk of inhaled nickel compounds in humans.

In the absence of results from epidemiological studies, such extrapolation can be attempted from results of animal inhalation studies, provided differences in deposition and retention of the nickel compound between the animal and human respiratory tract are considered. Aims of such extrapolation modeling include the calculation of a dose to the lung and an equivalent human exposure EHE (EHE) based on the animal study. The term is defined here as the chronic exposure concentration of a compound inhaled by humans that will result in the same dose at a target site of the respiratory tract in man as a known chronic exposure concentration did at these sites in experimental animals. The underlying principles of lung dosimetry to derive an EHE are discussed in the following sections.

2.   LUNG DOSIMETRY: DEPOSITION AND RETENTION

Inhalation exposure to the same concentration of a nickel compound will lead to different uptakes of this compound into target cells of the respiratory tract of animals and man due to differences in deposition and retention within the respiratory tract of the two species. Thus, it is not possible to simply take the exposure concentration of an animal inhalation study as EHE (Figure 32.1). Likewise, although the inhaled dose in an animal resulting from the exposure could be transferred into a human inhaled dose, it should not be used to calculate the EHE, and even the deposited dose of a compound in different regions of the respiratory tract in an animal is not necessarily suitable for calculating an EHE. In contrast, the long-term retained or accumulated dose in target cells of the respiratory tract is most likely the crucial dose, which, if it is the same in animal and man, can be assumed to lead to the same effect. This assumption can be challenged on the grounds that it does not consider a time factor for tumor development: Once the accumulated dose has reached a steady-state level during chronic constant exposure, it will not change with additional constant exposure. However, the time necessary to reach this steady-state level depends on the retention halftime ($T_{1/2}$) of the inhaled compound in the respiratory tract; that is, the $T_{1/2}$ is different in animal and man, the exposure time to reach steady state will be different, too, thus introducing a time factor. From the accumulated dose, the human-specific deposited dose and inhaled dose can be calculated—provided retention and deposition data are available—and finally the EHE can be derived (Fig. 32.1), which in turn can be used for a risk estimation.

The inhaled dose depends solely on the minute ventilation and the exposure concentration, whereas the deposited dose takes additionally into account airflow characteristics, particle characteristics, and airway geometry. Well-established lung models for predicting deposition of different particles in some experimental animal species and humans are available, for example, the typical path whole lung models for rat and man by Yeh and Schum (1980) and Schum and Yeh (1980). Yet only few data are available for describing retention ($T_{1/2}$) of nickel compounds in the respiratory tract of experimental animals, and not much is known about the retention of inhaled nickel compounds in the human respiratory tract. However, such data are necessary to calculate an accumulated dose in the respiratory tract. The accumulated dose could be expressed as micrograms of nickel per gram of lung or per square centimeter of surface area of the respiratory tract. The latter considers the fact that the tumors originate from the epithelial cells and assumes that accumulation occurs in these cells.

The extrapolation model based on respiratory tract dosimetry, which is outlined in Figure 32.1, presupposes that results of an animal inhalation study on the carcinogenicity of an inhaled nickel compound are available; ideally, this should be in the form of an exposure response curve with several inhaled concentrations in order to derive a dose–response curve and to perform an estimate of the tumor risk associated with the respective EHEs. Lacking such ideal data, the aforementioned study by Ottolenghi et al. (1974) with only one concentration of inhaled Ni_3S_2 will be used as a practical example for extrapolation modeling of the EHE.

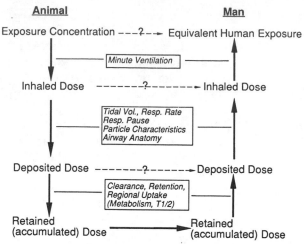

Figure 32.1. Extrapolation to EHE based on lung dosimetry.

3.   EHE FOR Ni₃S₂ DERIVED FROM RAT INHALATION STUDY

In the study by Ottolenghi et al. (1974), male and female F344 rats had been exposed for 78 weeks to 970 µg Ni_3S_2/m^3, 6 hr/day, 5 days/week. From the information on particle sizes given in the paper (70% of the mass smaller than 1 µm, 25% between 1 and 1.5 µm geometric diameter d_g), the following particle parameters can be deduced, assuming that the particle size distribution was lognormal and that the density ρ of Ni_3S_2 is 5.82 g/cm³ (i.e., a particle with geometric diameter d_g of 1 µm would have an aerodynamic diameter d_{ae} of 2.41 [$d_{ae} = d_g \times \rho$]). Then the mass median aerodynamic diameter (MMAD) becomes 2.0 µm and the geometric standard deviation (GSD) is 1.47 (Hatch–Choate equation; Mercer, 1973).

Under the exposure conditions of the study, both benign and malignant tumors of the respiratory tract developed to a significant degree in the animals. There was no clear distinction between tumors of bronchial origin and tumors of alveolar origin; tumors at both sites were induced. In contrast, workers exposed to nickel-containing particles (Ni_3S_2) developed tumors in the nasal cavity and in the lung (EPA, 1986). The latter possibly consisted also of bronchogenic and alveologenic tumors, although histological typing was not always performed. Although inhaled Ni_3S_2 can cause tumors in the nasal region as well as in the lower respiratory tract of humans, the following model calculations are directed to the lower respiratory tract only. However, if respective kinetic data for nickel in the nasal regions become available, the calculations could as well be applied to this region.

For calculating deposition in the respiratory tract the lung model described by Schum and Yeh (1980) and Yeh and Schum (1980) and modified by Wojciak (1988) was used. In this model, deposition of inhaled particles is modeled occurring by diffusion, sedimentation, and impaction during inhalation and by diffusion and sedimentation during exhalation, whereas nasopharyngeal deposition is based on empirically derived data. As pointed out by the authors of the model, the predicted depositions from the model agreed quite well with experimental studies in rats (Raabe et al., 1977) and humans (Lippmann, 1977).

Respiratory rate and tidal volumes, which are needed for deposition calculations, were calculated using the predictive formulas by Stahl for the rat of 300 g body weight (1967); for humans, values for reference man were used (ICRP, 1975). Thus, a tidal volume of 2 mL and respiratory rate of 73 min^{-1} for a 300-g rat with a lung weight of 1.6 g were used, and for man a tidal volume of 500 mL, respiratory rate of 15 min^{-1}, and a lung weight of 950 g were assumed.

Table 32.1 summarizes the data that were used for these model calculations for rat and man. The epithelial surface area of the respiratory tract was calculated from the data given by Yeh and Schum (1980) in their typical path whole lung model with correction for surface area overlap at the bifurcation as described by Wojciak (1988).

For calculation of the accumulated dose, the pulmonary retention of Ni₃S₂ has to be known. However, no specific data are available for rat or man. After a single exposure in rats, inhaled NiO (particle size 0.4 μm) had been found to be retained in the lungs with a retention half-time of 36 days (Hochrainer et al., 1980). The Ni₃S₂ is solubilized faster in serum than is NiO₂ (Kuehn and Sunderman, 1982) and may therefore have a shorter retention half-time. However, this could be offset by the bigger particle size of the Ni₃S₂ particles (2.0 μm) in the study by Ottolenghi et al. (1974) compared to the study by Hochrainer et al. [1980 (0.4 μm)]. Therefore, it is assumed that pulmonary retention of Ni₃S₂ in the rat also has a retention half-time of 36 days.

Under the assumption that solubility rates of Ni₃S₂ are the same among species (e.g., metal solubility in human and rabbit alveolar macrophages; Lundborg et al., 1985), a pulmonary $T_{1/2}$ of Ni₃S₂ in man may be estimated

Table 32.1 Data Used to Calculate EHE for Ni₃S₂[a]

Parameter	Rat	Man
Respiratory Rate, min^{-1}	73	15
Tidal volume, mL	2	500
Lung weight, g	1.6	950
Surface area of tracheobronchial region, cm^2	31	3,420
Surface area of alveolar region, cm^2	3,755	479,000

[a] From rat inhalation study by Ottolenghi et al., (1974); MMAD = 2.0 μm, GSD = 1.5.

from differences between mechanical and solubility pulmonary clearance rates.

The retention half-time of insoluble particles in the rat lung is about 50 days, as found for highly insoluble Fe_2O_3, Fe_3O_4, or TiO_2 particles (Bellman et al., 1983; Oberdörster et al., 1984; Ferin, 1982). The effective pulmonary retention half-time ($T_{1/2,eff}$) of Ni_3S_2 in rats of 36 days is due to both mechanical clearance as well as clearance by solubilization:

$$\frac{1}{T_{1/2,eff}} = \frac{1}{T_{1/2,m}} + \frac{1}{T_{1/2,s}} \tag{1}$$

where $T_{1/2,m}$ is the retention half-time due to mechanical clearance (50 days) and $T_{1/2,s}$ the retention half-time due to solubility clearance [129 days (Eq. 1)].

The pulmonary retention half-time in man of insoluble particles can be approximated to be about 500 days (Bailey et al., 1985). Assuming that $T_{1/2,s}$ is the same in rat and man, an effective pulmonary retention half-time for Ni_3S_2 in humans of 103 days can be calculated [Eq. (1)].

Assuming first-order clearance kinetics, the amount A_t of Ni accumulated at time t during continuous exposure is determined by

$$A_t = \frac{a}{b}(1 - e^{-bt}) \tag{2}$$

(Task Group on Metal Accumulation, 1973), where a is the amount being deposited each day and b is the elimination rate ($b = \ln 2/T_{1/2}$).

Since the rats were only exposed for 5 days each week, a was adjusted to

$$a' = a \cdot \frac{5}{7} \tag{3}$$

to account for the two days of no exposure. This was also applied to humans in these model calculations assuming occupational exposure conditions. An equilibrium or steady-state value A_e is reached after approximately five half-times have elapsed:

$$A_E' = a'/b \tag{4}$$

Table 32.2 summarizes the data on deposited and accumulated doses, normalized for tracheobronchial and pulmonary surface area and for lung weight. Included in the table is also the inhaled dose expressed per kilogram body weight (for man, 70 kg). This comparison shows that in most cases lower normalized doses of Ni_3S_2 are reached in man than in the rat under the conditions of the Ottolenghi study.

Table 32.2   Inhaled, Deposited, and Retained Dose of Ni₃S₂ in Rat and Man After Exposure to 970 μg/m³ at Particle Size of 2.0 μm (MMAD) with GDS of 1.5

	Rat	Man
Inhaled Dose (during 6 hr), μg/kg body weight	168	37
Deposited dose (during 6 hr):		
On tracheobronchial surface, ng/cm²	62	33
On alveolar surface, ng/cm²	0.9	0.8
Per pulmonary mass, ng/g¹	2040	420
Accumulated pulmonary dose		
(when equilibrium is reached, 6 h/day, 5 days/		
week exposure)		
Per pulmonary mass, μg/g¹	105	59
Per alveolar surface, ng/cm²	45	113

Table 32.3 shows equivalent human exposure concentrations leading to the same doses of Ni₃S₂ in man as in rats when the latter are exposed to 970 μg Ni₃S₂/m³ for 6 hr/day, 5 days/week at a particle size of 2.0 μm (MMAD) and a GSD of 1.5. Depending on the basis for comparison—inhaled dose, deposited dose, or accumulated dose—the EHEs range from 390 to 4710 μg/m³ for normal resting conditions. The ratios of the exposure concentrations between rats (970 μg/m³) and man range from 0.4 to 4.86.

As discussed in Section 1, an EHE is probably best derived from the accumulated dose since this reflects the long-term dose to the lung. Since tumors will develop from certain epithelial cells, the surface area of the

Table 32.3   Equivalent Human Exposure Concentrations Leading to Same Respective Doses of Inhaled Ni₃S₂ in Man as in a Rat Exposed to 970 μg/m³[a]

Basis for EHE	EHE (μg/m³)	Man (EHE)/Rat (970)
Inhaled dose/kg body weight	4400	4.5
Deposited dose/cm² total body surface	1800	1.85
Deposited dose/cm² alveolar surface	1120	1.15
Deposited dose/g lung weight	4710	4.86
Accumulated dose/g lung weight	1730	1.78
Accumulated dose/cm² alveolar surface	390	0.4

[a]Particle size: MMAD = 2.0; GSD = 1.5; exposure: 6 hr/day, 5 days/week.

epithelium might be most appropriate as a basis, which gives an EHE of 390 $\mu g/m^3$ (Table 32.3). This means that the rat exposure concentration of 970 μg Ni_3S_2/m^3 is equivalent to a human exposure of 390 μg Ni_3S_2/m^3 in terms of the equilibrium dose of Ni_3S_2 accumulating in epithelial cells of the pulmonary region during chronic constant exposure. Ideally, it would be desirable to base the EHE also on the accumulated dose in the epithelial cells of the tracheobronchial tree. However, no data on retention of Ni_3S_2 or other nickel compounds in this region are available, and although it is conceivable that a healthy epithelium of the conducting airways will clear Ni_3S_2 particles by mucociliary action very rapidly, a long-term retention in this region has to be considered, as was described for other insoluble particles (Stahlhofen et al., 1986). In addition, chronic exposure under occupational conditions may lead to ciliotoxic effects, thus affecting particulate clearance in the tracheobronchial region. Therefore, lacking appropriate data, we can at present only estimate the EHE from the accumulated dose per unit pulmonary surface area. The respective EHE of 390 $\mu g/m^3$ applies to exposure of 6 hr/day, 5 days/week (Table 32.3). This EHE is based on a pulmonary retention half-time of Ni_3S_2 that was derived indirectly since no direct respective experimental data for Ni_3S_2 are available. If such data become available in the future and if they are different from the presently assumed value, they will change the EHE but not the concept of deriving the EHE. The possibility of such change should be kept in mind when making a comparison between the lung tumor risk associated with the presently calculated EHE and the risk derived from epidemiological data, as is done in the following.

In the rat exposure study, 15% of the male rats had developed lung tumors and 12% of the female rats (Ottolenghi et al., 1974). Since the rats had only been exposed for about half of their life span (78 weeks), an adjustment should be made for lifetime exposure. The adjustment could take into consideration the product of accumulated dose and time during which this dose is present. Alternatively, the dose rate could be considered. However, since we do not know whether a linear relationship exists between the product of dose and time and the tumor incidence rate, we will only consider the EHE of 390 $\mu g/m^3$ (Table 32.3) for this tentative comparison of risks derived from the animal and the epidemiological study. Based on epidemiological studies (EPA, 1986), the midpoint of the estimates of unit risk for lung tumors from exposure to Ni_3S_2 has been calculated to be 4.8×10^{-4}. The unit risk refers to a continuous exposure to 1 $\mu g/m^3$. The EHE of 390 $\mu g/m^3$, based on exposure of 6 hr/day, 5 days/week (Table 32.3), would be equivalent to a continuous exposure of 70 $\mu g/m^3$ (390 × $\frac{6}{24}$ × $\frac{5}{7}$). This translates into a risk of 3.4×10^{-2}, based on the EPA (1985) risk estimate cited above, whereas the rat study predicts $12 \times 10^{-2} - 15 \times 10^{-2}$. This difference can be regarded as small considering the range of estimates in the epidemiological studies and may be resolved with the availability of new retention data, as discussed above.

3.1. Influence of *in Vivo* Solubility on EHE

Nickel compounds of greater solubility in the lung than Ni_3S_2 will obviously have a different EHE. This section examines the influence of solubility of the nickel compound in the lung on the accumulated dose by using retention data of inhaled $NiCl_2$ particles in the lung.

The pulmonary clearance of $NiCl_2$ has been studied by English et al. (1981) and by Menzel et al. (1987). In the former study, $NiCl_2$ was instilled intratracheally into the lungs of rats so that the initial lung burden was about 1.3 µg Ni. Over the first 3 days about 95% of the instilled dose was cleared with a retention half-time of 0.7 days (calculated from data of English et al., 1981), and less than 5% was retained with a retention half-time of about 4 days. Menzel and co-workers (1987) suggested from the results of their study a carrier-mediated transport for $NiCl_2$ clearance from the lung, which can be saturated. Consequently, they analyzed their data assuming Michaelis–Menten kinetics and modeled nickel accumulation in the lung during chronic exposure based on this assumption. However, their experimental data for nickel accumulation in the rat lung after 9 days of exposure to 400 µg $NiCl_2$/m^3 (2 hr/day) showed that an equilibrium concentration of nickel in the lung was reached after about 7 days (about 2.1 µg Ni per lung). This would also be consistent with first-order clearance kinetics with a retention half-time of 1.26 days. Possibly, higher lung burdens of $NiCl_2$ could affect and slow down its clearance from the lung, which would be consistent with observed Michaelis–Menten-type clearance kinetics and may thus explain the retention data of Menzel et al. (1987) at high lung burdens of nickel. Additional studies would be needed to clarify this point. Taking together the data of English et al. (1981) and of Menzel et al. (1987), we will assume for the present discussion that reasonably low lung burdens of $NiCl_2$ are cleared from the lung of rats with a retention half-time of 1 day.

Assuming that the accumulated dose (at equilibrium) of nickel in the lung should reach the same value as that predicted from Ni_3S_2 exposure of the Ottolenghi et al. 1974 study [45 ng Ni_3S_2/cm^2 (Table 32.2 rat, alveolar surface), which is equivalent to 33 ng Ni/cm^2], the respective deposited and retained doses and finally the exposure concentration for the rat can be calculated. Figure 32.2 shows the result of this calculation for a 300-g rat and a particle size of inhaled $NiCl_2$ aerosol of 2.0 µm (MMAD) with a GSD of 1.5 for exposure of 6 hr daily, 5 days/week, which were the conditions of the Ottolenghi study. Figure 32.2 shows that in order to reach the same accumulated dose (at steady state) of nickel on the epithelial surface of the alveolar region after chronic inhalation of $NiCl_2$ as that after chronic exposure to 970 µg Ni_3S_2/m^3, the exposure concentration would have to be 51.1 mg $NiCl_2$/m^3. This is a concentration that would lead to acute toxic effects in the lung, including the surfactant system, and would not be tolerated by the animals over an extended period of time (Wiernik et al., 1983; Johansson et al.,

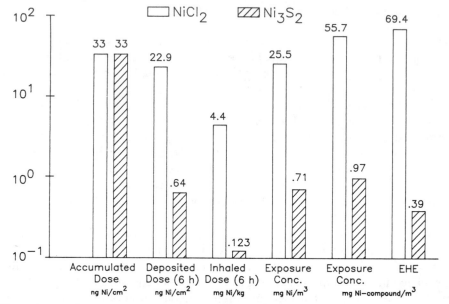

Figure 32.2. Influence of low (Ni$_3$S$_2$) and high (NiCl$_2$) in vivo solubility of nickel compounds on deposited dose, inhaled dose, and exposure concentration during chronic exposure that will lead to the same accumulated dose of 33 ng Ni/cm^3 surface area in the lungs of rats. The last two columns show the EHE necessary to reach the same accumulated lung dose (33 ng Ni/cm^2) in the human lungs.

1983). Thus, it might be hypothesized that it is difficult or even impossible to induce lung tumors with inhalation of highly soluble nickel compounds because acute toxic effects occur long before an accumulated carcinogenic dose is reached. Obviously, more research is needed, in particular on the importance of the dose rate and on the carcinogenic mechanism of nickel, to support this hypothesis.

It follows from the simple relationships between accumulated dose and daily deposited dose [Eq. (4)] and between deposited dose, inhaled dose, and exposure concentration: An x-fold shorter pulmonary retention half-time of a nickel compound requires an x-fold higher exposure concentration in the inhaled air in order to reach the same accumulated dose in the lung. In the above example this factor is 36 (Figure 32.2; estimated retention half-time is 36 days for Ni$_3$S$_2$ and 1 day for NiCl$_2$). The same relationship applies to the EHE: The assumed pulmonary retention half-time for Ni$_3$S$_2$ in humans is 103 days, and 1 day for NiCl$_2$; that is, an EHE of 308 μg Ni/m^3 for Ni$_3$S$_2$ would change into 31.7 mg Ni/m^3 for NiCl$_2$ or 69.4 mg NiCl$_2$/m^3 (Fig. 32.2). If the actual retention half-time for NiCl$_2$ in humans is longer than the assumed 1 day, then the EHE would decrease correspondingly. Thus, although it is obvious that in vivo solubility of nickel compounds has a major impact on the EHE the magnitude of this impact needs to be studied further.

3.2. Influence of Particle Size on EHE

A further factor of possible influence for the EHE is the particle size of the inhaled nickel compound. This factor has two aspects. First, solubilization rates of particles of low in vivo solubility depend on their surface area (Mercer, 1967), thus affecting their retention in the lung; the second factor concerns the effect on deposition efficiency within the respiratory tract. The latter will be discussed in the following paragraphs.

An x-fold higher (lower) daily deposition will increase (decrease) the accumulated dose by the same factor x [Eq. (4)] provided the retention half-time does not change. The particle size of Ni_3S_2 in the rat study of Ottolenghi et al. (1974) was estimated to be 2.0 μm (MMAD) with a GSD of 1.5. In the following example, the particle size is changed to 0.5 μm (MMAD) with a GSD of 1.5 for both rat and human exposure to Ni_3S_2. The predictive lung deposition models of Schum and Yeh (1980) and Yeh and Schum (1980) as modified by Wojciak (1988) are used for the deposition calculations.

Table 32.4 shows the result, comparing the deposition for the 2.0- and 0.5-μm particles in the nasopharyngeal, tracheobronchial, and pulmonary regions of rat and man. As expected, nasopharyngeal deposition is very high in both rat and man for the bigger particle size. This underlines the possible importance of the particle size of the inhaled nickel-containing particles for the development of nasal cancer, although it remains to be studied how these particles get into the nasal sinuses, the sites of these tumors.

Pulmonary deposition of the smaller particles is lower than for the bigger particles, by almost a factor of 2 in the human lung (Table 32.4). Based on these data, respective exposure concentrations for Ni_3S_2 in rat and man were calculated that give the same accumulated dose. As can be seen from Figure 32.3, the resulting changes in exposure concentration for the rat and for man (EHE) are not nearly as big as they are when the in vivo solubility of the

Table 32.4   Relative Regional Deposition of 2.0- and 0.5-μm Particles (GSD 1.5) in Rat and Man[a]

	Rat			Man		
	2.0 μm	0.5 μm	ratio[b]	2.0 μm	0.5 μm	ratio[b]
Nasopharyngeal	23.3	3.8	6.1	31.9	<1	>31.9
Tracheobronchial	4.2	1.8	2.3	5.2	2.3	2.3
Pulmonary	6.6	4.7	1.4	14.5	7.8	1.9

[a]Numbers are percentage of inhaled dose deposited.
[b]Ratio of deposition 2.0 μm/0.5 μm.

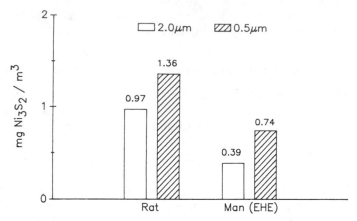

Figure 32.3. Change of exposure concentration of Ni_3S_2 to reach the same accumulated dose of 33 ng Ni/cm^2 surface area in the lungs of rat and man when inhaled particles size changes from 2.0 to 0.5 μm (MMAD).

nickel compound changes (Fig. 32.2). Moreover, it can be argued that the smaller particle size of the Ni_3S_2 particles (0.5 vs. 2.0) may increase their solubility rate, thus additionally decreasing their pulmonary retention. This may more than offset a decrease in EHE based on the smaller particle size.

However, while the particle size of the inhaled Ni_3S_2 does not seem to have the same impact as their in vivo solubility, it should still be considered since it is the determining factor for initial deposition, e.g., nasal or deep lung. Also, within a given region of the respiratory tract the deposition efficiencies can vary quite significantly depending on the inhaled particle size. The significance of such regional deposition differences underlines the importance of the knowledge of the deposited dose. In contrast, the inhaled dose does not take into account differences in particle size, breathing pattern, or solubility; this dose is the same for all nickel compounds as long as the exposure concentration is the same and is therefore not very relevant or useful for establishing dose-response relationships.

4. CONCLUDING DISCUSSION AND FUTURE RESEARCH NEEDS

For extrapolating results of chronic inhalation studies with nickel compounds in rats to humans, both deposition and retention of the inhaled nickel compound in the respiratory tract have to be considered. It is not appropriate to use the inhaled dose. Predictive models for calculating the deposition of inhaled nickel compounds are sufficiently well developed. However, modeling of retention requires more studies, in particular of a comparative nature between rodents and primates, since respiratory tract retention is dependent both on the chemical species and on the animal species. Table 32.5 lists research areas in which data are needed to improve extrapolation modeling and risk estimation of nickel carcinogenicity.

Foremost, nickel retention kinetics in different regions of the respiratory tract should be investigated. While such retention studies need to be performed in both rodents and nonhuman primates, it may be sufficient to study the clearance mechanisms in rodents only. Such studies are necessary to understand better the bioavailability of nickel compounds. For example, a long retention half-time may be due to either low solubility or binding of nickel to cellular components. The site of retention and the biological and toxicological impact could be quite different in both situations. It should be mentioned in this context that the contribution of ingested nickel to the lung burden of nickel should be studied, as recently pointed out by Sunderman et al. (1988).

In a chronic exposure situation, such as occupational exposures, the pulmonary clearance of nickel compounds could be very different from a single acute exposure. For example, Tanaka et al. (1985, 1988) estimated a $T_{1/2}$ for NiO of 7.7, 11.5, and 21 months in rats under chronic exposure conditions, obviously correlated to particle size. Kodama et al. (1985) found an increasing pulmonary $T_{1/2}$ with increasing NiO exposure concentration. Such long $T_{1/2}$ are quite different from the 36 days found after a single low-concentration exposure to NiO (Hochrainer et al., 1980), which was used in the present model calculations. Oberdörster and Hochrainer (1980) reported that long-term exposure of rats to 50 μg NiO/m^3 resulted in a retardation of alveolar macrophage mediated test particle clearance from 58 days (control) to 520 days. The result of their study offers an explanation for the long pulmonary $T_{1/2}$ observed by Tanaka et al. (1985, 1988) in that alveolar macrophages are affected in their clearance function by chronic NiO exposure: They can no longer effectively clear both inert test particles and the nonsolubilized fraction of the NiO particles from the alveolar space. This effect was not considered in the present model calculations since we do not know whether it also occurs with Ni_3S_2. However, it could easily be built into the model. If such prolongation of pulmonary nickel retention does occur during chronic Ni_3S_2 exposure in rats and if a proportionally similar retardation of alveolar clearance occurs in humans, the calculation of the EHE would give the same result as derived in this chapter. It can be concluded from this brief discussion

Table 32.5   Data Needed for Extrapolation Modeling and Risk Estimation of Nickel Carcinogenicity

Kinetics of inhaled nickel compounds in rats and primates.
Mechanisms of clearance of different nickel compounds from the respiratory tract.
Respiratory tract carcinogenicity of nickel compounds in rats
Mechanisms of nickel carcinogenicity.

that it is important to know more about (i) nickel compound specific retention kinetics for different inhaled particle sizes, (ii) different exposure concentrations and durations, and (iii) the mechanisms of pulmonary clearance for these compounds.

Studies on the mechanism of nickel carcinogenicity will shed led on the question of the carcinogenicity of the nickel ion per se; that is, is the carcinogenic response dependent on the bioavailability of ionic nickel? If so, is it necessary that the nickel ion be delivered to target cells at a certain dose rate over an extended period of time? Are secondary effects involved, for example, chronic release of oxidants by inflammatory cells due to phagocytosis of nickel-containing particles? Both carefully designed mechanistic in vitro and in vivo studies are necessary to answer these questions. As pointed out by Doll (1984), such mechanistic studies are essential to assure that extrapolation from one species to another is appropriate. As long as answers to the above questions are not available, it would be prudent and is certainly conservative to regard all nickel compounds in general as having a carcinogenic potential.

It could be argued that the deposited dose rate rather than the accumulated equilibrium lung dose is better suited for extrapolation modeling from rat to man. During chronic inhalation exposure, when the nickel concentration in the lung has reached an equilibrium, the clearance rate—which is proportional to the solubility rate and mechanical transport rate of nickel particles—is equal to the deposition rate of the particles. Thus, it could be reasoned that the deposition rate alone is the most crucial parameter for these dosimetric considerations and not the equilibrium concentration, as proposed in this chapter. This in turn would imply that highly soluble nickel compounds (e.g., $NiCl_2$) have the same carcinogenic potency as poorly soluble compounds (like NiO) provided that the aerodynamic particle size and nickel concentration of the inhaled particles is the same, thereby resulting in the same deposition rate. Indeed, a recent reevaluation of epidemiological studies seems to indicate that highly soluble nickel compounds are as carcinogenic as poorly soluble nickel compounds (Chapters 29 and 31). This cannot easily be explained knowing the rapid clearance rate of soluble nickel compounds from the lung (English et al., 1981; Menzel et al., 1987), which implies that the equilibrium nickel concentration in the lung of the two compounds will be orders of magnitude apart when inhaled at the same concentration and particle size, as can be judged from Figure 32.2. Moreover, results of recent animals studies (Chapter 37) with intraperitoneally injected nickel compounds as well as of recent in vitro studies on the cell mutagenic potency of different nickel compounds (Chapter 23 and 24) would not support a conclusion of equal carcinogenic potency of soluble and insoluble nickel compounds. Further investigations of the carcinogenic mechanism of nickel are needed. Conceivably, during the combined exposure to soluble and insoluble nickel compounds occurring at the workplace in nickel refineries, the soluble nickel compounds could contribute to the carcinogenic action of particulate nickel compounds of lower in vivo solubility via an adverse effect on pulmonary clearance mechanisms, thus prolonging nickel retention. Therefore, studies on both carcinogenic mechanisms and mechanisms of lung clearance are necessary (Table 32.5) to solve the controversial question of carcinogenicity of highly soluble nickel compounds. This will also allow us to refine the dosimetric model described in this chapter and to develop a more appropriate mechanistically oriented dosimetric model.

When the results of studies listed in Table 32.5 become available, extrapolation modeling can be considerably improved, although the principles outlined in this chapter would remain unchanged. The calculation of the EHE as discussed here could then be better justified, possibly even the cellular level (microdosimetry) could be included; a carcinogenic risk associated with inhalation of different nickel compounds could then also be derived.

ACKNOWLEDGMENT

This chapter was partially based on work supported by grants ESO-1247 and ESO-1248 from the National Institutes of Health.

REFERENCES

Bailey, M. R., Fry, F. A., and James, A. C. (1985). Long-term retention of particles in the human respiratory tract. *J. Aerosol Sci.* **16,** 295–305.

Bellmann, B., Muhle, H., and Heinrich, U. (1983). Lung clearance after long time exposure of rats to airborne pollutants. *J. Aerosol Sci.* **14,** 194–196.

Doll, R. (1984). Occupational cancer: Problems in interpreting human evidence. *Ann. Occup. Hyg.* **28,** 291–305.

English, J. C., Parker, R. D. R., Sharma, R. P., and Oberg, S. G. (1981). Toxicokinetics of nickel in rats after intratracheal administration of a soluble and insoluble form. *Am. Ind. Hyg. Assoc. J.* **42,** 486–492.

Environmental Protection Agency (EPA) (1986). *Health Assessment Document for Nickel and Nickel Compounds.* EPA/600/8-83/012FF, Office of Health and Environmental Assessment, Washington, DC.

Ferin, J. (1982). Alveolar macrophage mediated pulmonary clearance suppressed by drug-induced phospholipidosis. *Experim. Lung Res.* **4,** 1–10.

Hochrainer, D., Oberdörster, G., and Mihm, U. (1980). Generation of NiO aerosols for studying lung clearance and their effects on lung function. *Gesellschaft f. Aerosolforschung* **8,** 259–264.

International Agency for Research on Cancer (IARC) (1987). Nickel and nickel compounds. In *IARC Monographs on the Evaluation of the Carcinogenic Risk of Chemicals to Humans,* Suppl. 7. IARC, WHO, Lyon, France, p. 440.

International Commission on Radiological Protection. (ICRP) (1975). *Report of the Task Group on Reference Man,* No. 23. Pergamon Press, Oxford, pp. 151–173.

Johansson, A., Curstedt, T., Robertson, B., and Camner, P. (1983). Rabbit lung after inhalation of soluble nickel. II. Effects on lung tissue and phospholipids. *Environ. Res.* **31,** 399–412.

Kodama, Y., Ishimatsu, S., Matsuno, K., Tanaka, I., and Tsuchiya, K. (1985). Pulmonary deposition and clearance of a nickel oxide aerosol by inhalation. *Biol. Trace Elem. Res.* **7,** 1–9.

Kuehn, K., and Sunderman, F. W., Jr. (1982). Dissolution halftimes of nickel compounds in water, rat serum, and renal cytosol. *J. Inorg. Biochem.* **17,** 19–39.

Lippmann, M. (1977). Regional deposition of particles in the human respiratory tract. In *Handbook of Physiology*. American Physiol. Society, Bethesda, MD, Chapter 14, pp. 213–239.

Lundborg, M., Eklund, A., Lind, B., and Camner, P. (1985). Dissolution of metals by human and rabbit alveolar macrophages. *Br. J. Ind. Med.* **42,** 642–645.

Menzel, D. B., Deal, D. L., Tayyeb, M. I., Wolpert, R. L., Boger, J. R., Shoaf, C. R., Sandy, J., Wilkinson, K., and Francovitch, R. J. (1987). Pharmacokinetic modeling of the lung burden from repeated inhalation of nickel aerosols. *Toxicol. Lett.* **38,** 33–43.

Mercer, T. T. (1967). On the role of particle size in the dissolution of lung burdens. *Health Phys.* **13,** 1211–1217.

Mercer, T. T. (1973). *Aerosol Technology in Hazard Evaluation*. Academic Press, New York.

Oberdörster, G., and Hochrainer, D. (1980). Effect of continuous nickel oxide exposure on lung clearance. In *Nickel Toxicology*. Academic Press, New York, pp. 125–128.

Oberdörster, G., Green, F.H.Y., and Freedman, A. P. (1984). Clearance of $^{59}Fe_3O_4$ particles from the lungs of rats during exposure to coal mine dust and diesel exhaust. *J. Aerosol Sci.* **15,** 235–237.

Ottolenghi, A. D., Haseman, J. K., Payne, W. W. Falk, H. L., and MacFarland, H. N. (1974). Inhalation studies of nickel subsulfide in pulmonary carcinogenesis of rats. *J. Natl. Cancer Inst.* **54,** 1165–1172.

Raabe, O. G., Yeh, H. C., Newton, G. J., Phalen, R. F., and Velasquez, D. J. (1977). Deposition of inhaled monodisperse aerosols in rodents. In W. H. Walton and B. McGovern (Eds.), *Inhaled Particles IV*. Pergamon Press, Oxford, pp. 3–21.

Schum, G. M., and Yeh, H. C. (1980). Theoretical evaluation of mammalian airways. *Bull. Math. Biol.* **42,** 1–15.

Stahl, W. R. (1967). Scaling of respiratory variables in mammals. *J. Appl. Physiol.* **22,** 453–460.

Stahlhofen, W., Gebhart, J., Rudolf, G., and Scheuch, G. (1986). Clearance from the human airways of particles of different sizes deposited from inhaled aerosol boli. In *Aerosols, Formation and Reactivity*. Pergamon Press, Oxford, pp. 192–194.

Sunderman, F. W., Jr., An, L. L., Zaharia, O., Wong, S.H.Y., and Hopfer, S. M. (1988). Acute depletion of pulmonary lavage cells, inhibition of 5′-nucleotidase activity, and enhanced lipid peroxidation in alveolar macrophages of rats following parenteral injection of nickel chloride. *Toxicologist* **8,** 192.

Tanaka, I., Ishimatsu, S., Matsuno, K., Kodama, Y., and Tsuchiya, K. (1985). Biological half-time of deposited nickel oxide aerosol in rat lung by inhalation. *Biol. Trace Elem. Res.* **8,** 203–210.

Tanaka, I., Horie, A., Haratake, J., Kodama, Y., and Tsuchiya, K. (1988). Lung burden of green nickel oxide aerosols and histopathological findings in rats after continuous inhalation. *Biol. Trace Elem. Res.* **16,** 19–26.

Task Group on Metal Accumulation (1973). Accumulation of toxic metals with special reference to their absorption, excretion and biological half-times. *Environ. Physiol. Biochem.* **3,** 65–107.

Wiernik, A., Johansson, A., Jarstrand, C., and Camner, P. (1983). Rabbit lung after inhalation of soluble nickel. I. Effects on alveolar macrophages. *Environ. Res.* **30,** 129–141.

Wojciak, J. F. (1988). Theoretical and experimental analyses of aerosol deposition in the lung: Implications for human health effects. Ph.D. Dissertation, Department of Chemical Engineering, Pediatrics and Biophysics, University of Rochester.

Yeh, H.-C., Schum, G. M. (1980). Models of human lung particle deposition. *Bull. Math. Biol.* **42,** 461–480.

33

INHALATION TOXICITY OF NICKEL COMPOUNDS

J. K. Dunnick and M. R. Elwell

National Toxicology Program, National Institute of Environmental
Health Sciences, Research Triangle Park, North Carolina 27709

*J. B. Benson, C. H. Hobbs, Y. S. Cheng, F. F. Hahn,
P. J. Haley, and A. F. Eidson*

Lovelace Inhalation Toxicology Research Institute,
U.S. Department of Energy, Albuquerque, New Mexico 87185

Nickel and Human Health: Current Perspectives, Edited by Evert Nieboer and
Jerome O. Nriagu.
ISBN 0-471-50076-3   © 1992 John Wiley & Sons, Inc.

1. INTRODUCTION

Epidemiology studies of workers exposed to nickel compounds have demonstrated an excess risk of cancer of the nasal cavity and lung, although the chemical form of nickel responsible for the carcinogenic effect is not known (IARC, 1976, 1982, 1987). The National Toxicology Program is conducting inhalation studies to evaluate and compare the carcinogenic potential of nickel oxide, nickel sulfate, and nickel subsulfide.

Other investigators have reported the effects of aerosol exposure of rodents to solitary compounds, but comparative toxicology has not been extensively studied. The primary target organ after inhalation exposure is the lung. Inhalation studies in the F344/N rat (Ottolenghi et al., 1974) have shown that nickel subsulfide exposure at 1 mg/m^3 (6 hr/day, 5 days/week for 108 weeks) caused an increase in lung tumors. Twelve to fourteen percent of exposed animals (208 animals examined histologically) had lung tumors compared to less than 0.5% of control animals (215 animals examined histologically). At the end of the 108-week exposure period fewer than 5% of animals in treated groups were alive compared with survival of 31% in control groups. The exposed animals had a high incidence of nonneoplastic lung lesions that might have contributed to the decreased survival.

Nickel oxide exposure of hamsters at 53 mg/m^3 for two years did not cause an increase in lung tumors (Wehner et al., 1975). Several authors have suggested that the hamster is less sensitive than the rat to the carcinogenic effects of nickel (Furst and Schlauder, 1971). No lung tumors were seen in rats exposed to nickel metal powder at 60 or 200 μg/m^3 for 18 months (Glaser et al., 1986). Sunderman and coworkers (1959, 1966) have investigated the carcinogenicity of nickel carbonyl in Wistar rats and found a low incidence of lung tumors in exposed animals. Early studies by Hueper and Payne (1962) showed that inhalation exposure to metallic nickel caused adenomatoid lung lesions in rats. Tanaka et al. (1986) have shown that the water-insoluble nickel salt, nickel oxide, is retained in the lungs, while nickel is cleared after exposure to the more water-soluble salts such as nickel chloride (Menzel et al., 1987).

National Toxicology Program (NTP) studies are designed to expand our knowledge on the mechanism of metal carcinogenicity and the relative potency of water-soluble and water-insoluble metal salts and to compare carcinogenicity data in rodents with effects in humans. This chapter describes 13-week toxicity studies of nickel sulfate, nickel oxide, and nickel subsulfide in the F344/N rat and B6C3F$_1$ mouse, and the use of the data to select exposure levels for two-year toxicity and carcinogenicity studies.

2. METHODS AND MATERIALS

2.1. Animals

Male and female F344/N rats and B6C3F$_1$ mice were obtained from NTP breeding colonies. At the start of the 13-week inhalation exposure periods, animals were 7–8 weeks of age. Animals were housed in individual cages in multitiered inhalation chambers throughout the course of the studies. Water was available ad libitum, and NIH-07 diet (Zeigler Brothers, Gardners, PA) was available during the nonexposure hours. Temperature in the exposure chambers was maintained at $23 \pm 2°C$, and average humidity was 17–45%. The light cycle was maintained at 12 hr on (6 a.m.) and 12 hr off (6 p.m.).

2.2. Chemicals

Nickel oxide (NiO; molecular weight (MW) 74.71; CAS No. 1313-99-1; formed at 1350°C; Boldt, 1967) and nickel subsulfide (Ni_3S_2; MW 240.25; CAS No. 12035-72-2; low-temperature form; Sharma and Change, 1980) were obtained from International Nickel Company (INCO), Toronto, Ontario, Canada, and nickel sulfate hexahydrate ($NiSO_4 \cdot 6H_2O$; MW 262.86; CAS No. 10101-97-0) was obtained from Aldrich Chemical Co., Milwaukee, WI. The purity estimation of these compounds was based on elemental analyses and spark source mass spectroscopy. The nickel oxide and nickel sulfate were greater than 99% pure, and the nickel subsulfide was greater than 97% pure (Dunnick et al., 1985).

The aerosol system used for nickel oxide and nickel subsulfide was a fluid-bed aerosol generator. Nickel sulfate aerosol was generated using a nebulizer. The aerosol was passed through a K-85 discharger to neutralize electrical charge and was mixed with diluting air to achieve the desired concentration in the chambers.

Multitiered inhalation chambers of two sizes were used in each study (H-1000 and H-2000, Hazleton Systems, Aberdeen, MD). The H-1000 has a total volume of $1.0 \, m^3$ and H-2000 has a total volume of $1.7 \, m^3$. The flow through each chamber provided 12 ± 2 air changes per hour. Target concentrations for nickel sulfate hexahydrate (22.3% nickel) were 0, 0.12, 0.25, 0.5, 1.0, and 2.0 mg compound/m^3 (0, 0.02, 0.05, 0.1, 0.2, and 0.4 mg Ni/m^3); for nickel subsulfide (73.3% nickel), 0, 0.15, 0.3, 0.6, 1.2, and 2.5 mg compound/m^3 (0, 0.11. 0.2, 0.4, 0.9, and 1.8 mg Ni/m^3); and for nickel oxide (78.6% nickel), 0, 0.6, 1.2, 2.5, 5.0, and 10.0 mg compound/m^3 (0, 0.4, 0.9, 2.0, 3.9, 7.9 mg Ni/m^3).

The aerosol concentration in the exposure chambers was monitored by taking three 2-hr filter samples during the a 6-hr exposure day and found to be within $\pm 10\%$ of target levels. The mean daily concentration in the exposure chambers was calculated from the filter samples. Aerosol size distribution was determined using cascade impactors (Table 33.1).

2.3. Experimental Design

Groups of 10 male and 10 female F344/N rats and B6C3F$_1$ mice were exposed to filtered air (controls) or to one of five concentrations of each of the nickel compounds. These animals comprised the basic study group. Endpoints evaluated were body weight gain, clinical observations, and histopathological

Table 33.1 Exposure Levels for Nickel Sulfate, Nickel Subsulfide, and Nickel Oxide

Compound	Target Concentration		Actual Concentration[a] (mg/m^3)	Mass Median Aerodynamic Diameter (μm)	Geometric Standard Deviation
	mg Ni/m^3	mg cpd/m^3			
NiSO$_4$	0.027	0.12	0.12 (0.01)	2.3	2.4
	0.11	0.50	0.52 (0.03)	2.3	2.4
	0.45	2.0	2.01 (0.09)	2.3	2.4
Ni$_3$S$_2$	0.11	0.15	.15 (0.02)	2.4	2.2
	0.45	0.60	.61 (0.05)	2.4	2.2
	1.8	2.5	2.52 (0.21)	2.4	2.2
NiO	0.47	0.60	.61 (0.05)	2.8	1.8
	2.0	2.5	2.46 (0.17)	2.8	1.8
	7.9	10.0	9.95 (0.73)	2.8	1.8

[a]Values are the mean of 68 exposure days; numbers in parentheses are standard deviation.

change. Organ weights taken at necropsy were lung, liver, kidney, testes, brain, and thymus.

Animals were sacrificed while under halothane anesthesia by exsanguination from the heart or axial arteries. A complete necropsy was performed on each animal, and tissues were preserved in 10% neutral buffered formalin, embedded in paraffin, sectioned at 5 μm, and stained with hematoxylin and eosin. Complete histopathology was performed on high-exposure and control group animals and to a no-effect level in target tissues in lower dose group animals (Dunnick et al., 1988).

Groups of five male rats from the control group and from the groups exposed to the high, middle, and lowest nickel aerosol concentrations were sacrificed for quantitation of nickel concentrations in lungs. Methods for tissue digestion and nickel analysis have been described previously (Benson et al., 1987, 1988). Limits of detection and quantitation of the analytical method were calculated for each species and for each sacrifice period. Limits of detection ranged from 0.216 to 0.301 mg Ni/sample, while limits of quantitation ranged from 0.597 to 0.826 mg Ni/sample. For purposes of statistical analyses, measured nickel concentrations below the limits of detection or quantitation were considered to be zero.

2.4. Statistical Evaluation

The significance of differences between doses and control groups means was assessed using multiple-comparison procedures (two-tail test) (Dunn, 1964; Shirley, 1977; Williams, 1986).

3. RESULTS

No exposure-related mortality was seen in rats or mice. The only treated animals that had a decrease in final body weight relative to controls were male rats in the nickel subsulfide high-exposure level (Table 33.2). There was an exposure-related increase in lung weight after nickel exposure (Table 33.3). There were no exposure-related changes in organ weights in the thymus, liver, kidney, brain, or testis.

Inflammation was seen in lungs of both species and sexes after exposure to all three nickel compounds and in the nasal cavity after exposure to nickel sulfate and nickel subsulfide (Table 33.4). No exposure-related lesions were seen in other organs. Exposed rats and mice had increased numbers of macrophages within the alveolar spaces throughout the lung (also see Chapter 34). At the lower exposure levels the only change present was a minimal increase in alveolar macrophages. The incidence and severity of inflammation (classified as chronic-active) in the lung increased with the level of nickel exposure. These foci of inflammation were predominantly adjacent to terminal bronchioles, but some were subpleural and consisted of a mild thickening

Table 33.2 Final Body Weight (g) of F344/N Rats and B6C3F₁ Mice after 13 weeks Exposure to Nickel Sulfate, Nickel Subsulfide, or Nickel Oxide[a]

Exposure Concentrations (mg/m³ Ni)	Male Rats			Female Rats			Male Mice			Female Mice		
	N	Mean	SE	N	Mean	SE	N	Mean	SE	N	Mean	SE
Nickel Sulfate												
0	10	328	4	10	197	4	6	30.7	0.5	7	25.9	0.8
0.02	10	324	4	10	189	4	8	32.1	0.9	10	27.2	0.5
0.05	10	339	5	10	193	3	10	30.6	0.4	10	27.0	0.6
0.1	10	317	5	10	192	3	10	31.9	0.6	10	27.3	0.3
0.2	10	334	5	10	198	3	10	31.9	0.7	10	26.6	0.5
0.4	9	311	5	10	187	3	10	31.4	0.5	10	25.2	0.6
Nickel Subsulfide												
0	10	353	4	10	200	3	8	29.8	0.5	10	25.1	0.5
0.11	10	354	6	10	203	3	10	30.5	0.5	8	25.4	0.5
0.2	10	337[b]	4	10	208	4	10	31.6	0.6	10	25.2	0.4
0.4	10	338[b]	7	10	202	2	8	30.8	0.5	9	25.4	0.4
0.9	10	350	4	10	201	3	9	30.2	0.8	9	25.3	0.4
1.8	10	330[b]	7	10	198	3	10	29.0	0.3	7	24.9	0.5
Nickel Oxide												
0	10	306	7	10	190	5	10	32.4	0.4	9	28.8	0.5
0.4	10	317	6	10	192	3	10	32.6	0.6	10	27.9	0.6
0.9	10	319	7	10	192	3	10	32.0	0.3	7	28.7	1.0
2.0	9	303	6	10	186	4	10	31.4	0.5	10	27.6	0.8
3.9	10	314	5	10	187	4	10	31.6	0.9	10	27.0	0.7
7.9	10	308	6	10	190	4	9	31.3	0.6	9	28.1	0.5

[a] N = final survival.
[b] Significantly different from the control group (P = 0.05).

Table 33.3 Lung Weight (mg) of F344/N Rats and B6C3F$_1$ Mice after 13 Weeks Exposure to Nickel Sulfate, Nickel Subsulfide, or Nickel Oxide

Exposure Concentrations (mg/m^3 Ni)	Male Rats		Female Rats		Male Mice		Female Mice	
	Mean	SE	Mean	SE	Mean	SE	Mean	SE
Nickel Sulfate								
0	1350[a]	43	1022	22	200	0	200	0
0.02	1254	26	1017	33	200	0	200	0
0.05	1509	54	1162[a]	16	200	0	200	0
0.1	1641[a]	47	1335[a]	51	210	10	200	0
0.2	2137[a]	37	1715[a]	40	250[a]	17	220	13
0.4	2217[a]	51	1722[a]	44	310[a]	10	270[a]	15
Nickel Subsulfide								
0	1330	15	1010	28	188	12	190	10
0.11	1740[a]	52	1290[a]	24	200	0	175	16
0.2	1830[a]	37	1390[a]	35	220	20	200	0
0.4	2300[a]	67	1820[a]	44	212	12	211	11
0.9	2630[a]	30	1850[a]	37	233[b]	17	260[a]	22
1.8	2420[a]	68	1810[a]	38	280[a]	13	288[a]	12
Nickel Oxide								
0	1180	37	983	26	214	9	195	12
0.4	1349[a]	24	1027	23	215	11	195	18
0.9	1474[a]	41	1134[a]	24	206	12	192	8
2.0	1698[a]	53	1550[a]	38	213	8	213[b]	8
3.9	1906[a]	32	1610[a]	47	243	10	223[b]	8
7.9	2467[a]	45	2111[a]	81	288[a]	19	271[a]	7

[a] Significantly different from the control group ($P = 0.01$).
[b] Significantly different from the control group ($P = 0.05$).

Table 33.4 Lung and Nasal Lesions after Inhalation Exposure of Rats and Mice to Nickel Compounds for 13 Weeks[a]

Dose mg/m³ (mg Ni/m³)	Nickel Sulfate					Nickel Subsulfide						Nickel Oxide					
	.12 (.02)	.25 (.05)	.50 (.1)	1.0 (.2)	2.0 (.4)	0	.15 (.1)	.30 (.2)	.60 (.4)	1.2 (.9)	2.5 (1.8)	0	.6 (.4)	1.2 (.9)	2.5 (2.0)	5.0 (3.9)	10.0 (7.9)
Male Rats																	
Lung																	
Alveolar macrophage hyperplasia	−	+ (1.0)	+ (1.0)	+ (2.4)	+ (3.6)	−	+ (1.1)	+ (1.5)	+ (1.6)	+ (3.4)	+ (3.8)	−	+ (1.0)	+ (1.0)	+ (1.0)	+ (1.5)	+ (2.5)
Chronic active inflammation	−	−	+ (1.0)	+ (1.5)	+ (1.4)	−	+ (1.0)	+ (1.3)	+ (1.8)	+ (2.9)	+ (3.7)	−	−	−	+ (1.0)	+ (1.4)	+ (3.0)
Nose																	
Olfactory epithelial atrophy	−	*	+	+	+	−	−	+	+	+	+	−	*	*	*	*	−
Female Rats																	
Lung																	
Alveolar macrophage hyperplasia	−	+ (1.0)	+ (1.1)	+ (2.2)	+ (3.6)	−	+ (1.0)	+ (1.7)	+ (1.8)	+ (2.9)	+ (3.8)	−	+ (1.0)	+ (1.0)	+ (1.0)	+ (1.4)	+ (2.4)
Chronic active inflammation	−	+ (1.0)	+ (1.0)	+ (1.3)	+ (1.0)	−	+ (1.0)	+ (1.0)	+ (1.9)	+ (2.6)	+ (3.8)	−	−	−	+ (1.0)	+ (1.3)	+ (2.4)
Nose																	
Olfactory epithelial atrophy	−	*	+	+	+	−	−	+	+	+	+	−	*	*	*	*	−

444

Lung															
Alveolar macrophage hyperplasia	–	–	+(1.0)	+(1.0)	+(1.0)	–	–	+(1.0)	+(1.0)	+(1.0)	+(1.0)	+(2.0)	+(2.2)	+(1.0)	+(1.0)
Chronic active inflammation	–	–	+(1.0)	+(1.0)	+(1.5)	–	–	+(1.0)	+(1.2)	+(2.0)	–	–	–	–	+(1.0)
Fibrosis	–	–	–	+	+	–	–	–	+	+	–	–	–	–	–
Nose															
Olfactory epithelial atrophy	*	*	–	+	+	*	–	+	+	+	*	*	*	*	–

Female Mice

Lung															
Alveolar macrophage hyperplasia	–	–	+(1.0)	+(1.0)	+(1.0)	–	–	+(1.0)	+(1.0)	+(1.0)	+(1.0)	+(2.4)	+(2.6)	+(1.0)	+(1.1)
Chronic active inflammation	–	–	+(1.0)	+(1.0)	+(2.1)	–	–	+(1.0)	+(1.5)	+(2.0)	–	–	–	+(1.0)	+(1.0)
Fibrosis	–	–	–	+	+	–	–	–	+	+	–	–	–	–	–
Nose															
Olfactory epithelial atrophy	*	*	*	+	+	*	–	+	+	+	*	*	*	*	–

[a] Numbers in parentheses represent mean severity of macrophage hyperplasia and chronic active inflammation graded minimal (1) to severe (4); +, lesion present; –, lesion absent; *, group not examined.

of the alveolar septae by a mononuclear (lymphocytes and macrophages) cell infiltrate. In these focal lesions, macrophages and a few neutrophils were also present within the alveolar spaces. Interstitial infiltrate, consisting primarily of lymphocytes around vessels, was also present in the higher exposure groups. Mice exposed to the two highest levels of nickel subsulfide and nickel sulfate also had focal areas of fibrosis in the alveolar septae.

Both species and sexes of rats and mice exposed to the higher levels of nickel subsulfide and nickel sulfate had atrophy of the olfactory epithelium. This change was minimal and was characterized by a decreased number of sensory cell nuclei, particularly in the olfactory epithelium of the dorsal meatus. There was no evidence of an inflammatory or regenerative response in the olfactory mucosa.

Concentrations of nickel in lungs of rats and mice increased with exposure concentrations. Males and females of each species had similar nickel concentrations in lung. Concentration of nickel in lungs of animals exposed to 0.4 mg Ni/m^3 was lowest for animals exposed to $NiSO_4$ and greatest for those exposed to NiO. Equilibrium levels of nickel in the lung were reached by 13 weeks after nickel sulfate and nickel subsulfide exposures, while the lung levels of nickel after nickel oxide exposure continued to increase (Table 33.5).

4. DISCUSSION

As has been seen in other nickel toxicology studies (Ottolenghi et al., 1974; Taneka et al., 1986; Menzel et al., 1987), the primary target organ after inhalation exposure to nickel was the lung. The order of toxicity was nickel sulfate $>>$ nickel subsulfide $>$ nickel oxide. Rats were more sensitive than mice to nickel lung toxicity. Toxicity to the lung paralleled the water solubility of nickel compound with the most water-soluble compound (nickel sulfate) being the most toxic, even though nickel sulfate was more readily cleared from the lung than were nickel subsulfide or nickel oxide.

Exposure levels for two-year toxicity and carcinogenicity studies were selected to give maximal exposure without causing severe body weight effects, increase in mortality, or life-threatening target organ toxicity (Haseman, 1985) and to maximize the sensitivity of a study in which a small number of animals is used to determine increases or decreases in tumor incidences (Office of Science and Technology Policy, 1984; NTP, 1984).

In these nickel studies the selection of exposure levels was based primarily on the severity of lung toxicity (Table 33.6). Use of exposure levels with inflammatory lesions of greater than moderate severity was considered to be too toxic for a two-year study. The current threshold limit value (TLV) for nickel metal and insoluble salts is a 1 mg/m^3, and for water-soluble nickel compounds it is 0.1 mg/m^3 (ACGH, 1985). The high exposure level in the nickel subsulfide two-year study is comparable to the LTV for insoluble salts

Table 33.5 Analysis of Nickel Quantities in Whole Lungs (μg Ni/g sample) after 13 Weeks Exposure to Nickel Sulfate, Nickel Subsulfide, or Nickel Oxide

Exposure Concentrations (mg/m³ Ni)	Male Rats		Female Rats		Male Mice		Female Mice	
	Mean	SE	Mean	SE	Mean	SE	Mean	SE
Nickel Sulfate								
0	0.0	0.0	0.0	0.0	0.0	0.0	0.0	0.0
0.02	0.1	0.1	0.0	0.0	0.0	0.0	0.0	0.0
0.1	1.1[a]	0.1	1.2[a]	0.1	0.0	0.0	0.0	0.0
0.4	3.4[a]	0.1	3.7[a]	0.3	0.8	0.5	2.2[a]	0.4
Nickel Subsulfide								
0	0.0	0.0	0.0	0.0	0.0	0.0	0.0	0.0
0.1	4.9[a]	0.2	4.8[a]	0.1	4.2[a]	0.3	5.7[a]	0.4
0.4	7.5[a]	0.4	6.9[a]	0.1	11.4[a]	0.8	13.4[a]	1.0
1.8	17.6[a]	0.5	17.4[a]	0.7	17.2[a]	1.6	22.7[a]	1.4
Nickel Oxide								
0	0.0	0.0	[b]	[b]	0.0	0.0	[b]	[b]
0.4	80.1[a]	5.5	[b]	[b]	41.5[b]	2.9	[b]	[b]
1.0	181.5[a]	25.6	[b]	[b]	201.7[b]	18.3	[b]	[b]
7.9	524.1[a]	38.2	[b]	[b]	736.2[b]	123.0	[b]	[b]

[a] Significantly different from the control group ($P = 0.01$).
[b] Sample not measured.

Table 33.6   Two-Year Exposure Concentrations[a] (mg/m^3)

Nickel Sulfate (22.3% Ni)	
Rats	
Amount of compound	0, 0.125, 0.25, 0.5
Amount of nickel	0, 0.3, 0.06, 0.12
Mice	
Amount of compound	0, 0.25, 0.5, 1.0
Amount of nickel	0, 0.06, 0.12, 0.24
Nickel Subsulfide (73.5% Ni)	
Rats	
Amount of compound	0, 0.15, 1.0
Amount of nickel	0, 0.11, 0.74
Mice	
Amount of compound	0, 0.6, 1.2
Amount of nickel	0, 0.44, 0.9
Nickel Oxide (78.6% Ni)	
Rats	
Amount of compound	0, 0.62, 1.25, 2.5
Amount of nickel	0, 0.5, 1.0, 2.0
Mice	
Amount of compound	0, 1.25, 2.5, 5.0
Amount of nickel	0, 1.0, 2.0, 4.0

[a] Occupational exposure Limits in the United States: 1 mg Ni/m^3 for nickel metal, 0.1 mg Ni/m^3 for soluble nickel compounds.

(1 mg/m^3). The high exposure level in the nickel oxide two-year study is two (rats) to four (mice) times the TLV level. The high exposure level for the nickel sulfate two-year rat studies is approximately equal to the TLV level for soluble nickel compounds (0.1 mg/m^3), and the high exposure level for mice is approximately twice the TLV level.

ACKNOWLEDGMENTS

We gratefully acknowledge contributions to this study in toxicology by Dr. R. Chhabra; in chemistry by Dr. C. W. Jameson and Dr. T. Goehl; in animal care by Dr. G. Rao; in health and safety by Dr. D. Walters; and the technical assistance of J. Holmes, J. Kim, D. Brown, and A. Stevens. The research was conducted under Interagency Agreement YO1-ES-30108 between the U.S. Department of Energy (Contract NO. DE-AC04-76 EV01013) and the National Institute of Environmental Health Sciences. The facilities used for this research were fully accredited by the American Association for the Accreditation of Laboratory Animal Care.

REFERENCES

American Conference of Governmental Industrial Hygienists (ACGIH) (1987). *Threshold Limit Values (TLV) and Biological Exposure Indices for 1986–1987*. ACGIH, Cincinnati.

Benson, J. M., Carpenter, R. L., Hahn, F. F., Haley, P. J., Hanson, R. L., Hobbs, C. H., Pickrell, J. A., and Dunnick, J. K. (1987). Comparative inhalation toxicity of nickel subsulfide to F344/N rats and B6C3F$_1$ mice exposed for twelve days. *Fund. Appl. Toxicol.* **9**, 251–265.

Benson, J. M., Burt, D. G., Carpenter, R. L., Eidson, A. F., Hahn, F. F., Haley, P. J., Hanson, R. L., Hobbs, C. H., Pickrell, J. A., and Dunnick, J. K. (1988). Comparative inhalation toxicity of nickel sulfate to F344/N rats and B6C3F$_1$ mice exposed for twelve days. *Fund. Appl. Toxicol.* **10**, 164–178.

Boldt, J. R., Jr. (1967). *The Winning of Nickel, Its Geology, Mining, and Extraction Metallurgy*. Longmans Canada Limited, Toronto, Canada, pp. 392–398.

Dunn, O. J. (1964). Multiple comparisons using rank sums. *Technometrics* **6**, 241–252.

Dunnick, J. K., Jameson, C. W., and Benson, J. M. (1985). Toxicology and carcinogenesis studies of nickel oxide, nickel sulfate, and nickel subsulfide: Designs of study and characterization of nickel compounds. In S. S. Brown and F. W. Sunderman, Jr. (Eds.), *Progress in Nickel Toxicology*, Blackwell, Oxford, England, pp. 49–52.

Dunnick, J. K., Benson, J. H., Hobbs, C. H., Hahn, F. F., Cheng, Y. S., and Eidson, A. F. (1988). Comparative toxicity of nickel oxide, nickel sulfate hexahydrate, and nickel subsulfide after 12 days of inhalation exposure to F344/N rats and B6C3F$_1$ mice. *Toxicology* **50**, 145–156.

Furst, A., and Schlauder, M. C. (1971). The hamster as a model for metal carcinogenesis. *Proc. West. Pharmacol. Sec.* **14**, 68–71.

Glaser, U., Hockrainer, D., Oldiges, H., Takenaka, S. (1986). Long-term inhalation studies with NiO and As$_2$O$_3$ aerosols in Wistar rats. *Int. Congr. Ser.—Excerpta Med.* **676**, 325–328.

Haseman, J. K. (1985). Issues in carcinogenicity testing; dose selection. *Fundam. Appl. Toxicol.* **5**, 66–78.

Hueper, W. C., and Payne, W. W. (1962). Experimental studies in metal carcinogenesis. *Arch. Environ. Health* **5**, 445–462.

International Agency for Research on Cancer (IARC) (1976). In Vol. 11. Cadmium, nickel, some epoxides, miscellaneous industrial chemicals and general considerations on volatile anesthetics. *IARC Monographs on the Evaluation of Carcinogenic Risk of Chemicals to Man*, IARC, Lyon, pp. 75–112.

International Agency for Research on Cancer IARC (1982). Chemicals, industrial processes and industries associated with cancer. *IARC Monographs on the Evaluation of the Carcinogenic Risk of Chemicals to Humans*, suppl. 4. IARC, Lyon, pp. 167–170.

International Agency for Research on Cancer (IARC) (1987). Overall evaluations of carcinogenicity. *IARC Monographs on the Evaluation of Carcinogenic Risks to Humans*, Suppl. 7, IARC, pp. 264–269.

Menzel, D. B., Deal, D. L., Tayyeb, M. I., Wolpert, R. L., Boger, J. R., Shoaf, C. R., Sandy, J., Wilkinson, K., and Francovitch, R. J. (1987). Pharmacokinetic modeling of the lung burden from repeated inhalation of nickel aerosols. *Toxicol Lett.* **38,** 35–43.

National Toxicology Program (NTP) (1984). Report of the NTP Ad Hoc Panel on chemical carcinogenesis testing and evaluation. Board of Scientific Counselors, NIEHS. Research Triangle Park, NC: U.S. Department of Health and Human Services pp. 165–188.

Office of Science and Technology Policy (1984). Executive Office of the President. Chemical Carcinogens; Review of the Science and its Associated Principles. *Fed. Regis.* **49**(250), 21594–21661, November 23.

Ottolenghi, A. D., Haseman, J. K., Payne, W. W., Falk, H. L., and MacFarland, H. N. (1974). Inhalation studies of nickel sulfide in pulmonary carcinogenesis of rats. *J. Natl. Cancer Inst.* **54,** 1165–1172.

Sharma, R. C., and Chang, Y. A. (1980). Thermodynamics and phase relationships of transition metal-sulfur systems: IV. Thermodynamic properties of the Ni-S liquid phase and the calculation of the Ni-S phase diagram. *Metallurgical Trans.* **11,** 139–146.

Shirley, E. (1977). A non-parametric equivalent of William's Test for contrasting increasing dose levels of a treatment. *Biometrics* **33,** 386–389.

Sunderman, F. W., and Donnelly, A. J. (1966). Studies of nickel carcinogenesis metastasizing pulmonary tumors in rats induced by the inhalation of nickel carbonyl. *Am. J. Pathol.* **46,** 1027–1041.

Sunderman, F. W., Donnelly, A. J., West, B., and Kincaid, J. F. (1959). Nickel poisoning. IX. Carcinogenesis in rats exposed to nickel carbonyl. *Arch. Industr. Hlth.* **20,** 36–41.

Tanaka, I., Ishimatsu, S., Matsuno, K., Kodama, Y., and Tsuchiya K. (1986). Retention of nickel oxide (green) aerosol in rat lungs by long-term inhalation. *Bio. Trace Elements Res.* **9,** 187–195.

Wehner, A. P., Busch, R. H., Olson, R. J., and Craig, D. K. (1975). Chronic inhalation of nickel oxide and cigarette smoke by hamsters. *Am. Ind. Hyg. Assoc.* **36,** 801–810.

Williams, D. A. (1986). A note on Shirley's nonparametric test for comparing several dose levels with a zero-dose control. *Biometrics* **42,** 183–186.

34

EFFECTS OF INHALED NICKEL ON LUNG BIOCHEMISTRY

Janet M. Benson, David G. Burt, Yung Sung Cheng, Fletcher F. Hahn, Patrick J. Haley, Rogene F. Henderson, Charles H. Hobbs, and John A. Pickrell

*Inhalation Toxicology Research Institute,
Lovelace Biomedical and Environmental Research Institute,
Albuquerque, New Mexico 87185*

June K. Dunnick

*National Toxicology Program, National Institute of Environmental
Health Sciences, Research Triangle Park, North Carolina 27709*

Nickel and Human Health: Current Perspectives, Edited by Evert Nieboer and
Jerome O. Nriagu.
ISBN 0-471-50076-3   © 1992 John Wiley & Sons, Inc.

1. INTRODUCTION

Pulmonary exposure to a variety of nickel compounds, including nickel subsulfide (Ni_3S_2), nickel sulfate ($NiSO_4$), and nickel oxide (NiO), resulted in development of inflammation in monkeys, rabbits, rats, hamsters, and mice; fibrosis in mice; and emphysema, alveolar proteinosis, and cancer in rats (Bingham et al., 1972; Mastromatteo, 1967; Wehner et al., 1975; Wehner and Craig, 1972; Ottolenghi et al., 1974; Takenaka et al., 1985; Benson et al., 1986, 1987, 1988; Haley et al., 1987; Dunnick et al., 1988). The purpose of this study was to determine the relative toxicities of Ni_3S_2, $NiSO_4$, and NiO to lungs of rats and mice exposed subchronically to occupationally relevant aerosol concentrations of these compounds. The aerosol concentrations of Ni_3S_2 and NiO used in this study bracketed the current threshold limit value (TLV) of 1 mg Ni/m^3 set for nickel sulfide–roasting fumes and dust and for nickel metal, while concentrations of $NiSO_4$ employed closely bracketed the TLV of 0.1 mg Ni/m^3 set for soluble nickel compounds (ACGIH, 1986). Toxic effects were evaluated by biochemical analyses of bronchoalveolar lavage fluid and by histopathology. Biochemical parameters evaluated in bronchoalveolar lavage fluid (BALF) included lactate dehydrogenase (LDH; cytoplasmic enzyme used as a measure of cell injury), β-glucuronidase (BG; released from phagocytic cells by active processes or cell death), collagenous peptides (CPs; indicator of turnover of extracellular collagenous matrix), total protein (TP; measure of increased permeability of the alveolar–capillary barrier), and total and differential cell counts (measure of influx and nature of inflammatory cells). Collagen concentration in lung was also determined as a possible indicator for the development of fibrosis. We evaluated the above biochemical and cytological changes, compared these changes with histopathological changes, and ranked the compounds by toxicity.

2. METHODS

2.1. Test Compounds

Nickel subsulfide (α-Ni_3S_2; CAS No. 12035-72-2) and nickel oxide (green oxide, calcined at 1200°C; CAS No. 1313-99-1) were supplied by the International Nickel Company (INCO, Toronto, Ontario, Canada). Nickel sulfate hexahydrate (referred to as nickel sulfate, CAS No. 10101-97-0) was from Aldrich Chemical Company (Milwaukee, WI). Elemental analyses conducted on these compounds by the Midwest Research Institute (Kansas City, MO) indicated overall purities of 97.0, 99.5, and 99.5% for Ni_3S_2, $NiSO_4$, and NiO, respectively (Dunnick et al., 1988).

2.2. Aerosol Characteristics

The exposure systems and methods for aerosol generation and characterization have been described in detail (Benson et al., 1987, 1988; Dunnick et al.,

1988). The Ni_3S_2 and NiO aerosols were generated using 2- and 4-in. fluid bed generators, respectively. Nickel sulfate aerosols were generated by nebulization of nickel sulfate solutions (62 g $NiSO_4 \cdot 6H_2O$/L water). Aerosol concentration was determined by taking three 2-hr filter samples through the exposure day. Real-time determination of aerosol concentration was made using real-time aerosol monitor (RAM) units (Cheng et al., 1988). Aerosol size was determined using cascade impactors. Aerosol characteristics are given in Table 34.1.

2.3. Experimental Design

Male and female F344/N rats and $B6C3F_1$ mice were obtained from the Frederick Cancer Research Facility (Frederick, MD) and from Simonsen Laboratories (Gilroy, CA), respectively. Animals were acclimatized for 20 days in Hazleton 2000 multitiered inhalation chambers (Hazleton Systems, Aberdeen, MD).

Groups of 6 male and 6 female rats and groups of 8 male and 8 female mice were exposed 6 hr/day 5 days/week for 13 weeks to $NiSO_4$ (0, 0.02, 0.1, or 0.4 mg Ni/m^3), Ni_3S_2 (0, 0.4, or 1.8 mg Ni/m^3), or NiO (0, 0.4, 0.9, or 7.9 mg Ni/m^3) for evaluation of biochemical and cytological changes in bronchoalveolar lavage fluid. In addition, groups of 10 male and 10 female rats and mice per compound were exposed similarly for evaluation of histopathological changes.

2.4. Biochemical and Cytological Evaluation of Lavage Fluid

Animals were anesthetized using 4% halothane in O_2, sacrificed by exsanguination from the axillary arteries and the heart–lung block was removed. Rat lungs were lavaged four times with 6 mL (male) or 5 mL (female) of physiological saline. Cells recovered by centrifugation ($300g$ for 10 min at 5°C) of lavage fluid from all four washes were pooled for cytological evaluation. Fluid from the first two lavages was pooled for evaluation of biochemical changes after cell removal. Mouse lungs were lavaged four times with 1 mL (male and female) of physiological saline. Fluid from all four washes from two mice of the same sex within the same exposure group was pooled to

Table 34.1 Summary of Experimental Design and Exposure Conditions

| Compound | Aerosol Concentration | | Aerosol Size |
	mg Compound/m^3	mg Ni/m^3	MMAD (σg)
$NiSO_4$	0, 0.12, 0.5, 2.0	0, 0.02, 0.1, 0.4	1.9 (2.1)
Ni_3S_2	0, 0.6, 2.5	0, 0.4, 1.8	2.8 (2.2)
NiO	0, 0.6, 2.5, 10	0, 0.4, 2.0, 7.9	3.0 (1.9)

form single samples. Cells recovered from lavages from two mice were also pooled to form single samples. The BALF from each rat and pooled BALF from pairs of mice were analyzed for LDH, BG, and TP as described previously (Henderson et al., 1985a). Aliquots of BALF supernatant were hydrolyzed (6 N HCl, 110°C, 18 hr) for analysis of collagenous peptides (as hydroxyproline). Lung tissue was frozen in liquid nitrogen, pulverized, and extracted (0.5 M acetic acid, 4°C, overnight) prior to hydroxyproline analyses. Hydroxyproline was measured according to the method of Grant (1965). Concentration of collagenous peptides in BALF and lung were obtained by multiplying corresponding concentrations of hydroxyproline by 7.46 (Neuman and Logan, 1950).

Total nucleated cells recovered in BALF were determined using a hemocytometer. Cell preparations were then transferred to glass slides using a cytocentrifuge and stained with Wright-Giemsa, and differential cell counts were performed.

2.5. Necropsy

Animals for histopathological examination were sacrificed as described above and lungs were removed. Lungs were fixed at constant pressure by intratracheal instillation of a 10% neutral-buffered formalin solution, embedded in paraffin, sectioned at 5 μm, stained with hematoxylin and eosin, and examined microscopically.

2.6. Statistical Evaluation

Data for males and females were combined because they were no significant differences in results obtained for the two sexes. Dunn's test or William's modification of Shirley's multiple-comparison procedure was applied based on the occurrence of an exposure concentration–related response in the data (Dunn, 1964; Shirley, 1977; Williams, 1986). Differences at significance levels of 0.05 and 0.01 are reported.

3. RESULTS

Lavage fluid obtained from rats exposed to all nickel compounds was somewhat milky in appearance. Centrifugation produced a clear supernatant on which all biochemical assays were performed. Pellets resulting from centrifugation of BALF from nickel-exposed rats increased in size with increasing exposure concentration. The effect was greatest for NiO and least for NiSO$_4$. Microscopically, the pellet consisted of nucleated cells, cellular debris, and some amorphous material. This effect occurred to a much lesser extent in mice.

Inhalation of Ni$_3$S$_2$, NiSO$_4$, and NiO resulted in exposure concentration–dependent increases in LDH, BG, and TP in both rats and mice (Figures

34.1–34.3). Concentrations of collagenous peptides were increased in rats exposed to NiSO$_4$, Ni$_3$S$_2$, and NiO and in mice exposed to Ni$_3$S$_2$ and NiO (Fig. 34.4). Collagen concentration was significantly increased in both rats and mice exposed to the highest concentration of NiSO$_4$ and Ni$_3$S$_2$ (Table 34.2).

The biochemical parameter increasing most in rats and mice was BG. It increased the most in rats exposed to 7.9 mg Ni/m^3 (as NiO), the highest

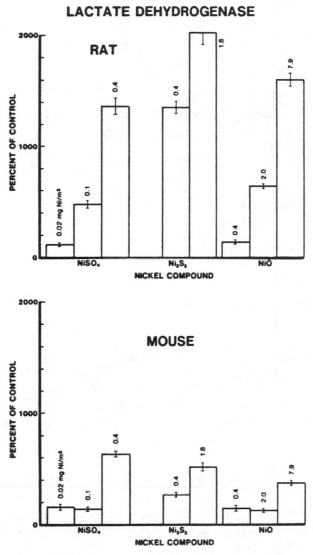

Figure 34.1. Effect of nickel compound exposure on LDH activity in BALF. Error bars represent 1 SE.

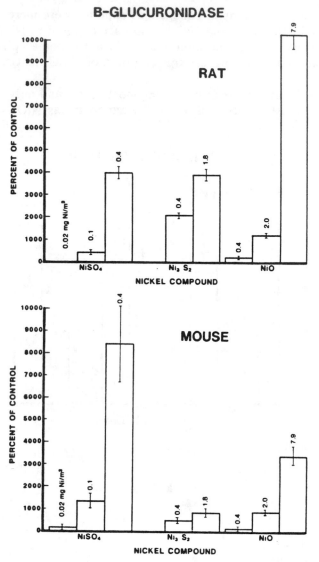

Figure 34.2. Effect of nickel compound exposure on BG activity in BALF. Error bars represent 1 SE.

nickel aerosol concentration used in this series of studies. Increases in BALF biochemical parameters were greater in rats than mice. When effects produced by each compound at an equivalent nickel aerosol concentration (0.4 mg Ni/m³) were compared, increases were consistently greatest for NiSO₄ and least for NiO (Fig. 34.5).

Total nucleated cell (TNC) and differential cell counts are given in Table 34.3. Total cell counts are not reported for Ni₃S₂-exposed rats because cellular debris might have been erroneously included in the cell counts for this study. Total nucleated cell counts in rats increased with increasing $NiSO_4$ and NiO exposure concentration. This effect was paralleled in $NiSO_4$-, Ni_3S_2-, and NiO- exposured mice. Greatest overall increases in TNC occurred in rats exposed to 0.4 mg Ni/m^3 as $NiSO_4$.

An influx of neutrophils occurred in nickel-exposed rats and mice. This influx occurred at a lower nickel aerosol concentration for $NiSO_4$-exposed

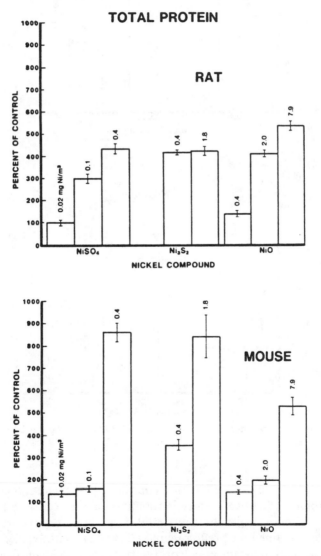

Figure 34.3. Effect of nickel compound exposure on TP content of BALF. Error bars represent 1 SE.

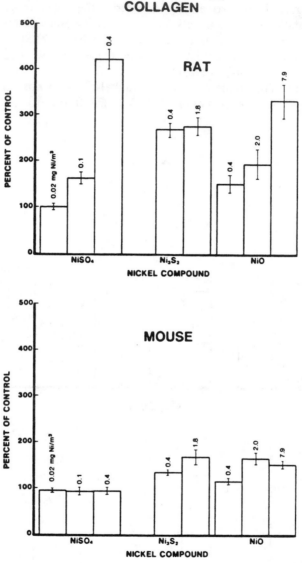

Figure 34.4. Effect of nickel compound exposure on CP concentration in BALF. Error bars represent 1 SE.

animals than for those exposed to Ni_3S_2 or NiO. Furthermore, the magnitude of the neutrophil influx was greater in rats than in mice exposed to the same aerosol conditions. Despite the neutrophil influx in nickel-exposed animals, macrophages remained the predominant cell type present. Nickel compound exposure had no significant effect on lymphocyte and eosinophil numbers in BALF (data not shown).

Lesions in rats and mice exposed to either Ni_3S_2, $NiSO_4$, or NiO for 13 weeks included chronic active inflammation, macrophage hyperplasia, and interstitial infiltrates. Lesions occurring in lungs of rats and mice exposed to the three compounds were quite similar in histological appearance with the following exceptions. Dark, basophilic staining particles were present within alveolar macrophages of animals exposed to NiO. Chronic active inflammation was more prominent in rats than in mice and was present in animals from all three studies. However, in the highest doses of NiO, granulomatous inflammation was also present in some animals. This was characterized by the aggregates of multinucleated giant cells, macrophages, and neutrophils in alveolar spaces. When the three nickel compounds are compared to the basis of milligrams of nickel per cubic meter, the morphological changes were least severe in mice and rats exposed to NiO. An additional important difference was that the mice exposed to $NiSO_4$ and Ni_3S_2, but not to NiO, developed pulmonary fibrosis. This lesion was absent in rats exposed to the nickel compounds.

4. DISCUSSION

Subchronic exposure of F344/N rats and B6C3F$_1$ mice to Ni_3S_2, $NiSO_4$, and NiO by inhalation resulted in increases in LDH and BG activity and in TP

Table 34.2   Effect of Nickel Inhalation on Total Lung Collagen Concentration[a]

Compound (mg Ni/m^3)	mg Collagen/g Control Lung, Mean	
	Rat	Mouse
$NiSO_4$		
0	17.9 (0.49)	9.82 (0.29)
0.02	17.1 (0.30)	10.0 (0.31)
0.1	18.4 (0.96)	9.87 (0.35)
0.4	24.1 (0.74)[b]	13.7 (0.28)[b]
Ni_3S_2		
0	15.0 (0.43)	6.93 (0.24)
0.4	17.3 (0.62)	7.87 (0.73)
1.8	22.7 (1.0)[b]	11.3 (0.56)[b]
NiO		
0	15.8 (1.1)	7.04 (0.29)
0.4	14.7 (0.73)	7.16 (0.18)
2.0	14.5 (0.36)	6.12 (0.44)
7.9	16.5 (1.22)	8.42 (0.10)

[a] Results represent the mean ± SE of 12 values for rats and 8 values for mice. Numbers in parentheses are standard errors.
[b] Mean significantly different from control, $p \leq 0.05$.

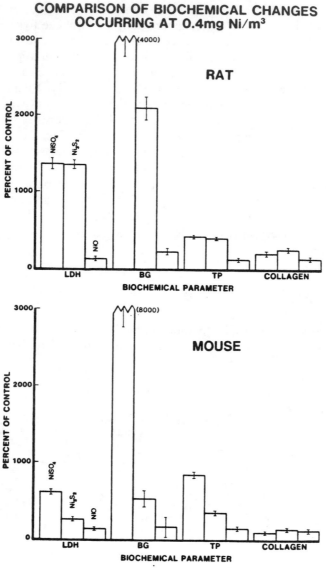

Figure 34.5. Comparison of changes in LDH, BG, and TP and CP in lavage fluid from rats and mice exposed to 0.4 mg Ni/m³ as $NiSO_4$, Ni_3S_2, or NiO. Error bars represent 1 SE.

and CP content, indicating damage or death of cells within the lung, increased phagocytic activity, increased vascular permeability, and turnover of the extracellular collagen matrix, respectively. The magnitude of increase in BG was greater than that of LDH, indicating that increases in BG were due to the presence of active phagocytic cells and not due solely to release to enzyme from dying or damaged phagocytic cells. The chronic inflammation,

Table 34.3 Differential Cell Counts in Lavage Fluid[a]

Compound (mg Ni/m³)	Rat			Mouse		
	Total Nucleated, ×10⁻⁶	Percent Neutrophils	Percent Macrophages	Total Nucleated, ×10⁻⁶	Percent Neutrophils	Percent Macrophages
NiSO₄						
0.0	6.04 ± 0.57	0.08 ± 0.1	95 ± 1	1.17 ± 0.18	0.34 ± 0.3	91 ± 1
0.02	10.6 ± 1.1[b]	0.0 ± 0.0	96 ± 1	1.50 ± 0.15	2.4 ± 0.6[b]	90 ± 2
0.1	22.2 ± 4.0[b]	19 ± 3[b]	77 ± 3[b]	2.62 ± 0.26[b]	14 ± 4[b]	82 ± 4
0.4	54.4 ± 7.3[b]	24 ± 2[b]	74 ± 3[b]	9.48 ± 0.82[b]	4.5 ± 2[b]	91 ± 2
Ni₃S₂						
0.0	[c]	1.1 ± 0.4	88 ± 2	2.17 ± 0.51	0.50 ± 0.4	90 ± 1
0.4	[c]	16 ± 2[b]	80 ± 2[b]	7.48 ± 0.96[b]	10 ± 2[b]	84 ± 2
1.8	[c]	12 ± 2[b]	82 ± 2[b]	8.46 ± 0.99[b]	2.2 ± 0.8[b]	92 ± 1
NiO						
0.0	6.69 ± 0.46	0.29 ± 0.1	94 ± 1	1.13 ± 0.08	0.07 ± 0.07	94 ± 2
0.4	11.2 ± 1.4[b]	6.6 ± 1[b]	91 ± 1	1.41 ± 0.14	1.1 ± 0.5[b]	95 ± 1
2.0	19.7 ± 1.2[b]	21 ± 2[b]	78 ± 2[b]	0.68 ± 0.16[b]	6.7 ± 1[b]	83 ± 3[b]
7.9	38.8 ± 4.8[b]	28 ± 2[b]	70 ± 2[b]	2.05 ± 0.27[b]	16 ± 2[b]	81 ± 3[b]

[a] Results represent the $\bar{X}$ ± SE of 12 values for rats and 8 values for mice.
[b] Results significantly different from control values.
[c] Data suspect; see Section 3.

461

macrophage hyperplasia, and interstitial infiltrates observed microscopically qualitatively paralleled the biochemical and cytological changes. The large increases in BG activity in animals correlated well with the extent of inflammatory response seen. The magnitude of increase in total nucleated cells recovered from lungs of rats and mice exposed to NiO was less than might be expected from the severity of inflammatory response observed histopathologically. This may have been because inflammatory cells in foci of granulomatous inflammation are more adherent and resistant to removal by lavage than other macrophages.

Of interest was the development of fibrosis in mice, but not in rats exposed to $NiSO_4$ and Ni_3S_2, despite the fact that the collagen concentration in lung was significantly elevated in both species. The difference may be explained by the higher degree of collagen turnover in rats than in mice (as indicated by the higher concentration of collagenous peptides in BALF).

The development of an inflammatory response in rats and mice exposed to Ni_3S_2, $NiSO_4$, and NiO in this study is consistent with results we have obtained previously with these compounds in 12-day inhalation studies (Benson, et al., 1986, 1987; Dunnick et al., 1988). In the 12-day studies, the lowest aerosol concentrations producing an inflammatory response were 0.46 mg Ni/m^3 (rats) and 1.8 mg Ni/m^3 (mice) for Ni_3S_2, 0.77 mg Ni/m^3 (rats and mice, lowest concentration tested) for $NiSO_4$, and 7.8 mg $Ni/m;^3$ (rats) and 23.4 mg Ni/m^3 (mice) for NiO. It is significant that in the subchronic inhalation study, inflammation was produced at aerosol concentrations four to five times lower than those producing inflammation in the 12-day studies. This suggests that the products of nickel exposure concentration and time for the 12- and 13-week (65-exposure-day) studies were similar for the production of inflammation. The effect of subchronic exposure to NiO in the 13-week study was much greater than would have been predicted based on the results of earlier studies. The presence of granulomatous inflammatory response in rats and mice exposed to NiO for 13 weeks may have been related to the greater persistence of this compound within the lung, due to its low solubility, rather than to its inherent toxicity as compared to the other two nickel compounds.

The increases in LDH and BG in rats exposed to $NiSO_4$ (0.4 mg Ni/m^3), Ni_3S_2 ($\geq$0.4 mg Ni/m^3), and NiO ($\geq$2.0 mg Ni/m^3) were greater than those occurring in rats exposed to α-quartz (35 mg/m^3) 6 hr/day 5 days/week for 1 month (Henderson et al., 1985b). The response of TP was similar in nickel- and quartz-exposed rats, while the neutrophil response was greater for quartz- than for nickel-exposed rats. Biochemical changes in rats exposed to all nickel aerosol concentrations were greater than those occurring in rats exposed to coal combustion fly ash (35 mg/m^3) 5 days/week for 4 weeks. As with quartz, the ash-exposed rats had a greater influx of neutrophils (Henderson et al., 1985b). Quartz and, to a much lesser extent, fly ash were fibrogenic in rats 11 months after cessation of the inhalation exposures. The magnitudes of biochemical changes in rats and mice exposed to nickel compounds were much greater than those observed in rats and mice exposed to diesel exhaust (7.0 mg particles/m^3) 7 hr/day 5 days/week for up to 2 years. Fibrosis developed in rats but not in mice exposed to diesel exhaust (Henderson et al., 1988). This is in direct contrast to our results where mice, but not rats, exposed to NiSO$_4$ and Ni$_3$S$_2$ developed fibrosis.

The toxicity ranking of the three compounds was NiSO$_4$ > Ni$_3$S$_2$ > NiO, based on the magnitude of biochemical, cytological, and morphological changes produced in lung at equal nickel exposure concentrations. This ranking parallels the solubilities of these compounds in biological fluid, with NiSO$_4$ being readily soluble and Ni$_3$S$_2$ and NiO having dissolution half-times of 23 days and >11 years in rat serum, respectively (Sunderman et al., 1987). The ranking is also similar to that determined on the basis of results obtained following intratracheal instillation of the three compounds in rats (Benson et al., 1988). As expected, the magnitude of the biochemical, cytological, and morphological changes were greater in the case of the subchronic inhalation exposures than to the changes occurring upon intratracheal instillation of the compounds. This was most evident in the case of NiO, which did not cause an inflammatory response upon intratracheal instillation but did produce a significant response when administered by inhalation. This difference may again have been due to the persistence of NiO particles within the lung upon repeated exposures.

These results indicate that exposure to occupationally relevant aerosol concentrations of Ni$_3$S$_2$, NiSO$_4$, or NiO can produce toxic effects in lung. Changes in biochemical indicators of lung damage measured in BALF paralleled the nature and incidence of morphological changes. The toxicities of these three compounds closely parallel their solubilities in biological fluids.

ACKNOWLEDGMENTS

The present research was conducted under interagency agreement YO1-ES-30108 between the U.S. Department of Energy (Contract No. DE-AC04-76EV01013) and the National Institute of Environmental Sciences as part of the National Toxicology Program. The facilities used for this research were fully accredited by the American Association for the Accreditation of Laboratory Animal Care. The authors gratefully acknowledge J. Holmes, F. Straus, J. Kim, J. Waide, V. Sanchez, and C. Dickey for their excellent technical assistance.

REFERENCES

American Conference of Governmental Industrial Hygenists (ACGIH) (1986). *Threshold Limit Values for 1986.* ACGIH, Cincinnati, OH.

Benson, J. W., Henderson, R. F., McClellan, R. O., Hanson, R. L., and Rebar, A. H. (1986). Comparative acute toxicity of four nickel compounds to F344 rat lung, *Fundam. Appl. Toxicol.* **7**, 340–347.

Benson, J. M., Carpenter, R. L., Hahn, F. F., Haley, P. J., Hanson, R. L., Hobbs, C. H., Pickrell, J. A., and Dunnick, J. K. (1987). Comparative inhalation toxicity of ickel subsulfide to F344/N rats and B6C3F₁ mice exposed for 12 days, *Fundam. Appl. Toxicol.* **9**, 251–265.

Benson, J. M., Burt, D. G., Carpenter, R. L., Eidson, A. F., Hahn, F. F., Haley, P. J., Hanson, R. L., Hobbs, C. H., Pickrell, J. A., and Dunnick, J. K. (1988). Comparative inhalation toxicity of nickel sulfate to F344/N rats and B6C3F₁ mice exposed for 2 days, *Fundam. Appl. Toxicol.* **10**, 164–178.

Bingham, E., Barkley, W., Zerwas, M., Stemmer, K., and Taylor, P. (1972). Response of alveolar macrophages to metals. I. Inhalation of lead and nickel, *Arch. Environ. Health* **25**, 406–414.

Cheng, Y. S., Barr, E. B., Benson, J. M., Damon, E. G., Medinsky, M. A., Hobb, C. H., and Goehl, T. J. (1988). Evaluation of a real time aerosol monitor (RAM-S) for inhalation studies. *Fund. Appl. Toxicol.* **10**, 321–328.

Dunn, O. J. (1964). Multiple comparisons using rank sums, *Technometrics* **6**, 241–252.

Dunnick, J. K., Benson, J. M., Hobbs, C. H., Hahn, F. F., Cheng, Y. S., and Eidson, A. F. (1988). Comparative toxicity of nickel oxide, nickel sulfate hexahydrate, and nickel subsulfide after 12 days of inhalation exposure to F344/N rats and B6C3F₁ mice. *Toxicology* **50**, 145–156.

Grant, R. A. (1965). Estimation of hydroxyproline by the autoanalyzer, *J. Clin. Pathol.* **17**, 685–686.

Haley, P. J., Bice, D. E., Muggenburg, B. A., Hahn, F. F., and Benjamin, S. A. (1987). Immunopathologic effects of nickel subsulfide on the primate pulmonary immune system. *Toxicol. Appl. Pharmacol.* **88**, 1–12.

Henderson, R. F., Benson, J. M., Hahn, F. F., Hobbs, C. H., Jones, R. K., Mauderly, J. L., McClellan, R. O., and Pickrell, J. A. (1985a). New approaches for the evaluation of pulmonary toxicity: bronchoalveolar lavage fluid analysis, *Fundam. Appl. Toxicol.* **5**, 451–458.

Henderson, R. F., Hobbs, C. H., Hahn, F. F., Benson, J. M., Pickrell, J. A., and Silbaugh, S. A. (1985b). A comparison of *in vitro* and *in vivo* toxicity of mineral dusts. In E. G. Beck and J. Bigman (Eds.), *In Vitro Effects of Mineral Dusts*. Springer-Verlag, Berlin, NATO ASI Series, Vol. G3, pp. 521–527.

Henderson, R. F., Pickrell, J. A., Jones, R. K., Sun, J. D., Benson, J. M., Mauderly, J. L., and McClellan, R. O. (1988). Response of rodents to inhaled diluted diesel exhaust: biochemical and cytological changes in bronchoalveolar lavage fluid and in lung tissue. *Fundam. Appl. Toxicol.*, **11**, 546–567.

Mastromatteo, E. (1967). Nickel: a review of its occupational health aspects, *Occup. Med.* **9**, 127–136.

Neuman, R. E., and Logan, M. A. (1950). The determination of hydroxyproline, *J. Biol. Chem.* **184**, 299–306.

Ottolenghi, A. D., Haserman, J. K., Payne, W. W., Falk, H. L., and MacFarland, H. N. (1974). Inhalation studies of nickel sulfide in pulmonary carcinogenesis of rats. *J. Natl. Cancer Inst.* **54**, 1165–1172.

Shirley, E. (1977). A nonparametric equivalent of William's test for contrasting increasing dose levels of a treatment, *Biometrics* **33**, 386–389.

Sunderman, W. F., Jr., Hopfer, S. M., Knight, J. A., McCully, K. S., Cecutti, A. G., Thornhill, P. G., Conway, K., Miller, C., Patierno, S. R., and Costa, M. (1987). Physiochemical characteristics and biological effects of nickel oxides carcinogenesis, *Carcinogenesis* **8**, 305–313.

Takenaka, S., Hochrainer, D., and Oldiges, H. (1985). Alveolar proteinosis induced in rats by long-term inhalation of nickel oxide. In S. S. Brown and F. W. Sunderman, Jr. (Eds.), *Progress in Nickel Toxicology*. Blackwell Scientific Publications, Oxford, pp. 89–92.

Wehner, A. P., and Craig, D. K. (1972). Toxicology of inhaled NiO and CiO in golden hamsters, *Am. Ind. Hyg. Assoc. J.* **33,** 146–155.

Wehner, A. P., Busch, R. H., Olson, R. J., and Craig, D. K. (1975). Chronic inhalation of nickel oxide and cigarette smoke by hamsters, *Am. Ind. Hyg. Assoc. J.* **36,** 801–810.

Williams, D. A. (1986). A note on Shirley's nonparametric test for comparing several dose levels with a zero-dose control, *Biometrics* **42,** 183–186.

35

CHRONIC EFFECTS OF INTRATRACHEALLY INSTILLED NICKEL-CONTAINING PARTICLES IN HAMSTERS

H. Muhle, B. Bellmann, S. Takenaka, R. Fuhst, and U. Mohr

Fraunhofer-Institut für Toxikologie und Aerosolforschung, Hannover, Germany

F. Pott

Medizinisches Institut für Umwelthygiene, Düsseldorf, Germany

Nickel and Human Health: Current Perspectives, Edited by Evert Nieboer and Jerome O. Nriagu.
ISBN 0-471-50076-3   © 1992 John Wiley & Sons, Inc.

1. INTRODUCTION

Only limited information is available concerning whether a potential carcino-genic risk is associated with exposure to particles of nickel alloys. In a chronic study, Syrian golden hamsters were treated by intratracheal instillations to evaluate the carcinogenic potency of grinding dusts derived from stainless steel and particles from the mineral pentlandite, both containing nickel. Nickel powder and nickel subsulfide (α-Ni_3S_2) were used as reference materials. In a previous study the acute and subchronic effects were investigated to determine the maximum tolerated dose (Muhle et al., 1986).

2. MATERIALS AND METHODS

2.1. Test Materials

A characterization of test materials nickel powder, nickel subsulfide, pen-tlandite, and grinding dusts is given in Table 35.1. The diameters of particles were determined by electron microscopy. For irregular shaped particles (grinding dusts), the median geometric diameter was estimated whereas for more regular shaped particles the mass median diameter was calculated. From these data the values of the mass median aerodynamic diameters were estimated.

A respirable fraction of stainless steel grinding dust was collected in a factory in which metal pieces were ground. For the separation of coarse and fine dust a high-volume heavy-grain-load impaction sampler was used (Koch et al., 1986; Muhle et al., 1986). Animals were treated with two types of stainless steel grinding dusts, one originating from a normal 18/10 chromium-nickel steel and the other originating from a ferritic chromium steel. The latter was used to investigate biological effects of particles with a very low nickel content (negative control). The composition of the grinding dusts is presented in Table 35.2.

The chromium–nickel steel dust was also investigated by powder diffraction analysis. It was composed of a phase mixture of austenite and a cubic spinel $(FeCrNi)_3O_4$. Additionally, about 100 single randomly selected particles were investigated in a scanning electron microscope by energy-dispersive X-ray analysis. This investigation showed no evidence for the presence of nickel or nickel oxide in pure form. The range of the nickel content per particle was 4–10%.

The chromium steel dust was composed of a phase mixture of ferrite and a cubic spinel $(FeCrAl)_3O_4$.

2.2. Animal Model

Syrian golden hamsters were chosen as experimental animals because in this species inflammatory reactions in the lung are less pronounced after intra-

Table 35.1 Characterization of Test Materials

Test Material	Median Geometric Diameter (μm)	Mass Median Diameter (μm)	Estimated Mass Median Aerodynamic Diameter (μm)	Chemical Characterization	Source [a]
Ni Powder	—	3.1	9	99.9% Ni	INCO, Clydach, GB, Experimental Nickel Powder No. 384
Ni_3S_2	—	2.2	5	αNi_3S_2; 72.9% Ni, 27.1% S	INCO; ground sample
Pentlandite	—	2.2	5	34.3% Ni, 32.2% S, 28.8% Fe, 0.028% Cu, 0.05% Al_2O_3	INCO
Cr–Ni stainless steel grinding dust	1.1	—	3–5	See Table 35.2	Grinding machine
Cr stainless steel grinding dust	0.8	—	3–5	See Table 35.2	Grinding machine

[a] INCO, International Nickel Company.

Table 35.2   Chemical Analysis of Stainless Steel Grinding Dusts

Constituent	Cr–Ni Steel (%)	Cr Steel (%)
Total Fe	59.2	68.0
Fe met[a]	17.2	28.5
Cr	13.92	12.87
Cr^{6+}	<0.01	0.04
Ni	6.79	0.54
Mn	1.0	0.3
As	0.016	0.018
S	0.13	0.12
Al	2.14	1.99
Si	0.65	0.28
Zn	0.03	0.04
Pb	<0.01	<0.01
Cd	<0.01	<0.01
C	3.90	3.43
Be	<0.001	<0.001
Co	0.10	0.03
Ti	0.03	0.02
Mo	0.22	0.09

[a] Fraction of iron crystallized in a metallic iron lattice.

tracheal instillation compared to rats (Della Porta et al., 1958; Saffiotti et al., 1968; Nettesheim, 1972). Persistent inflammatory reactions may aggravate other chronic lung diseases. A further reason for selecting the hamster was the low spontaneous tumor rate of 0.1–0.5% (Saffiotti et al., 1968; Dontenwill et al., 1973; Mohr et al., 1984). Syrian golden hamsters (Cpb-ShGa 51) from TNO, The Netherlands, were used. At the start of the treatment, animals were 10–12 weeks old and weighed about 80 g. Hamsters were randomized.

2.3.   Sacrifices and Histopathology

The observation period was 26 months for females and 30 months for males. The median lifetime is presented with the results. At termination of the study, hamsters were sacrificed by an overdose of ether. Tissues and organs were fixed in 10% neutral buffered formalin and processed for routine histology (H&E stain). Additionally, in animals bearing lung tumors the main organs were examined histopathologically to exclude metastatic lung tumors.

3.   RESULTS

3.1.   Retention of Nickel-containing Particles in Lung

In satellite groups at 14 days and 6 and 12 months after the last intratracheal instillation two hamsters of each sex were sacrificed. Exempted from this protocol was the group treated by chromium steel dust. Lungs were ashed, and the nickel content was analyzed. Table 35.3 shows the dose used and the retained mass, calculated from the nickel content of the instilled particles.

For calculation of the clearance kinetics of the particles an exponential curve was fitted to the values of the three retention dates. The half-time of the pulmonary nickel clearance was estimated as 152 days for nickel powder, <14 days for Ni_3S_2, 120 days for pentlandite, and about 450 days for chromium–nickel stainless steel dust. The number of animals for this evaluation was relatively low. Values of male and female hamsters were pooled.

3.2. Carcinogenicity Study

3.2.1. Study Design

Thirty hamsters per sex were treated by intratracheal instillation a total of 12 times at 14-day intervals and with 0.15 ml per treatment. The particles were suspended in saline. The single and the cumulative doses are shown in Table 35.4.

3.2.2. Clinical Observations

The number of animals that died during the treatment period of 170 days is listed in Table 35.4. Most of these hamsters died during the period corresponding to the first two treatments. The reason for the high mortality rate in the chromium stainless steel dust group could be particle agglomeration, which may have caused partial airway obstruction. To prevent this effect, a more intensive ultrasonic treatment of the particle suspension was performed in the subsequent instillations. The mortality rate during the treatment period was larger in males (uneven-group numbers) than in females (even-group numbers). Beside airway obstruction, acute toxic effects may have caused some mortality. Fourteen days after the first instillation additional hamsters were added to those groups indicating losses as a result of the treatment (see Table 35.4). The reserve hamsters came from the same delivery as the original animals. Animals that died during the treatment period were not included in assessing the carcinogenic risk because of their short life span and were excluded from the calculation of the median lifetime. The listed lifetime refers to the period since treatment was initiated. For the total lifetime, the span of 10 weeks before treatment has to be added.

Body weights of female hamsters in group 6 (pentalandite) and group 10 (chromium–nickel stainless steel dust) were lower than those of the control groups at about day 500 of the study. Body weights of other groups were comparable to those of controls.

3.2.3. Results of Histopathological Investigations

Results of histopathological investigations of the hamsters surviving the treatment period of 170 days are listed in Table 35.5. After the treatment to nickel subsulfide, particle deposition was detected in only a few animals. This is due to the low amount of mass used for the treatment and the relatively

Table 35.3 Results of Retention Measurements of Nickel in Lungs

Test Material	Cumulative Dose (mg/animal)	Lung Retention 14, 182, and 365 Days after Last Treatment (mg/lung)[b]	Number of Investigated Hamsters	Estimated Half-Time of Nickel Clearance (days)
Ni powder	9.6 (12 × 0.8 mg)	0.367 ± 0.172	3	152
		0.195 ± 0.094	4	
		0.075 ± 0.029	4	
Ni$_3$S$_2$	1.2 (12 × 0.1 mg)	0.051 ± 0.0196	3	<14
		a	3	
		a	3	
Pentlandite	36 (12 × 3 mg)	13.9 ± 4.1	4	120
		2.2 ± 1.6	4	
		1.8 ± 1.7	4	
Cr/Ni stainless steel dust	36 (12 × 3 mg)	21.3 ± 4.5	4	481
		14.5 ± 6.2	4	
		12.8 ± 2.9	4	
Cr/Ni stainless steel dust	108 (12 × 9 mg)	71.8 ± 25.8	4	415
		51.0 ± 19.8	4	
		39.9 ± 14.5	4	

[a]Below the detection limit.
[b]Mean and standard deviation.

Table 35.4 Mortality Development and Median Lifetime in Chronic Study

Test Material	Cumulative Dose (mg/animal)	Group Number[a]	Number of Dead Hamsters Added after 14 Days	Number of Hamsters by End of Treatment	Number of Hamsters after Treatment Period	Median Lifetime (weeks)
Ni Powder	9.6 (12 × 0.8 mg)	1	2	5	27	111
		2	1	0	31	100
Ni_3S_2	1.2 (12 × 0.1 mg)	3	3	3	30	114
		4	2	0	32	101
Pentlandite	36 (12 × 3 mg)	5	1	2	29	109
		6	2	1	31	91
Cr–Ni stainless steel dust	36 (12 × 3 mg)	7	2	1	31	130
		8	2	0	32	102
Cr–Ni stainless steel dust	108 (12 × 9 mg)	9	5	4	31	112
		10	2	1	31	96
Cr stainless steel dust	108 (12 × 9 mg)	11	10	12	28	119
		12	2	4	28	90
Vehicle control (NaCl solution)	—	13	0	0	30	117
		14	0	0	30	103

[a]Groups with odd number, males; groups with even number, females.

Table 35.5 Histopathological Changes in the Lungs of Hamsters Exposed to Intratracheal Nickel Compounds

Compounds	Group Number	Sex	Number of Hamsters Examined Histologically	Number of Hamsters with Stated Findings[a]																
				Deposition of Pigments					Granulomatous Changes				Bronchiolo-alveolar Hyperplasia				Squamous Metaplasia			Primary Tumors
				T	(1+	2+	3+	4+)	T	(1+	2+	3+)	T	(1+	2+	3+)	T	(1+	2+)	
Ni Powder	1	M	26	9	3	6			0				15	8	7		1	1		
	2	F	30	21	10	11			0				20	7	11	2	4	1	3	1[b]
Ni₃S₂	3	M	30	3		3			0				2	1	1		1	1		
	4	F	32	0					0				0				0			
Pentlandite	5	M	29	27			27		15	7	8		22	13	9		0			
	6	F	31	31			31		20	3	17		18	10	8		0			1[c]
Cr/Ni Stainless Steel	7	M	30	30		1	29		2	2			12	10	2		0			
	8	F	32	32			32		12	4	8		10	9	1		1	1		
Cr/Ni Stainless Steel	9	M	31	31			9	22	26	3	22	1	15	14	1		1	1		
	10	F	31	31			14	17	27	4	21	2	10	8	2		0			
Cr Stainless Steel	11	M	28	28			26	2	27	11	15	1	24	4	20		0			
	12	F	28	28			21	7	28	1	27		14	10	4		0			
NaCl Solution	13	M	30	0					0				5	4	1		1	1		
	14	F	30	1	1				0				1	1			0			

[a]T = Total number; breakdown is given in parentheses (1+, very slight; 2+ slight; 3+, moderate; 4+, severe)

[b]Adenosquamous carcinoma

[c]Bronchiolo-alveolar adenoma

474

fast pulmonary clearance of this material. In all other groups, distinct particle deposits were found, with the smallest amounts occurring in the group treated with nickel powder. In this latter group, the administered mass was also relatively low.

Granulomatous changes, which reflect inflammatory processes, were only observed in groups treated with pentlandite and stainless steel dusts. These were the groups with the highest lung burden of particles. Bronchiolo-alveolar hyperplasia was detected for the treatments with nickel powder, pentlandite, and stainless steel dusts, while squamous metaplasia was slightly increased in the nickel group. Only two primary tumors were observed: one adeno-squamous carcinoma in the group exposed to nickel powder and one bronchiolo-alveolar adenoma after pentlandite exposure. This indicates that no statistical increase in tumor rates was found.

In the groups to which nickel powder and nickel subsulfide were administered, the larynx, trachea, and kidneys were also examined histologically. There were no indications of treatment-related effects.

4. DISCUSSION

In this study, the calculation of the half-times for the pulmonary clearance of the applied dusts was based on the retention data of nickel. For nickel subsulfide the half-time was relatively short, being less than 14 days. This value was of the same magnitude as the one published by Fisher et al. (1986) and Finch et al. (1987), namely 5 days for mice. The fast clearance of nickel subsulfide could be one of the reasons why little response was detected in this treatment group.

The half-times of the pulmonary clearance of nickel powder and pentlandite were 152 and 120 days, respectively. The corresponding value for chromium–nickel stainless steel dusts was about 450 days. The significant retardation of lung clearance after treatment with stainless steel dust is probably caused by an overloading of the alveolar clearance. This effect is produced by lung burdens higher than 1 mg/g lung and is reported also for other dusts like plastic powder (Muhle, 1988, 1990). It is therefore not specific for stainless steel grinding dusts.

None of the treatment groups had a statistically significant number of tumors. Five cases of squamous cell metaplasia and one case of an adeno-squamous carcinoma after nickel treatment indicate that nickel powder has only a very weak carcinogenic effect in hamsters, if at all. After treatment with pentlandite, one bronchiolo-alveolar adenoma was observed in 60 animals. A ranking of the relative carcinogenic potential of nickel, nickel compounds, and nickel alloys is not possible as a result of this study.

No inflammatory reactions were observed after administration of nickel powder and Ni_3S_2. On the other hand, inflammation was visible after treatment with pentlandite, chromium–nickel steel, and chromium steel dusts. This

Table 35.6 Carcinogenicity of Various Substances in Human, Rat, and Hamsters[a]

Test Material	Human		Rat			Hamster		
	Carcinogenicity[b]	Author	Lung Tumors[c]	Route[e]	Author	Lung Tumors[d]	Route[e]	Author
Nickel powder	S[f]	IARC (1987)	+	i.tr.	Pott et al. (1987)	−	i.tr.	this study
Nickel subsulfide	S[f]	IARC (1987)	+	inh.	Ottolenghi et al. (1974)	−	i.tr.	this study
			+	i.tr.	Pott et al. (1987)			
Cadmium	L	IARC (1987)	+	inh.	Takenaka et al. (1983)	−	inh.	Heinrich et al. (1988)
			+	i.tr.	Pott et al. (1987)			
α-Quartz	L	IARC (1987)	+	inh.	Holland et al. (1986)	−	i.tr.	Holland et al. (1983)
			+	inh.	Muhle et al. (1989)	−	i.tr.	Renne et al. (1985)
Diesel engine exhaust	L	IARC (1989)	+	inh.	Brightwell et al. (1986)	−	inh.	Brightwell et al. (1986)
			+	inh.	Heinrich et al. (1986b)	−	inh.	Heinrich et al. (1986b)
			+	inh.	Mauderly et al. (1987)			
PAH-containing exhaust	S	IARC (1984)	+	inh.	Heinrich et al. (1986a)	−	inh.	Heinrich (1988)

[a]Selected results which show clear differences between rats and hamsters after similar exposure.
[b]S, sufficient evidence for carcinogenicity; L, limited evidence for carcinogenicity.
[c]+: Induction of lung tumors.
[d]−: No statistically significant increase of tumor rate.
[e]i.tr., intratracheal instillation; inh., inhalation.
[f]Chemical characterization by IARC: nickel and nickel compounds.

effect is mostly a nonspecific reaction of the type commonly found after application of many particulates with low solubility. The particles are phago-cytosed by alveolar macrophages and induce inflammatory granulomatous changes (Saffiotti, 1986).

Since the beginning of the design of this study in 1982, various other carcinogenicity studies were completed in which hamsters and rats were investigated concurrently (see Table 35.6). These studies show that after inhalation or intratracheal instillation of various test materials, rats did develop lung tumors but not hamsters. The data in Table 35.6 represent only a small section of the numerous carcinogenicity studies. A more complete overview has been compiled by Pepelko (1984).

In humans, carcinogenic effects are associated with exposure to coke oven exhaust. For this material, the rat is obviously a more suitable species for investigating the carcinogenic potency on humans than the hamster. For other materials (diesel motor exhaust, cadmium, and α-quartz), there is limited evidence for carcinogenicity in humans (IARC, 1987). In the rat, these materials and also nickel and some nickel compounds induced lung tumors. Consequently, the absence of statistically significant carcinogenic effects in hamsters after treatment with nickel-containing particles does not exclude a carcinogenic effect of the tested substances in humans. A better knowledge on the mechanism of nickel-induced tumors in various species will provide a better data base for risk extrapolation.

ACKNOWLEDGMENTS

This study was sponsored by the Nickel Producers Environmental Research Association, Industrieverband Schneidwaren und Bestecke e.V., and the Industrieverband Haushalt-, Küchen-, und Tafelgeräte e.V., West Germany.

We thank Krupp Stahl AG, Bochum, for chemical analysis of the dust samples.

REFERENCES

Brightwell, J., Fouillet, X., Cassano-Zoppi, A.-L., Gatz, R., and Duchosal, F. (1986). Neoplastic and functional changes in rodents after chronic inhalation of engine exhaust emissions. In N. Ishinishi, A. Koizumi, R. O. McClellan, and W. Stöber (Eds.), *Carcinogenic and Mutagenic Effects of Diesel Engine Exhaust*. Elsevier, New York, pp. 471–488.

Della Porta, G., Kolb, L., and Shubik, H. P. (1958). Induction of tracheobronchial carcinomas in the Syrian golden hamster. *Cancer Res.* **18**, 592–597.

Dontenwill, W., Chevalier, H. J., Harke, H. P., Lafrenz, U. Reckzeh, G., and Schneider, B. (1973). Spontantumoren des syrischen Goldhamsters. *Z. Krebsforsch.* **80**, 127–158.

Finch, G. L., Fisher, G. L., and Hayes, T. L. (1987). The pulmonary effects and clearance of intratracheally instilled Ni_3S_2 and TiO_2 in mice. *Environ. Res.* **42**, 83–93.

Fischer, G. L., Chrisp, C. E., and McNeill, D. A. (1986). Lifetime effects of intratracheally instilled nickel subsulfide on B6C3F1 mice. *Environ. Res.* **40**, 313–320.

Heinrich, U. (1988). Personal communication.

Heinrich, U, Pott, F., and Rittinghausen, S. (1986a). Comparison of chronic inhalation effects in rodents after long-term exposure to either coal oven flue gas mixed with pyrolized pitch or diesel engine exhaust. In N. Ishinishi, A. Koizumi, R. O. McClellan, and W. Stöber (Eds.), *Carcinogenic and Mutagenic Effects of Diesel Engine Exhaust.* Elsevier Science (Biomed. Div.), New York, pp. 441–457.

Heinrich, U., Muhle, H., Takenaka, S., Ernst, H., Fuhst, R., Mohr, U., Pott, F., and Stöber, W. (1986b). Chronic effects of the respiratory tract of hamsters, mice and rats after long-term inhalation of high concentrations of filtered and unfiltered diesel engine emissions. *J. Appl. Toxicol.* **6**, 383–396.

Heinrich, U., Peters, L., Rittinghausen, S., Ernst, H., Dasenbrock, C., and Mohr, U. (1988). Long-term inhalation exposure of Syrian golden hamster and NMRI-mice to various cadmium compounds. Presented at the 4th IUPAC Cadmium Workshop, September 11–13, 1988 in Schmallenberg-Grafschaft, Germany.

Holland, L. M., Gonzales, M., Wilson, J. S., and Tillery, M. I. (1983). Pulmonary effects of shale dusts in experimental animals. In W. L. Wagner, W. N. Rom, and J. A. Merhants (Eds.), *Health Issues Related to Metal and Non-Metalic Mining,* Ann Arbor Sciences, Ann Arbor, pp. 485–496.

Holland, L. M., Wilson, J. S., Tillery, M. I., and Smith, D. M. (1986). Lung cancer in rats exposed to fibrogenic dusts. In D. Goldsmith, D. Winn, and C. Shy (Eds.), *Silica, Silicosis and Cancer.* Praeger Scientific, New York, pp. 267–279.

International Agency for Research on Cancer (IARC) (1984). *Polynuclear Aromatic Compounds.* Part 3, Monograph. Vol. 34, IARC, Lyon, France.

International Agency for Research on Cancer (IARC) (1987). *Monograph on the Evaluation of Carcinogenic Risks to Humans,* Supplement 7. IARC, Lyon, France. p. 440.

International Agency for Research on Cancer (IARC) (1989). *Diesel and Gasoline Engine Exhausts and Some Nitroarenes,* Monograph. Vol. 46, IARC, Lyon, France.

Koch, W., König, H. P., Lödding, H., and Holländer, W. (1986). Sampling and characterization of metal grinding dusts. In *Aerosols: Formation and Reactivity.* Pergamon Press, Oxford, pp. 1208–1211.

Mauderly, J. L., Jones, R. K., Griffith, W. C., Henderson, F. R., and McClellan, R. O. (1987). Diesel exhaust is a pulmonary carcinogen in rats exposed chronically by inhalation. *Fundam. Appl. Toxicol.* **9**, 1–13.

Mohr, U., Heinrich, U., Fuhst, R., Ketkar, M., and Muhle, H. (1984). The suitability of the Syrian golden hamster for chronic inhalation experiments. In P. Grosdanoff, R. Bass, U. Hackenberg, D. Henschler, D. Müller, and H.-J. Klimish (Eds.), *Problems of Inhalatory Toxicity Studies.* MMV Medizin Verlag, München, pp. 145–152.

Muhle, H., Koch, W., and Bellmann, B. (1986). Acute and subchronic effects of intratracheally instilled nickel containing particles in hamster. In R. M. Stern, A. Berlin, A. C. Fletcher, and J. Jörvisalo (Eds.), *Health Hazards and Biological Effects of Welding Fumes and Gases.* Excerpta Medica, Amsterdam, pp. 337–340.

Muhle, H., Creutzenberg, O., Bellmann, B., Heinrich, U., Mermelstein, R. (1990). Dust overloading of lungs: investigations of various materials, species differences, and irreversibility of effects. *J. Aerosol Med.* **3**, S111–S128.

Muhle, H., Bellmann, B., and Heinrich, U. (1988). Overloading of lung clearance during chronic exposure of experimental animals to particles. *Ann. Occup. Health,* **32**, pp. 141–147 Supplement 1.

Muhle, H., Takenaka, S., Mohr, U., Dasenbrock, C., Mermelstein, R. (1989). Lung tumor induction upon long-term low-level inhalation of crystalline silica *Am. J. Ind. Med.* **15**, 343–346.

Nettesheim, P. (1972). Respiratory studies with the Syrian golden hamster: A review. *Prog. Exp. Tumor Res.* **16,** 185–200.

Ottolenghi, A. D., Haseman, J. K., Payne, W. W., Falk, H. L., and MacFarland, H. N. (1974). Inhalation studies of nickel subsulphide in pulmonary carcinogenesis of rats. *J. Nat. Cancer Inst.* **54,** 1165–1172.

Pepelko, W. E. (1984). Experimental respiratory carcinogenesis in small laboratory animals. *Environ. Res.* **33,** 144–188.

Pott, F., Ziem, U., Reiffer, F.-J., Huth, F., Ernst, H., and Mohr, U. (1987). Carcinogenicity studies on fibres, metal compounds, and some other dusts in rats. *Exp. Pathol.* **32,** 129–152.

Renne, R. A., Eldridge, S. R., Lewis, T. R., and Stevens, P. L. (1985). Fibrogenic potential of intratracheally instilled quartz, ferric oxide, fibrous glass, and hydrated alumina in hamsters. *Toxicol. Pathol.* **13,** 306–314.

Saffiotti, U. (1986). The pathology induced by silica in relation to fibrogenesis and carcinogenesis. In D. F. Goldsmith, D. M. Winn, and C. M. Shy (Eds.), *Silica, Silicosis and Cancer*, Praeger Publishers, New York, pp. 287–307.

Saffiotti, U., Cefis, F., and Kolb, L. H. (1968). A method for experimental induction of bronchogenic carcinoma. *Cancer Res.* **28,** 104–124.

Takenaka, S., Oldiges, H., König, H., Hochrainer, D., and Oberdörster, G. (1983). Carcinogenicity of cadmium chloride aerosols in W rats. *J. Nat. Cancer Inst.* **70,** 367–371.

36

DURABILITY OF VARIOUS KINDS OF NICKEL COMPOUNDS IN RATS ADMINISTERED BY INHALATION

Isamu Tanaka, Sigeko Ishimatsu, Yasushi Kodama, Joji Haratake, Akio Horie*, and Kenzaburo Tsuchiya

University of Occupational and Environmental Health, Japan, Yahatanishi, Kitakyushu, 807 Japan.

Shuji Cho

Faculty of Home Life Science, Fukuoka Women's University, Fukuoka 83, Japan

1. **Introduction**
2. **Methods**
3. **Results**
    - 3.1. Solubility of Nickel Compound Aerosol
    - 3.2. Body and Organ Weights after Exposure
    - 3.3. Nickel Concentrations in Rat Organs and in Whole Blood after Exposure
    - 3.4. Body and Organ Weights after Clearance Period
    - 3.5. Nickel Concentration in Rat Organs and in Blood after Clearance Period

* Deceased

Nickel and Human Health: Current Perspectives, Edited by Evert Nieboer and Jerome O. Nriagu.
ISBN 0-471-50076-3 © 1992 John Wiley & Sons, Inc.

---

1. INTRODUCTION

Nickel sulfide roasting fumes and dust are recognized as having carcinogenic or cocarcinogenic potential and have been assigned a threshold limit value (TLV) of 1 mg/m^3 (ACGIH, 1987). The IARC (Saracci, 1981; IARC, 1987) also considers that nickel and nickel compounds as a group should be regarded as potential occupational carcinogens. Nevertheless, the identity of the nickel compounds that may induce cancers of the nasal cavities and lungs in nickel refinery workers is uncertain. Therefore, principal attention has been focused on possible differences in carcinogenic activities of specific nickel compounds to which workers have been exposed and upon the identification of physical, chemical, or biological factors that could be used to predict the carcinogenicity of nickel species. Considering these possible differences and the main exposure routes of the workers in an occupational environment, an inhalation study is obviously important in order to find out the specific nickel compounds that induce cancers of the nasal cavities and lungs.

The penetration and deposition of particles in the nasal cavities and lungs and the clearance of those particles from the lungs are of primary importance in the development of a tumor in the case of chronic exposure.

Tanaka et al. (1985, 1986, 1988a,b) investigated the biological half-time of green nickel oxide [NiO(G)] and in nickel monosulfide (amorphous) aerosols deposited in rat lungs as well as the distribution of nickel in rats by altering the exposure time, clearance period, exposure concentration, and aerosol diameter.

This chapter describes the differences of nickel distribution in rat organs (lungs, liver, kidneys, and spleen) due to various kinds of nickel compounds administered by aerosol inhalation.

2. METHODS

In the experiment described, the nickel compounds used (Wako Pure Chemical, Japan) were amorphous nickel monosulfide [NiS(A)], nickel metal [Ni(M)], and Ni(III) oxide [Ni$_2$O$_3$] and green nickel oxide [NiO(G)] (produced by Soekawa Chemical Co. Japan). Respirable aerosols were transported to exposure chambers by a fluidized-bed-type dust generator (Tanaka and Akiyama, 1984). Crystalline or amorphous structure was verified by powder X-ray diffraction. The exposure systems have been fully reported elsewhere (Tanaka et al., 1982, 1985; Kodama et al., 1985). The aerosol concentration of the nickel compounds in the exposure chamber was monitored continuously by a light-scattering method and was measured gravimetrically at daily intervals by the isokinetic suction of air through a glass fiber filter. The size distribution of the aerosol in the exposure chambers was determined by using an Andersen cascade impactor.

The exposure period, exposure concentrations, mass median aerodynamic diameter, and geometric standard deviation are shown in Table 36.1.

Wistar male rats were exposed to aerosols of the mentioned nickel compounds for 7 hr/day, 5 days/week for 1–3 months. Controls were exposed to clean air in adjacent, identical chambers, under similar flow (150 L/min), temperature (24 ± 1°C), and humidity (50 ± 5%) conditions.

Some rats were sacrificed 24 hr after the termination of the final exposure sequence and others were exposed for 1–3 months and then kept for a clearance period of 6–12 months before sacrifice. The control rats were also sacrificed at the same time. The nickel contents in the organs (lungs, liver, spleen, and kidneys) and in whole blood were measured by flameless atomic absorption spectrometry (Kodama et al., 1985).

3. RESULTS

3.1. Solubility of Nickel Compound Aerosol

All nickel compounds were equilibrated for 1 week at 37°C employing a magnetic stirrer with distilled water and a saline solution (100 mg/20 mL). The solubilities of nickel compounds are shown in Table 36.2.

3.2. Body and Organ Weights after Exposure

Table 36.3 shows the body and organ weights of rats used after the termination of the inhalation period. There was no significant difference in body and organ weights, except for lung weights, between exposed rats and controls. The wet-lung weights in the exposed rats were heavier than those in the controls.

3.3. Nickel Concentrations in Rat Organs and in Whole Blood after Exposure

Table 36.4 summarizes the nickel concentrations in rat organs and in blood after the 1- or 3-month exposure period.

The nickel concentrations in exposed rat lungs were much higher than those in controls for all nickel compounds. Nickel concentrations in kidneys and liver were also significantly higher than those in the controls for NiS(A),

Table 36.1 Experimental Conditions

| Parameter | NiO(G) | | NiS(A) | | Ni(M) | | Ni$_2$O$_3$, |
	Run 1	Run 2	Run 3	Run 4	Run 5	Run 6	Run 7
Exposure and clearance period,[a] months	1	1–12	1	1–6	1	1–12	3
Total exposure time,[b] hr	147	140	147	147	140	140	470
Exposure concentration, mg/m^3 ± SD	7.0 ± 3.9	8.0 ± 1.8	8.8 ± 2.8	8.8 ± 2.8	7.9 ± 1.3	7.9 ± 1.3	11.0 ± 2.9
MMAD[c]	1.2 (2.2)	1.2 (2.2)	3.9 (2.6)	3.9 (2.6)	5.8 (2.0)	5.8 (2.0)	4.0 (1.8)

[a]1, 1-month exposure; 3, three-month exposure; 1–6 and 1–12 denote a clearance period of 6 or 12 months after 1-month exposure.
[b]7 hr/day, 5 days/week.
[c]Numbers in parentheses are geometric standard deviations.

Table 36.2 Solubility (μ/mL)

	NiO(G)	NiS(A)	Ni(M)	Ni_2O_3
Distilled water	0.32	1990	0.13	2.54
Saline solution	2.32	7610	0.73	5.13

Ni(M), and Ni_2O_3. On the other hand, there were no significant differences in the case of exposure to NiO(G). Neither were there differences in the nickel concentration of the spleen between exposed rats and controls for all nickel compounds in this experiment. By contrast, the concentration differences in whole blood were statistically significant for NiS(A) and Ni(M) but not for Ni_2O_3 and NiO(G).

3.4. Body and Organ Weights after Clearance Period

Table 36.5 shows the body and organ weights of rats after the clearance observation interval of 6 or 12 months. There were no significant differences in body and organs weights, except for the lung. The wet-lung weights of the exposed rats exceeded those of the controls.

3.5. Nickel Concentration in Rat Organs and in Blood after Clearance Period

Table 36.6 presents the various nickel concentrations. The nickel concentration in the lungs remained elevated only in the case of NiO(G), for which most of the nickel remained after 12 months. Similarly, there were no differences in the nickel levels in other organs, except for liver in the NiO(G) case, for which a slight elevation was observed. The clearance for Ni_2O_3 was not evaluated.

4. DISCUSSION

4.1. Clearance Rates

In Table 36.4, it is seen that the lung clearance rate was different for the various nickel compounds. For NiS(A), Ni_2O_3, and Ni(M), the nickel may be assumed to be deposited and cleared simultaneously, since nickel was detected in liver, kidneys, and blood immediately after exposure. This is in contrast to NiO(G), for which nickel was detected only in the lung tissue after the exposure.

Table 36.7 presents the actual amount of nickel remaining in rat lung after the exposure and clearance periods. The residual nickel content after clearance was very small for Ni(M) or NiS(A). From this result, we deduce that the

Table 36.3 Body and Organ Wet Weight after Exposure (g ± SD)

Compound	Group[a]	N[b]	Body	Lung	Liver	Kidney	Spleen
NiO(G)	C	5	357 ± 19	1.2 ± 0.1	12.5 ± 0.7	2.02 ± 0.25	0.70 ± 0.06
(run 1)	E	5	354 ± 18	1.7 ± 0.2[c]	12.2 ± 1.2	1.91 ± 0.08	0.49 ± 0.07
NiS(A)	C	4	280 ± 16	1.3 ± 0.1	11.4 ± 0.9	2.03 ± 0.07	0.78 ± 0.06
(run 3)	E	5	265 ± 13	1.8 ± 0.2[c]	10.9 ± 0.7	2.05 ± 0.19	0.66 ± 0.7[d]
Ni(M)	C	6	264 ± 23	1.2 ± 0.1	8.2 ± 0.8	2.15 ± 0.23	0.72 ± 0.10
(run 5)	E	6	238 ± 21	1.9 ± 0.4[c]	7.3 ± 0.9	1.92 ± 0.27	0.60 ± 0.07
Ni₂O₃	C	3	447 ± 51	1.5 ± 0.1	11.4 ± 1.9	2.58 ± 0.29	0.69 ± 0.08
(run 7)	E	5	432 ± 32	3.3 ± 0.3[c]	10.6 ± 0.9	2.87 ± 0.12	0.84 ± 0.18

[a]C, Control; E, experimental.
[b]Number of rats sacrificed.
[c]Significance level: $P < 0.01$, as compared with controls by t-test.
[d]Significance level: $P < 0.05$, as compared with controls by t-test.

Table 36.4 Nickel Concentration in Organ after Exposure (µg/g organ wt ± SD)

Compound	Group[a]	Lung	Liver	Kidney	Spleen	Whole Blood[b]
NiO(G)	C	0.14 ± 0.03	0.05 ± 0.02	0.15 ± 0.04	0.21 ± 0.07	ND
(run 1)	E	577 ± 57[c]	0.06 ± 0.02	0.15 ± 0.05	0.23 ± 0.07	ND
NiS(A)	C	0.09 ± 0.05	0.03 ± 0.01	0.10 ± 0.04	0.25 ± 0.09	ND
(run 3)	E	1.14 ± 0.19[c]	0.09 ± 0.03[c]	0.96 ± 0.25[c]	0.25 ± 0.27	0.31 ± 0.06[c]
Ni(M)	C	0.24 ± 0.04	0.04 ± 0.3	0.06 ± 0.02	0.22 ± 0.06	ND
(run 5)	E	13.5 ± 4.5[c]	0.23 ± 0.16[d]	3.63 ± 2.19[d]	0.28 ± 0.11	0.37 ± 0.13[c]
Ni₂O₃	C	0.10 ± 0.04	0.02 ± 0.01	0.07 ± 0.03	0.11 ± 0.05	ND
(run 7)	E	54 ± 10[c]	0.1 ± 0.05[d]	1.09 ± 0.54[c]	0.14 ± 0.03	ND

[a]C, Control; E, experimental.
[b]ND :lt 0.1 µg/g.
[c]Significance level: ** $P < 0.01$, as compared with controls by t-test.
[d]Significance level: $P < 0.05$, as compared with controls by t-test.

Table 36.5 Body and Organ Wet Weight after Clearance Period (g ± SD)

Compound	Group[a]	N	Body	Lung	Liver	Kidney	Spleen
NiO(G)	C	5	618 ± 46	2.7 ± 0.7	19.5 ± 1.4	3.35 ± 0.56	1.1 ± 0.14
(Run 2)	E	5	653 ± 29[b]	3.5 ± 0.3[c]	18.8 ± 1.6	3.05 ± 0.20	1.0 ± 0.13
NiS(A)	C	2	448 ± 11	1.8 ± 0.4	10.1 ± 0.1	2.57 ± 0.02	0.86 ± 0.10
(Run 4)	E	5	516 ± 53	2.6 ± 0.3[b]	11.4 ± 1.3	3.01 ± 0.27[b]	1.03 ± 0.11
Ni(M)	C	18	703 ± 86	1.9 ± 0.2	18.1 ± 3.2	3.68 ± 0.39	1.10 ± 0.19
(Run 6)	E	18	704 ± 92	2.1 ± 0.3[b]	17.5 ± 2.5	3.79 ± 0.56	1.10 ± 0.29

[a] C, Controls; E, experimental.
[b] Significance level: $P < 0.05$, as compared with controls (C) by t-test.
[c] Significance level: $P < 0.01$, as compared with controls by t-test.

Table 36.6 Nickel Concentration in Organ after Clearance Period (μg/g organ ± SD)

Compound	Group[a]	Lung	Liver	Kidney	Spleen	Blood[b]
NiO(G)	C	0.10 ± 0.02	0.02 ± 0.01	0.10 ± 0.02	0.23 ± 0.23	ND
(Run 2)	E	277 ± 98[c]	0.11 ± 0.07[d]	0.10 ± 0.09	0.55 ± 0.22	ND
NiS(A)	C	0.07 ± 0.01	0.02 ± 0.01	0.04 ± 0.03	0.13 ± 0.06	ND
(Run 4)	E	0.09 ± 0.08	0.03 ± 0.02	0.05 ± 0.0	0.2 ± 0.05	ND
Ni(M)	C	0.05 ± 0.02	0.03 ± 0.02	0.04 ± 0.02	0.04 ± 0.02	ND
(Run 6)	E	0.2 ± 0.33	0.03 ± 0.01	0.04 ± 0.02	0.05 ± 0.02	ND

[a] C, Controls; E, experimental.
[b] ND < 0.1 μg/g.
[c] Significance level: $P < 0.01$, as compared with controls by t-test.
[d] Significance level: $P < 0.05$, as compared with controls by t-test.

Table 36.7 Nickel Content in Rat Lungs (μg ± SD)^a

	NiO(G)		NiS(A)		Ni(M)		Ni$_2$O$_3$
	Run 1 (Exp.)	Run 2 (Clear)	Run 3 (Exp.)	Run 4 (Clear)	Run 5 (Exp.)	Run 6 (Clear)	Run 7 (Exp.)
Control	0.17 ± 0.03	0.27 ± 0.05	0.11 ± 0.05	0.12 ± 0.01	0.39 ± 0.07	0.1 ± 0.04	0.14 ± 0.06
Exposed	1018 ± 31^b	984 ± 373^b	2.0 ± 0.4^b	0.21 ± 0.15	251 ± 8.7^b	0.4 ± 0.64^c	176 ± 40^b
Apparent deposition fraction, %	14.5		0.04		0.4		0.57

^{a}Exp., after exposure period; Clear., after clearance period.
^{b}Significance level: $P < 0.01$, as compared with controls by t-test.
^{c}Significance level: $P < 0.05$, as compared with controls by t-test.

clearance rate was fast for NiS(A) and Ni(M). Corroborating evidence for this is provided by the lack of elevation of nickel in liver, kidneys, spleen, and blood after clearance. By contrast, for NiO(G), the nickel content in the lungs decreased sightly during the 12-month clearance period, indicating a slow clearance rate. Slow removal is also consistent with the slight elevation of the liver nickel at this time (Table 36.6).

4.2. Solubility and Durability

Solubility of the nickel compounds in distilled water and saline solution had the following ranking: $NiS(A) >> Ni_2O_3 > NiO(G) > Ni(M)$ (Table 36.2). The term *solubility* as used here relates to the behavior of aerosols in various fluids. In general, this term is more appropriate for *in vitro* than for *in vivo* studies, because in tissue the degradation of aerosols involves more phenomena than solubility. We call the degradation of the aerosol *in vivo durability*.

From the data in Tables 36.4, 36.6, and 36.7, it is clear that the NiS(A) aerosol initially persisted in the lungs, dissolved easily into the blood, and transferred rapidly to the kidneys, from which it is subsequently excreted in the urine. Tanaka et al. (1988b) showed that the biological half-time in rat lungs was only 20 hr for a single 4-hr exposure to NiS(A).

The NiO(G) in rat lungs was present for a long time, and after exposure there was no nickel in any other organ (Table 36.4). Most of the nickel was still present in the lungs after 12 months clearance (Table 36.7). One of the main reasons for this is that its solubility is extremely low. Although the solubility of Ni(M) is also small, most of the nickel in the lungs disappeared in 12 months. This indicates that the length of time that aerosols of nickel compounds persist in the lungs is not only a function of their solubility but also a function of their durability. Durability is directly related to chemical composition and physical characteristics. For example, Tanaka et al. (1985, 1986) showed that the retention time in the lungs increased with increasing particle diameter: half-times of 12 months [mass median aerodynamic diameter (MMAD) 1.2 μm] and 21 months (MMAD 4.0 μm) in rats exposed for 140 hr to 0.24–55 mg Ni/m^3 as NiO(G).

Since comparable nickel levels were found in lungs, kidneys, and liver after exposure to Ni_2O_3 and Ni(M), the clearance rate of Ni_2O_3 (MMAD 4.0 μm seemed to be similar to that of Ni(M) (MMAD 5.8 μm). The absence of nickel elevation in blood for Ni(M) but not for Ni_2O_3 suggests that the retention time of Ni(M) is likely shorter than that of Ni_2O_3. The biological half-time (inverse of durability) in the lungs may be estimated to be in the order $NiO(G) >> Ni_2O_3 > Ni(M) >> Ni(S)$. This is illustrated by the apparent deposition fraction given in Table 36.7.

It is interesting to speculate whether the short retention time of NiS(A) is the reason for its noncarcinogenicity. Costa et al. (1982) reported that water-insoluble crystalline nickel sulfide compounds consistently induce high incidences of cancers in experimental animals, independent of the site of administration. Similar treatment with NiS(A) does not induce cancer at any of a variety of injection sites. Sunderman (1987) also found that the crystalline nickel sulfides and oxides, which slowly dissolve in body fluids and can readily enter cells by phagocytosis, tend to be most carcinogenic, while water-soluble nickel salts have not been shown to initiate carcinogenesis in rodents (but see Pott et al., Chapter 28.4).

In conclusion, we suggest that persistence and durability are important factors in the development of tumors or fibrosis and should therefore be evaluated. Histopathological findings for the studies described will be reported at some future date, as done previously for NiO(G) (Horie et al., 1985; also see Chapter 39).

REFERENCES

American Conference of Governmental Industrial Hygenists (ACGIH) (1987). *Threshold Limit Values and Biological Exposure Indices for 1987–1988*. ACGIH, Cincinnati, p. 39.

Costa, M., Heck, J. D., and Robinson, S. H. (1982). Selective phagocytosis of crystalline metal sulfide particles and DNA strand breaks as a mechanism for the induction of cellular transformation. *Cancer Res*. **42**, 2757–2763.

Horie, A., Haratake, J., Tanaka, I., Kodama, Y., and Tsuchiya, K. (1985). Electron microscopical findings with special reference to cancer in rats caused by inhalation of nickel oxide. *Biol. Trace Element Res*. **7**, 223–239.

International Agency for Research on Cancer (IARC) (1987). *Overall Evaluation of Carcinogenicity on Updating of IARC Monographs*, Volumes 1 to 42, IARC Monographs on the Evaluation of Carcinogenic Risks to Humans, Supplement 17. IARC, Lyon, France.

Kodama, Y., Ishimatsu, S., Matsuno, K., Tanaka, I., and Tsuchiya, K. (1985). Pulmonary deposition and clearance of a nickel oxide aerosol by inhalation. *Biol. Trace Element Res*. **7**, 1–9.

Saracci, R. (1981). Quantification of Occupational Cancer (Banbury report 9), Cold Spring Harbor Lab., London, pp. 165.

Sunderman, F. W. Jr. (1987). Physiochemical and biological attributes of nickel compounds in relationship to carcinogenic activities. *J. UOEH* **9**, Sup. 84–94.

Tanaka, I., and Akiyama, T. (1984). A new dust generator for inhalation toxicity studies. *Ann. Occup. Hyg*. **28**, 157–162.

Tanaka, I., Hayashi, H., Horie, A., Kodama, Y., ad Tsuchiya, K. (1982). Toxicity of a nickel oxide aerosol by inhalation. *UOEH* **4**, 441–449.

Tanaka, I., Ishimatsu, S., Matsuno, K., Kodama, Y., and Tsuchiya, K. (1985). Biological half-time of deposited nickel oxide aerosol in rat lung by inhalation. *Biol. Trace Element Res*. **8**, 203–210.

Tanaka, I., Ishimatsu, S., Matsuno, K., Kodama, Y., and Tsuchiya, K. (1986). Retention of nickel oxide (green) aerosol in rat lungs by long-term inhalation. *Biol. Trace Element Res*. **9**, 187–195.

Tanaka, I., Horie, A., Haratake, J., Kodama, Y., and Tsuchiya, K. (1988a). Lung burden of green nickel oxide aerosol and histopathological findings in rats after continuous inhalation. *Biol. Trace Element Res*. **16**, 19–26.

Tanaka, I., Ishimatsu, S., Haratake, J., Horie, A., and Kodama, Y. (1988b). Biological half-time in rats exposed to nickel monosulfide (amorphous) aerosol by inhalation. *Biol. Trace Element Res*. **17**, 237–243.

37

CARCINOGENICITY OF NICKEL COMPOUNDS AND NICKEL ALLOYS IN RATS BY INTRAPERITONEAL INJECTION

F. Pott, R. M. Rippe, M. Roller, M. Csicsaky, and M. Rosenbruch

Medical Institute for Environmental Hygiene at the University of Düsseldorf, Germany

F. Huth

Institute for Pathology, Municipal Hospital, Hildesheim, Germany

Nickel and Human Health: Current Perspectives, Edited by Evert Nieboer and Jerome O. Nriagu.
ISBN 0-471-50076-3 © 1992 John Wiley & Sons, Inc.

1. INTRODUCTION

Regulation of carcinogens at the workplace and in the environment should be adapted as closely as possible to their risk for humans. Various nickel compounds showed marked differences in the incidences of sarcomas in rats after parenteral administration (Sunderman and Maenza, 1976). Therefore, it became necessary to quantify these differences and establish their causes. Sunderman (1984) tested 18 nickel compounds by intramuscular injection and reported the carcinogenic activities of particulate nickel compounds correlated well with the nickel mass fraction.

In our previous investigation, lung tumors were induced in rats by intratracheal instillation of nickel subsulfide, nickel powder, and nickel oxide (Pott et al., 1987). Nickel subsulfide showed the most toxic and carcinogenic activity; nickel oxide showed the least. Corresponding results were found after intraperitoneal injection. Further experiments were, therefore, carried out using the intraperitoneal test model for establishing the ranking order of the carcinogenicity of 10 dusts containing nickel. This test is simple to perform and is sensitive for detecting the carcinogenicity of metals and especially of mineral fibers.

2. MATERIALS AND METHODS

2.1. Substances Used

Nickel(II)acetate [Ni(OCOCH$_3$)$_2$·4H$_2$O], 23.6% nickel per unit weight of the compounds, was obtained from Fluka Feinchemikalien GmbH, Neu-Ulm, Germany, No. 72225-24 28 63, purity > 99%. Nickel(II)carbonate (alkaline) [NiCO$_3$·2 Ni(OH)$_2$·4 H$_2$O], 51.2% nickel per unit weight of the compound, was obtained from Fluka Feinchemikalien GmbH, Neu-Ulm, Germany, No. 72245-24 16 48, purity >97%. Nickel(II)chloride (NiCl$_2$·6 H$_2$O), 24.7% nickel per unit weight of the compound, was obtained from Fluka Feinchemikalien GmbH, Neu-Ulm, Germany, No. 72247-23 90 30, purity >98%. Nickel(II)oxide (NiO; green-grey), 78.6% nickel per unit weight of the compound, was provided by Aldrich-Chemie GmbH & Co. KG, Steinheim, Germany, purity 99.99%. Nickel powder (Ni), 100% nickel per unit weight of the compound, was provided by INCO Metals Company, Toronto, Ontario, Canada (purity not stated). Nickel subsulfide (Ni$_3$S$_2$), 73.3% nickel per weight of the compound, was provided by INCO Metals Company, Toronto, Ontario, Canada, purity >99%. Nickel(II)sulfate heptahydrate (NiSO$_4$·7H$_2$O), 20.9% nickel per unit weight of the compound, was obtained from Ega-Chemie GmbH & Co. KG., Steinheim, Germany, purity 99.99%. Nickel aluminum alloy, content 50.4% Ni detected by photometry or 52.2% detected by atomic absorption spectrometry, was obtained from Fluka Feinchemikalien GmbH, Neu-Ulm, Germany, No. 72240-22 81 74, purity >97%. This alloy is used for the production of Raney nickel. Nickel alloy 29, content 29.2% Ni detected by photometry or 28.0% detected by atomic absorption spectrometry (before milling, the analysis resulted in 32% Ni, 21% Cr, 0.8% Mn, and 55% Fe), was prepared by Bundesanstalt für Arbeitsschutz, Dortmund, Germany, and milled to particles of <10 μm. This alloy is a heat-resistant steel used for the production of tubes. Nickel alloy 66, a powder containing 66.0% Ni detected by photometry or 66.6% Ni detected by atomic absorption spectrometry (before milling, the analysis resulted in 74% Ni, 16% Cr, and 7% Fe), was prepared by Bundesanstalt für Arbeitsschutz, Dortmund, Germany, and milled to particles of <10 μm. The alloy is heat resistant and resistant to inflammation. Iron(III)oxide (Fe_2O_3) was obtained from Merck, Darmstadt, Germany, No. 3924-1179599, purity >99%. Magnetite (Fe_3O_4) was obtained under the name Ferroso Ferric Oxide from Research Organic/Inorganic Chemical (Bellville, NJ). The particles are very small and cannot be measured by light microscopy.

2.2. Laboratory Animals and Animal Keeping

Female Wistar rats (Wistar-WU/Kißlegg) were obtained from the breeding farm S. Ivanovas (Kißlegg, Allgäu, Germany) and randomly allocated to plastic cages on wood granule bedding (T-grob, Buntenbach, Solingen, Germany). The animals were maintained under conventional conditions. A standard pelleted laboratory diet (Altromin GmbH & Co. KG, Lage, Germany) and water were provided ad libitum. The average body weight of the animals used in the long-term nickel experiment was about 200 g at the time of the first injection; this corresponds to an age of 18 weeks. The rats (same strain and sex) of the parallel experiment with two iron oxides (and other dusts not described in this chapter) were lighter (about 160 g) and younger (11 weeks).

2.3. Method of Administration

The dusts were suspended in 0.9% sodium chloride (buffered with Na_2HPO_4–NaH_2PO_4) solution with an ultrasonic mixer for about 1–5 min. The dose of the nickel compounds and alloys per animal was calculated on the basis of their nickel content. In the nickel study, 1 mL of the suspension was injected intraperitoneally per administration; in the parallel study with two iron oxides 2 mL were applied. If the test substance was highly toxic, then a lower concentration was selected and more than one injection was given in weekly intervals up to 25 administrations or twice weekly when 50 injections were applied (see Table 37.1).

2.4. Method of Examination and Tumor Evaluation

A postmortem examination was made of the abdominal cavity of the animals that died spontaneously or were killed when in bad health. Parts of tumors

Table 37.1 Carcinogenicity after Intraperitoneal Injection of Various Nickel and Iron Compounds and Nickel Alloys

Substance	Intraperitoneal Dose[a] (no. of injections × mg)	Ratio of Rats with Tumors[b] to Rats Examined[c]		Number of Mesothelioma	Number of Sarcoma	Life Span after First Treatment of Rats (weeks)					
		Number	%			Percentage of All Rats				Rats with Tumors	
						≤20	≤50	≤80	≤100	First to Die	Average
Ni chloride	50 × 1	4/32	12.5[d]	1	3	84	103	119	132	90	103
Ni sulfate	50 × 1	6/30	20.0[e]	1	5	47	100	132	132	47	90
Ni acetate	25 × 1	3/35	8.6	1	2	86	114	131	132	47	86
Ni acetate	50 × 1	5/31	16.1[d]	2	4	61	100	132	132	55	74
Ni carbonate	25 × 1	1/35	2.9	0	1	88	109	132	132	86	86
Ni carbonate	50 × 1	3/33	9.1	2	1	84	113	130	132	51	90
Ni powder	1 × 6	4/34	11.8[d]	2	2	64	104	130	132	22	72
Ni powder	2 × 6	5/34	14.7[d]	3	2	50	97	132	132	31	44
Ni powder	25 × 1	25/35	71.4[f]	9	16	41	50	97	118	27	48
Ni subsulfide	1 × 6	20/36	55.6[f]	4	16	27	45	105	131	24	40
Ni subsulfide	2 × 6	23/35	65.7[f]	8	15	27	36	92	131	22	41
Ni subsulfide	25 × 1	25/34	73.5[f]	13	13	33	40	53	131	26	42

Compound	Dose[a]									
Ni oxide	1×25	12/34	35.3[f]	1	11	64	86	120	131	45	79
Ni oxide	1×100	15/36	41.7[f]	3	13	59	87	117	131	42	81
Ni–Al alloy	1×50	8/35	22.9[f]	1	7	74	97	122	132	53	83
Ni–Al alloy	3×50	13/35	37.1[f]	3	11	50	82	113	132	34	81
Ni alloy 29	1×50	2/33	6.1	1	1	82	106	129	132	108	113
Ni alloy 29	2×50	1/36	2.8	0	1	84	114	124	131	53	53
Ni alloy 66	1×50	12/35	34.3[f]	0	12	65	92	125	131	58	81
Ni alloy 66	3×50	22/33	66.7[f]	5	19	38	69	97	131	25	63
NaCl solution	3×1 mL	1/33	3.0	0	1	76	101	120	131	81	81
NaCl solution	50×1 mL	0/34	0.0	0	0	87	105	130	132	—	—
NaCl solution	4×2 mL	3/66	4.5	1	2	91	114	130	130	93	118
Magnetite	1×40	0/36	0.0	0	0	96	124	130	130	—	—
Magnetite	4×40	2/34	5.9	0	2	73	102	130	130	101	116
Fe(III)oxide	1×40	1/36	2.8	1	0	95	118	130	130	101	116
Fe(III)oxide	4×40	0/33	0.0	0	0	98	122	130	130	—	—

[a] Dose of nickel compounds and alloys refers to the injected mass of nickel.

[b] Animals with mesothelioma and/or sarcoma in the abdominal cavity, excluding animals with malignant uterine tumors even when both tumor types were diagnosed.

[c] Minimum life span longer than 16 weeks after first injection.

[d] Significance versus vehicle controls according to Fisher's exact test: $p < 0.05$.

[e] Significance versus vehicle controls according to Fisher's exact test: $p < 0.01$.

[f] Significance versus vehicle controls according to Fisher's exact test; $p < 0.001$.

or organs with macroscopically suspected tumor tissue were fixed in formalin (6%) and prepared for histological examination in paraffin-embedded H&E stained sections. The rats examined (listed in Table 37.1) comprise all autopsied animals, including those that died relatively early but excluding those lost through cannibalism. The percentage of dead rats also includes those lost through cannibalism and is related to the number of rats treated. For calculation of tumor incidences, animals that developed a malignant tumor of the uterus simultaneously with a mesothelioma or sarcoma in the abdominal cavity were excluded. Our experience with Wistar rats would indicate that a malignant tumor of the uterus seems to metastasize fairly frequently into the abdominal cavity. In 102 rats injected only with sodium chloride solution, 6 of 15 animals with a malignant tumor of the uterus also showed mesothelioma or sarcoma of the peritoneum, but in only 2 rats, a mesothelioma or sarcoma was found without a malignant uterine tumor (Pott et al., 1989). On the other hand, it may be possible that in the groups treated with dusts a tumor could have been classified as a metastasis of a spontaneous uterine tumor even though it was induced by the substance administered and occurred accidentally alongside a malignant neoplasm of the uterus.

3. RESULTS

3.1. Toxicity

For ethical reasons, the observations on acute and subchronic toxicity made before planning the long-term experiment were limited to a small numbers of rats. Four nickel compounds (acetate, carbonate, chloride, and sulfate) showed a relatively high acute toxicity with mortality on the first day after intraperitoneal injection of milligram quantities of nickel in female Wistar rats (body weight more than 170 g). The mortality rate after injection of 2 mg nickel as a component of the four mentioned compounds was estimated to be between 10 and 40%. All rats (body weight 200–220 g, 10 per group) survived twice-weekly intraperitoneal injections of the four highly toxic compounds at the level of 1 mg nickel. The total number of injections of sulfate or of the other three compounds was 22 or 35 injections, respectively. However, in the long-term experiment some rats died early during the treatment period.

Nickel powder and nickel subsulfide were much less toxic than nickel acetate, carbonate, chloride, and sulfate. After intraperitoneal injection of 25 mg nickel as nickel powder or nickel subsulfide into rats, a death rate of about 20% can be expected within two weeks. The maximum tolerated dose for either single and repeated intraperitoneal injections may be about 10 mg nickel for the samples examined.

No acute or subchronic mortality has been detected after intraperitoneal injection of nickel oxide, three nickel alloys, and two iron oxides. A maximum total dose of 1000 mg nickel in nickel oxide divided into two intraperitoneal injections did not result in an acute mortality. Out of 47 rats, no animal died within the first five months after treatment (Pott et al., 1987).

Six months after the start of the experiment, the average body weight in all groups treated with nickel ranged from 229 to 246 g, which did not differ statistically from the two saline control groups (234 and 242 g). Also the median life span of the groups injected 50 times with the four highly toxic nickel compounds did not differ significantly from that of the two controls (Table 37.1). The sparingly soluble dusts (nickel oxide, nickel alloys) were found in the abdominal cavity after more than two years. The amount corresponded roughly to the total injected mass, however, within a certain range in each group. In the groups treated with nickel acetate, carbonate, chloride, and sulfate, no residues were detected macroscopically.

3.2. Carcinogenicity

The tumor incidences found after two and a half years, when the experiment was finished, are listed in Table 37.1. All results were significantly different from the saline controls for most nickel compounds in at least one dose group; nickel chloride, nickel carbonate, and nickel alloy 29 (Fisher's exact test) were exceptions. However, when compared with the larger number of historic controls (Pott et al., 1987, 1989), the carcinogenicity of nickel chloride is also statistically significant.

Examples of the survival times of tumor-bearing animals in some groups are shown in Figures 37.1 and 37.2.

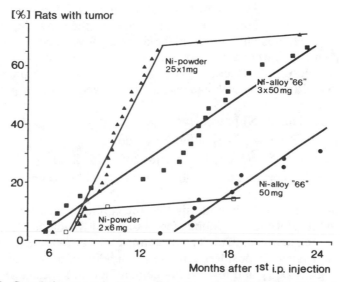

Figure 37.1. Cumulative tumor incidence after intraperitoneal injection of nickel powder (2 × 6 mg Ni or 25 × 1 mg Ni) and of nickel alloy 66 (1 × 50 mg Ni or 3 × 50 mg Ni) over a 2-year period. Two types of tumor–time relationships are indicated (see Section 4).

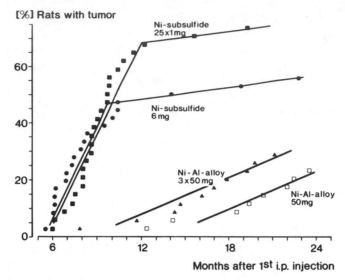

Figure 37.2. Cumulative tumor incidence after intraperitoneal injection of nickel subsulfide (1 × 6 mg Ni or 25 × 1 mg Ni) and of nickel aluminum alloy (1 × 50 mg Ni or 3 × 50 mg Ni) over a 2-year period. Two types of tumor–time relationships are indicated (see Section 4).

4. DISCUSSION

The final results (Table 37.1) show that a wide range of the tumor incidences was induced by the different nickel compounds. In the group injected with 6 mg Ni in nickel subsulfide, 56% of the animals developed tumors, while 50 mg Ni as nickel carbonate or 100 mg Ni as nickel alloy 29 did not lead to a clear carcinogenic effect. Regarding their carcinogenic potency, the ranking order of the tested substances can be divided into four groups:

High: Ni subsulfide, Ni powder.
Medium: Ni alloy 66, Ni–Al alloy, Ni oxide.
Low: Ni acetate, Ni sulfate, Ni chloride (questionable).
Not proven: Ni carbonate, Ni alloy 29, Fe(III)oxide, magnetite.

The results confirm the hypothesis that the carcinogenicity is associated with the solubility of the compounds (see Furst and Radding, 1984). This is clearly illustrated in Figure 37.3; it also demonstrates an association between solubility and toxicity. Considering the low solubility of nickel carbonate in water, the high toxicity and low carcinogenicity are remarkable. However, residues of the 50 mg nickel carbonate injected were unequivocally not visible in the abdominal cavity. It can be concluded that there is a great difference between the solubility of this compound in water and in the body.

If the solubility hypothesis is correct, absolute durable nickel alloys do not induce tumors. This conclusion would be most relevant for evaluation of the large number and amount of alloys that play an important role in daily life. However, it must be acknowledged that low solubility of a special nickel compound may be too small to induce tumors within the short life span of a rat. Yet it is not necessarily inactive in humans. It is plausible that slow dissolution of particles retained in the lung maintains significant internal exposure to nickel for many years.

Two types of relationships between tumor rates and lifespan may be derived from Figure 37.1. One can be called the *nickel powder type* and the other the *nickel alloy type*. The first shows a short life span of tumor-bearing rats for both dose groups, with most of the animals dying during the first year after treatment. By contrast, the second type shows a much slower increase of the cumulative tumor rates in both dose groups, with no evidence of levelling off of the curves and an extension of the survival time for the low-dose group by about half a year. Nickel subsulfide also belongs to the first type and nickel aluminum alloy to the second (see Fig. 37.2). The nickel alloy type shows a similar pattern to that observed with mineral fibers.

Sunderman (1984) found that the carcinogenic activities of nickel compounds correlated with the nickel mass fraction injected intramuscularly in rats. Our results also show a good correlation between the two parameters if the data for nickel subsulfide are excluded (see Fig. 37.4).

The surface properties of the particles may also be relevant for tumor induction. This possibility is currently under discussion to explain the mechanism of the carcinogenicity of mineral fibers. Therefore, in a parallel experiment, nickel powder was injected together with polyvinylpyridine-*N*-oxide (PVNO); the rats were treated again with PVNO intraperitoneally after

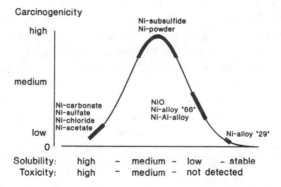

Figure 37.3. Interpretation of experimental results in terms of a hypothesis relating the degree of toxicity and solubility in vivo of nickel compounds and their carcinogenic activity. The carcinogenic potency of the highly soluble nickel compounds is considered to be low, but not zero. The carcinogenic potency of absolute stable nickel alloys may be zero.

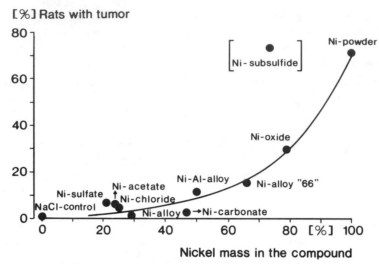

Figure 37.4. Relationship between nickel mass fractions of injected compounds and tumor incidences observed or estimated 2 years after intraperitoneal injection of 25 mg nickel contained in injected compounds. [The tumor incidences of Ni chloride, Ni sulfate, Ni acetate, and Ni alloys were estimated to be half as high as the tumor incidences found after injection of the compound containing 50 mg nickel (see Table 37.1). These results refer to 2 years after the start of the experiment.] The solid line corresponds to the logistic function: $y = 0.7149/\{1 + 6.933\ [x/(1 - x)]^{-1.161}\}$, fitted to the experimental data (excluding Ni subsulfide) by means of nonlinear regression techniques (x, nickel mass fraction; y, ratio of tumor-bearing animals).

4, 8, 12, 16, and 20 months. This kind of treatment inhibits the silicotic activity of quartz and prevents the development of an experimental silicosis (Schlipköter and Brockhaus, 1960). However, we did not find a decreased carcinogenicity of nickel powder after the treatment with PVNO.

Two reviews on the questionable carcinogenicity of iron oxides have come to the same conclusion: There are no data that could seriously confirm this long-held suspicion (Stokinger, 1984; Henschler, 1984). The tumors observed after intraperitoneal injection of two samples of gamma-ferric oxide hydrate were explained by the fibrous shape of the dust particles (Pott et al., 1987, 1989). However, large amounts of nonfibrous magnetite instilled intratracheally (15 × 15 mg) led to an unexpected lung tumor incidence of 69% (Pott et al., 1987). This raises the question of whether an unspecific reaction of the rat lung to the large surface of the very fine particles induced the tumors. The possibility of an overloading effect in the lung by magnetite is supported by the negative intraperitoneal test.

This consideration leads to the general question as to what extent the carcinogenic response observed at the serosa after intraperitoneal injection of substances is qualitatively and quantitatively similar to their carcinogenic activity on the epithelial cells of the airways. From the available results with mineral fibers and nickel compounds, it can be concluded that qualitatively there is an analogy between the carcinogenic reaction of the two types of tissues. There are also indications of a similar ranking order of the carcinogenic potency of various nickel compounds or identical numbers of durable mineral fibers in the lung and the serosa. But there is obviously a great difference between the dose–response relationship of carcinogenic dusts calculated for the inhaled dose and those calculated for the injected dose. The effects of particle deposition and lung clearance have to be considered carefully (Oberdörster, 1988). Certainly, the intraperitoneal injection of dusts is a nonrealistic exposure route, but the carcinogenicity observed with this or with similar test models is not necessarily irrelevant. Conversely, the predominant negative carcinogenicity studies with cigarette smoke and with crocidolite asbestos show clearly that a physiological exposure route does not necessarily lead to results that are relevant for humans.

A more detailed description of this experiment with a comprehensive discussion was published in German (Pott et al., 1991).

ACKNOWLEDGMENT

This work was supported by the Bundesanstalt für Arbeitsschutz, Dortmund, Germany.

REFERENCES

Furst, A., and Radding, S. B. (1984). New developments in the study of metal carcinogenesis. *J. Environ. Sci. Health*, **C2**(1), 103–133.

Henschler, D. (Ed). (1984). Gesundheitsschädliche Arbeitsstoffe. Toxikologisch-arbeitsmedizinische Begründung von MAK-Werten (Maximale Arbeitsplatz-Konzentrationen). Bearb. von den Arbeitsgruppen "Aufstellung von MAK-Werten" und "Festlegung von Grenzwerten für Stäube" der Kommission zur Prüfung gesundheitsschädlicher Arbeitsstoffe der Deutschen Forschungsgemeinschaft. *Eisenoxide*. Lfg. 10, pp. 1–10.

Oberdörster, G. (1988). Lung clearance of inhaled insoluble and soluble particles. *J. Aerosol. Med.*, **1**, 289–330.

Pott, F., Ziem, U., Reiffer, F.-J., Huth, F., Ernst, H., and U. Mohr (1987). Carcinogenicity studies on fibres, metal compounds, and some other dusts in rats. *Exp. Pathol. (Jena)* **32**, 129–152.

Pott, F., Roller, M., Ziem, U., Reiffer, F.-J., Bellmann, B., Rosenbruch, M., and Huth, F. (1989). Carcinogenicity studies on natural and man-made fibres with the intraperitoneal test in rats. In J. Bignon, J. Peto, and R. Saracci (Eds.), *Non-occupational Exposure to Mineral Fibres*. IARC Scientific Publ. No. 90. IARC, Lyon, pp. 173–179.

Pott, F., Rippe, R. M., Roller, M., Rosenbruch, M., and Huth, F. (1991). Vergleichende Untersuchungen über die Kanzerogenität verschiedener Nickelverbindungen und Nickellegierungen. Bremerhaven: Wirtschaftsverlag NW, Verl. für neue Wissenschaft. (= Schriftenreihe der Bundesanstalt für Arbeitsschutz. Forschung Fb 638.)

Schlipköter, H.-W., and Brockhaus, A. (1960). Die Wirkung von Polyvinylpyridin auf die experimentelle Silikose. *Dtsch. Med. Wochenschr.* **85,** 920–923, 933–934.

Stokinger, H. E. (1984). A review of world literature finds iron oxides noncarcinogenic. *Am. Ind. Hyg. Assoc. J.* **45,** 127–133.

Sunderman, F. W. (1984). Carcinogenicity of nickel compounds in animals. In F. W. Sunderman, Jr. (Ed.), *Nickel in the Human Environment.* IARC Scientific Publ. No. 53. IARC, Lyon, pp. 127–142.

Sunderman, F. W., and Maenza, R. M. (1976). Comparisons of carcinogenicities of nickel compounds in rats. *Res. Comm. Chem. Pathol. Pharmacol.* **14,** 319–330.

38

MECHANISM OF MALIGNANT TUMOR INDUCTION BY NICKEL SUBSULFIDE

George Lumb

Department of Pathology, Hahnemann University College of Medicine, Philadelphia, Pennsylvania 19102 and Department of Pathology, Duke University Medical Center, Durham, North Carolina 27710

F. William Sunderman, Sr.

Institute for Clinical Science, Pennsylvania Hospital, Philadelphia, Pennsylvania 19107

Nickel and Human Health: Current Perspectives, Edited by Evert Nieboer and Jerome O. Nriagu.
ISBN 0-471-50076-3   © 1992 John Wiley & Sons, Inc.

1. INTRODUCTION

In previously reported experimental induction of tumors with nickel subsulfide (Ni_3S_2) in rats, striking similarities were observed with human malignant fibrous histiocytoma (MFH). These have been seen in primary development, after transplantation, and also following inoculation of tumor cells grown in tissue culture (Lumb et al., 1985a,b, 1987; Sunderman et al., 1984).

Other workers have described these experimental tumors as rhabdomyosarcomas (Gilman, 1962; Sunderman, 1981, 1983), and some investigators have used the designation leiomyosarcoma (Bruni, 1985). These differences in terminology seem to suggest the development from pluripotential mesenchymal cells, with a variety of appearances possible in the fully developed final tumor.

In our studies the presence of two tumor cell types was observed (Lumb et al., 1987). These have been described as fibroblastlike, with the characteristics of myofibroblasts, and histiocytelike. Myofibroblasts have been described and reviewed by many authors (Lipper et al., 1980). Their relationship to wound healing in animals and humans (Gabbiani et al., 1976; Ryan et al. 1973), neoplastic development in animals associated with methylcholanthrene (Katenkamp and Stiller, 1975) and dimethyl nitrosamine (Hard and Butler, 1971), as well as their presence in human MFH have also been noted (Churg and Kahn, 1977; Taxy and Battifora, 1977).

In our earlier experiments with Ni_3S_2-induced tumors, the nickel compound was implanted into striated muscle (Lumb et al., 1985a; Sunderman et al., 1984). In more recent experiments (Lumb et al., 1987) tumors were induced by subcutaneous implantations, thereby making tumor cytology easier to determine, without the confusion of interpretation of damaged and regenerating muscle fibers in the developing neoplasms.

The difference in the time required for tumor development when Ni_3S_2 was inoculated directly into rat tisues (i.e., per primum) versus those resulting from transplantation or tissue culture inoculation was impressive. The tumors initiated per primum developed after an average latent period of seven to eight months, versus transplants in five and a half weeks, and tumors induced from cell culture in three to four weeks, despite the fact that all three types had essentially similar cytological and histological appearances (Lumb et al., 1987).

The present studies were undertaken to make a preliminary investigation of the hypothesis that in the tumors arising per primum, a latent period was required for cellular "irritation" by Ni_3S_2, which has an extremely low solubility in water. It is conceivable that the time required for tumor formation might be correlated with this low solubility. During this latent period the chemical might bind to macromolecules, leading to cell necrosis. Subsequently *reparative hyperplasia* either in the muscle cells themselves or in pluripotential mesenchymal cells with development of mutations and ultimate malignant tumor formation might result. Such a manifestation has been described in other tumors as *epigenetic carcinogenicity of the cytoxic variety* (Andersen, 1986; Bosan et al., 1987; Newberne et al., 1987a,b).

For our present investigation, it was decided to return to the intramuscular implantation model in order to take advantage of the previously acquired knowledge (Lumb et al., 1985a,b) that muscle fiber damage and apparent regeneration occurred following the introduction of the Ni_3S_2.

2. MATERIALS AND METHODS

A previously described methodology (Lumb et al., 1987) was used for the implantation of Ni_3S_2. The anesthetic was a mixture of acepromazine (Azeco Co., West Fort Dodge, IA) and ketamine (Ketaset Bristol Labs, Syracuse, NY). A 00 gelatin capsule containing 10 mg Ni_3S_2 was implanted through a small skin and fascia incision into the body of the quadriceps muscle of the right hind limb of seven male F344 rats. Previous experiments had shown no sex differences in tumor histology and cytology, and empty control capsules had caused no significant irritation and no tumor formation (Lumb et al., 1985a, 1987).

Minimal swelling occurred after inoculation, which subsided after 5–8 days so that the capsule could be palpated. After 2–3 weeks the capsules were absorbed and nothing could be felt. Every 2 weeks careful light palpation was performed until 5 months, when daily observations were made until the earliest swelling could be detected. When these were large enough that it was considered possible to make a small biopsy without removing the entire lesion (approximately 1–3 mm), biopsies were performed, and then observations were continued until tumors reached reasonable proportions (5–7 mm) when further biopsies were performed. Then the tumor was allowed to grow until examined at sacrifice at 10–12 months following the initial inoculation. At this time tumors were approximately 0.8–1.2 cm in diameter, firm in consistency, and without significant necrosis. Tumors developed in all of the seven animals, and the first observations of nodules were made in the whole group within 3–8 days of each other, at approximately 7 months after implantation. Once nodules were palpated, their size increased quite rapidly, and the first biopsies (1–3 mm) were performed within 10 days of each other at approximately 8 months after Ni_3S_2 implantation. The same held true for the second set of biopsies (5–7 mm) at 9 months.

Light microscopy (LM) was performed on tumor material fixed in buffered formalin (pH 7.0) and embedded in paraffin wax and methacrylate for routine and ultrathin sections (Lumb, 1983). Material fixed first in 2.5% gluteraldehyde and then in 1% percent osmium tetroxide was embedded in Spurr for electron microscopy (EM). Immunohistochemical (IH) studies were undertaken using the peroxidase antiperoxidase method (PAP) (Sternberger, 1979; Taylor, 1986) in paraffin-embedded material, with antibodies directed against vimentin, desmin, myoglobin, chymotrypsin (Dako, Santa Barbara, CA), and macro-phage-specific and muscle-specific actin (Enzo Diagnostics, New York, NY).

3. RESULTS

3.1. First Biopsies (Nodules 1–3 mm)

Light microscopy from the early nodules showed groups of indeterminate cells scattered among degenerating muscle fibers. These cells were mostly spheroidal with occasional spindle forms (Fig. 38.1). They had no special features on which to make a specific cytologic diagnosis. They resembled histiocytes or activated mesenchymal cells. They tended to have nuclei with large nucleoli but were not specific enough to make malignancy a certainty. There were occasional mitoses. Inflammatory cells were not a feature.

In adjacent areas where there were normal muscle fibers, scattered isolated degenerating fibers were found, which, when seen in transverse section, were intimately surrounded by cells of the same type as those seen in the large groups already described (Fig. 38.2). In five of the seven biopsies taken at this period, acellular material was seen scattered between cells and also was apparently intracellular (Fig. 38.3). This was confirmed by EM. The material stained black with H and E and showed a positive staining reaction (blue-violet) with rubeanic acid (Thompson, 1966). Rubeanic acid stains copper, cobalt, and nickel compounds in tissues and gives a different staining color reaction with each of them. Nickel compounds produce a blue-violet color, copper a greenish-black color, and cobalt a yellowish-brown color. A blue-violet color reaction was obtained when the tissues were stained with

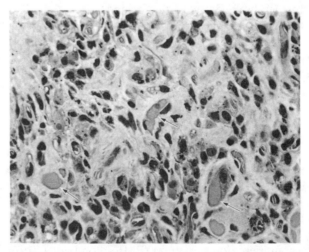

Figure 38.1. Scattered pleomorphic cells, mostly spheroidal in shape, among degenerating muscle fibers (arrow). (H&E stain, ×250)

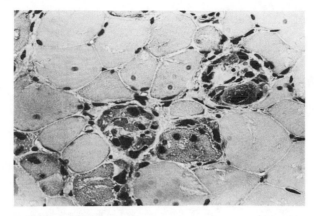

Figure 38.2. Isolated degenerating and destroyed muscle fibers seen in transverse section and surrounded by cells with hyperchromatic nuclei similar to those seen in Figure 38.1 (H&E stain, ×400)

rubeanic acid; therefore, it is reasonable to assume that nickel-containing fragments were being stained.

Electron microscopy from these biopsies showed degenerating muscle cells with the cells (noted already in LM) surrounding them (Fig. 38.4) These cells were usually spheroidal, with occasional spindle forms, some of which had many of the appearances of myofibroblasts, but at this stage most of them were not quite specific enough for a positive diagnosis (Fig. 38.5). Some of the spindle cells, however, had indented nuclei and very well defined rough endoplasmic reticulum, with frequently dilated cysternae. Scattered

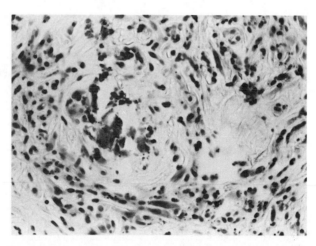

Figure 38.3. Scattered fragments of acellular material that stained blue-violet with rubeanic acid. The fragments are scattered among proliferating cells and destroyed muscle cells (H&E stain, ×100)

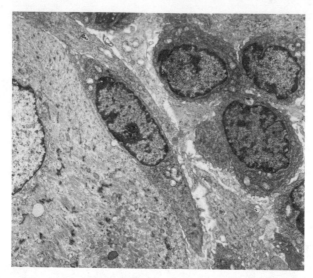

Figure 38.4. Degenerating muscle fiber (left) encircled by indeterminate cells, one of which has a spindle form and shows dilated rough endoplasmic reticulum. The remainder are spherical. (×6050)

microfilaments with focal "dense bodies" were seen. Such microfilaments are characteristically seen in muscle cells, both smooth and striated. These cells, therefore, were characteristic of myofibroblasts (Churg and Kahn, 1977; Fu et al., 1975; Lipper et al., 1980) (Fig. 38.6). In some of these cells acellular material with suggestion of a crystalline-type pattern was seen. It seems reasonable to assume that these deeply osmium-staining fragments

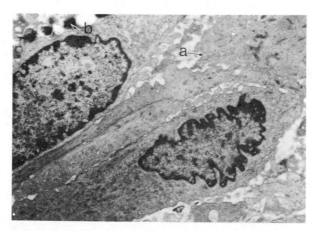

Figure 38.5. Spindle shaped cell with indented nucleus and rough endoplasmic reticulum in abundance, suggestive of an early form of myofibroblast (lower right), close to a degenerating muscle cell (arrow *a*). Fragments of osmiphilic material (probably nickel) can also be seen (arrow *b*). (×9250)

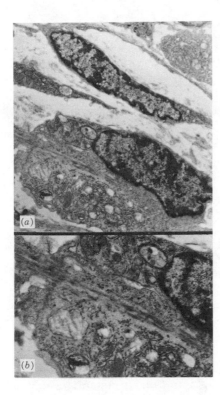

Figure 38.6. Myofibroblast lying close to a typical fibroblast (upper part of a). (*a*) Intercellular collagen is seen (×23,300). (*b*) In the enlarged area the distended cysternae and microfilaments with dense bodies are more clearly seen.

were, in fact, nickel. The material was seen in the cytoplasm and in some instances was found inside the nuclei. In many cells total destruction had occurred (Fig. 38.7).

Immunohistochemical study showed only myoglobin positivity in obviously degenerating and normal muscle fibers. Although other scattered cells showed positive reaction to vimentin, there was insufficient characteristic staining to make a positive conclusion.

3.2. Second Biopsies (Nodules 5–7 mm)

Light microscopy showed mainly spindle-shaped tumor cells. There was some suggestion of the beginning of storiform arrangement (Fig. 38.8). Scattered spheroidal cells were also found. Mitoses were frequent. These cells predominated over degenerating muscle cells, which in these biopsies were relatively few in number. The nuclei of the spindle-shaped cells were indented with large nucleoli, and chromatin was arranged mainly around the nuclear margin.

Electron microscopy showed large numbers of the myofibroblasts previously described (Fig. 38.6).

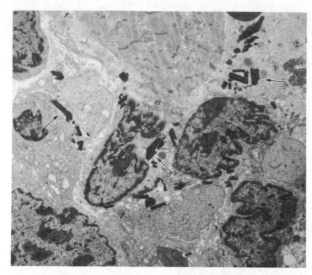

Figure 38.7. Remnant of degenerating muscle fiber (upper segment) with indeterminate spheroidal cells around it. Numerous fragments of acellular material that is almost certainly nickel subsulfide are seen both between cells and in cells (arrows). (×6050)

Some cells showed fine microfilaments suggestive of actin scattered among the filaments with dense bodies (Fig. 38.9). Intracellular membrane–bound collagen could be identified. Mitoses were seen, and there was considerable extracellular collagen between the spindle-shaped tumor cells (Figs. 38.6 and 38.9). Occasional cell fragments and nickel particles were seen in these second-stage biopsies.

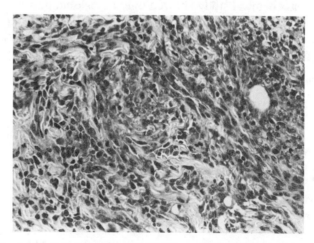

Figure 38.8. Mixed cells including spindle cells and spheroidal cells with the beginning of a storiform pattern taken from the second set of biopsies. (H&E stain, ×250)

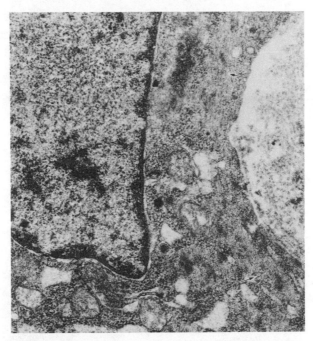

Figure 38.9. Part of malignant cell of myofibroblast type showing microfilaments with dense bodies and scattered finer filaments suggestive of actin (arrows). Extracellular collagen is also seen. (×43,100)

Immunohistochemical stains showed positivity for vimentin (Fig. 38.10) and muscle-specific actin (Figs. 38.11a,b) in most of the tumor cells. Myoglobin was positive only in the few remaining degenerating muscle cells.

3.3. Biopsies of Well-defined Tumors at Time of Sacrifice (Nodules 0.8–1.2 cm)

Light microscopy showed the typical storiform-type appearance with the two cell types previously described as fibroblastlike (myofibroblasts) and histiocytelike (Lumb et al., 1987). Electron microscopy confirmed the two cell types, and in addition, large undifferentiated cells were seen (Fig. 38.12). The myofibroblasts showed well-formed rough endoplasmic reticulum with distended cisternae, intracellular and extracellular collagen, microfilaments scattered through the cytoplasm with dense bodies, and typical indented nuclei with peripherally arranged clumps of chromatin. The spherical cells had well-defined Golgi and frequent microvilli at the cell surface. Microfilaments were not a feature, though occasional very fine filaments near the periphery of the cells suggestive of Actin fibrils were seen. All of these changes have been well documented previously (Lumb et al., 1987).

Immunohistochemical stains from these tumors showed intense staining with vimentin and muscle-specific actin in all the tumor cells (Figs. 38.10a,b

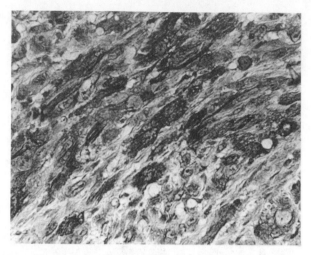

Figure 38.10. Intense diffuse positivity for vimentin mostly in spindle-shaped cells. (×400)

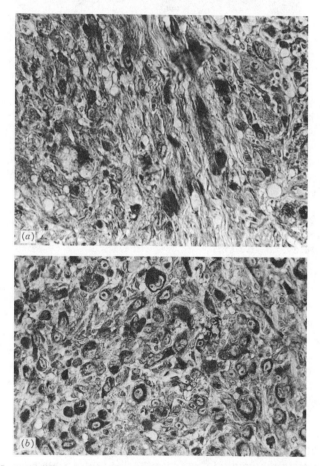

Figure 38.11. Intense diffuse staining for muscle-specific actin seen both in (*a*) spindle-shaped cells and (*b*) spheroidal cells. (×400)

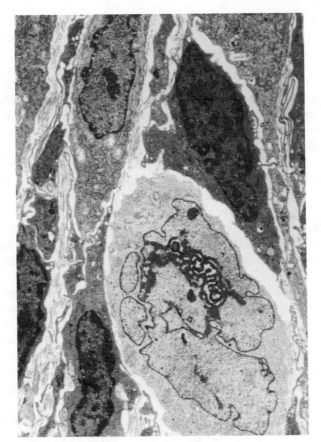

Figure 38.12. Typical fibroblast-type (upper left) and histiocyte-type (upper right) cells lying adjacent to each other. Also seen is a spherical undifferentiated cell (lower right). (×9250)

and 38.11*a,b*). Other IH stains were unproductive from the diagnostic point of view. Occasional cells showed positivity for desmin and some stained for chymotrypsin. However, these two results were insufficiently constant to be helpful. Antibodies designed to be reactive against histiocytes showed no positivity in this material.

4. DISCUSSION

These experiments demonstrate the time lag of several months between the implantation of Ni$_3$S$_2$ into muscle and the development of nodules and ultimately of malignant tumors. These tumors are invasive, and they undergo ultimate necrosis and metastasize to local lymph nodes, but principally to the lungs (Lumb et al., 1987).

In the initial stages of development, it has been demonstrated that Ni_3S_2 is an irritant and destroys the muscle fibers. Degenerating muscle fibers were easy to demonstrate, and other cells that are difficult to differentiate between reacting mesenchymal cells and regenerating muscle cells can also be seen in the area of destruction. No typical inflammatory reaction was seen, but as time progressed, development of myofibroblasts was clearly demonstrated, and these appear to arise from the proliferating and/or regenerating cells. The LM and EM appearances were typical of myofibroblasts, and further credence to the diagnosis was obtained by the IH results of positivity in reactions to vimentin and muscle actin-specific monoclonal antibody (HHF35). Although positive reactions to vimentin are rather nonspecific in many connective tissue tumors, the association of positivity with HHF35 makes the diagnosis of myofibroblasts more certain. The monoclonal antibody HHF35 has been characterized biochemically as recognizing isotypes of actin (alpha and gamma). These isotypes are specific to muscle cells, but Tsukada et al. (1987b), who identified HHF 35, have shown that the antibody localizes to myofibroblasts (Tsukada et al., 1987a,b).

The close resemblance of the morphological features of these experimental tumors with the human MFH is noteworthy, and the similar appearances of the nickel-induced tumor with one produced by 4-(hydroxyamino)quinoline-1-oxide in rats and designated by the authors as MFH (Maruyama et al., 1983) is also striking. The relationship of myofibroblasts to human MFH remains controversial, with some workers strongly in favor of the association (Churg and Kahn, 1977; Taxy and Battifora, 1977), while others feel that a histiocytic origin is more likely (Fu et al., 1975; Weiss, 1982). Our feeling is that the tumors produced in these experiments arise from primitive mesenchymal cells with pluripotentiality. These could be altered or regenerating myocytes when the tumor arises in the body of a muscle. However, when the tumor arises subcutaneously as was previously demonstrated (Lumb et al., 1987), the likelihood of an undifferentiated mesenchymal cell seems to be more probable. In the experiments reported here, cellular irritation, degeneration, necrosis, and regeneration were present in all cases. This raises the question of whether or not this tumor development falls into the category of an epigenetic carcinogenesis of cytotoxic variety. Most of the investigations of this type of tumor development have been performed in relation to the liver (Andersen, 1986; Bosan et al., 1987; Newberne, 1987a,b). The findings of these experiments are presented as further evidence of the epigenetic possibility of tumor induction, but with a tumor model differing from those already studied. This may give opportunities for increased knowledge of this type of cytotoxic damaging phenomenon.

5. CONCLUDING SUMMARY

Experiments were performed to study the earliest changes and their sequence in the development of malignant tumors following implantation of 10 mg of nickel subsulfide (Ni$_3$S$_2$) into the right quadriceps muscle of seven male Fischer rats. Biopsies were performed when nodules reached 1–3 mm, later when they were 5–7 mm, and finally at sacrifice to confirm the fully developed tumor pattern. Light microscopy of the earliest samples showed groups of cells clumped and scattered among degenerating muscle fibers. Mitoses were seen and inflammatory cells were not a feature. Electron microscopy showed individual degenerating muscle fibers, but also cells with characteristic features of myofibroblasts. In many cells, osmium-dense fragments suggestive of crystalline material were seen in the cytoplasm and nuclei. In the second set of biopsy material myofibroblasts with well-defined and dilated rough endoplasmic reticulum, intracellular membrane-bound collagen, and micro-filaments with focal dense bodies were numerous. Mitoses were frequent. Immunohistochemistry showed strongly positive reaction to vimentin and muscle-specific actin in the tumor cells. In the fully developed tumors the previously described typical storiform cell pattern with spindle and spheroidal cells with frequent mitoses was seen (Lumb et al., 1985a; Sunderman et al., 1984). Vimentin and muscle-specific actin stains were strongly positive. The long latent period, the evidence of cell degeneration, necrosis, foreign material (probably of nickel composition), cell invasion, and subsequent rapid myofibroblast-type cell development, proliferating to malignant tumors highly suggestive of malignant fibrous histiocytoma, seem to suggest an epigenetic form of carcinogenicity of cytotoxic variety. Whether the tumor cells derive from transformed myofibrils or from activated pluripotential mesenchymal cells, or from both, remains in doubt.

REFERENCES

Andersen, M. (1986). Mechanistic considerations in chemical carcinogenesis: pharmacokinetics. *Proc. Tox. Forum* 68–73.

Bosan, W. S., Shank, R. C., MacEwen, J. D., Gaworski, C. L., and Newberne, P. M. (1987). Methylation of DNA guanine during the course of induction of liver cancer in hamsters by hydrazine or dimethylnitrosamine. *Carcinogenesis* **8**, 439–444.

Bruni, C. (1985). In S. S. Brown and F. W. Sunderman, Jr. (Eds.), Electron Microscopy Observations on Leiomyosarcomas Induced with Nickel Subsulfide. *Progress in Nickel Toxicology.* Blackwell Scientific, Oxford, U.K., pp. 45–47.

Churg, A. M., and Kahn, L. B. (1977). Myofibroblasts and related cells in malignant fibrous and fibrohistiocytic tumors. *Human Pathol.* **8**, 205–218.

Fu, Y. S., Gabbiani, G., Kaye, G. I., and Lattes, R. (1975). Malignant soft tissue tumors of probable histiocytic origin (malignant fibrous histiocytomas): General considerations and electron microscopic and tissue culture studies. *Cancer* **35**, 176–197.

Gabbiani, G., Le Lous, M., Bailey, A. J., Bazin, S., and Delaunay, A. (1976). Collagen and myofibroblasts of granulation tissue. *Virchows Arch. (Cell Pathol.)* **21**, 133.

Gilman, J.P.W. (1962). Metal carcinogenesis. II. A study on the carcinogenic activity of cobalt, copper, iron and nickel compounds. *Cancer Res.* **22**, 158–162.

Hard, G. C., and Butler, W. M. (1971). Ultrastructural study of the development of interstitial lesion leading to mesenchymal neoplasia induced in the rat renal cortex by dimethylnitrosamine. *Cancer Res.* **31**, 337.

Katenkamp, D., and Stiller, D. (1975). Structural patterns and histological behaviour of experimental sarcomas. II. Ultrastructural cytology. *Exp. Pathol. (Jena)* **11,** 190.

Lipper, S., Kahn, L. B., and Reddick, R. L. (1980). The myofibroblast. *Path. Ann.* **18,** 409–441.

Lumb, G. (1983). Plastic embedding as a tool in surgical pathology diagnosis. *Ann. Clin. Lab. Sci.* **13,** 393–399.

Lumb, G., Sunderman, F. W., Sr., and Schneider, H. P. (1985a). Nickel-induced malignant tumors. *Ann. Clinical Lab. Sci.* **15,** 374–379.

Lumb, G., Sunderman, F. W., Sr., and Schneider, H. P. (1985b). In S. S. Brown and F. W. Sunderman, Jr. (Eds.), Nickel-Induced Malignant Tumors. *Progress in Nickel Toxicology.* Blackwell Scientific, Oxford, U.K., pp. 33– 44.

Lumb, G., Sunderman, F. W. Sr., Schneider, H. P., and Chou, R. H. (1987). Histogenesis of subcutaneous malignant tumors resulting from nickel subsulfide implantation. *Ann. Clin. Lab. Sci.* **17,** 286–299.

Maruyama, H., Mii, Y., Emi, Y., Masuda, S., Miyauchi, Y., Masuhara, K., and Konishi, Y. (1983). Experimental studies on malignant fibrous histiocytomas. II. Ultrastructure of malignant fibrous histiocytomas induced by 4-(hydroxyamino)-quinoline 1-oxide in rats. *Lab. Invest.* **48,** 187–198.

Newberne, P. M., Punyarit, J. de Camargo, and Suphakarn, V. (1987a). The role of necrosis in hepatocellular proliferation and liver tumors. Mouse liver tumors. *Arch. Toxicol.,* Suppl. 10, 54–67.

Newberne, P. M., Suphakarn, V., Punyarit, P., and de Camargo, J. (1987b). Nongenotoxic Mouse Liver Carcinogens. Banbury Report 25: Nongeneotoxic Mechanisms in Carcinogenesis Cold Spring Harbor Laboratory.

Ryan, G. B., Cliff, W. J., Gabbiani, G., et al. (1973). Myofibroblasts in an avascular fibrous tissue. *Lab. Invest.* **29,** 197.

Sternberger, L. A. (1979). *Immunocytochemistry,* 2nd ed., Wiley, New York, pp. 104–169.

Sunderman, F. W. Jr. (1981). Recent research on nickel carcinogenesis. *Environ. Health Perspect.* **40,** 131–144.

Sunderman, F. W., Jr. (1983). In R. Langenbach, S. Nesnow, and J. Rice (Eds.), Organ and Species Specificity in Nickel Subsulfide Carcinogenesis. Organ and Species Specificity in Chemical Carcinogenesis. Plenum Publishing, New York, pp. 107–126.

Sunderman, F. W. Sr., Schneider, H. P., and Lumb, G. (1984). Sodium diethyldithiocarbamate administration in nickel-induced malignant tumors. *Ann. Clin. Lab. Sci.* **14,** 1–9.

Taxy, J. B., and Battifora, H. (1977). Malignant fibrous histiocytoma. An electron microscopic study. *Cancer* **40,** 254.

Taylor, C. R. (1986). *Immunomicroscopy: A Diagnostic Tool for the Surgical Pathologist.* W. B. Saunders, Philadelphia, pp. 32–37.

Thompson, S. W. (1966). Rubeanic acid method for cooper, cobalt, nickel. In S. W. Thompson (Ed.), *Selected Histochemical and Histopathological Methods.* C. C. Thomas, Springfield, Ill., pp. 604–606.

Tsukada, T., Tippens, D., Godron, D., Ross, R., Gown, A. M. (1987a). HHF35: A muscle-actin-specific monoclonal antibody: I. Immunocytochemical and biochemical characterization. *Am. J. Pathol.* **127,** 51–60.

Tsukada, T., McNutt, M. A., Ross, R., Gown, A. M. (1987b). HHF35: A muscle-actin-specific monoclonal antibody: II. Reactivity in normal, reactive, and neoplastic human tissues. *Am. J. Pathol.* **127,** 389–402.

Weiss, S. W. (1982). Malignant fibrous histiocytoma. *Am. J. Surg. Pathol.* **6,** 773–784.

39

HISTOPATHOLOGICAL ASPECTS OF INHALATION EXPERIMENTS WITH NiO AEROSOLS AND CIGARETTE SMOKE

*A. Horie** and J. Haratake*

*Department of Pathology and Oncology, School of Medicine,
University of Occupational and Environmental Health, Japan,
Kitakyushu 807, Japan*

I. Tanaka

*Department of Environmental Health Engineering, Institute of
Industrial Ecological Sciences, University of Occupational and
Environmental Health, Japan, Kitakyushu 807, Japan*

Y. Kodama

*Department of Environmental Health, School of Medicine,
University of Occupational and Environmental Health, Japan,
Kitakyushu 807, Japan*

K. Tsuchiya

*President, School of Medicine, University of Occupational
and Environmental Health, Japan, Kitakyushu 807, Japan*

S. Cho

*Faculty of Home Life Science,
Fukuoka Women's University, Fukuoka 803, Japan*

* Died in 1991.

Nickel and Human Health: Current Perspectives, Edited by Evert Nieboer and
Jerome O. Nriagu.
ISBN 0-471-50076-3   © 1992 John Wiley & Sons, Inc.

1. INTRODUCTION

Epidemiological reports have confirmed an increased incidence of cancer of the respiratory tract among nickel workers involved in production and refining facilities. The identification by inhalation studies of carcinogenic nickel compounds is limited, except for nickel carbonyl vapor (Sunderman and Donnelly, 1965) and nickel sulfide (Ottolenghi et al., 1974). One of the insoluble nickel compounds, nickel oxide (NiO), was selected for the present inhalation study. Our previous report (Horie et al., 1985) indicated that inhalation of NiO aerosol by rats had produced hyperplasia of type II pneumocytes and Clara cells. However, clear carcinogenicity of NiO aerosol has not been established. The objective of the current experiment is to demonstrate whether the carcinogenicity of NiO can be promoted by tobacco smoke.

2. MATERIAL AND METHODS

Respirable aerosols of NiO(G) (green) were obtained from the Nickel Producers Environmental Research Association (NiPERA). The median aerodynamic size was 4.1 μm, with a geometric standard deviation of 2.2. The exposure concentration of NiO(G) was 0.96 ± 0.51 mg/m^3. All the animals exposed were male Wistar rats about two months of age and around 250 g in body weight and were supplied by the animal center of our university. They were kept in exposure chambers 0.48 m^3 in size, five to a cage, and were provided with commercial pellet food. The exposure system for the generation of NiO aerosol has been reported in detail by Tanaka et al. (1982). The rats were exposed to NiO(G) for 6 hr/day 5 days/week, for up to one month.

For the tobacco smoke inhalation studies, the exposure conditions are summarized in Table 39.1. The rats were exposed to the side streams of 10 cigarettes ("Peace" brand, made in Japan) for 2 hr a day for up to one month. The daily exposure condition corresponded to the consumption of 13.5 ± 3.7 mg/m^3 of tobacco.

A hundred and thirty rats were divided into four groups as follows.

Group 1, NIOGT[NiO(+), Tobacco (+)]. Inhalation of NiO(G) for 1 month followed by inhalation of tobacco smoke for 1 month and a clearance period without treatment for 16 months.

Table 39.1   Exposure Conditions

Volume m^3	0.1
Airflow rate, m^3/min	0.05
Differential pressure from atmospheric pressure, Pa	+10 to +50
Temperature, °C	27 ± 2
Relative humidity, %	52 ± 6

Group 2, NiOGTC[NiO(+), Tobacco (−)]. Inhalation of NiO(G) for 6 months followed by a clearance period without treatment for 1 year.

Group 3, NiOGCT[NiO(−), Tobacco (+)]. Inhalation of tobacco smoke for 1 month followed by a clearance period without treatment for 17 months.

Group 4, NiOGCTC[NiO(−), Tobacco (−)]. Nontreated control, sacrificed after 1½ years.

We examined the tissues of the upper respiratory organs including paranasal sinuses, lungs, liver, pancreas, kidneys, and spleen by both light microscopy (LM) and electron microscopy (EM) as well as by a careful gross examination. For LM studies, the tissue slices were fixed in 10% formalin. For EM work, 1-mm cubes of the same tissue were fixed in 2.5% phosphate-buffered glutaraldehyde overnight, postfixed in 1% phosphate-buffered osmium tetroxide, dehydrated in a graded ethanol series, and embedded in Epon 812. Ultrathin sections were cut with a diamond knife and stained with uranyl acetate and lead citrate. The thick–thin correlative technique was used to relate the osmiophilic material (surfactant), which was visible in thick sections of light microscopy, and its corresponding ultrastructural continuum (i.e., the adjacent thin sections of the electron micrograph). A JEM 100 CX electron microscope was used for the investigation of ultrastructure.

3. RESULTS

Experimental results are summarized in Table 39.2. An increased incidence of neoplastic lesions in the respiratory tract could not be related to the inhalation of NiO(G) with or without tobacco smoke. All the animals failed to demonstrate any neoplastic lesions in the upper respiratory tract, not withstanding the careful investigation of paranasal sinuses (Figs. 39.1 and 39.2).

The number of lesions of pulmonary adenomatosis due to alveolar cell hyperplasia, mainly of type II pneumocytes, was also not increased by the inhalation experiments (Fig. 39.3). However, hyperplasia of terminal bronchiolar epithelium, mainly due to Clara cell proliferation, was prominent in

Table 39.2 Experimental Yield of Tumors

	Pulmonary Adenomatosis		Pancreatic Adenoma		Other Tumors, N	No Neoplasm	Total
	N	%	N	%			
Group 1: NiOGT [NiO(+), Tob(+)]	2	9	4	17	0	19	23
Group 2: NiOGTC [NiO(+), Tob(−)]	1	2	2	4	Fibrosarcoma, 1; renal carcinoma, 1	42	47
Group 3: NiOGCT [NiO(−), Tob(+)]	3	14	1	5	Fibrosarcoma, 1: osteosarcoma 1	16	22
Group 4: NiOGCTC [NiO(−), Tob(−)]	3	8	1	3	0	34	38
Total	9	7	8	6	4	111	130

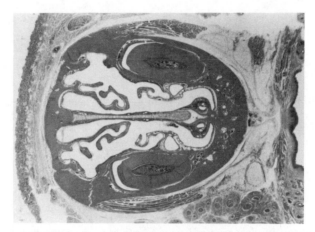

Figure 39.1. Paranasal sinuses, anterior region. Group 3. No neoplastic lesion is evident. (H&E stain, ×10)

NiO(G) inhalation groups (groups 1 and 2) when compared with the other groups without NiO inhalation (groups 3 and 4). As shown in Figure 39.4, group 1 demonstrated Clara cell hyperplasia, appearance of a macrophage, and peribronchial lymphatic permeation of osmiophilic materials. The electron micrographs in Figures 39.5–39.7 reveal prominent lysosomal granules in the ciliated epithelial cells, projection of Clara cells, and accumulation of tubular myelin in the peribronchial lymphatic lumen. An enlarged view of the tubular myelin and phagocytosis by endothelial cells is shown in Figure 39.6.

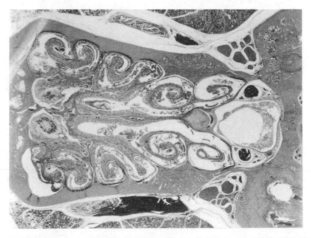

Figure 39.2. Paranasal sinuses, posterior region. Group 3 (same rat as Fig. 39.1). No neoplastic lesion is indicated. (H&E stain, ×10)

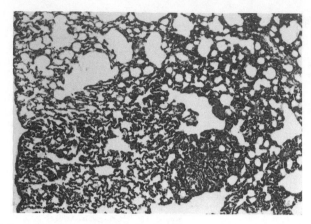

Figure 39.3. Pulmonary adenomatosis. Group 3. Note hyperchromatic epithelium in the lower two-thirds of the plate. (H&E stain, ×100)

One of the other neoplastic lesions noted in all four groups is pancreatic adenoma of the acinar cell type (Fig. 39.8). The pancreas of both treated and nontreated rats equally contained islets that were three- to fourfold larger than normal and also occurred in greater numbers per unit area of pancreas. In addition, malignant neoplastic tumors possibly of spontaneous origin listed in Table 39.2 are fibrosarcomas (groups 2 and 3), a renal carcinoma (group 2), and an osteosarcoma (group 3).

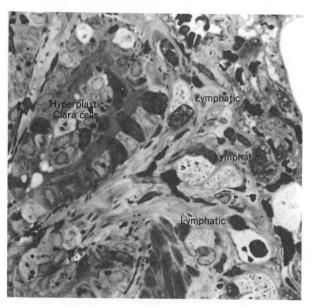

Figure 39.4. Clara cell hyperplasia (left) and osmiophilic materials in the peribronchiolar lymphatics (right). Group 1. Toluidine blue (plastic section). (×1000)

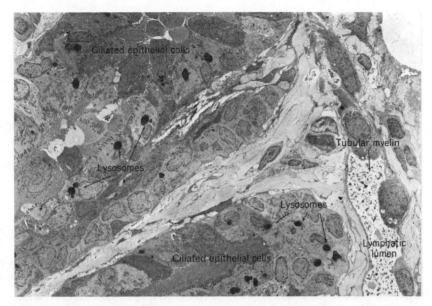

Figure 39.5. Electron micrograph revealing prominent lysosomal granules (round black granules) in the ciliated epithelial cells, projection of Clara cells (upper part) and accumulation of tubular myelin in the peribronchial lymphatic lumen (right lower part). The thick–thin correlation specimen of Fig. 39.4. Group 1. Double stain. (×1500)

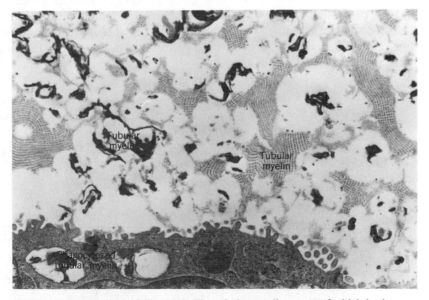

Figure 39.6. Enlarged view of Fig. 39.5. The tubular myelin, some of which is phagocytosed by endothelial cells. Group 1. Double stain. (×15,000)

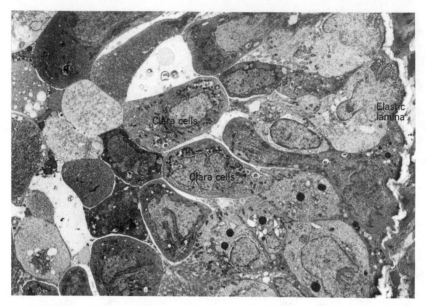

Figure 39.7. Projection and apocrine secretion of Clara cells in the bronchiolar epithelium (central part). Group 1. Double stain. (×1500)

4. DISCUSSION

The experiments illustrate that insoluble NiO(G) particles activated the epithelial cells of bronchioles and type II pneumocytes, which confirmed our previous report (Horie et al., 1985). As before, some of the NiO(G) particles were retained in the cells and also in the lysosomes of macrophages for more than one year.

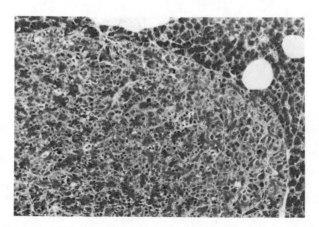

Figure 39.8. Pancreatic adenoma of acinar cell type. Group 1. (H&E, ×250)

It is concluded that NiO(G) is not carcinogenic to rats under the conditions of the present inhalation study. Surprisingly, tobacco smoke did not potentiate the carcinogenicity of NiO(G). Carcinogenic substances derived from tobacco smoke are expected to cling to mineral particles. However, the amount of tobacco smoke inhaled by the rats might be insufficient to provoke carcinoma in the respiratory epithelium. When considering the insignificant differences in cancer incidences between the experimental and control groups, the overall neoplastic lesions might be associated with the aging process (Goodman et al., 1979). As for the hyperplasia of pancreatic islets, these features are known in aged fatty rats (Boder and Johnson, 1972).

5. CONCLUDING REMARKS

The experimental data slightly increased the numbers of neoplastic lesions, namely pulmonary adenomatosis, pancreatic adenomas, and sarcomas in treated groups compared to controls. However, the data are not statistically significant, and particles of NiO(G) do not appear to constitute a carcinogenic substance.

ACKNOWLEDGMENTS

We wish to thank the Nickel Producers Environmental Research Association for their encouragement throughout these experiments.

REFERENCES

Boder, G. B., and Johnson, I. S. (1972). Summary of discussion at the Fiftieth Anniversary Insulin Symposium. *Diabetes* **21** (Suppl. 2), 535–537.

Goodman, D. G., Ward, J. M., Squire, R. A., Chu, K. C., and Linhart, M. S. (1979). Neoplastic and nonneoplastic lesions in aging F344 rats. *Toxicol. Appl. Pharm.* **48**, 237–248.

Horie, A., Haratake, J., Tanaka, I., Kodama, Y. and Tsuchiya, K. (1985). Electron microscopical findings with special reference to cancer in rats caused by inhalation of nickel oxide. *Bio. Trace Element Res.* **7**, 223–239.

Ottolenghi, A. D., Haseman, J. K., Payne, W. W., Falk, H. L., and MacFarland, H. N. (1974). Inhalation studies of nickel sulfide in pulmonary carcinogenesis of rats. *J. Natl. Cancer Inst.* **54**, 1165–1172.

Sunderman, F. W., and Donnelly, A. J. (1965). Studies of nickel cancinogenesis. Metastasizing pulmonary tumors in rats induced by the inhalation of nickel carbonyl. *Am. J. Path.* **46**, 1027–1041.

Tanaka, I., Hayashi, H., Horie, A., Kodama, Y., Akiyama, T., and Tuchiya, K. (1982). Toxicity of a nickel oxide aerosol by inhalation. *J. UOEH* **4**, 441–449.

40

EXPERIMENTAL STUDIES ON METAL CARCINOGENESIS: A COMPARATIVE ASSESSMENT OF NICKEL AND LEAD

Y. Teraki

Department of Histology, Nippon Dental University, School of Dentistry at Niigata, 951 Japan

A. Uchiumi

National Chemical Laboratory for Industry, Tsukuba Research Center, Tsukuba, Ibaragi, 305 Japan

Nickel and Human Health: Current Perspectives, Edited by Evert Nieboer and Jerome O. Nriagu.
ISBN 0-471-50076-3   © 1992 John Wiley & Sons, Inc.

1. INTRODUCTION

This chapter constitutes an investigation, through comparative assessments, of the carcinogenicity of Ni(II) and Pb(II) salts by mutagenicity tests in bacteria and animal injection studies. Concentrations and distribution of these and other elements in soft and bone tissues are also reported.

2. MUTAGENICITY STUDIES

Assessments for mutagenicity using the Ames test and rec-assay were carried out. The former test was performed with *Salmonella typhimurium* strains TA-100 and TA-98 in the usual manner, with and without metabolic S-9 activation. Ames tests without incorporation of the S-9 preparation were negative with both nickel nitrate and lead acetate. With metabolic activation, the nickel salt was negative, while lead acetate was positive. The rec-assay was conducted on *Bacillus subtilis* strains H-17 (Rec$^+$) and M-45 (Rec$^-$) by the streak method of Kada et al. (1972). Assay paper discs in the latter procedure were impregnated with 50 μL of each sample solution containing 1 μmol of the metal. Results of the rec-assay were recorded as the difference in distance, in millimeters, between the edge of an assay disc and that of streaks and the diameter of inhibition zones for Rec$^+$ and Rec$^-$ organisms on agar plates. For interpretation of the assay results, the following criteria were employed: a difference of at least 3 mm, positive (+); at most 3 mm, doubtfully positive (±); or 0 mm, negative (−). The Ni(NO$_3$)$_2$·6H$_2$O was noted to have a negative effect and Pb(COOCH$_3$)$_2$·3H$_2$O a positive effect in the test (Table 40.1). For comparison, data for other metal salts are also summarized in Table 40.1.

3. ANIMAL STUDIES

Nickel acetate and lead acetate were tested in vivo in rats for carcinogenicity. The type, incidence, and histological features of tumors induced, if any, and distribution and concentration of inorganic metals in organs were evaluated.

3.1. Procedures

Male F344 rats 5–6 weeks old were injected with nickel acetate (30 mg/kg) or lead acetate (60 mg/kg) subcutaneously into the dorsal back once or twice a week for a period of 5–8 weeks. Controls received equal volumes (0.2 ml per animal) of physiological saline. The animals were followed by daily observation for clinical signs and by periodic hematological, blood chemical, and urine examinations over 80 weeks. Those with positive signs of illness and/or abnormal laboratory values were sacrificed and necropsied.

Table 40.1   Rec-Assay Results[a]

Experiment number	Elements	Compound	S-9	Rec Effect
1	Fe	$FeSO_4 \cdot 7H_2O$	−	±
			+	±
2	Fe	$FeCl_3$	−	±
			+	±
3	Co	$CoCl_2$	−	+++
			+	+++
4	Ni	$Ni(NO_3)_2 \cdot 6H_2O$	−	−
			+	−
5	Ru	$RuCl_3$	−	±
			+	±
6	Zn	$ZnSO_4$	−	++++
			+	++++
7	Zn	Zn protein (Zn 6.58%)	−	−
			+	−
8	Sn	$SnCl_2 \cdot 2H_2O$	−	±
			+	±
9	Sn	$(C_6H_5)_3SnCOOCH_3$	−	++++
			+	++++
10	Sn	$(C_4H_9)_3SnCl$	−	±
			+	±
11	Pb	$Pb(COOCH_3)_2 \cdot 3H_2O$	−	±
			+	+
12	As	$Na_2HAsO_4 \cdot 12H_2O$	−	+
			+	+
13	As	$KAsO_2$	−	+++
			+	+++

[a] Symbols: − , $Rec^- - Rec^+ = 0$ mm; ± , $Rec^- - Rec^+ = 0$ [X] 3 mm; + , $Rec^- - Rec^+ = 3$ mm; + , 10 μ mol/disc; ++ , 1 μ mol/disc; +++ , 0.1 μ mol/disc; ++++ , 0.05 μ mol/disc.

Tumor tissues and organs from rats were subjected to chemical analyses for inorganic elements by inductively coupled plasma emission spectrometry (ICP-ES) and other methods. Tumor tissue was further examined by X-ray fluorescence element mapping spectrometry (XEMS) to delineate the distributions of selective inorganic elements.

3.2. Induction of Tumors

Of 16 rats injected with nickel acetate, 5 were found to have developed subcutaneous tumors between weeks 22 and 66 after injections. In 8 of 14 rats receiving lead acetate, tumors were detected in the subcutaneous tissue between weeks 32 and 78. All these subcutaneous tumors were microscopically diagnosed as fibrosarcomas (Table 40.2; Figs. 40.1 and 40.2). Fourteen control

Table 40.2 Tumors Induced in Rats at Site of Subcutaneous Injection of Lead Acetate or Nickel Acetate[a]

Pathology Number	Dosage (mg/kg)	Total Dose	Period Tumor Detected (weeks after injection)	Size and Location of Tumors	Microscopic Diagnosis
N-5	30 mg/kg	150 mg/kg	91	4×3×3 cm dorsal back/s.c.	Fibrosarcoma
N-1	30 mg/kg	240 mg/kg	22	3×3×2 cm, posterior abdomen	Fibrosarcoma
N-4	30 mg/kg	240 mg/kg	43	4×4×3 cm, dorsal back/s.c.	No microscopic examination
N-12	30 mg/kg	240 mg/kg	47	3.5×3.5×1.8 cm, dorsal back and abdomen s.c.	Fibrosarcoma
N-8	30 mg/kg	240 mg/kg	66	4×4×4 cm, dorsal back/s.c.	Fibrosarcoma
Pb-1	60 mg/kg	300 mg/kg	39	Dorsal back	Fibrosarcoma
Pb-2	60 mg/kg	300 mg/kg	66	Dorsal back	Fibrosarcoma
Pb-3	60 mg/kg	300 mg/kg	78	Dorsal back	Fibrosarcoma
Pb-5	60 mg/kg	300 mg/kg	52	Dorsal back	No microscopic examination
Pb-13	60 mg/kg	400 mg/kg	62	Dorsal back	Fibrosarcoma
Pb-14	60 mg/kg	400 mg/kg	35	Dorsal back	Fibrosarcoma
Pb-15	60 mg/kg	400 mg/kg	35	Dorsal back	Fibrosarcoma
Pb-16	60 mg/kg	400 mg/kg	53	Dorsal back	Fibrosarcoma

[a]At start of injection rats are 5-weeks old, mean body weight is 152 g.; s.c., subcutaneous

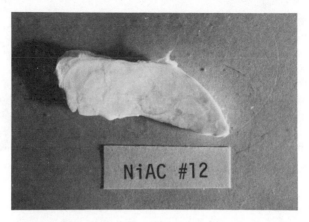

Figure 40.1. Subcutaneous tumor in dorsal back of rat following eight 30-mg/kg subcutaneous doses of nickel acetate (3.5 × 3.5 × 1.8). Diagnosis: fibrosarcoma.

rats injected with physiological saline in the same manner showed no abnormalities, nor evidence of tumor growth.

Gross and histopathological findings of the heart, lungs, spleen, liver, kidneys, pancreas, and skeletal muscle and hematogical findings were unremarkable in both groups. Organs and tissues of the tumor-bearing rats were quantitatively analyzed for inorganic elements such as Ca, Cu, Zn, P, Pb, Ni, Mg, and Mn. Blood of tumor-bearing rats of the nickel-injected group showed no detectable level of nickel. In the liver of these rats, 0.14 µg/g wet weight (or 0.49 µg/g dry weight) of Ni was detected; the concentration

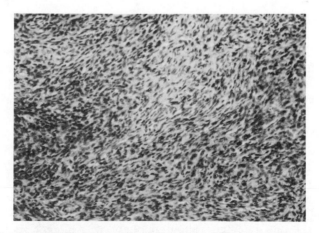

Figure 40.2. Photomicrograph of subcutaneous tumor in rat following five 60-mg/kg subcutaneous doses of lead acetate (H&E stain, ×100). Diagnosis: fibrosarcoma. Note the slender, fusiform tumor cells arranged irregularly in intricate bundles, with prominent anisocytosis and frequent mitotic figures.

of P being highest, followed in order by Mg, Fe, Ca, Zn, Cu, Mn, and Pb. Other tissues examined included the hair, tongue, brain, heart, spleen, kidneys, stomach, duodenum, pancreas seminal vesicle, skeletal muscles, femora, mandibula, incisors, and neoplastic tissue, among which the femur was the only tissue formed to contain Ni (0.42 µg/g). In the tumors of rats treated with nickel acetate, P showed the highest concentration, followed in order by Ca, Mg, Fe, Zn, Cu, and Mn. In tumor-bearing rats of the lead-injected group, both the soft and hard tissues showed high Pb contents (Table 40.3) and the tumor tissue Pb, Ca, and Fe levels were significantly higher than control levels, namely 265-, 153-, and 30-fold increases, respectively.

3.3. Tumor Tissue Analysis by XEMS

A new data-processing unit using a commercially available 16-bit personal computer was developed for XEMS. With this unit, elemental distribution can be presented as maps or bird's-eye views. Maps can feature either color codes or shaded patterns.

The subcutaneous neoplasms induced by lead acetate were further analyzed by XEMS for localization of inorganic elements. The tumor tissues exhibited markedly increased distribution of Ca, Fe, and Pb compared to those in normal tissues of controls (Fig. 40.3). As can be seen, lead was detected predominantly at the center of tumor, while iron accumulated in necrotized areas, surrounded by localization of calcium. This pattern seemed to reflect a process of formation of the tumor and coincided with ICP-ES analytical findings. Analyses of tumors from nickel-injected rats by the same procedures failed to reveal any evidence for occurrence of nickel.

4. CONCLUSIONS

When assessed by the Ames test, a commonly used procedure, the nickel and lead compounds as well as other metals gave negative results in the *S. typhimurium* tester strains without metabolic activation. With the incorporation of hepatic microsomal S-9 fraction, however, lead acetate was positive while nickel nitrate remained negative. The nickel and lead compounds, as well as compounds of other metals such as arsenic, tin, and cobalt, were further examined for mutagenic potential by a rec-assay (again, with or without S-9 activation). The nickel compound gave negative results while the lead compound was positive in replicate assays. From these results of the tests in vitro, experiments were designed to examine the carcinogenic potential of nickel and lead acetate in mammals.

Fischer 344 rats injected subcutaneously with nickel acetate or lead acetate solution into the dorsal back developed subcutaneous neoplastic growths in the region of injection site at week 22 or 35 posttreatment. Fourteen control rats injected with physiological saline in the same manner showed

Table 40.3 Inorganic Element Levels in Tissues of Rats Bearing Lead Acetate–induced Tumors[a]

Tissue	K	Na	Cu	Ca	Zn	P	Pb	Mg	Mn	Fe
Liver	1:1.21	1:1.3	1:1.44[b]	1:2.2[b]	1:2[b]	1:1.3	1:24[b]	1:1.1	1:0.63	1:1.6
Kidney	1:1.1	1:1.15	1:0.47	1:2.8[b]	1:1	1:0.87	1:31[b]	1:0.8	1:0.53	1:0.60
Spleen	1:2.3	1:30[b]	1:1.57[b]	1:10[b]	1:4.7[b]	1:3[b]	1:4[b]	1:34[b]	1:36[b]	1:1.03
Duodenum	1:0.9	1:1.1	1:0.77	1:1.27	1:0.7	1:0.9	1:0.9	1:0.12	1:2.3	1:0.47
Muscles	1:0.87	1:225	1:1.6[b]	1:2.2[b]	1:1.76[b]	1:1.2	1:11[b]	1:1.26	1:1.44	1:12[b]
Lung	1:2.75	1:2.1	1:2.7[b]	1:4.5[b]	1:2.0[b]	1:2.7[b]	1:0.28	1:2.4	1:1.56	1:0.78
Brain	1:1.1	1:1.0	1:1.32	1:5[b]	1:1.1	1:1.15	1:1.0	1:1.18	1:1.05	1:0.81
Testes	1:1.05	1:1.35	1:1.64[b]	1:17[b]	1:1.28	1:1.7[b]	1:1.08	1:1.1	1:1.1	1:2.3
Femur	1:1.5	1:5.6[b]	1:1.28	1:3885[b]	1:12[b]	1:54[b]	1:216[b]	1:12.4	1:0.8	1:0.92
Mandibular bone	1:0.95	1:1.1	1:1.1	1:0.97	1:1.02	1:0.95	1:69.8	1:0.97	1:0.93	1:1.2
Incisors	1:1.07	1:24[b]	1:2.7[b]	1:0.8	1:1.26	1:1.23	1:44[b]	1:0.93	1:1.74	1:1.0
Subcutaneous	1:2.0	1:2.16	1:2.6[b]	1:153[b]	1:1.9[b]	1:9.0[b]	1:265[b]	1:4.04	1:2.43	1:30[b]

[a]Values are expressed as a ratio relative to concentrations in control tissues.
[b]$p < 0.01$.

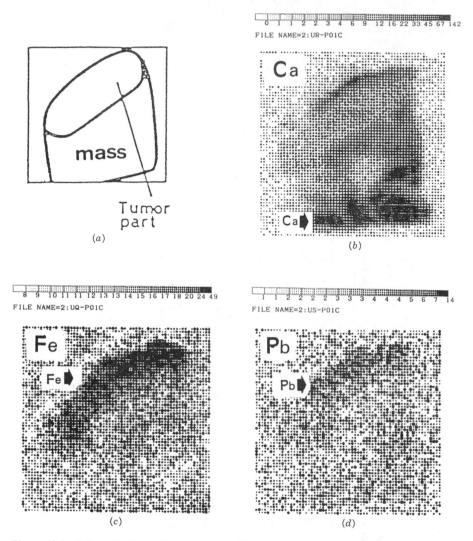

Figure 40.3. Schematic illustration of section of lead-induced tumor (*a*) and inorganic element distribution in the section irradiated (*b* to *c*). Arrows indicate localization of metals.

no abnormalities and no evidence of tumor growth. The tumors, microscopically fibrosarcomas, occurred with an incidence of 31.3% with the nickel acetate and 57.1% with the lead acetate. Chemical analysis of tissues revealed the presence of low concentrations of nickel in the liver and femora 76 weeks after dosing, but not in any other organs or tissues examined of the rats bearing nickel-induced tumor. The results suggest that nickel is absorbed from the tissue at the site of injection into organs and is cleared relatively rapidly. By contrast, in the rats with lead-induced tumors, remarkably high concentrations of lead were demonstrated in both the soft and hard tissues. To follow the process of tumor development in the rats locally injected with metal compounds, we examined the metal distribution in sections of the induced fibrosarcoma in toto by means of XEMS developed at this laboratory. The observation failed to reveal the occurrence of nickel in any of the nickel-induced tumors examined. In the tumors induced with lead, on the other hand, lead and iron localized centritumorally were demonstrated with peri-tumoral calcium and potassium deposition. The finding is of considerable interest when viewed in conjunction with the process of tumor formation in this experimental system.

REFERENCE

Kada, T., Sadaie, Y., and Tutikawa, K. (1972). *Mutat. Res.* **16,** 165–174.

41

EFFECT OF INHALED GREEN NICKEL OXIDE ON THE LIPID LEVELS OF RAT LUNGS

Shuji Cho

Faculty of Home Life Science, Fukuoka Women's University, Fukuoka 813, Japan

Isamu Tanaka, Yasushi Kodama, Akio Horie, Joji Haratake, and Kenzaburo Tsuchiya

School of Medicine, University of Occupational and Environmental Health, Kitakyushu 807, Japan

Nickel and Human Health: Current Perspectives, Edited by Evert Nieboer and Jerome O. Nriagu.
ISBN 0-471-50076-3   © 1992 John Wiley & Sons, Inc.

1. INTRODUCTION

Health effects of nickel metal and its compounds have been reviewed extensively (Kodama and Ishinishi, 1980; Brown and Sunderman, 1980; Sunderman, 1984). However, there are very few reports on the inhalation toxicity of these compounds. Considering the main exposure route of workers in an occupational environment, inhalation studies are very important.

The effects of inhalation of nickel metal dust on the lipids of male rabbits have been investigated. Chronic exposure to respirable nickel dust resulted in a significant increase in the concentration of lipids in the lungs of rabbits, specifically phosphatidylcholine (Bruce et al., 1980). By contrast, studies on the inhalation of nickel oxide (NiO) and lung lipids formation are not known. This chapter describes such work in inhalation experiments in which rats were exposed to green nickel oxide [NiO(G)].

2. EXPERIMENTAL PROCEDURE

2.1. Test Substance and Exposure System

Respirable aerosols of NiO(G) were generated and transported to exposure chambers. The green nickel oxide powder (Soekawa Chemicals, Japan) was ground in a ball mill to reduce the size of the particles and then fed into a continuous fluidized bed generator (Tanaka et al., 1985).

2.2. Experimental Design

The NiO(G) aerosol concentration in the exposure chamber was monitored continuously by a light-scattering method and was measured gravimetrically at daily intervals by the isokinetic suction of air through a glass fiber filter. The daily average aerosol concentrations as NiO were 1.2 ± 0.3 mg/m^3 in the high-exposure group and 0.3 ± 0.2 mg/m^3 in the low-exposure group (mean $\pm$ SD). The size distribution of the aerosol in the exposure chambers was determined by using an Andersen cascade impactor. The mass median aerodynamic diameter and geometric standard deviation were 0.6 μm and 1.6, respectively (Tanaka et al., 1985).

2.3. Experimental Animals

Male Wistar rats (2 months old) fed on a standard laboratory chow were exposed to NiO(G) aerosols for 7 hr/day, 5 days/week, for up to 12 months. Controls were exposed to clean air in adjacent and identical chambers under similar flow (150 L/min), temperature ($24 \pm 1°C$), and humidity ($50 \pm 5\%$) conditions (Tanaka et al., 1985). The present study is composed of two components. In experiment 1, rats were exposed for 3, 6, or 12 months and then sacrificed immediately after the termination of the exposure; in experiment 2, rats were exposed for 12 months and were then kept for an 8-month clearance period before sacrifice. The control rats were also killed at the same time as the experimental animals.

2.4. Measurement of Nickel and Lipid Levels in the Lung

The nickel content in rat lung was measured by a flameless atomic absorption spectrometer (Hitachi, model 170-70, Japan) (Tanaka et al., 1985). The lung lipids were extracted and purified (Folch et al., 1957) and were finally dissolved in chloroform. Triglyceride (Fletcher, 1968), cholesterol (Sperry and Webb, 1950), and phospholipid (Fiske and Subbarow, 1925) were determined on an aliquot of this solution. Individual phospholipid fractions were separated with thin-layer chromatography (Sugano et al., 1970) and the fatty acid composition of lung phosphatidylcholine was assessed (Sugano et al., 1970).

3. RESULTS

3.1. Body and Lung Weights of Rats

In Table 41.1, the body and lung weights of rats are summarized. There were no significant differences in the end body weight in experiment 1 of exposed and control groups within each exposure period. Lungs of rats exposed to NiO were gray in color, in contrast to the normal pink color of control rats. Lung weight of rats tended to be elevated as the air nickel concentration increased for various exposure times. Lungs corresponding to high levels of nickel oxide were approximately twice as heavy as control lungs after both the 6- and 12-month exposure periods. In experiment 2, lung weight remained high in rats that continuously inhaled 1.2 mg/m³ of NiO dust for 12 months and were subsequently kept for the 8-month clearance period.

3.2. Nickel Content of Rat Lung

The nickel concentrations in the lungs of exposed rats in experiment 1 were much higher than those of the controls (Table 41.2). Relative to controls, there was a 900–2200-fold increase in the concentration of lung nickel in the low-exposed groups (L) and 2600–3700-fold in the high-exposed groups (H).

Table 41.1 Body and Lung Weight of Rats[a]

Exposure Period	Number of Rats	Body (g)	Lung (g)	Lung (g/100 g body weight)
		Experiment 1		
3M				
C	5	421 ± 39^a	1.39 ± 0.16^a	0.33 ± 0.02^a
L	5	435 ± 22^a	1.54 ± 0.15^a	0.35 ± 0.02^a
H	4	443 ± 11^a	2.01 ± 0.17^b	0.45 ± 0.03^b
6M				
C	5	498 ± 35^a	1.54 ± 0.08^a	0.31 ± 0.02^a
L	5	495 ± 60^a	1.91 ± 0.20^b	0.39 ± 0.01^b
H	5	497 ± 54^a	3.27 ± 0.46^c	0.66 ± 0.06^c
12M				
C	5	563 ± 39^a	1.62 ± 0.12^a	0.29 ± 0.01^a
L	5	568 ± 88^a	1.96 ± 0.17^b	0.35 ± 0.02^b
H	4	608 ± 27^a	4.15 ± 0.81^c	0.68 ± 0.08^c
		Experiment 2		
12M/8M				
C	1	465	1.93	0.42
L	3	594 ± 48^a	1.87 ± 0.10^a	0.31 ± 0.01^a
H	5	654 ± 49^a	4.19 ± 1.69^b	0.64 ± 0.12^b

[a] Mean $\pm$ SE. Values in a column not followed by the same alphabetic superscript are significantly different ($p < 0.05$). Abbreviations: C, control group; L, low-exposure group; H, high-exposure group. Exposure period is in months; 12M/8M: an 8-month clearance period following 12 months of exposure.

Table 41.2 Nickel Concentrations in Rat Lung (μg/g lung)[a]

Exposure Group	Experiment 1			Experiment 2, 12M/8M[b]
	3 months	6 months	12 months	
C	0.18 ± 0.08^a	0.12 ± 0.01^a	0.20 ± 0.03^a	0.15
L	169 ± 17^b	258 ± 33^b	293 ± 147^b	157 ± 7^a
H	471 ± 44^c	447 ± 140^c	614 ± 237^c	451 ± 238^b

[a] Mean $\pm$ SE. See footnotes to Table 41.1 for explanation of alphabetic superscripts.
[b] An 8-month clearance period following 12 months of exposure.

Table 41.3   Lipid Concentrations of Rat Lung (mg/g lung)[a]

Exposure Period	TC	TG	PL
Experiment 1			
3M			
C	4.8 ± 0.4^a	5.8 ± 0.3^a	28.7 ± 0.3^a
L	6.2 ± 0.4^b	6.8 ± 0.2^b	37.9 ± 1.5^b
H	6.7 ± 0.5^b	7.7 ± 0.7^b	54.9 ± 3.6^c
6M			
C	4.6 ± 0.4^a	5.0 ± 0.8^a	27.4 ± 1.0^a
L	6.7 ± 0.7^b	5.5 ± 0.6^a	44.3 ± 1.1^b
H	9.5 ± 1.3^b	6.1 ± 1.5^a	64.5 ± 7.6^c
12M			
C	7.9 ± 1.3^a	26.8 ± 4.2^a	25.5 ± 1.8^a
L	6.3 ± 0.6^a	31.0 ± 6.9^a	32.3 ± 1.9^b
H	6.6 ± 0.7^a	19.6 ± 3.0^a	99.1 ± 10.3^c
Experiment 2			
12M/8M			
C	4.1	—	22.5
L	4.5 ± 0.4^a	15.1	23.4 ± 4.2^a
H	7.7 ± 1.1^a	13.0 ± 1.5	65.5 ± 18.0^b

[a] Mean $\pm$ SE. See the footnotes to Table 41.1. Abbreviations: TC, total cholesterol; TG, triglyceride; PL, phospholipid.

In experiment 2, the concentrations of lung nickel in the low-exposed group was about 1000 times the control values, whereas in the high-exposed group it was elevated about 3000-fold.

3.3. Lipid Concentrations in Rat Lung

Inhalation of NiO(G) caused a significant increase in the levels of phospholipid (PL) in the lung in experiment 1, proportional to the aerosol concentration (Table 41.3). There were smaller increases in the amounts of total cholesterol (TC) and triglyceride (TG) in the lung of NiO-exposed animals, which reached significance only after 3-month (TC and TG) or 6-month (TC only) exposure. These data suggest that the enlargement of rat lung by NiO inhalation was due in part to the accumulation of lipids, especially phospholipids. The levels of pulmonary PL in high-exposed rats in experiment 2 were higher than those of the low-exposed and control groups, while there were no changes in TC and TG (Table 41.3).

3.4. Concentrations of Lung Phospholipids

Accompanying the increase in PL (Table 41.3), there were increases in the individual PL components (Table 41.4). In both experiments, the fraction

Table 41.4   Concentrations of PC and PE in Rat Lung (mg/g lung)[a]

Exposure Period	PC	PE
Experiment 1		
3M		
C	12.8 ± 1.1[a]	8.4 ± 0.5[a]
L	16.6 ± 2.6[a]	7.9 ± 1.7[a]
H	29.4 ± 3.1[b]	10.5 ± 3.6[a]
6M		
C	19.6 ± 3.6[a]	2.2 ± 0.9[a]
L	29.7 ± 1.3[a]	10.2 ± 3.7[a,b]
H	47.4 ± 7.9[b]	9.7 ± 2.3[b]
12M		
C	11.5 ± 0.4[a]	3.8 ± 0.2[a]
L	16.5 ± 2.0[a]	4.8 ± 1.8[a]
H	50.5 ± 6.6[b]	17.3 ± 3.9[b]
Experiment 2		
12M/8M		
C	10.4	3.2
L	8.4 ± 3.7[a]	1.6 ± 0.6[a]
H	43.1 ± 15.8[b]	4.5 ± 1.4[b]

[a] Mean ± SE. See the footnotes to Table 41.1. Abbreviations: PC, phosphatidylcholine; PE, phosphatidylethanolamine.

of phosphatidylcholine (PC) and phosphatidylethanolamine (PE) increased with exposure (except PE in the 3-month exposrue). From these data, it is concluded that accumulation of pulmonary phospholipids was due, in large part, to PC formation. In Figure 41.1, a relationship between the nickel and PL contents in the lung is depicted. It is clear that the amount of phospholipids increased in proportion to the accumulation of nickel ($r = 0.98$).

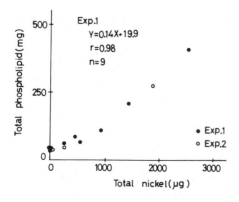

Exp.1
Y=0.14X+19.9
r=0.98
n=9

• Exp.1
○ Exp.2

Figure 41.1. Relationship between total nickel and phospholipid concentrations in rat lung.

Table 41.5   Palmitic Acid Composition of PC in Rat Lung (%)[a]

Exposure	Experiment 1		
	3 months	6 months	12 months
C	54.5 ± 0.4^a	51.6 ± 0.8^a	53.0 ± 1.3^a
L	57.1 ± 1.0^b	59.6 ± 0.8^b	$58.4 \pm 2.8^{a,b}$
H	61.8 ± 0.6^c	63.7 ± 1.2^c	65.5 ± 2.5^b

[a] Mean $\pm$ SE. See the footnotes to Table 41.1 for explanation of alphabetic superscripts. Abbreviation: PC, phosphatidylcholine.

3.5.   Palmitic Acid Composition of Lung PC

The fatty acid composition of lung PC was measured, and palmitic acid was the major component (Table 41.5).

4.   DISCUSSION

In this study, the effect of the exposure to NiO(G) on lipid accumulation in the lungs of male Wistar rats was investigated. Chronic inhalation by rats of NiO(G) caused significant increases in lung weight and in lipid concentrations. Although there were small increases in the amounts of cholesterol and triglyceride in the lung of NiO-exposed rats, the major increase was in the PL fraction. It is known that chronic exposure to nickel dust (Bruce et al., 1980), asbestos (Tetley et al., 1976), and diesel particulates (Eskelson et al., 1987) also increased pulmonary lipid concentrations. We found that pulmonary phospholipidosis was accompanied by increased PC and palmitate contents. Therefore, by analogy to diesel fumes (Eskelson et al., 1987), it is concluded that dipalmitoylphosphatidylcholine (DPPC) is a major component of the PLs that accumulate in rat lung on exposure to NiO(G). The DPPC is a primary constituent of lung surfactants and is necessary to maintain alveolar opacity during respiration. It is of interest that asbestos has been shown to bind large quantities of surfactant, most of which is PC, and that nickel powder has also been found to bind PC (Bruce et al., 1980). Such an interaction between PC and inhaled materials could be a part of the pulmonary clearance and/or defense mechanism. The good correlation observed between PL and nickel content suggests that PL level may serve an an indicator of the lung burden of toxic particulate materials.

REFERENCES

Brown, S. S., and Sunderman, F. W. (1980). *Nickel Toxicology*. Academic Press, New York.

Bruce, M. C., Camner, P., and Curstedt, T. (1980). Influence of chronic inhalation exposure to nickel dust on accumulation of lung lipids. In C. L. Sanders et al. (Eds.), *Pulmonary Toxicology of Respirable Particles*. Technical Information Center, Washington, D.C., pp. 357–366.

Eskelson, C. D., Chvapil, M., Strom, K. A., and Vostal, J. J. (1987). Pulmonary phospholipidosis in rats respiring air containing diesel particulates. *Environ. Res.* **44,** 260–271.

Fiske, C. H., and Subbarow, Y. (1925). The colorimetric determination of phosphorus. *J. Biol. Chem.* **66,** 375–380.

Fletcher, M. J. (1968). A colorimetric method for estimating serum triglycerides. *Clin. Chim. Acta* **22,** 393–398.

Folch, J., Lees, M., and Sloane Stanley, G. H. (1957). A simple method for the isolation and purification of total lipids from animal tissues. *J. Biol. Chem.* **226,** 497–509.

Kodama, Y., and Ishinishi, N. (1980). Nickel toxicology and carcinogenicity on nickel. *J. Univ. Occup. Environ. Health* **2,** 99–108.

Sperry, W. M., and Webb, M. (1950). A reversion of the Schoenheimer and Sperry method for cholesterol determination. *J. Biol. Chem.* **187,** 97–106.

Sugano, M., Cho, S., Imaizumi, K., and Wada, M. (1970). Changes in phosphatidylcholine and phosphatidylethanolamine during hepatic injury caused by carbon tetrachloride. *Biochem. Pharmacol.* **19,** 2325–2333.

Sunderman, F. W. Jr. (Ed.-in-Chief) (1984). *Nickel in the Human Environment*, IARC Sc. Publ. 53, Oxford University Press, Oxford.

Tanaka, I., Ishimatsu, S., Matsuno, K., Kodama, Y., and Tsuchiya, K. (1985). Biological half time of deposited nickel oxide aerosol in rat lung by inhalation. *Biol. Trace Element Res.* **8,** 203–210.

Tetley, T. D., Hext, P., Richard, R., and McDermott, M. (1976). Chrysotile induced asbestosis: Changes in the free cell population, pulmonary surfactant and whole lung tissue of rats. *Br. J. Exp. Pathol.* **57,** 505–514.

42

INHIBITORY EFFECTS OF ZINC, MAGNESIUM, AND IRON ON NICKEL SUBSULFIDE CARCINOGENESIS IN RAT SKELETAL MUSCLE

Kazimierz S. Kasprzak and Ricardo E. Rodriguez

Laboratory of Comparative Carcinogenesis, National Cancer Institute, FCRDC, Frederick, Maryland 21702

Nickel and Human Health: Current Perspectives, Edited by Evert Nieboer and Jerome O. Nriagu.
ISBN 0-471-50076-3   © 1992 John Wiley & Sons, Inc.

1. INTRODUCTION

Although nickel in various chemical and physical forms is carcinogenic to humans and animals, the mechanisms involved in the process of tumor induction by this metal remain unclear. Some theoretical and experimental data indicate that among other possible ways, the carcinogenic and other toxic effects of nickel may be mediated through its interaction with some divalent essential metals, for example, calcium, magnesium, manganese, zinc, and others (Sunderman and McCully, 1983; Kasprzak and Waalkes, 1986; Kasprzak et al., 1987a; Sunderman and Barber, 1988). In other words, depending on physicochemical similarities, nickel might substitute for one or more of the essential metals at binding sites of certain biomolecules (e.g., structural proteins, enzymes, nucleic acids) whose conformation and physiological function(s) depend on the metal bound and thus alter their normal function. This may lead to cell death or transformation and cancer. If this scenario were true, then we should be able to counteract nickel action by supplementing the animal with high doses of the essential metal(s). Indeed, Mn(II) supplementation was shown by Sunderman and McCully (1983) to inhibit nickel carcinogenesis in rats. As found in our laboratory, calcium and magnesium prevented nickel-caused increase in the pulmonary adenoma incidence in strain A mice (Poirier et al., 1984), and magnesium, but not calcium, inhibited nickel carcinogenesis in rats (Kasprzak et al., 1985). Following these positive results, we decided to broaden our research on magnesium and extend our investigations on two more essential metals, zinc and iron, which along with the other essential metals seemed to offer some practical applications in prevention of nickel tumors (Waalkes et al., 1985; Kasprzak, 1988; Kasprzak and Poirier, 1985; Kasprzak et al., 1986a,b, 1987b, 1988; Miki et al., 1987). In this chapter, the results of various experiments on nickel interactions with zinc, magnesium, and iron in carcinogenesis have been analyzed comparatively and some general conclusions are drawn.

2. METHODOLOGY

Male Fischer F334/NCr rats from the FCRDC-APA colony (FCRDC, Frederick, MD), 50–100 g of body weight, were randomly divided into treatment groups of 20 rats per group. The rats were injected in the thigh musculature of both hind limbs with 2.5 mg of nickel subsulfide (Ni₃S₂) per site suspended in 0.1–0.2 mL of water or 0.15 M saline. This dose is equal to 31 μmol Ni per site. The experimental groups of rats were given Ni₃S₂ alone or in combination with different metal–nickel molar proportions of the investigated physiological metal compounds listed and characterized in Table 42.1. Both water-soluble and water-insoluble forms of the essential metals were examined. The control groups were injected intramuscularly with the highest tested doses of the physiological metal compounds alone or the injection vehicle. In order to

Table 42.1 Characteristics of Investigated Compounds

Compound	Chemical Formula	Abbreviation	Producer and Purity Grade	Solubility in Water	Injected Dose per Site
Nickel subsulfide	Ni_3S_2	Ni_3S_2	INCO, Toronto, Canada	Insoluble particles, < 10 μm diameter	2.5 mg = 10.4 μmol Ni_3S_2 (31 μmol Ni)
Zinc oxide	ZnO	ZnO	Fisher Scientific, Fair Lawn, NJ; ACS grade	Insoluble particles, ~ 1 μm diameter	0.6–4.9 mg ZnO (7.5–60 1 μmol Zn)
Zinc acetate	$Zn(CH_3COO)_2 \cdot 2H_2O$	ZnAcet	J. T. Baker Chemical, Phillipsburg, NJ; Baker Analyzed Reagent	Soluble	3.4 mg ZnAcet (15.5 μmol Zn)
Magnesium basic carbonate	$4MgCO_3 \cdot Mg(OH)_2 \cdot n\text{-}H_2O$	MgCarb	J. T. Baker Chemical, Phillipsburg, NJ; Baker Analyzed Reagent, containing 40.6 wt % Mg expressed as MgO	Insoluble particles, ~ 4 μm diameter	0.75–25 mg MgCarb (7.5–120 μmol Mg)
Magnesium acetate	$Mg(CH_3COO)_2 \cdot 4H_2O$	MgAcet	J. T. Baker Chemical, Phillipsburg, NJ; Baker Analyzed Reagent	Soluble	13.5 mg MgAcet (60 μmol Mg)
Iron, metallic	Fe	Fe^0	Aldrich Chemical, Milwaukee, WI; Gold Label	Insoluble particles, 325 mesh (<45 μm diameter)	0.9–3.4 mg Fe^0 (15.5–60 μmol Fe)
Iron(III) sulfate	$Fe_2(SO_4)_3 \cdot n\text{-}H_2O$	Fe(III)	J. T. Baker Chemical, Phillipsburg, NJ; Baker Analyzed Reagent, containing 77.1 wt % of anhydrous formula	Soluble	3.9–15.6 mg Fe(III) (15.5–60 μmol Fe)

establish the local effect of and the dose response to a given essential metal, Ni$_3$S$_2$ was injected intramuscularly in a common suspension with that metal compound. Systemic effects were tested by injecting Ni$_3$S$_2$ intramuscularly as before and the other metal subcutaneously at the nape of the neck.

All rats had free access to drinking water and Purina 5001 or NIH formula 31 diet. The rats were weighed and the injection sites examined by palpation every second week until the termination of the experiments at 66–79 weeks. They were sacrificed earlier if a firm mass 0.5–2 cm in diameter developed at any of the injection sites. The tumors, the injection site muscles, and any other tissues with gross lesions were fixed in formalin for histological examination. The incidence of the injection site tumors and the differences among treatments were analyzed statistically with a computer program developed by Thomas et al. (1977).

The retention of nickel, zinc, magnesium, or iron in the injected muscles was determined in separate groups of rats treated as in the bioassay experiments and sacrificed at times from 5 hr to 22 weeks postinjection, at least three rats per time point. One injected muscle of each rat was solubilized in concentrated HCl and H$_2$O$_2$ (Ultrex, J.T. Baker Chemical Co, Phillipsburg, NJ; 5 mL HCl and 0.2 mL 30% H$_2$O$_2$/g wet muscle) for 5 days at room temperature; the second muscle was fixed for histological examination. The recovery of metals from the digests was determined by addition of a series of known amounts of a metal derivative (listed in Table 42.1) to intact muscle samples prior to the digestion. A typical digest contained some fine particulate residue that was separated by centrifugation at 2000 g. The metals were determined by flame atomic absorption spectrometry. The recovery of the added metal derivatives exceeded 90% $\pm$ 5% SD.

3. RESULTS

No acute toxic effects of any of the treatments with Ni$_3$S$_2$ alone or combined with the essential metal compounds were observed. Body weight gain data did not indicate any significant treatment-related depression of growth. None of the animals given the injection vehicle developed a muscle tumor. In the control groups, only one local sarcoma was found at week 48 in 20 rats given 60 μmol ZnO per site (Table 42.1).

The local tumors were predominantly unilateral sarcomas. Histologically most of the tumors were rhabdomyosarcomas (average of all experiments, 72%), the rest being undifferentiated sarcomas (average 22%), fibrosarcomas (average 4%), and histiocytic sarcomas (average 2%). No differences were noticed in the distribution of various histological types of sarcomas among the treatment groups. Bilateral injection site tumors were quite infrequent (0–16%) without significant differences among most of the treatments except for the intramuscular Ni$_3$S$_2$–MgAcet group in which bilateral sarcoma incidence of 42% was significantly higher than in the other tumor-bearing rats. By time of sacrifice, the local sarcomas metastasized readily to the lungs and lymph nodes and occasionally to the kidney, pancreas, and spleen. Primary tumors at locations other than the injection site were found only in few cases with no apparent relationship to the treatments.

3.1. Zinc

The bioassay was terminated after 66 weeks when the incidence of local tumors in the experimental groups exceeded 85%, that is, became statistically identical. The results are shown in Figure 42.1. Addition of different local doses of ZnO to Ni_3S_2 had no influence on the final tumor yield; the negative slope of the dose–effect curve for ZnO illustrated in Figure 42.2 did not differ significantly from zero. However, all doses of ZnO did significantly diminish the tumor occurrence rates ($p < 0.001$, 0.0074, 0.0016, 0.0004 for the Ni_3S_2 group vs. groups receiving Ni_3S_2–ZnO at the Zn–Ni molar ratios of 0.25, 0.5, 1, and 2, respectively). Local administration of water-soluble ZnAcet was equally effective as the administration of insoluble ZnO at the same dose despite their very different retention times in the injected muscle

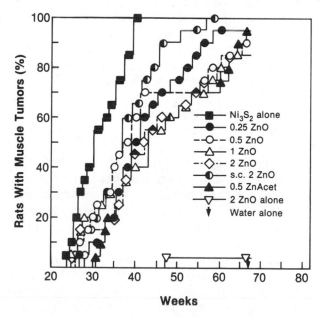

Figure 42.1. Cumulative incidences of local sarcomas in rats injected with Ni_3S_2 alone or combined with different doses of ZnO or ZnAcet. All Ni_3S_2 doses were 2.5 mg per site, equal to 31 μmol Ni per site. The zinc doses are marked as numbers indicating molar proportions of zinc to nickel (0.25–2). All injections of the zinc compounds, except one (subcutaneous), were made intramuscularly together with intramuscular Ni_3S_2.

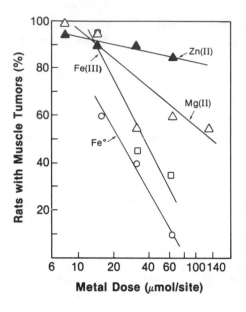

Figure 42.2. Dose effects of essential metals on the final incidence of muscle tumors induced in rats by a constant dose of Ni_3S_2 (31 μmol Ni) after 66 weeks (zinc) or 79 weeks (magnesium and both iron forms).

(Table 42.2). Both compounds delayed the occurrence of the Ni_3S_2-induced sarcomas to the same extent ($p < 0.007$ for Ni_3S_2 with either compound vs. Ni_3S_2 alone). As can be seen from Figure 42.1, the same dose of ZnO admixed to Ni_3S_2 retarded the occurrence of muscle tumors more effectively than when the same dose was injected separately at a site distant from Ni_3S_2 ($p < 0.004$, 0.007 vs. Ni_3S_2 alone, respectively)

Addition of ZnO to Ni_3S_2 also tended to diminish the frequency of metastatic tumors in lungs and kidneys. However, that tendency reached a statistically significant level only in rats treated with the highest dose of ZnO ($p < 0.045$ vs. Ni_3S_2 alone by the χ^2 test).

Histological examination of muscles of rats injected with Ni_3S_2 revealed, at days 1–3, a necrotizing inflammatory reaction around the Ni_3S_2 particles with a dense infiltration by neutrophils and some macrophages. From day 3, there were foci of proliferation of cells of undetermined origin among the muscle fibers surrounding the necrotic areas. Those cells resembled histiocytes or connective tissue stromal cells. The necrotizing inflammatory response slowly subsided with time, persisting in some rats up to week 4. The predominant morphological changes around the inflammatory foci after day 3 consisted of continuous proliferation of the histiocytic–stromal cells with increasing cytologic atypia (high nucleus–cytoplasm ratio, hyperchromasia, pleomorphism, multiple nucleoli) and new muscle fiber formation. Admixture of ZnO or ZnAcet to Ni_3S_2 or distant subcutaneous administration of ZnO following intramuscular Ni_3S_2 did not substantially alter that pattern.

Determination of metals in the injected muscle at days 7, 21, and 35 after treatment showed that nickel was eliminated from the injection site with an approximate half-time of 20 days (Table 42.2) regardless of the presence of the zinc compounds. Therefore, the local ZnO or ZnAcet had no detectable effect on the mobilization of nickel from the injection site. Zinc as ZnO persisted in the muscle longer than nickel and much longer than zinc administered as ZnAcet (Table 42.2).

3.2. Magnesium

The cumulative muscle tumor incidence curves for rats injected with Ni_3S_2 alone or with different combinations of MgCarb or MgAcet are shown in Figure 42.3. As can be seen in this figure, MgCarb significantly suppressed the carcinogenic activity of Ni_3S_2 by decreasing both the tumor occurrence rates and final tumor incidences at the termination of this bioassay at 79 weeks postinjection. For example, at week 45, when Ni_3S_2 alone produced the injection site sarcomas in 100% of the rats, the same Ni_3S_2 dose with added MgCarb induced tumors in 55% of the animals when the magnesium–nickel molar ratio in the mixture was 0.25 and in only 15% of the rats when the magnesium–nickel ratio was 1.0. The magnitude of the suppression depended on the dose of local MgCarb as illustrated in Figure 42.2. The MgCarb administered subcutaneously at a site distant from the Ni_3S_2 injection site was ineffective; the slight retardation of the tumor occurrence

Table 42.2   Retention of Metals in Injected Muscle

Metal Compound Injected[a]	Half-time (days ± SE) ($n = 9$–18)[b]
Ni_3S_2 [c]	20 ± 1.5
MgCarb[d]	0.2 ± 0.03
ZnO	24 ± 3.5
ZnAcet	2.5 ± 0.4
Fe^0	72 ± 5.5
Fe(III)	33 ± 4.0

[a] See Table 42.1 for chemical formulas.
[b] Determined for the first one [ZnO, Fe(III); Fe^0 by extrapolation] or two (MgCarb, ZnAcet, Ni_3S_2) half-life periods from the corresponding time/retention graphs based on metal determinations at 2–24 hr and 1–35 days after a single intramuscular injection.
[c] 10.4 μmol Ni_3S_2/site (equivalent to 31 μmol Ni) injected alone or in common mixtures with an essential metal compound amount equivalent to 62 μmol of the metal.
[d] Retention of the essential metals was determined for compound doses equivalent to 62 μmol of the metal.

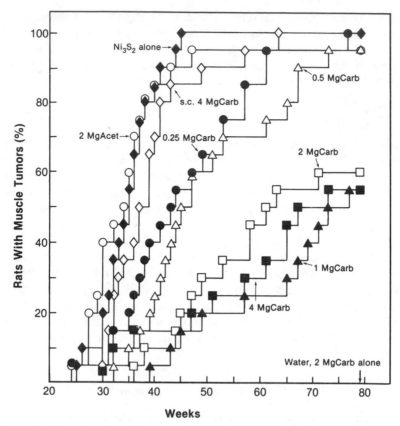

Figure 42.3. Cumulative incidences of local sarcomas in rats injected with Ni₃S₂ alone or combined with different doses of MgCarb or MgAcet. All Ni₃S₂ doses were 2.5 mg per site, equal to 31 μmol Ni per site. The magnesium doses are marked as numbers indicating molar proportions of magnesium to nickel (0.25–4). All injections of the magnesium compounds, except one (subcutaneous), were made intramuscularly together with intramuscular Ni₃S₂.

rate compared to the Ni₃S₂-only group, visible in Figure 42.3, lacks statistical significance. Likewise ineffective was local addition of the water-soluble MgAcet to Ni₃S₂ (Fig. 42.3).

There was no statistically significant influence of the magnesium salts on the frequency of metastatic tumors observed in the lungs, lymph nodes, kidneys, and other sites.

Histological examination of the injection site between day 3 and week 22 postinjection showed marked differences in the intensity and sequence of tissue reactions between the rats administered with Ni₃S₂ alone and/or with MgCarb. The course of the local necrotizing inflammatory response to Ni₃S₂ alone was the same as described in Section 3.1. Admixture of MgCarb to Ni₃S₂ significantly attenuated the initial necrosis and dramatically changed the response by macrophages. The latter became predominating over the few neutrophils present as early as day 3. In the Ni_3S_2–MgCarb rats, the inflammatory reaction lasted up to 3 weeks postinjection with a vigorous infiltration of the dense macrophage and multinuclear giant cell population. Those infiltrates progressed toward resolution with extensive mineralization and fibrosis. Although the injection site lesion in that group appeared more benign than in rats given Ni_3S_2 alone, small foci of the proliferating histiocytic–stromal cells among intact muscle fibers close to the inflammatory site (described in Section 3.1) could be found in some rats from week 12 onward. There was no myoblast formation up to week 22, in contrast to the Ni_3S_2-only rats. In the muscles injected with MgCarb alone, at day 3 there was a foreign body granulomatous myositis that completely resolved by week 6 after treatment.

The results of determination of nickel and magnesium in the injected muscles at times ranging from 5 hr to 112 days after injection revealed no influence of MgCarb on the retention of nickel whose biological half-life remained at approximately 20 days (Table 42.2). The clearance of magnesium from intramuscular MgCarb was much faster, with 50% of the dose removed in approximately 5 hr (Table 42.2).

3.3. Iron

The bioassays for the effects of iron on Ni_3S_2 carcinogenesis lasted 1.5 years. This metal was tested in two forms, as water-insoluble powdered metallic Fe^0 and as soluble Fe(III) sulfate (Table 42.1). The cumulative tumor incidence curves for Ni_3S_2 and Fe^0 are shown in Figure 42.4. As can be easily concluded

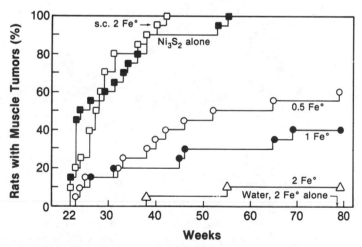

Figure 42.4. Cumulative incidences of local sarcomas in rats injected with Ni_3S_2 alone or combined with different doses of iron powder, Fe^0. All Ni_3S_2 doses were 2.5 mg per site, equal to 31 μmol Ni per site. The Fe^0 doses are marked as numbers indicating molar proportions of iron to nickel (0.5–2). All injections of Fe^0, except one (subcutaneous), were made intramuscularly together with intramuscular Ni_3S_2.

from this figure, Fe^0 was a very effective local antagonist of Ni_3S_2 carcinogenesis in the rat muscle. When administered together with Ni_3S_2 at the iron–nickel molar dose ratio of 2.0, Fe^0 delayed the onset of tumors to 38 weeks and decreased the final tumor incidence to 10% compared to 22 weeks and 100% for Ni_3S_2 alone. As shown in Figure 42.2, that inhibition was dose dependent. Unlike the local injection, the distant subcutaneous administration of Fe^0 had no effect on the carcinogenic activity of Ni_3S_2 (Fig. 42.4).

Figure 42.5 presents the results of the bioassay with Fe(III). This form of iron was also a strong antagonist of the Ni_3S_2 action, though not as effective as Fe^0. It delayed the onset of tumors by at most 9 weeks and decreased the final tumor yield to 35% relative to Ni_3S_2 alone. Its effect was strictly local [see the subcutaneous Fe(III) in Fig. 42.5] and depended on the dose (Fig. 42.2).

Histological examination of the injection site at early stages revealed tissue reactions to Ni_3S_2 plus Fe^0 or Fe(III) resembling those described for MgCarb (see Section 3.2). In the presence of either iron form, however, there was more severe necrosis and more vigorous infiltration by macrophages and giant cells than around Ni_3S_2 alone or Ni_3S_2 combined with MgCarb. Also, the granulomas formed in the injected muscles were relatively larger and more persistent than in the other experiments. The intramuscular injection of Fe^0 or Fe(III) caused foreign body response resembling that produced by MgCarb but with more extensive and longer lasting necrosis and stronger response by the phagocytes. Although more detailed comparative histological

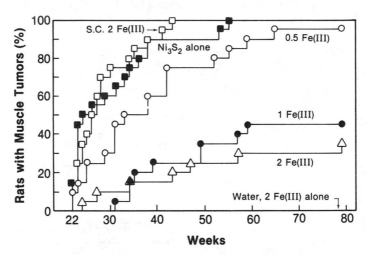

Figure 42.5. Cumulative incidences of local sarcomas in rats injected with Ni_3S_2 alone or combined with different doses of iron sulfate, Fe(III). All Ni_3S_2 doses were 2.5 mg per site, equal to 31 μmol Ni per site. The Fe(III) doses are marked as numbers indicating molar proportions of iron to nickel (0.5–2). All injections of Fe(III), except one (subcutaneous), were made intramuscularly together with intramuscular Ni_3S_2.

evaluation was not made, the necrotizing inflammatory response to Fe(III) seemed to be more severe than that to Fe^0.

Determination of the injected metals in the muscles at days 7, 21, and 35 revealed that both forms of iron were retained substantially longer than nickel, the half-lives being approximately 20, 33, and 72 days for nickel, Fe(III), and Fe^0, respectively. Neither Fe^0 nor Fe(III) had any measurable effect on the retention of nickel.

4. DISCUSSION

The results of our experiments demonstrate that the development of muscle tumors induced by Ni_3S_2 in the rat can be markedly inhibited by a simultaneous administration of one of the physiological metals, zinc, magnesium, or iron. The magnitude of that inhibition depends on the metal, its chemical form and dose, and in most cases the mode of administration. At comparable doses, the capacity of Fe^0 and the salts of the three essential metals tested to inhibit nickel carcinogenesis could be ranked as Fe^0 > Fe(III) > MgCarb > ZnO, ZnAcet. The effect is predominantly local. Only zinc appeared to exert some systemic inhibition of nickel tumors. Magnesium and iron were inhibitory only if injected in a common mixture with Ni_3S_2 but completely ineffective if injected away from the carcinogen. Most interestingly, none of the investigated essential metals had any detectable effect on the gross retention of nickel at the injection site. Thus, the inhibition was not due to any enhanced mobilization of the nickel dose [the carcinogenic activity of Ni_3S_2 is dose dependent (Sunderman, 1982)]. The same result was found for manganese by Sunderman et al. (1976). An important finding of the present experiments is that the inhibition of nickel carcinogenesis by the essential metals is not simply related to the metals' retention times in the muscle; prolonged retention is not necessary. In contrast, a long persistence of Ni_3S_2 in the rat muscle is obligatory for the induction of tumors (Kasprzak, 1978). Indeed, ZnO and ZnAcet, having very different retention half-lives, were equally effective in counteracting Ni_3S_2 carcinogenesis. The inhibition by Fe^0 and Fe(III) at higher doses differed only slightly from each other despite very different retention times in the muscle. And finally, MgCarb, whose $T_{1/2}$ in the muscle was only about 5 hr, was also very effective. However, water-soluble MgAcet, which is absorbed much more readily, did not inhibit Ni_3S_2 action. Hence, there seems to exist a *Minimum effective time* for the presence of a certain elevated concentration of the metal inhibitor to express its activity. This time period looks very short (hours to a few days) relative to the latency of Ni_3S_2 carcinogenesis (>5 months). The shortness of this period indicates that some critical events, possibly involving initiation, in muscle carcinogenesis by nickel subsulfide take place in a few hours or days after the injection of Ni_3S_2. Consequently, the obligatory further presence of Ni_3S_2 at the injection site would be necessary for promotion of the initiated cells. We may thus conclude that magnesium can act only at the tumor initiation stage while the other metals, owing to their long retention, would be able to influence the carcinogenic process also at its later stages. However, some delayed indirect effects of this short magnesium presence mediated through the immune cells discussed below must also be considered.

The results of the present experiments focused our attention on two cell populations that responded to the metals in a most conspicuous way, that is, the target cells (the cells that give rise to nickel tumors: myoblasts, fibroblasts, histiocytes) and the inflammatory/immune cells. Generally, a strong phagocytic reaction is typical for strains less susceptible to nickel carcinogenicity (Daniel, 1966). Nickel-induced sarcoma cells are antigenic (Corbeil, 1968). On the other hand, nickel is known to suppress cellular immunity by affecting macrophages (Benson et al., 1986; Haley et al., 1987), T cells (Smialowicz et al., 1984; Kasprzak et al., 1988), and the natural killer (NK) cells (Smialowicz et al., 1986a,b; Kasprzak et al., 1987b; Judde et al., 1987), which constitute the first-defense line against newly arising tumor cells. Therefore, protection of these cells would lead to the same result as would protection of the target cells alone—fewer tumors. Indeed, manganese, which is a potent inhibitor of nickel carcinogenesis (Sunderman and McCully, 1983), was also found to prevent nickel-caused suppression of the NK cells (Smialowicz et al., 1987b; Judde et al., 1987). Likewise, magnesium counter-acted suppression by nickel of NK cells (Kasprzak et al., 1987b) and T cells (Kasprzak et al., 1988). Essential roles for magnesium and zinc in immune system function are established (Kasprzak and Waalkes, 1986).

The exact sites and mechanisms involved in the essential metals' inhibitory action toward nickel effects on the target and the immune cells cannot be derived from the present experiment. The metals may compete for transport channels and at a variety of intracellular binding molecules including enzymes and nucleic acids in various types of cells. Magnesium, for example, is known to counteract nickel toxicity by inhibiting nickel transport across cell membranes in microorganisms and mammals (Nielsen, 1980; Kasprzak et al., 1986a). It also competes strongly with nickel for binding to the cell nuclei in vivo (Kasprzak and Poirier, 1985) and to DNA in vitro (Kasprzak et al., 1986b). The strength of that competition is consistent with the effectiveness of magnesium's inhibition of nickel carcinogenesis. The mode of zinc's interaction with nickel is less clear. Unlike magnesium, zinc seems to affect nickel toxicity on the pharmacodynamic rather than pharmocokinetic level (Waalkes et al., 1985). Zinc's inability to inhibit nickel transport across cell membranes (and to prevent Ni₃S₂-caused cell necrosis) and very limited ability to antagonize nickel–DNA binding (Kasprzak et al., 1986b) would be consistent with its weak activity against nickel carcinogenesis. The systemic character of the zinc action may indicate involvement of mechanisms other than a direct zinc–nickel interaction at the target cell. Although no morphological indications of alteration by zinc of the cellular response to Ni₃S₂ were observed in the present study, protection of at least some of the immune cells including the NK cells (Fraker et al., 1986) seems possible. Iron, like magnesium and zinc, prevents some toxic effects of nickel in various organisms; interaction of these two metals in biological systems is apparently competitive, but its exact nature remains unknown (Nielsen, 1980). The local action of iron against nickel carcinogenesis, like that of magnesium, is also accompanied by a vivid response from macrophages and giant cells. This effect stresses again the possible involvement of phagocytes in the observed inhibition of nickel carcinogenesis, but a possible direct interaction of nickel and iron at the target cells cannot be excluded. Iron differed from the other two metals by enhancing and prolonging the initial necrosis at the site of Ni_3S_2 injection. Iron compounds alone were also more necrotizing than the compounds of magnesium and zinc tested alone. Whether or not the severity of that necrotizing reaction contributed to the inhibition of tumor induction remains to be determined. Iron is known to generate a highly reactive hydroxyl radical in the Fenton and Haber–Weiss reactions with H_2O_2, which makes it cytotoxic and/or mutagenic in some systems (Loeb et al., 1988). Nickel, in turn, decreases the activity of the H_2O_2-decomposing enzyme glutathione peroxidase (Athar et al., 1987), which in the presence of large numbers of neutrophils will assist in a H_2O_2 buildup in the Ni_3S_2-injected tissue. These two facts put together would explain the severe necrosis by iron and nickel. They also make us pose the following questions:

1. What is the role of the inflammatory and immune system cells in nickel carcinogenesis?
2. Does nickel affect enzymes handling the active oxygen species generated by inflammatory cells?
3. Do magnesium and iron counteract nickel effects on these enzymes?

The results of first experiments designed to answer these questions are presented in Chapter 30

5. CONCLUSIONS

The following conclusions can be drawn from the present study:

1. The essential metals zinc, magnesium, and iron, administered in a common mixture with the carcinogen, inhibit Ni_3S_2-induced muscle carcinogenesis in rats.
2. The inhibitory effect is not produced by faster mobilization of Ni_3S_2 from the injection site in the presence of the administered essential metals.
3. Systemic administration of magnesium and iron does not prevent the action of Ni_3S_2 while systemic administration of zinc is effective.

4. The inhibition of the Ni_3S_2 carcinogenesis by magnesium, iron, and possibly zinc seems to be mediated through a protection and/or activation of cellular immune response to Ni_3S_2.

ACKNOWLEDGMENTS

Valuable comments on this chapter by Drs. L. Anderson and J. Rice and editorial help of Ms. K. Breeze are gratefully acknowledged.

REFERENCES

Athar, M., Hasan, S. K., and Srivastava, R. C. (1987). Evidence for the involvement of hydroxyl radicals in nickel mediated enhancement of lipid peroxidation: implications for nickel carcinogenesis. *Biochem. Biophys. Res. Commun.* **147,** 1276–1281.

Benson, J. M., Henderson, R. F., and McClellan, R. O. (1986). Comparative cytotoxicity of four nickel compounds to canine and rodent alveolar macrophages *in vitro. J. Toxicol. Environ. Health* **19,** 105–110.

Corbeil, L. B. (1968). Antigenicity of rhabdomyosarcomas induced by nickel sulphide, Ni_3S_2. *Cancer* **21,** 184–189.

Daniel, M. R. (1966). Strain differences in the response of rats to the injection of nickel sulphide. *Br. J. Cancer* **20,** 886–895.

Fraker, P. J., Gershwin, M. E., Good, R. A., and Prasad, A. A. (1986). Interrelationship between zinc and immune function. *Federation Proc.* **45,** 1474–1480.

Haley, P. J., Bice, D. E., Muggenburg, B. A., Hahn, F. F., and Benjamin, S. A. (1987). Immunopathologic effects of nickel subsulfide on the primate pulmonary immune system. *Toxicol. Appl. Pharm.* **88,** 1–12.

Judde, J. G., Breillout, F., Clemenceau, C., Poupon, M. F., and Jasmin, C. (1987). Inhibition of rat natural killer cell function by carcinogenic nickel compounds: preventive action of manganese. *J. Natl. Cancer Inst.* **78,** 1185–1190.

Kasprzak, K. S. (1978). *Problems of Metabolism of the Carcinogenic Nickel Compounds.* Technical University of Poznan Press, Poznan, Poland, pp. 1–65.

Kasprzak, K. S. (1988). Inhibitory effect of iron on the carcinogenicity of nickel subsulfide in F344/NCr rats. *Toxicologist* **8,** 193.

Kasprzak, K. S., and Poirier, L. A. (1985). Effects of calcium and magnesium on nickel uptake and stimulation of thymidine incorporation into DNA in the lungs of strain A mice. *Carcinogenesis* **6,** 1819–1821.

Kasprzak, K. S., and Waalkes, M. P. (1986). The role of calcium, magnesium, and zinc in carcinogenesis. In L. A. Poirier, P. M. Newberne, and M. W. Pariza (Eds.), *Essential Nutrients in Carcinogenesis.* Plenum Press, New York, pp. 497–515.

Kasprzak, K. S., Quander, R. V., and Poirier, L. A. (1985). Effects of calcium and magnesium salts on nickel subsulfide carcinogenicity in Fischer rats. *Carcinogenesis* **8,** 1161–1166.

Kasprzak, K. S., Waalkes, M. P., and Poirier, L. A. (1986a). Effects of magnesium acetate on the toxicity of nickelous acetate in rats. *Toxicology* **42,** 57–68.

Kasprzak, K. S., Waalkes, M. P., and Poirier, L. A. (1986b). Antagonism by essential divalent metals and amino acids of nickel(II)-DNA binding *in vitro. Toxicol. Appl. Pharm.* **82,** 336–343.

Kasprzak, K. S.,Waalkes, M. P., and Poirier, L. A. (1987a). Effects of essential divalent metals on carcinogenicity and metabolism of nickel and cadmium. *Biol. Trace Elem. Res.* **13**, 253–273.

Kasprzak, K. S., Ward, J. M., Poirier, L. A., Reichardt, D. A., Denn, C. A., III, and Reynolds, C. W. (1987b). Nickel-magnesium interactions in carcinogenesis: dose effects and involvement of natural killer cells. *Carcinogenesis* **8**, 1005–1011.

Kasprzak, K. S., Kiser, R. F., and Weislow, O. S. (1988). Magnesium counteracts nickel-induced suppression of T-lymphocyte response to concanavalin A. *Magnesium*, **7**, 166–172.

Loeb, L. A., James, E. A., Waltersdorph, A. M., and Klebanoff, S. J. (1988). Mutagenesis by the autooxidation of iron with isolated DNA. *Proc. Natl. Acad. Sci. USA* **85**, 3918–3922.

Miki, H., Kasprzak, K. S., Kenney, S., and Heine, U. I. (1987). Inhibition of intercellular communication by nickel(II): antagonistic effect of magnesium. *Carcinogenesis* **8**, 1757–1760.

Nielsen, F. (1980). Interactions of nickel with essential minerals. In O. Nriagu (Ed.), *Nickel in the Environment*, Wiley-Interscience, New York, pp. 611–634.

Poirier, L. A., Theiss, J. C., Arnold, L. J., and Shimkin, M. B. (1984). Inhibition by magnesium and calcium acetates of lead subacetate- and nickel acetate-induced lung tumors in strain A mice. *Cancer Res.* **44**, 1520–1522.

Smialowicz, R. J., Rogers, R. R., Riddle, M. M., and Stott, G. A. (1984). Immunologic effects of nickel. I. Suppression of cellular and humoral immunity. *Environ. Res.* **33**, 413–427.

Smialowicz, R. J., Rogers, R. R., Rowe, D. G., Riddle, M. M., and Luebke, R. W. (1987a). The effects of nickel on immune function in the rat. *Toxicology* **44**, 271–281.

Smialowicz, R. J., Rogers, R. R., Riddle, M. M., Rowe, D. G., Luebke, R. W., and Fogelson, L. D. (1987b). The effects of manganese, calcium, magnesium, and zinc on nickel-induced suppression of murine natural killer cell activity. *J. Toxicol. Environ. Health* **20**, 67–80.

Sunderman, F. W., Jr. (1982). Organ species specificity in nickel subsulfide carcinogenesis. In R. Langenbach, S. Nesnow, and J. Rice (Eds.), *Organ and Species Specificity in Chemical Carcinogenesis*. Plenum Press, New York, pp. 107–127.

Sunderman, F. W., Jr., and Barber, A. M. (1988). Finger-loops, oncogenes, and metals. *Ann. Clin. Lab. Sci.* **18**, 267–286.

Sunderman, F. W., Jr., and McCully, K. S. (1983). Effects of manganese compounds on carcinogenicity of nickel subsulfide in rats. *Carcinogenesis* **4**, 461–465.

Sunderman, F. W., Jr., Kasprzak, K. S., Lau, T. J., Minghetti, P. P., Maenza, R. M., Becker, N., Onkelinx, C., and Goldblatt, P. J. (1976). Effects of manganese on carcinogenicity and metabolism of nickel subsulfide. *Cancer Res.* **36**, 1790–1800.

Thomas, D. G., Breslow, D., and Gart, J. J. (1977). Trend and homogeneity analyses of proportions and life-table data. *Comput. Biomed. Res.* **10**, 373–381.

Waalkes, M. P., Kasprzak, K. S., Ohshima, M., and Poirier, L. A. (1985). Protective effects of zinc acetate toward the toxicity of nickelous acetate in rats. *Toxicology* **34**, 29–41.

43

NICKEL-INDUCED DERANGEMENTS OF THERMOREGULATION

Sidney M. Hopfer and F. William Sunderman, Jr.

Departments of Laboratory Medicine and Pharmacology,
University of Connecticut School of Medicine,
Farmington, Connecticut 06032

Nickel and Human Health: Current Perspectives, Edited by Evert Nieboer and
Jerome O. Nriagu.
ISBN 0-471-50076-3   © 1992 John Wiley & Sons, Inc.

---

1. CLINICAL OBSERVATIONS OF NICKEL-INDUCED HYPOTHERMIA

In 1883, Da Costa observed nickel-induced hypothermia during therapeutic
trials of nickel salts in epilepsy and other neurological diseases, noting that
a patient developed hypothermia and bradycardia following ingestion of five
grains (325 mg) of NiSO$_4\cdot$6H$_2$O. Diminished body temperatures were recently
observed in several electroplating workers who accidently ingested a solution
of NiSO$_4$ and NiCl$_2$ (Sunderman et al., 1988).

2. STUDIES OF NICKEL-INDUCED HYPOTHERMIA IN EXPERIMENTAL ANIMALS

2.1. Investigations by Gordon and Co-workers

Gordon and Stead (1986), Gordon et al. (1989), and Gordon (1989) reported
immediate reductions of body temperature following intraperitoneal administration of NiCl$_2$ to rats and mice based on measurements of colonic temperature by a thermocouple probe inserted via the anus. The NiCl$_2$ treatment
resulted in hypothermia of rats that lasted more than 1 hr, with reductions
of colonic temperature by 3–4°C when the ambient temperature was 20°C.
The Ni^{2+}-induced hypothermia was accentuated at lower ambient temperature
(e.g., 10°C) and ameliorated at higher ambient temperature (e.g., 30°C). The
metabolic rate, assayed by O$_2$ consumption, was diminished when NiCl$_2$-
treated rats were kept at 10° or 20°C but not at 30°C.
   Gordon et al. (1983) designed a chamber with a temperature gradient that
allowed rodents to chose their preferred thermal environment. When NiCl$_2$-
treated rats and mice were placed in the chamber, they consistently selected
a reduced ambient temperature (Gordon and Stead, 1986; Gordon, 1989).
These studies suggest that Ni^{2+}-induced hypothermia involves derangements
of behavioral as well as autonomic thermoregulation in rodents.

2.2. Investigations by the Present Authors

Hopfer and Sunderman (1988) evaluated the immediate hypothermic response
of rats following intraperitoneal or subcutaneous injection of NiCl$_2$ and monitored the delayed effects of NiCl$_2$ upon diurnal rhythms of core body tem-

perature and physical activity using radiotelemetry from a thermistor probe implanted in the peritoneal cavity.

2.2.1. Experimental Design

In the study of Hopfer and Sunderman (1988), male Fischer 344 rats (body weight 226 ± 11 g) were fed laboratory rat chow and tap water ad libitum, housed individually in polypropylene cages with wood-chip bedding, and maintained at 20 ± 1°C in a quiet room with 12-hr light–dark cycles, changing at 0600 and 1800. Injections were performed at 1500 or 1600, approximating the end of the diurnal trough of core body temperature of rats (De Castro, 1978). Test rats were given single injections of $NiCl_2$ (62 or 100 μmol/kg intraperitoneal; 250 or 375 μmol/kg subcutaneous) in NaCl vehicle, and control rats were given similar injections of NaCl vehicle (140 mmol/L, 0.1–0.2 mL per rat). Core body temperature and physical activity of the rats were monitored by telemetry, as described by De Castro and Brower (1977) and Thorne et al. (1987). Six to 9 days prior to the injection, a radiothermistor capsule was implanted in the peritoneal cavity via a midline abdominal incision. Data for body temperature and physical activity of three rats were collected in each experiment via a Dataquest telemetry system (Data Sciences, Roseville, MN) and recorded at 5-min intervals, 24 hr/day, during 120 hr preinjection and 152 hr postinjection.

2.2.2. Statistical Techniques

The immediate responses to $NiCl_2$ vs. NaCl vehicle were compared by Student's t-test and the Mann–Whitney U test (Sachs, 1984), based upon the means and standard deviations of the telemetry data collected from rats during 8 hr postinjection. Statistical analyses of circadian rhythms were performed by the least-squares cosine regression technique (cosinor analysis) (Fort and Mills, 1970) using definitions of circadian parameters according to Minors and Waterhouse (1981). The cosinor analyses were computed from hourly averages of data collected during three study periods as follows: (a) the 120-hr period preinjection, (b) the 72-hr period from 8 to 80 hr postinjection, and (c) the 72-hr period from 80 to 152 hr postinjection. The 8-hr postinjection period was excluded from the cosinor analyses. The acrophase and amplitude of the diurnal cycles of body temperature and physical activity were computed for each study period, and the means and standard deviations of these parameters in $NiCl_2$-treated and control rats were compared by Student's t-test.

2.2.3. Body Temperature Response after Intraperitoneal Injection of $NiCl_2$

In the *first and second experiments*, rats were given intraperitoneal injections at 1500 of (a) NaCl vehicle, (b) $NiCl_2$, 62.5 μmol/kg, or (c) $NiCl_2$, 100 μmol/ kg. Prompt reductions of core body temperature ensued in four $NiCl_2$-treated rats but not in two controls. Illustrative plots of telemetry data (Fig. 43.1)

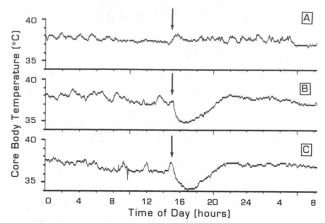

Figure 43.1. Computer plots of telemetry data for body temperature in a control rat (panel A) that received intraperitoneal injection of NaCl vehicle and test rats (panels B and C) that received intraperitoneal injection of $NiCl_2$ (100 μmol/kg). The times of the injections are indicated by arrows. [From Hopfer and Sunderman (1988) with permission.]

show that core body temperatures diminished 3.3 and 2.6°C, respectively, in two rats that received $NiCl_2$ (100 μmol/kg), reaching the nadir at 1.5 hr postinjection and returning to the baseline within 5–6 hr postinjection. Core body temperatures diminished 2.2 and 2.5°C, respectively, in two rats that received a lower dosage of $NiCl_2$ (62.5 μmol/kg), reaching the nadir at 1.5 hr postinjection and returning to the baseline within 4–5 hr.

2.2.4. *Immediate and Delayed Responses after Subcutaneous Injection of NiCl₂*

In the *third and fourth experiments*, rats were given subcutaneous injections at 1600 of NaCl vehicle or $NiCl_2$ at a dosage of 375 μmol/kg. The Ni^{2+}-treated rats evidenced hypothermia immediately postinjection, as described later, but the $NiCl_2$ dosage proved to be excessive, since three of four rats died on the third day postinjection. Accordingly, in the *fifth and sixth experiments*, rats were given subcutaneous injections at 1600 of NaCl vehicle or $NiCl_2$ at a dosage of 250 μmol/kg, which did not cause any deaths. Typical plots of telemetry data for body temperature (Fig. 43.2) show that the diurnal temperature cycle was unaffected by administration of NaCl vehicle to a control rat (panel A), whereas administration of $NiCl_2$ (250 μmol/kg, subcutaneous) to a test rat caused prompt hypothermia ($-1.8°C$ from baseline) followed by brief rebound and dampening of the diurnal temperature cycles during subsequent days (panel B). Telemetry data for physical activity in the rats showed that the diurnal activity cycle was unaffected by NaCl injection in control rats but was markedly dampened in $NiCl_2$-treated rats for at least 3 days postinjection.

2.2.5. *Quantitative Description of Acute Hypothermia Induced by Subcutaneous NiCl₂*

Table 43.1 summarizes the changes of core body temperature observed in experiments 3–6, immediately following subcutaneous injections at 1600 of NaCl vehicle or NiCl₂. During 8 hr after injection of the NaCl vehicle, the control rats maintained their usual circadian rhythm of core body temperature, with an average temperature peak of 1.1 ± 0.1°C above the predose value. In contrast, during 8 hr after injection of NiCl₂ at dosages of 250 or 375 μmol/kg, the core body temperatures of the test rats became markedly diminished, reaching nadirs that averaged 3.0 ± 0.5°C and 3.4 ± 0.6°C, respectively, below the simultaneous values in control rats. The interval from NiCl₂ injection to the nadir of core body temperature was longer and the duration of the hypothermic response was prolonged in rats that received the higher dosage of NiCl₂ compared to those that received the lower dosage.

2.2.6. *Effects of Subcutaneous NiCl₂ on Circadian Rhythm of Core Body Temperature*

Table 43.2 summarizes the parameters for circadian rhythms of core body temperature in experiments 3–6 during the 120-hr interval preinjection and the intervals 8–80 and 80–152 hr after subcutaneous injection of NaCl vehicle or NiCl₂. Data for rats that received NiCl₂ at the higher dosage (375 μmol/ kg) are omitted, since three of the four rats died before 80 hr postinjection. No significant change of average values for core body temperature occurred during the specified intervals after injection of NiCl₂, but the amplitude of the diurnal cycle of body temperature during the period from 8 to 80 hr postinjection was dampened to 59 ± 12% of the preinjection amplitude.

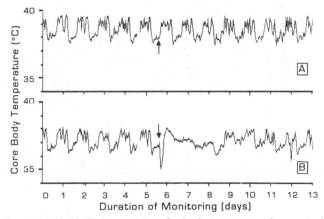

Figure 43.2. Computer plots of telemetry data for body temperature in a control rat (panel A) that received subcutaneous injection of NaCl vehicle and a test rat (panel B) that received subcutaneous injection of NiCl₂ (250 μmol/kg). The times of the injections are indicated by arrows. [From Hopfer and Sunderman (1988) with permission.]

Table 43.1 Core Body Temperature Responses of Rats 0–8 hr after Injection of NiCl$_2$ or NaCl Vehicle (Controls)^a

NiCl$_2$ Dosage (μmol/kg)^b	Number of Rats	Maximum Temperature Change from Preinjection Value (°C)	Interval from NiCl$_2$ Dose to Nadir of Body Temperature (min)	ΔT^c at Nadir of Body Temperature (°C)	Duration of Hypothermic Response to NiCl$_2$ (min)
0	4	+1.1 ± 0.1	—	—	—
250	4	−2.0 ± 0.4^d	84 ± 17	3.0 ± 0.5^d	235 ± 10
375	4	−2.3 ± 0.6^d	138 ± 25^e	3.4 ± 0.6^d	419 ± 87^e

^{a}Data are given as means ± SD.
^{b}Subcutaneous.
^{c}Controls − NiCl$_2$-dosed rats.
$^d P < 0.05$ vs. controls by t-test.
$^e P < 0.05$ vs. results at 250 μmol/kg dosage, by Mann–Whitney U-test.

Table 43.2 Circadian Rhythm of Core Body Temperature of Rats after Injection of NiCl$_2$ or NaCl Vehicle (controls)[a]

NiCl$_2$ Dosage (μmol/kg)[b]	Number of Rats	Interval Pre- or Postinjection (hr)	Average Core Body Temperature (°C)	Amplitude of Diurnal Temperature Cycle (°C)	Acrophase of Diurnal Temperature Cycle[c]
0	4	−120 to 0	37.1 ± 0.1	0.58 ± 0.06	2212 ± 0046
		8 to 80	37.0 ± 0.1	0.54 ± 0.05	2324 ± 0030
		80 to 152	37.0 ± 0.1	0.59 ± 0.06	2244 ± 0021
250	4	−120 to 0	37.1 ± 0.1	0.56 ± 0.08	2232 ± 0016
		8 to 56	37.2 ± 0.1	0.33 ± 0.07[d]	0300 ± 0114[d]
		56 to 128	37.1 ± 0.1	0.47 ± 0.10	0022 ± 0021[d]

[a] Data are given as means ± SD.
[b] Subcutaneous.
[c] Listed as hours and minutes on a 24-hr clock.
[d] $P < 0.05$ vs. controls by t-test and $P < 0.01$ vs. preinjection values by paired sample t-test.

Concomitantly, the acrophase of the diurnal cycle of body temperature was delayed from 10:32 p.m. to 3:00 a.m. These parameters returned toward the preinjection and control values during the interval from 80 to 152 hr postinjection.

2.2.7. *Effects of Subcutaneous NiCl₂ on Circadian Rhythm of Physical Activity*

Table 43.3 summarizes the parameters for circadian rhythms of physical activity in the control and NiCl$_2$-treated rats, corresponding to the body temperature parameters in Table 43.2. The average physical activity of rats was significantly reduced during 8–80 hr after subcutaneous injection of NiCl$_2$ (250 μmol/kg), and the amplitude of the diurnal cycle of physical activity was dampened. However, no significant shift occurred in the acrophase of the diurnal cycle of physical activity in NiCl$_2$-treated rats, compared to their preinjection values or to the corresponding values in control rats (Hopfer and Sunderman 1989).

3.   POSSIBLE PATHOPHYSIOLOGICAL MECHANISM OF IMMEDIATE NiCl₂-INDUCED HYPOTHERMIA

Several studies suggest that Ni^{2+} crosses cell membranes via Ca^{2+} channels, competing with Ca^{2+} for specific receptors (Wang et al., 1984; Brommundt and Kavaler, 1987; Kavaler and Brommundt, 1987; Raffa et al., 1987). Since exposure of the hypothalamus to excess Ca^{2+} causes a sharp diminution of body temperature (Myers and Veale, 1971; Myers et al., 1971; Myers and Lee, 1984), Ni^{2+} may mimic the effect of Ca^{2+} upon the hypothalamic thermoregulatory center, provoking the immediate hypothermic response (Hopfer and Sunderman, 1988). Hypothermia induced by NiCl$_2$ resembles the acute hypothermic responses of rats and mice following parenteral administration of CdCl$_2$ (Johnson et al., 1970; Christensen and Fujimoto, 1984; Gordon and Stead, 1986), suggesting that thermoregulatory effects of Ni^{2+} and Cd^{2+} might share a common pathophysiological mechanism. Christensen and Fujimoto (1984) showed that direct intracerebroventricular injection of a miniscule dose of CdCl$_2$ induces hypothermia, pointing to a direct effect of Cd^{2+} on the brain. On the other hand, Johnson et al. (1970) suggested that CdCl$_2$ might induce hypothermia by interfering with effector pathways or thermoeffector organs (e.g., vasodilatation of dermal vasculature). Since nickel salts have been reported to inhibit thyroid function (Anbar and Inbar, 1964; Shvayko and Tsvetkova, 1972), the acute hypothermia induced by NiCl$_2$ could conceivably be mediated by reduced release of triiodothyronine (T$_3$) from the thyroid gland or blocked uptake of T$_3$ by peripheral tissues. Thus, the hypothetical mechanisms for nickel-induced hypothermia include (a) Ni^{2+} suppression of the thermoregulatory center of the hypothalamus, (b) vasodilatory effect of Ni^{2+} on dermal vasculature, and (c) acute diminution of body metabolic

Table 43.3 Circadian Rhythm of Physical Activity of Rats after Injection of NiCl$_2$ or NaCl Vehicle (Controls)[a]

NiCl$_2$ Dosage (µmol/kg)[b]	Number of Rats	Interval Pre- or Postinjection (Hr)	Average Level of Physical Activity (arbitrary units)	Amplitude of Physical Activity Cycle (arbitrary units)	Acrophase of Physical Activity Cycle[c]
0	4	−120 to 0	21 ± 2	13 ± 2	2216 ± 0100
		8 to 80	19 ± 2	13 ± 3	0040 ± 0110
		80 to 152	20 ± 4	15 ± 5	0052 ± 0232
250	4	−120 to 0	20 ± 2	12 ± 6	2320 ± 0055
		8 to 80	12 ± 5[d]	6 ± 3[d]	2232 ± 0106
		80 to 152	19 ± 4	13 ± 2	2322 ± 0213

[a] Data are given as means ± SD.
[b] Subcutaneous.
[c] Listed as hours and minutes on a 24-hr clock.
[d] $P < 0.05$ vs. controls by t-test and $P < 0.01$ vs. preinjection values by paired sample t-test.

rate, mediated, for example, by inhibition of thyroid release of T_3 or blockage of T_3 uptake by tissue receptors.

4. POSSIBLE PATHOPHYSIOLOGICAL MECHANISM FOR NiCl$_2$- INDUCED DERANGEMENTS OF THE CIRCADIAN CYCLE OF CORE BODY TEMPERATURE

The study of Hopfer and Sunderman (1988) showed that, in addition to causing prompt hypothermia that lasts for 4 hr, administration of NiCl$_2$ (250 μmol/kg subcutaneous) to rats deranged the circadian rhythm of thermoregulation for at least three days following recovery from hypothermia. During this period, the amplitude of the diurnal cycle of core body temperature was significantly dampened and the acrophase of the cycle was delayed 4–5 hr. Consistent with this finding, Gibbs (1981) showed that the circadian pacemaker of rats is highly temperature dependent, so that a transient bout of hypothermia can set back the biological clock of thermoregulation. Although the physical activity of rats was reduced for several days after injection of NiCl$_2$ and the amplitude of the diurnal cycle of physical activity was consequently diminished, the acrophase of the diurnal cycle of physical activity was *not* delayed in synchrony with the thermoregulatory cycle.

REFERENCES

Anbar, M., and Inbar, M. (1964). The effect of certain metallic cations on the iodide uptake in the thryroid gland of mice. *Acta Endocrinol.* **26**, 648–652.

Brommundt, G., and Kavaler, F. (1987). La^{3+}, Mn^{2+}, and Ni^{2+} effects on Ca^{2+} pump and on Na$^+$-Ca^{2+} exchange in bullfrog ventricle. *Am. J. Physiol.* **253**, C45–C51.

Christensen, C. W. and Fujimoto, J. M. (1984). Cadmium induced hypothermia in mice: dose dependent tolerance development. *Gen. Pharmac.* **15**, 263–266.

Da Costa, J. M. (1883). Observations on the salts of nickel, especially the bromide of nickel. *Med. News* **43**, 337–338.

De Castro, J. M. (1978). Diurnal rhythms of behavioral effects on core temperature. *Physiol. Behavior* **21**, 883–886.

De Castro, J. M., and Brower, E. (1977). Simple, reliable and inexpensive telemetry system for continuous monitoring of small animal core temperature. *Physiol. Behavior* **19**, 331–333.

Fort, A. and Mills, J. N. (1970). Fitting sine curves to 24 h urinary data. *Nature* **226**, 657–658.

Gibbs, F. P. (1981). Temperature dependence of rat circadian pacemaker. *Am. J. Physiol.* **241**, R17–R20.

Gordon, C. J. (1989). Effect of nickel chloride on body temperature and behavioral thermoregulation in the rat. *Neurotoxicology and Teratology*, **11**, 317–320.

Gordon, C. J., and Stead, A. G. (1986). Effect of nickel and cadmium chloride on autonomic and behavioral thermoregulation in mice. *Neurotoxicology* **7**, 97–106.

Gordon, C. J., Bailey, E. B., Kozel, W. M., and Ward, G. H. (1983). A device for monitoring position of unrestrained animals in a temperature gradient. *Physiol. Behavior* **31**, 265–268.

Gordon, C. J., Fogelson, L., and Stead, A. G. (1989). Temperature regulation following nickel intoxication in the mouse: effect of ambient temperatures. *Comp. Biochem. Physiol.*, **92**, 73–76.

Hopfer, S. M., and Sunderman, F. W., Jr. (1988). Hypothermia and deranged circadian rhythm of core body temperature in nickel-chloride treated rats. *Res. Commun. Chem. Pathol. Pharm.*, **62**, 495–505.

Johnson, A. D., Gomes, W. R., and VanDemark, N. L. (1970). Testis and body temperature changes caused by cadmium and zinc. *J. Reprod. Fert.* **21**, 383–393.

Kavaler, F., and Brommundt, G. (1987). Potentiation of contraction in bullfrog ventricle strips by manganese and nickel. *Am. J. Physiol.* **253**, C52–C59.

Minors, D. S., and Waterhouse, J. M. (1981). *Circadian Rhythms and the Human*. Wright PSG Press, Bristol.

Myers, R. D., and Lee, T. F. (1984). Are brain amines or calcium involved in the thermolytic action of peptides? In J.R.S. Hales (Ed.), *Thermal Physiology*, Raven Press, New York, pp. 119–124.

Myers, R. D., and Veale, W. L. (1971). The role of sodium and calcium ions in the hypothalamus in the control of body temperature of the unanesthetized cat. *J. Physiol.* **212**, 411–430.

Myers, R. D., Veale, W. L., and Yaksh, T. L. (1971). Changes in body temperature of the unanaesthetized monkey produced by sodium and calcium ions perfused through the cerebral ventricles. *J. Physiol.* **217**, 381–392.

Raffa, R. B., Bianchi, P., and Narayan, S. R. (1987). Reversible inhibition of acetylcholine contracture of molluscan smooth muscle by heavy metals: correlation to Ca^{++} and metal content. *J. Pharmacol. Exper. Therap.* **243**, 200–204.

Sachs, L. (1984). *Applied Statistics*, 2nd ed. Springer-Verlag, New York, pp. 293–303.

Shvayko, I. I., and Tsvetkova, I. N. (1972). Influence of various doses of cobalt, nickel and their combinations on the function of the thyroid gland in the rat." *Mikroelementy Meditsine* **3**, 210–212.

Sunderman, F. W. Jr., Dingle, B., Hopfer, S. M., and Swift, T. (1988). Acute nickel toxicity in electroplating workers who accidently ingested a solution of nickel sulfate and nickel chloride. *Am. J. Indust. Med.* **14**, 257–266.

Thorne, P. S., Yeske, C. P., and Karol, M. K. (1987). Monitoring guinea pig core temperature by telemetry during inhalation exposures. *Fund. Appl. Toxicol.* **9**, 398–408.

Wang, Z., Bianchi, C. P., and Narayan, S. R. (1984). Nickel inhibition of calcium release from subsarcolemmal calcium stores of molluscan smooth muscle. *J. Pharm. Exper. Therap.* **229**, 697–701.

44

PLACENTAL METABOLISM OF NICKEL

B. Sarkar, A. Mas, and H. Yeger

*Research Institute, The Hospital for Sick Children,
Toronto, Ontario, Canada, M5G 1X8, and the Departments of
Biochemistry and Pathology, University of Toronto, Toronto
Ontario, Canada, M5S 1A8*

1. INTRODUCTION

Placenta plays an important role in the maternofetal transfer of nutrients during fetal growth and development. Consequently, this organ provides an excellent opportunity to study the metabolism of different substrates and xenobiotics. The knowledge of the mineral metabolism in placenta is slender, and the understanding of the mechanisms of transfer and retention is not clear.

Nickel and Human Health: Current Perspectives, Edited by Evert Nieboer and
Jerome O. Nriagu.
ISBN 0-471-50076-3   © 1992 John Wiley & Sons, Inc.

The concept of a placental barrier has been developed from the ability of this organ to discriminate the transport of different compounds, basically *essential* from *toxic*. Although there are some exceptions (e.g., organic mercury), this is particularly true in metal metabolism, where essential metals such as zinc or copper are easily transferred to the fetus whereas toxic metals such as cadmium and nickel are retained in the placenta (Mas and Sarkar, 1989).

The close approximation of fetal capillaries to the absorptive trophoblast resembles other such arrangements as in lung, liver, and kidney, which require rapid exchange of cellular products. The synthetic functions of trophoblast are not dependent on the integrity of this blood supply. The villous interstitial tissue supports this vascular network. Oedema of this tissue is associated with fetal hydrops and can result in impaired exchange of precursors. There is no evidence that the interstitial tissue is involved directly in transport of nutrients, but it does constitute a potential sink for diffusable substrates. As in other epithelial organs, the basement membrane operates as a selective filtration system. The basement membrane is thickened in association with pregnancy disorders including preeclampsia and diabetes mellitus (Fox, 1978). How this process would affect transport is not known.

2. NICKEL TOXICITY

Although nickel has been considered essential for mammals, no physiological or biochemical role has been described so far for nickel in these organisms. Also, the amount of essential nickel is considered to be so minute for humans that it is extremely unlikely that a case of nickel deficiency will be reported. Instead, the nickel effects on human health are more related to toxicity.

The chemical form of nickel is one of the determinants of nickel toxicity. Soluble forms such as nickel chloride seem to be easily excreted through the kidney, and this organ is considered to be the main target organ for nickel. Insoluble forms such as nickel subsulfide or nickel oxides are internalized in the cells by phagocytosis and are considered more toxic than soluble forms because of their longer half-life in the body. However, the toxicity is considered to be due to ionic nickel, and the effect of insoluble forms of nickel would be related to a continuous release of ionic nickel directly inside the cell.

2.1. Nickel Toxicity during Gestation

Nickel toxicity may have special relevance during gestation. The slow release of nickel from insoluble forms coupled with the short pregnancy of rodents has restricted the studies with such forms of nickel. Most of the toxic, teratologic, and fetotoxic effects have been studied with soluble forms of nickel. During pregnancy the sensitivity to xenobiotics is increased, and this seems to be the case with nickel. Acute doses of $NiCl_2$ have shown an increased toxicity during gestation (Sunderman et al., 1978; Lu et al., 1979; Mas et al., 1985). The values of the median lethal dose (LD_{50}) for acute administration of nickel salts during gestation are shown in Table 44.1. The observed pattern for nickel toxicity during gestation is a decrease of the LD_{50}, which is independent of the route of administration, at the end of the gestation. The toxicity can be relevant at midgestation, although contradictory results have been observed depending on the animal strain used. This increased sensitivity to nickel might be due to physiological changes associated with pregnancy or uneven distribution of the metal between the maternofetal compartments.

The analysis of data presented in this chapter does not allow any definitive conclusion. It is known that the lack of an even distribution between the maternal body and the conceptus might result in a higher dose to the mother. However, the correction by the increase of weight, assuming no transfer of nickel to fetus, does not justify completely the decrease in the LD_{50}.

One of the effects of an acute dose of nickel is hyperglycemia (Clary, 1975; Horak and Sunderman, 1975) mediated by hyperglucagonemia. Hyperglucagonemia is increased during pregnancy (Mas et al., 1986), and this effect can account for part of the increased sensitivity to acute doses of nickel at the end of gestation.

2.2. Nickel Teratology and Fetotoxicity

One of the most important effects of a xenobiotic during pregnancy is the possibility of affecting the growth and development of the embryo (early and midgestation) or the fetus (late gestation). The toxic effects on the embryo can cause serious malformations and delays in development (teratologic effects), whereas these effects on the fetus produce only delays in development or fetal death but without producing morphological abnormalities (fetotoxicity).

Teratogenesis and fetotoxicity due to soluble nickel salts are shown in Table 44.2. Nickel chloride appears to be a weak teratogen in rats (Mas et

Table 44.1   Toxicity of Soluble Nickel Salts during Gestation

Species	Nickel Form and Route of Administration	Observation	Authors
Rat (Fisher)	$NiCl_2$, intramuscular	LD_{50}, day 8, 22 mg/kg; LD_{50}, day 18, 16mg/kg	Sunderman et al. (1978)
Mouse	$NiCl_2$, intraperitoneal	Increased mortality	Lu et al. (1979)
Rat (Wistar)	$NiCl_2$, intraperitoneal	LD_{50}, control, 9.3 mg/kg; LD_{50}, day 12, 6.3 mg/kg; LD_{50}, day 19, 6.0 mg/kg;	Mas et al. (1985)

Table 44.2 Teratogenicity of Soluble Nickel Salts

Species	Nickel Form, Route and Day of Administration[a]	Observation	Authors
Golden hamster	Ni acetate, i.p.	Teratogenic	Ferm (1972)
Rat	NiCl$_2$, i.m., day 8	Embryotoxic, low-weight fetuses	Sunderman et al. (1978)
Mouse	NiCl$_2$, i.m., day 18	No fetotoxicity	Lu et al. (1979)
Mouse	NiCl$_2$, i.p., days 7–11	Vertebrae and rib fusion, cleft palate, open eyelids	Storeng and Jonsen (1981)
Rat	NiCl$_2$, i.p., days 1–6	Low implantation, low litter size, abnormalities: neck haematomas, exenphaly	Mas et al. (1985)
	NiCl$_2$, i.p., days 8, 12, and 16	Low weight, haemorrhages, hydrocephalia; no effect on day 16	

[a]i.p., intraperitoneal; i.m., intramusclar.

al., 1985) but a stronger one in other species such as mice (Lu et al., 1979) or golden hamsters (Ferm, 1972). Nevertheless, the manifestation of the teratogenic malformations occurs when nickel is administered during the critical period of organogenesis and is maximal at dose levels that are also toxic to the dam. The presence of teratogenic malformations after nickel exposure during organogenesis may be due to the permeability of the yolk sac to nickel. At the end of gestation, however, the chorioallantoic placenta is more selective to the cation (Mas et al., 1985). The sensitivity of the developing embryo to toxic substances can also add an effect on the teratogenic action of nickel. However, the alteration of the glucose metabolism, due to the acute overload of nickel, has been pointed out as the most likely cause of nickel toxicity in that period (Mas et al., 1985).

Quite contrary to the above, soluble nickel seems to be almost harmless for the developing fetus in late gestation, when organogenesis has been completed (Sunderman et al., 1978; Mas et al., 1985). This effect may be due to the limited amount of the cation that reaches the fetus or to the higher levels of glucose if hyperglycemia is considered as a possible cause for fetotoxicity.

However, the effect of nickel administration in late gestation may have a very deleterious effect on the development of the offspring, as seen by a drastic reduction in the number of pups 5 days after their delivery by nickel-treated dams (Mas, 1984).

3. PLACENTAL METABOLISM OF NICKEL

Little or no data are available on nickel levels in human placenta or fetuses, probably due to the low levels of this metal and the limits of detection. In untreated rats, Kirchgessner et al. (1982) reported levels of nickel in amniotic fluid (referred to as dry mass) 30 times higher than those of fetus or placenta, although in all the tissues studied nickel accounted for less than 1 ppm expressed in dry mass.

The placenta seems to follow a maturation process regarding the ability to discriminate transfer of nickel. As can be seen in Table 44.3, the early placenta allows transfer of the metal to fetuses (Sunderman et al., 1978; Olsen and Jonsen, 1979; Mas et al., 1985), while later in the gestation, when the chorioallantoic placenta is fully developed, nickel is more selectively retained in placenta (Sunderman et al., 1978; Mas et al., 1985; Mas and Sarkar, 1989).

Administration of an acute dose of nickel to pregnant rodents near term showed a high retention of nickel in placenta and a low clearance rate in this organ with one of the longest biological half-lives (Mas et al., 1986). The long half-life of nickel in placenta and fetuses together with the rapid clearance of nickel through urine in the maternal body may explain the disagreement in results reported from a number of different laboratories.

Table 44.3 Distribution of Soluble Nickel Salts during Gestation

Species	Nickel Form, Route, Dose and Day of Administration[a]	Observations	Authors and Method
Mouse	NiCl$_2$, i.p., 3.5 mg/kg, day 16, time dependence	Peaks of placenta at 2 hr, fetus at 8 hr	Lu et al. (1976, 1979), tissue counting
Mouse	NiCl$_2$, i.p., trace, days 18–20	Similar amounts in fetus and mother	Jacobsen et al. (1978), tissue counting
Rat	NiCl$_2$, i.m., 12 mg/kg, days 8 and 18	High levels in fetuses and membranes on day 9; high in placenta on day 19; in fetus high in urinary bladder and yolk sac on day 19	Sunderman et al. (1978), tissue counting Autoradiography
Mouse	NiCl$_2$, i.p., trace, days 16–19	Visceral yolk sac and fetal accumulation in early placenta but not in late	Olsen and Jonsen (1979), autoradiography
Rat	NiCl$_2$, i.p., 4 mg/kg, days 12 and 19, time dependence	High in placenta and fetus on day 12, high in late placenta; placental retention	Mass et al. (1985), tissue counting Autoradiography
Rat	NiCl$_2$, i.p., trace and 4 mg/kg, half-life	Placenta and fetuses longest half-life	Mas et al. (1986), tissue counting
Rat	NiCl$_2$, i.v., trace, day 19	Limited placental transfer and high placental retention	Mas and Sarkar (1989), tissue counting

[a]i.p., intraperitoneal; i.m., intramuscular, i.v., intravenous.

Thus, Jacobsen et al. (1978) observed that nearly similar levels of nickel in the maternal and the fetal body can easily lead to misinterpretation about the placental transfer. However, the nickel analysis was carried out two days after the injection, providing sufficient time for an estimated clearance of 80–90% of the injected dose (Van Soestbergen and Sunderman, 1972; Tandon, 1982). The placental retention becomes obvious in the autoradiography shown in Figure 44.1.

No carrier or retention mechanisms have been discussed for metals besides the induction of placental metallothionein by cadmium, related to the retention of zinc and as a possible mechanism for cadmium toxicity (Samawickrama and Webb, 1981).

In an initial attempt to study the placental metabolism of nickel, homogenized human placentas were used, although the large amount of blood with respect

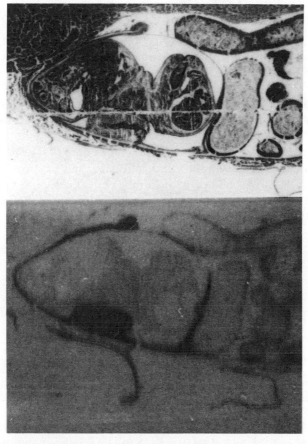

Figure 44.1. Detail of whole body autoradiography. Nineteen-day pregnant rat was injected intraperitoneally with 50 μCi of ^{63}NiCl$_2$ supplemented with 4 mg nickel/kg animal. Top, hematoxylin–eosin preparation; bottom, autoradiography. The animal was sacrificed 3 hr after the injection.

to the actual placental tissue (considered as trophoblast, fetal capillaries, and connective tissue) showed the approach to be unworkable. As a consequence, our experimental approach to the studies of placental metabolism of nickel had to be carried out with rat placentas in vivo (Mas and Sarkar, 1989) and with human trophoblast cells from placentas in vitro at term in primary cell culture.

3.1. An *in Vitro* System for Nickel Studies in Human Trophoblast

The chorionic villi of the human placenta undergo extensive branching as the placenta ages during gestation. The lining trophoblast layer becomes progressively deficient in proliferative cytotrophoblast and increasingly richer in the metabolically active secretory syncytiotrophoblast. The syncytiotrophoblast produces and secretes a remarkable array of polypeptide hormones (hCG, hPL) and steroids while also functioning as an active transporter of nutrients to the fetus. It has been possible in recent years to selectively isolate cytotrophoblast and syncytiotrophoblast cell populations from term placenta and demonstrate their functional activity as producers of these polypeptide hormones (Kliman et al., 1986; Zeiter et al., 1983) and steroids (Lobo et al., 1985).

In order to carry out nutrient uptake and transport studies, it was necessary to obtain large quantities of trophoblast cells that could be reproducibly cultured in vitro under controlled conditions.

Trophoblast cells were isolated from the chorionic villi of term placentas with a combination of enzymatic dissociation and purification after multiple steps of washing, low-speed centrifugation, and selective filtration through 62-μm nylon mesh (Yeger et al., 1988). The cells obtained by this procedure were identified as mostly (>95%) trophoblast cells by virtue of their immunocytochemical labeling with antibodies to low-molecular-weight cytokeratins the placental specific polypeptide hormones, chorionic gonadotropin, and placental lactogen. Contamination with mesenchymal cells was always >5% as determined by antibodies to vimentin and factor VIII antigen. Ultrastructurally, the cells were found to retain the complex organization of intracellular organelles and numerous microvilli typical of syncytiotrophoblast. Trophoblast cells dissociated from the underlying basement membrane organized into clusters and chains. Trophoblast cells were plated in Primaria (Falcon) culture dishes in DMEM/F12 (1 : 1) medium containing 10% fetal bovine serum. Cells rapidly attached, flattened out, and formed small epithelial looking colonies (Fig. 44.2). A small percentage of these cells appeared to be multinucleated, indicating the presence of syncytia.

For the metal uptake and binding studies trophoblast cells were cultured for 24 hr in the plating medium (10% FCS in DMEM/F12) and then switched to low-protein medium (0.5% in DMEM/F12) for overnight incubation. Subsequently, cultures were washed three times in EHG medium (Earle's salts, 10 mM HEPES and 1 g/L glucose) prior to adding the experimental incubation

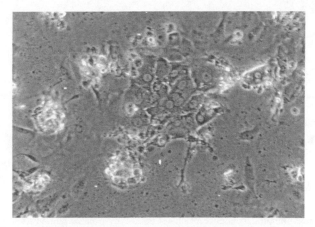

Figure 44.2. Phase contrast micrograph of human trophoblast cells cultured in Primaria plates in DMEM/F12 supplemented with 10% fetal bovine serum. The trophoblast cells have reaggregated prior to attachment and spreading. Trophoblast cells exhibit typical features of epithelium. (×200)

medium. Cells were incubated for periods up to 2 hr without appreciable loss of viability (as seen by trypan blue exclusion), and cell loss was negligible (as measured by protein and DNA content). Cells were detached by physical (rubber policeman) or chemical (0.1 M NaOH) means.

3.2. Nickel Metabolism in Isolated Human Trophoblast

As can be seen in the gross subcellular fractionation shown in Table 44.4, nickel is mostly present in the postmicrosomal fraction in pregnant rats and evenly distributed between nuclear and postmicrosomal fractions in cultured human trophoblast. However, as stated before, in rat placenta blood contamination was present, which may account for an increased amount of nickel in the soluble fraction. (Syncytiotrophoblast cells have a large proportion of nuclear fraction, accounting in our case for 60% of the total cell protein.)

Table 44.4 Subcellular Fractionation of Nickel in Rat Placenta and Human Cultured Trophoblast

Fraction	Rat in Vivo (Percent)	Human Trophoblast (Percent)
Nuclear	12.7	48
Mitochondrial	7.1	Trace
Microsomal	2.7	Trace
Postmicrosomal	77.4	45

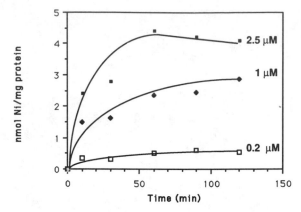

Figure 44.3. Concentration dependence of nickel uptake by isolated human trophoblast. Isolated cells in Primaria cell culture plates were incubated with $^{63}NiCl_2$ and the incubation was stopped at different times with cold phosphate buffer saline. Incubations were carried out at 37°C in CO_2–air mixture (5 : 95) in a shaking bath set at 60 oscillations/min. Results are the average of duplicate samples from two different experiments.

As in the previous in vivo experiments (Mas et al., 1986; Mas and Sarkar, 1989), nickel also showed, in vitro, a fast association with the placental cells. The uptake of nickel was rapid, reaching a plateau between 10 and 60 min of incubation (Fig. 44.3). The plateau can indicate a steady state between uptake and efflux or just a saturation of the uptake. The lack of efflux, as in the case of nickel, could be due to the long half-life of nickel in placenta.

In our earlier report (Mas and Sarkar, 1989) we showed the binding of the metal to low-molecular-weight components (Fig. 44.4). The resolution into two nickel-binding peaks of these low-molecular-weight components is evident by fast protein liquid chromatography (Superose 12) fractionation (Fig. 44.5). Fractionation of the cytosol from cultured human trophoblast

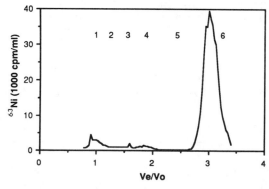

Figure 44.4. Sephadex G-150 elution profile of 100,000g supernatant of rat placenta homogenate 2 hr after ^{63}Ni injection.

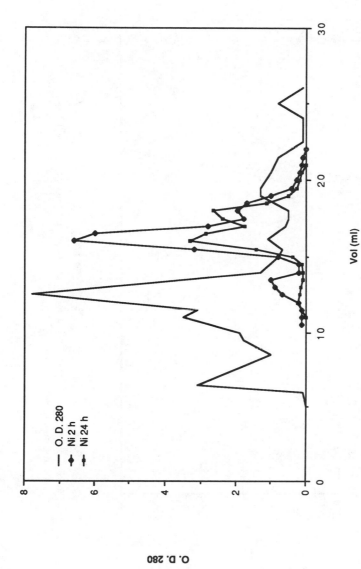

Figure 44.5. Superose 12 elution profile of 100,000g supernatant of rat placenta homogenate 2 and 24 hr after ^{63}Ni injection.

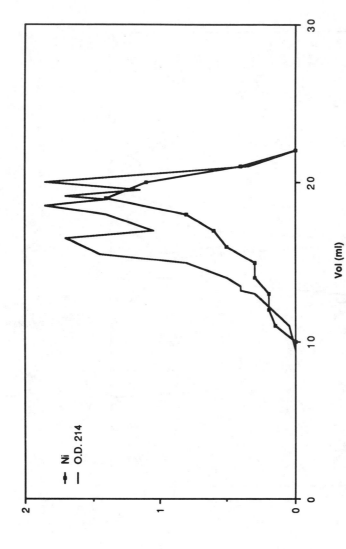

Figure 44.6. Superose 12 elution proflie of low-molecular-weight components (<10,000) from cytosol of trophoblast cells exposed to [63]Ni.

References 585

cells, previously incubated with ^{63}Ni, showed 60% of the total activity in the <10,000-dalton fraction, and further fractionation of this low-molecular-weight fraction through Superose 12 showed two peaks that superimpose with the peaks from the rat placenta in the in vivo experiments (Fig. 44.6).

The presence of these low-molecular-weight binding components in placenta are reminiscent of similar components found in kidney. The principal target organ for soluble nickel is kidney where most of the nickel is associated with low-molecular-weight binding components (Abdulwajid and Sarkar, 1983) comprised of an acidic polypeptide and sulfated oligosaccharides (Templeton and Sarkar, 1985). It is possible that similar nickel-binding components are also present in placenta. Work is currently underway to isolate and characterize these nickel-binding components in placenta.

ACKNOWLEDGMENTS

The research was supported by the Medical Research Council of Canada.

REFERENCES

Abdulwajid, A. W., and Sarkar, B. (1983). Nickel-sequestering renal glycoprotein. *Proc. Natl. Acad. Sci. USA* **80**, 4509–4512.

Clary, J. J. (1975). Nickel chloride-induced metabolic changes in the rat and guinea pig. *Toxicol. Appl. Pharm.* **31**, 55–65.

Ferm, V. H. (1972). The teratogenic effects of metals on mammalian embryos. *Adv. Teratol.* **5**, 51–75.

Fox, H. (1978). *Pathology of the Placenta*. W. B. Saunders, London.

Horak, E., and Sunderman, F. W., Jr. (1975). Effects of nickel upon plasma glucagon and glucose in rats. *Toxicol. Appl. Pharm.* **33**, 388–391.

Jacobsen, W., Alfheim, I., and Jonsen, J. (1978). Nickel and strontium distribution in some mouse tissues; passage through the placenta and mammary glands. *Res. Commun. Chem. Pathol. Pharm.* **29**, 571–584.

Kirchgessner, M., Roth-Maier, D. A., and Schnegg, A. (1982). Gehalte und Verteilung von Fe, Cu, Zn, Ni, und Mn in Foeten, Fruchtwasser, Placenta und Uterus von Ratten. *Res. Exp. Med.* **180**, 247–254.

Kliman, H. J., Nestler, J. E., Sermasi, E., Sanger, J. M., and Strauss, J. F. (1986). Purification, characterization and *in vitro* differentiation of cytotrophoblasts from human placentae. *Endocrinology* **118**, 1567–1582.

Lobo, J. O., Bellino, F. L., and Blankert, L. (1985). Estrogen synthetase activity in human term placental cells in monolayer culture. *Endocrinology* **116**, 889–895.

Lu, C. C., Matsumoto, M., and Iijima, S. (1976). Placental transfer of NiCl$_2$ to fetuses of mice. *Teratology* **14**, 245.

Lu, C. C., Matsumoto, M., and S. Iijima, S. (1979). Teratogenic effects of nickel chloride on embryonic mice and its transfer to embryonic mice. *Teratology* **19**, 137–142.

Mas, A. (1984). Distribució, Cinètica i efectes metabòlics del níquel durant la gestació de la rata. Ph. D. Dissertation, Universitat de Barcelona, Barcelona, Spain.

Mas, A., and Sarkar, B. (1988). The metabolism of metals in rat placenta. *Biol. Trace Elem. Res.*, **18**, 191–199.

Mas, A., Holt, D., and Webb, M. (1985). The acute toxicity and teratogenicity of nickel in pregnant rats. *Toxicology* **35**, 47–57.

Mas, A., Peligero, M. J., Arola, Ll., and Alemany, M. (1986). Distribution and kinetics of injected nickel in the pregnant rat. *Clin. Exp. Pharm. Physiol.* **13**, 91–96.

Olsen, I., and Jonsen, J. (1979). Whole-body autoradiography of Ni-63 in mice throughout gestation. *Toxicology* **12**, 165–172.

Samarawickrama, G. P., and Webb, M. (1981). The acute toxicity and teratogenicity of cadmium in the pregnant rat. *J. Appl. Toxicol.* **1**, 264–273.

Storeng, R., and Jonsen, J. (1981). Nickel toxicity in early embryogenesis in mice. *Toxicology* **20**, 45–51.

Sunderman, F. W., Jr, Shen, J. K., Mitchell, J. M., Allpass, P. R., and Damjanov, I. (1978). Embryotoxicity and fetal toxicity of nickel in rats. *Toxicol. Appl. Pharm.* **43**, 381–390.

Tandon, S. K. (1982). Deposition of Ni-63 in the rat. *Toxicol. Lett.* **10**, 71–73.

Templeton, D. M., and Sarkar, B. (1985). Peptide and carbohydrate complexes in human kidney. *Biochem. J.* **230**, 35–42.

Van Soestbergen, M., and F. W. Sunderman, F. W., Jr. (1972). ^{63}Ni complex in rabbit serum and urine after injection of ^{63}NiCl$_2$. *Clin. Chem.* **18**, 1478–1484.

Yeger, H., Lines, L. D., Wong, P.-Y., and Silver, M. M. (1989). Enzymatic isolation of human trophoblast and culture on various substrates: Comparison of first trimester with term trophoblast. *Placenta*, **10**, 137–151.

Zeitler, P., Markoff, E., and Handwerger, S. (1983). Characterization of the synthesis and release of human placental lactogen and human chorionic gonadotropin by an enriched population of dispersed placental cells. *J. Clin. Endocrinol. Metab.* **57**, 812–818.

45

MECHANISMS OF NICKEL-INDUCED CORONARY VASOCONSTRICTION IN ISOLATED PERFUSED RAT HEARTS

*Y. Edoute, P. M. Vanhoutte, and G. M. Rubanyi**

Department of Physiology and Biophysics, Mayo Clinic and Mayo Foundation, Rochester, Minnesota 55905

* Present address: Schering AG, Research Center, 1000 Berlin 65, Germany.

Nickel and Human Health: Current Perspectives, Edited by Evert Nieboer and Jerome O. Nriagu.
ISBN 0-471-50076-3   © 1992 John Wiley & Sons, Inc.

1. INTRODUCTION

Exogenous nickel chloride ($NiCl_2$) is a potent coronary vasoconstrictor in the in situ dog heart (Rubanyi et al., 1981, 1984) and isolated perfused rat heart (Rubanyi and Kovach, 1980; Rubanyi et al., 1981). In the in situ dog heart, nickel ions inhibit increases in coronary blood flow in response to ischemia and hypoxia (Rubanyi et al., 1984). A unique feature of nickel-induced coronary vasoconstriction is that myocardial hypoperfusion and hypoxia augment rather than inhibit it (Rubanyi et al., 1984). These findings suggested that inhibition of local vasodilator mechanisms (triggered by reduction in myocardial oxygen supply) may be (at least in part) responsible for the coronary vasoconstriction in response to low concentrations of nickel ion. Experiments were designed in isolated perfused rat hearts to analyze nickel-induced coronary vasoconstriction during stimulation of local metabolic vasodilator mechanisms (either by increasing myocardial oxygen demand or reducing myocardial oxygen supply) or in the presence of exogenous adenosine, the primary mediator of coronary vasodilatation triggered by disturbances in the balance between myocardial oxygen supply and demand (Belloni 1979; Berne, 1980).

2. METHODS

Wistar rats of either sex weighing 250–350 g were anesthetized with sodium pentobarbital (30 mg/kg intraperitoneally). After opening the thoracic cavity, the heart was removed by making a single cut with the scissors through the vessels arising from the heart. After excision, the still beating heart was put into a beaker containing ice-cold Krebs-Henseleit bicarbonate buffer solution of the following composition (mM): NaCl, 118; KCl, 4.7; $CaCl_2$, 2.5; $MgSO_4$, 1.2; KH_2PO_4, 1.2; calcium–disodium ethylenediaminetetraacetic acid (EDTA), 0.03; $NaHCO_3$, 25.0; and glucose, 11.1 (control solution). The prefiltered (Millipore) buffer was equilibrated with a 95 : 5 O_2–CO_2 gas mixture at 37°C. In some experiments the KCl concentration in the buffer solution was increased to 26 mM.

2.1. Retrograde Coronary Perfusion of Hearts by Langendorff Technique

The hearts were perfused in a modified version of the apparatus described earlier (Neely et al., 1967), which allows retrograde coronary (Langendorff preparation) or left atrial perfusion (working heart model; see Section 2.2) (Fig. 45.1). When the heart had cooled for 10–15 sec, the aortic stump was slipped onto a grooved perfusion cannula and held in place with a hemostat. After care was taken to position the tip of the cannula above the aortic valve, the aorta was secured with a ligature. Retrograde perfusion down the aorta was begun at a constant hydrostatic pressure of 85 cm H_2O (determined by the distance between the aortic bubble trap and the heart; see Fig. 45.1) and continued for 15 min to remove all blood and to allow the heart to recover from the period of hypoxia associated with excision and initiation of perfusion. The perfusion medium was pumped by a roller pump from the oxygenating chamber and reservoir to the overflow-type aortic bubble trap via a filter at a rate of 100 mL/min. The excessive medium (not used for perfusion of the heart) returned to the oxygenating chamber. From the bubble trap, the perfusion

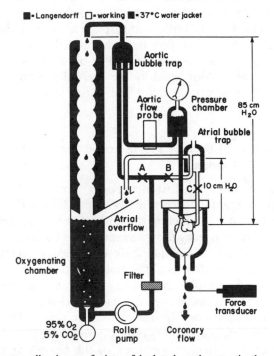

Figure 45.1. Apparatus allowing perfusion of isolated rat hearts via the aorta (Langendorff preparation; clamp A open; clamps B and C closed) or through the left atrium allowing the left ventricle to perform external work (working preparation; clamp A closed; clamps B and C open). For further details see text.

medium reached the aorta via a heat exchanger (37°C; not shown in Fig. 45.1). The perfusate leaving the heart was discarded.

2.2.   Perfusion of Working Rat Hearts

After the aorta was secured and retrograde perfusion initiated (see Section 2.1), the heart was rotated on the aortic cannula to position the openings in the left atrium to receive the second perfusion cannula. The left atrium was slipped onto this cannula and tied. Atrial perfusion was begun by clamping the tubes serving for retrograde aortic perfusion (clamp A; Fig. 45.1) and unclamping the tubes (clamps B and C; Fig. 45.1) supplying perfusate to the atrium via an overflow-type bubble trap and carrying perfusate back to the oxygenation chamber and reservoir. This design eliminated pressure fluctuations due to the pump and insured a constant 10 cm H_2O atrial filling pressure (determined by the distance between the atrial bubble trap and the left ventricle, which was fixed at 10 cm in all of the experiments). The buffer solution entering the atrium passed into the ventricle, and ventricular contraction forced the fluid into a pressure chamber attached to the aortic cannula. This chamber was one-third filled with air to provide some elasticity to an otherwise rigid system. When the left ventricle contracted, pressure developed in this chamber, which forced buffer out through a tube into the aortic bubble trap 85 cm above the heart. Overflow from the aortic bubble trap was returned inside the long central oxygenating chamber back to the lower reservoir, from which it was carried out again by the peristaltic pump. A water jacket maintained at 37°C was provided for the entire system, except the connecting tubing.

2.3.   Measurements of Coronary Flow, Aortic Output, and Ventricular Mechanical Performance

During retrograde coronary perfusion, coronary flow was continuously measured by an electromagnetic flow probe (Carolina Medical Electronics, model 322) placed into the tube connecting the aortic bubble trap and the aorta. In the working rat heart, coronary flow was measured by collecting the effluent from the heart in a graduated cylinder below the central opening of the heart chamber. Under these conditions aortic flow was continuously measured by the electromagnetic flow probe and also by collecting the overflow from the aortic bubble trap. The flow probe was calibrated before each experiment by known flow rates provided by the peristaltic pump. During retrograde coronary perfusion, left ventricular contractions were characterized by apex displacement. For this purpose the apex of the heart was connected to a force transducer (Grass FT03C) by a thread. After securing the thread to the apex of the heart, it was pulled through the central opening of the heart chamber and connected to the force transducer via a grooved wheel (Fig. 45.1). In the working preparation, left ventricular pressure development was measured by inserting a polyethylene tube (inside diameter 0.2 mm) into the ventricle through a side arm in the atrial cannula and connecting it to a pressure transducer (Statham P23). Apex displacement or ventricular pressure generation and the output of the flow probe (coronary or aortic flow) were recorded on a two-channel Honeywell recorder. Heart rate (beats per minute) was calculated from the apex displacement or pressure recordings. Stroke volume was calculated by dividing the value of cardiac output (aortic flow plus coronary flow) by heart rate. At the end of perfusion, hearts were cut from the cannulae, the atria and right ventricle removed and blotted on filter paper, and wet weight of the left ventricle measured. To correct for variations of initial coronary flow rates in hearts of various sizes, coronary flow data are expressed as milliliter per minute per gram of left ventricle wet weight, or the changes induced by various interventions expressed as percentage of initial flow rates.

2.4. Electrical Pacing of Hearts

In most experiments the hearts beat spontaneously. In some experiments the hearts were paced via a stainless steel wire fixed to the right atrium. Square-wave impulses (10 V; 2 msec; at frequencies exceeding the spontaneous heart rate by 50%) were provided by a Grass (SM6) stimulator.

2.5. Experimental Protocols

Hearts perfused by the Langendorff technique were first equilibrated for 30 min, during which all of the measured parameters (flow, contraction, and spontaneous heart rate) reached steady levels. In time control experiments, the hearts were perfused with control solution in the next 90 min. When concentration–response curves to nickel chloride ($NiCl_2$) were analyzed, nickel was added to the control buffer solution at appropriate concentrations, and the reservoir was first drained and then filled with the buffer containing $NiCl_2$. Perfusion with higher concentrations of $NiCl_2$ was started (without returning to control situation) after the effects of the previous concentrations reached steady state (which varied between 10 and 20 min). When the effects of KCl-induced arrest, pacing, and adenosine were analyzed on the concentration–response curve to $NiCl_2$, the cumulative increase of $NiCl_2$ concentration was started only after the effect of these interventions reached a steady level (approximately 5–15 min). Similarly, when the effects of $NiCl_2$ were tested on coronary vasodilatation, the vasodilator stimuli were applied only after the effects of $NiCl_2$ reached steady state. Reactive increases in coronary flow (during Langendorff perfusion) were induced by ischemia (60 sec cessation of perfusion) or by electrical pacing of the hearts. Adenosine (Sigma Chemical Co., St. Louis, MO) was added to the buffer solution at increasing concentrations (10^{-6}–10^{-4} M).

2.6. Calculations and Statistical Analysis

The mean and standard error of the mean ($\bar{x} \pm$ SEM) of grouped experimental data were calculated. The difference between means was estimated by Student's t-test for paired and unpaired observations. Their difference was denoted as statistically significant when $P \leqslant 0.05$.

3. RESULTS

3.1. Comparison of Effects of NiCl₂ in Langendorff Preparation and Working Rat Heart

After 30 min retrograde coronary perfusion at a fixed (85 cm H_2O) hydrostatic pressure, the coronary flow rate was 8.8 ± 0.7 mL/min g ($n = 12$), which did not change significantly in the next 90 min if the hearts were perfused with control solution only (time control; data not shown). Similarly, left ventricular contractions (characterized by measuring apex displacement) and spontaneous heart rate remained unchanged during this period (data not shown). Low concentrations ($10^{-8}-10^{-6}\ M$) of NiCl₂ had no significant effect on any of the parameters studied. At $10^{-5}\ M$, NiCl₂ induced almost complete cessation of coronary flow (Fig. 45.2; Table 45.1), which was accompanied by cessation of ventricular contractions (Table 45.1).

After 30 min atrial perfusion at a fixed atrial filling pressure of 10 cm H_2O, the working hearts had significantly higher coronary flow rate (16.8 ± 1.2 mL/min g; $p < 0.01$; $n = 6$) than the Langendorff preparations, which remained unchanged in the next 60-min perfusion period (time control; data not shown). Aortic flow at this atrial filling pressure exceeded coronary flow by a factor of 2.9, so it was more than sufficient for coronary perfusion. At $10^{-8}\ M$, NiCl₂ had no significant effects on the working hearts (Table 45.2), but at

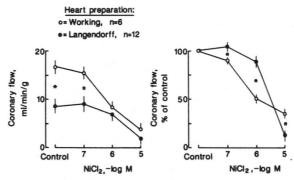

Figure 45.2. Effect of increasing concentrations ($10^{-7}-10^{-5}\ M$) of NiCl₂ on coronary flow (left, mL/min g wet heart weight; right, percentage of initial flow) in Langendorff and working rat heart preparations. Data shown as mean $\pm$ SEM. (*) Statistically significant differences between the effects of NiCl₂ in the Langendorff and working hearts.

Table 45.1 Effect of NiCl$_2$ on Langendorff Preparations[a]

	Control[b]	$10^{-8}\ M^c$	$10^{-7}\ M^c$	$10^{-8}\ M^c$	$10^{-6}\ M^c$	$10^{-5}\ M^c$
Coronary flow,[d] mL/min g	9.4 ± 0.5	9.8 ± 0.6	9.8 ± 0.5		9.8 ± 0.3	0.1 ± 0.08^d
Apex displacement, mm	4.5 ± 0.4	4.3 ± 0.4	4.6 ± 0.4		3.8 ± 0.5	0 ± 0^d
Heart rate	260 ± 14.5	251 ± 10.0	252 ± 10.9		259 ± 10.7	0 ± 0^d

[a]Data are shown as mean $\pm$ SEM of six experiments.
[b]Data are obtained after 30-min equilibration period at a fixed perfusion pressure of 85 cm H$_2$O.
[c]Data obtained 10–20 min after the start of NiCl$_2$ infusion when the effect reached steady state.
[d]Expressed as mL/min g left ventricular wet weight.
[e]Effect of NiCl$_2$ is statistically significant ($p < 0.05$).

Table 45.2 Effect of NiCl₂ on Isolated Working Heart Preparations[a]

	Control[b]	$10^{-8}\ M^c$	$10^{-7}\ M^c$	$10^{-6}\ M^c$	$10^{-5}\ M^c$
Coronary flow,[d] mL/min g	16.8 ± 1.4	16.4 ± 1.4	15.4 ± 1.5[e]	8.2 ± 1.0[e]	5.3 ± 0.8[e]
Aortic flow,[d] mL/min g	56.4 ± 4.5	55.8 ± 4.9	52.5 ± 4.1	22.2 ± 8.0[e]	7.8 ± 4.3[e]
Stroke volume,[d] mL/min g	0.37 ± 0.04	0.35 ± 0.03	0.33 ± 0.03	0.19 ± 0.07[e]	0.08 ± 0.04[e]
Left ventricular peak systolic pressure, mm Hg	129.7 ± 7.4	128.8 ± 6.9	124.2 ± 8.5	94.7 ± 9.4[e]	48.0 ± 15.8[e]
Left ventricular end-diastolic pressure mm Hg	2.6 ± 0.9	2.3 ± 0.8	2.8 ± 0.9	4.7 ± 0.9	6.8 ± 1.1[e]
Heart rate beats/min	206.2 ± 23.4	209.2 ± 19.1	179.7 ± 36.7	181.3 ± 18.1	115.8 ± 38.2[e]

[a] Data are shown as mean ± SEM of six experiments.
[b] Data are obtained after 30-min equilibration period at a fixed atrial filling pressure of 10 cm H₂O.
[c] Data obtained 10–20 min after the start of NiCl₂ infusion when the effect reached steady state.
[d] Flow data expressed as mL/min g left ventricular wet weight.
[e] Effect of NiCl₂ is statistically significant ($p < 0.05$).

10^{-7}–10^{-5} M, it reduced coronary flow rate significantly (Fig. 45.2, left; Table 45.2). At concentrations of 10^{-7} and 10^{-6} M, NiCl$_2$ was significantly more effective in reducing coronary flow in the working hearts than in the Langendorff preparations (Fig. 45.2, right). In contrast, the relative decrease of coronary flow at 10^{-5} M NiCl$_2$ was significantly less in the working (68.1 ± 6.2% of initial flow rate; n = 6) than in the Langendorff preparations (90.2 ± 5.6%; Fig. 45.2, right; n = 12; $p < 0.05$). During perfusion with 10^{-6} and 10^{-5} M NiCl$_2$, the significant difference in absolute coronary flow rate (observed under control conditions and also in the presence of 10^{-7} M NiCl$_2$) disappeared between the two preparations (Fig. 45.2, left). The decrease in coronary flow during perfusion of the working hearts with 10^{-7} M NiCl$_2$ did not affect aortic flow or the mechanical performance of the hearts (Table 45.2). Reduction of coronary flow by 10^{-6} M NiCl$_2$ was accompanied by significant reduction in aortic flow, stroke volume, and left ventricle peak systolic pressure, while spontaneous heart rate decreased and left ventricular end-diastolic pressure increased significantly only in the presence of 10^{-5} M NiCl$_2$ (Table 45.2).

3.2. Electrical Pacing of Langendorff Preparations

Langendorff preparations (n = 6) were paced electrically at a rate to exceed the initial spontaneous heart rate by 50%. The increase in heart rate was accompanied by a significant augmentation of coronary flow (inset in Fig. 45.3) but no significant change in ventricular contractile force (Table 45.3).

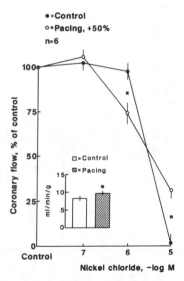

Figure 45.3. Effect of electrical pacing (at frequencies exceeding initial heart rate by 50%; see also Table 45.3) on coronary flow (inset) and on vasoconstrictor effect of NiCl$_2$. The effects of NiCl$_2$ are expressed as percentage of the initial coronary flow rate. Data are shown as mean ± SEM. (*) Statistically significant differences between the effects of NiCl$_2$ on control and paced hearts.

Table 45.3 Effect of Electrical Pacing, Adenosine, and KCl on Coronary Flow, Apex Displacement, and Heart Rate of Isolated Rat Hearts Perfused by Langendorff Technique[a]

	Electrical Pacing[b]		Adenosine (10^{-6} M)		KCl (26 mM)	
	Before[c]	During[d]	Before	During	Before	During
Coronory flow,[e] mL/min g	8.2 ± 0.6	11.4[f] ± 0.3	8.8 ± 0.8	12.7[f] ± 0.6	9.8 ± 0.6	6.8[f] ± 0.5
Apex displacement, mm	6.6 ± 0.9	4.7 ± 0.7	4.5 ± 0.4	4.2 ± 0.4	5.1	0[f]
Heart rate	254 ± 22	386[f] ± 20	230 ± 21	236 ± 15	212 ± 19	0[f]

[a] Data are shown as means ± SEM of six experiments.
[b] The hearts were stimulated at a rate exceeding the spontaneous heart rate by 50% (see also mean heart rate values before and during pacing).
[c] Data obtained after 30-min equilibration–perfusion period (before the various interventions were started).
[d] Data measured when the effect of the various interventions reached steady state (10–20 min).
[e] Coronary flow values (mL/min) were normalized for the wet weight (g) of the left ventricle.
[f] Difference between data obtained before and after the interventions is statistically significant ($p < 0.05$).

Pacing of the hearts significantly augmented the reduction of coronary flow induced by 10^{-6} M NiCl$_2$ but significantly depressed the coronary vasoconstrictor action of 10^{-5} M NiCl$_2$ (Fig. 45.3). Thus, pacing of the Langendorff preparation made the concentration–response curve of NiCl$_2$ similar to that obtained in the working rat hearts (see Fig. 45.2, right). In addition, as in the working heart, the initial difference between coronary flow rate in the spontaneously beating and paced Langendorff preparations (see inset in Fig. 45.3) disappeared in the presence of 10^{-6} MNiCl$_2$.

3.3. Correlation between Initial Coronary Flow Rate and Vasoconstrictor Effect of NiCl$_2$

The decrease of coronary flow (expressed as milliliters per minute per gram) induced by 10^{-6} M NiCl$_2$ was analyzed as a function of initial coronary flow rate in the three experimental groups studied (Fig. 45.4). Linear regression analysis of the data pairs revealed a strong positive correlation ($r = 0.90$; $p < 0.01$), indicating that under these experimental conditions the sensitivity of coronary arteries to the vasoconstrictor action of 10^{-6} M NiCl$_2$ increases with increased initial coronary flow rate.

3.4. Effect of Exogenous Adenosine

Addition of 10^{-6} M adenosine to the perfusate increased coronary flow significantly (inset in Fig. 45.5) but had no effect on ventricular contractile

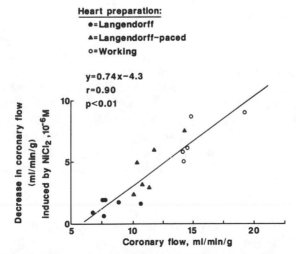

Figure 45.4. Correlation between initial flow rate and the decrease in coronary flow (expressed as ml/min g) induced by 10^{-6} M NiCl$_2$ in working preparations and in unpaced and paced Langendorff preparations. Note the significant positive correlation ($r = 0.90$; $p < 0.01$) between the two variables.

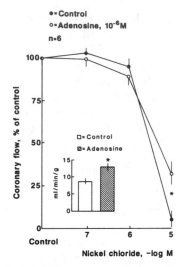

Figure 45.5. Effect of adenosine (10^{-6} M) on coronary flow rate (inset) and on the vasoconstrictor action of $NiCl_2$ in unpaced Langendorff preparations. Data are shown as mean ± SEM. (*) Statistically significant differences between the effects of $NiCl_2$ on control and adenosine-treated hearts.

force or spontaneous heart rate in Langendorff preparations (Table 45.3). Adenosine did not influence the effects of lower concentrations (10^{-7} and 10^{-6} M) of $NiCl_2$ but significantly depressed the coronary vasoconstrictor action of 10^{-5} M $NiCl_2$ (Fig. 45.5).

3.5. Effect of KCl Arrest

To determine whether myocardial contractile activity is essential for the coronary vasoconstrictor action of $NiCl_2$, the effects of the trace metal were studied in Langendorff preparations where ventricular contractions were inhibited by exposure to 26 mM KCl. Within 1 min after the start of high KCl perfusion, heart beat was stopped and coronary flow decreased significantly from 9.8 ± 0.6 mL/min g to a steady level of 6.8 ± 0.5 mL/min g (inset in Fig. 45.6; Table 45.3). Increasing concentrations of $NiCl_2$ produced a decrease in coronary flow in arrested hearts as well; at 10^{-6} M concentration, $NiCl_2$ evoked significantly greater reduction in flow than in control Langendorff preparations (Fig. 45.6).

3.6. Effect of $NiCl_2$ on Vasodilator Responses in Langendorff Preparation

Cessation of perfusion for 60 sec or electrical pacing of the hearts was followed by significant increases in coronary flow (42 ± 3% and 24 ± 11% of initial flow rate, respectively; Fig. 45.7). Twenty minutes after return to

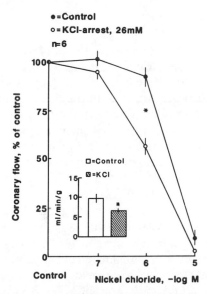

Figure 45.6. Effect of elevation of perfusate potassium ion (K⁺) concentration (to 26 mM) on coronary flow rate (inset) and on the vasoconstrictor action of NiCl₂ in unpaced Langendorff preparations. Data are shown as mean ± SEM. (*) Statistically significant differences between the effects of NiCl₂ in control and K⁺-arrested hearts.

control perfusion (when coronary flow rate returned to initial level), the hearts were perfused with buffer solution containing 10^{-6} M NiCl₂, which causes a 12 ± 4% decrease in flow. Reactive increases the flow following ischemia or electrical pacing were prevented in the presence of NiCl₂ (Fig. 45.7). Similarly, NiCl₂ inhibited the increases in coronary flow induced by 10^{-6} M adenosine (Fig. 45.7).

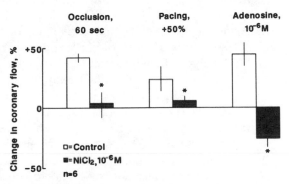

Figure 45.7. Effect of NiCl₂ (10^{-6} M) on reactive increases in coronary flow following cessation of perfusion (60 sec) or electrical pacing and on the vasodilatation induced by exogenously administered adenosine (10^{-6} M) in Langendorff preparations. Data are shown as mean ± SEM. (*) Effect of NiCl₂ on the responses to the vasodilator stimuli is statistically significant.

4. DISCUSSION

The present experiments confirm earlier observations that $NiCl_2$ induces coronary vasoconstriction in isolated perfused rat heart preparations (Rubanyi and Kovach, 1980; Rubanyi et al., 1981). They also demonstrate that an increase of mechanical performance of the heart (either by allowing the left ventricle to pump fluid against an external resistance or by electrical pacing of the nonpumping ventricle) produces coronary vasodilation (as determined by increases in flow at constant perfusion pressure) and has two distinct effects on nickel-induced coronary vasoconstriction: (a) it augments the effect of lower concentrations (10^{-7}–10^{-6} M) of $NiCl_2$ and (b) it reduces the vasoconstriction in response to 10^{-5} M $NiCl_2$. The finding that 10^{-6} M $NiCl_2$ reduced coronary flow rate in paced Langendorff preparations and in the working hearts to a level existing in the nonpaced Langendorff preparations suggests that $NiCl_2$ effectively antagonized the vasodilator mechanism(s) triggered by increases in myocardial mechanical performance. This suggestion was confirmed by (a) the positive correlation found between initial flow rate and vasoconstrictor action of 10^{-6} M $NiCl_2$ and (b) the prevention of pacing-induced vasodilatation by 10^{-6} M $NiCl_2$. Reactive hyperemia induced by increases in myocardial mechanical performance was shown to be mediated by local (metabolic) vasodilator mechanism triggered by augmented myocardial O_2 demand (Belloni, 1979; Berne, 1980; Sparks, 1980). The primary mediator of "exercise-hyperemia" is thought to be adenosine (Belloni, 1979; Berne, 1980), similar to the reactive hyperemia evoked by conditions (e.g., ischemia, hypoxia) that disturb the balance between myocardial O_2 demand and O_2 supply by reducing O_2 supply (Belloni, 1979; Berne, 1980). However, prevention of the vasodilatory action of endogenously produced adenosine cannot be the primary cause of augmented Ni^{2+} sensitivity in hearts with increased mechanical performance, since exogenously administered adenosine did not augment the vasoconstrictor action of 10^{-6} M $NiCl_2$. Alternatively, rather than interfering with its action, $NiCl_2$ may prevent the production/release of endogenous adenosine, a mechanism that cannot be observed during continuous infusion of exogenous adenosine. Earlier demonstration that exogenously administered $NiCl_2$ can penetrate the capillary wall and cell membrane of myocardial cells (Rubanyi et al., 1980) and can alter myocardial metabolism (Rubanyi and Kovach, 1980) may suggest that interference with myocardial metabolism may be (at least in part) responsible for the coronary vascular actions of the trace metal. Nickel chloride effectively prevented reactive increases in flow following 60 sec total coronary occlusion and electrical pacing and the vasodilator action of exogenous adenosine. This then suggests that one of the mechanisms of Ni^{2+} action on coronary vessels may be the indirect consequence of inhibition of local vasodilator mechanisms, potentially by inhibiting the production/release and vascular action of adenosine.

The increased Ni^{2+} sensitivity of coronary arteries of hearts with disturbed O_2 demand–O_2 supply balance may be the consequence of the "sensitizing" action of factors (e.g., K^+, pO_2, pH, prostaglandins, etc.) other than adenosine, which were also shown to contribute to local metabolic vasoregulation (Belloni, 1979; Sparks, 1980). The potential involvement of potassium ions [released from myocardial cells during electrical stimulation (Gilmore et al., 1971; Sybers et al., 1965) and hypoxia/ischemia (Murray and Sparks, 1976; Sarnoff et al., 1966; Sparks, 1980)] may be suggested on the basis of the present findings that similar to increased mechanical performance, increased extracellular K^+ concentration also augmented Ni^{2+} sensitivity. This latter observation is of particular interest, since Ni^{2+} sensitivity increased in the absence of cardiac mechanical activity (inhibited by high K^+). It is then obvious that contraction of myocardial cells play no essential role in Ni^{2+}-induced coronary vasoconstriction. This may also imply that the observed higher Ni^{2+} sensitivity of coronary vessels in isolated rat hearts (Rubanyi and Kovach, 1980; Rubanyi et al., 1981) and in situ dog hearts (Rubanyi et al., 1984) as compared to isolated large coronary arteries (Rubanyi et al., 1982) may be the consequence of differential responsiveness of large and small coronary arteries rather than due to the presence of the surrounding functional myocardium.

The selective depression by exogenous adenosine and by increased mechanical activity of the vasoconstrictor action of high (10^{-5} M) but not lower (10^{-7}–10^{-6} M) concentrations of $NiCl_2$ suggests that the trace metal may act via two different mechanisms on coronary vessels. It remains to be determined whether the adenosine-insensitive and adenosine-sensitive actions occur on the same or different segments of the coronary artery tree.

In conclusion, the present study suggests that nickel ions cause coronary vasoconstriction in isolated rat heart via multiple mechanisms: (1) adenosine-insensitive direct vascular action at lower concentrations (10^{-7}–10^{-6} M), which can be augmented by an increase in cardiac mechanical activity (possible by mediators such as K^+); (2) adenosine-sensitive direct action evoked by high (10^{-5} M) concentration of the trace metal, which is depressed rather than augmented by increases in cardiac mechanical activity; and (3) indirect action via inhibition of vasodilation triggered by local metabolic mechanisms and by exogenous adenosine.

REFERENCES

Belloni, F. L. (1979). The local control of coronary blood flow. *Cardiovasc. Res.* 13, 63–85.

Berne, R. M. (1980). The role of adenosine in the regulation of coronary blood flow. *Circ. Res.* 47, 807–813.

Gilmore, J. P., Nizolek, J. A., Jr., and Jacob, R. J. (1971). Further characterization of myocardial K^+ loss induced by change in contraction frequency. *Am. J. Physiol.* 221, 465–469.

Murray, P. A., and Sparks, H. V. (1976). The role of K^+ in the control of coronary vascular resistance. *Physiologist* 19, 307.

Neely, J. R., Liebermeister, H., Battersby, E. J., and Morgan, H. E. (1967). The effect of pressure development on oxygen consumption by isolated rat heart. *Am. J. Physiol.* 212, 804–814.

Rubanyi, G., and Kovach, A.G.B. (1980). The effect of nickel ions on cardiac contractility, metabolism and coronary flow in the isolated rat heart. In M. Anke, H.-J. Schneider, and Chr. Bruckner (Eds.), *Spurenelement Symposium, Ni.*, Vol. 3, Friedrich-Schiller-Universität, Jena, pp. 111–115.

Rubanyi, G., Balogh, I., and Kovach, A.G.B. (1980). The effect of nickel ions on ultrastructure of isolated rat heart. *J. Mol. Cell. Cardiol.* **12,** 609–619.

Rubanyi, G., Ligeti, L., and Koller, A. (1981). Nickel is released from the ischemic myocardium and contracts coronary vessels by a Ca-dependent mechanism. *J. Mol. Cell. Cardiol.* **13,** 1023–1026.

Rubanyi, G., Kalabay, L., Pataki, T., and Hajdu, K. (1982). Nickel stimulates isolated canine coronary artery contraction by a tonic calcium activation mechanism. *Acta Physiol. Acad. Sci. Hung.* **59,** 155–160.

Rubanyi, G., Ligeti, L., Koller, A., and Kovach, A.G.B. (1984). Possible role of nickel ions in the pathogenesis of ischemic coronary vasoconstriction in the dog heart. *J. Mol. Cell. Cardiol.* **16,** 533–542.

Sarnoff, S. J., Gilmore, J. P., Mcdonald, R. J., Jr., Daggett, W. M., Weisfeldt, P. B., and Mansfield, P. B. (1966). Relationship between myocardial K^+ balance, O_2 consumption, and contractility. *Am. J. Physiol.* **211,** 361–375.

Sparks, H. V., Jr. (1980). Effect of local metabolic factors on vascular smooth muscle. In D. H. Bohr, A. P. Somlyo, and H. V. Sparks, Jr. (Eds.), *Handbook of Physiology, Section 2: The Cardiovascular System,* Vol. II, Vascular Smooth Muscle, American Physiological Society, Bethesda, pp. 475–513.

Sybers, R. G., Sybers, H. D., Helmer, P. R., and Murphy, Q. R. (1965). Myocardial potassium balance during cardioaccelerator nerve and atrial stimulation. *Am. J. Physiol.* **209,** 699–701.

46

RESPIRATORY CANCER MORTALITY IN WELSH NICKEL REFINERS: WHICH NICKEL COMPOUNDS ARE RESPONSIBLE?

D. F. Easton and J. Peto

*Institute of Cancer Research, Sutton, Surrey,
United Kingdom SM2 5NG*

L. G. Morgan, L. P. Metcalfe, and V. Usher

INCO Europe, Clydach, Swansea, United Kingdom SA6 5QR

R. Doll

ICRF Cancer Studies Unit, Oxford, United Kingdom OX2 6PE

Nickel and Human Health: Current Perspectives, Edited by Evert Nieboer and
Jerome O. Nriagu.
ISBN 0-471-50076-3   © 1992 John Wiley & Sons, Inc.

---

1. INTRODUCTION

The excesses of nasal sinus and lung cancer in early Welsh nickel refiners was recognized over 50 years ago [Bridge, 1933; A. B. Hill (1939), quoted in Morgan, 1958], and well-documented excesses have also been observed in men employed in nickel refineries in several countries, notably Norway (Magnus et al., 1982; Chapter 47) and Canada (Roberts et al., 1984; Chapter 48). Despite this wealth of evidence that nickel refining can present a cancer hazard, however, it is still unclear which nickel compound or compounds are primarily responsible for the cancer risk.

Previous reports demonstrated that the cancer risk in a Welsh refinery was associated with employment in the calciners, the copper plant, and the furnaces and possibly with the nickel sulfate plant (Peto et al., 1984; Kaldor et al., 1986). Hitherto exposure in these different areas has been described only in general qualitative terms. Quantitative estimates of airborne levels of various nickel species for each period and area in the plant have now been provided by one of the authors (L.P.M.). No measurements were available prior to 1970, and these estimates have been synthesized from anecdotal evidence and extrapolation from more recent measurements.

In addition, we have constructed a new and more complete cohort consisting of 2524 workers employed at the refinery for five years or more before the end of 1969. This cohort includes a much larger number of men employed since 1930 than had previously been studied. Work histories of all men in the cohort have been linked to the exposure estimates, thus allowing intensity and duration of exposure to be related to mortality.

The cohort and exposure data reported here have formed part of a recent international collaborative study of nickel and human carcinogenesis (IWG, 1990). This study sought mortality and exposure data from most of the available cohorts of nickel-exposed workers in order to review the evidence on human carcinogenicity pertaining to the various nickel species.

In this chapter, we describe first the exposure estimates that have recently been obtained for the refinery and then the mortality experience of the updated cohort and the relationship between risk and employment in different areas of the plant. Finally, we consider the problem of quantifying the carcinogenic effect of the different nickel species and apply some simple dose –response models to the data.

2. PROCESS DESCRIPTION AND EXPOSURES

The refinery commenced production in 1902 and is still operating. The main refining process involves hydrogen reduction of impure nickel oxide followed by reaction of nickel with carbon monoxide to form nickel carbonyl gas and thermal decomposition of nickel carbonyl to produce pure nickel. These operations were carried out in the *nickel plants*, of which seven were built prior to 1931, although not all were in use at any one time. In 1958 the decomposer units in the nickel plants were shut down and a new *pellet plant* opened for nickel pellet production. In 1969 the nickel plants ceased operation except for some secondary processing, when the *kiln plants* were brought on line. In 1941 production of nickel powder started following the commissioning of the *powder plant* in a disused nickel plant.

Until 1948 the main refinery feed was a sulfide, which had to be roasted in the *calciners* to give the oxide before processing. The oxide feed used between 1948 and 1969 was also heat treated in the calciners before processing. The calciners also treated a wide range of refinery intermediates before further processing. Prior to 1936 all grinding of materials was done in small mills in the calciner building; subsequently grinding was carried out in a *central grinding plant*.

The sulfide feed up to 1933 was Bessemer matte, containing approximately equal proportions of nickel and copper. After calcination, the copper content was reduced in the *copper plants* by leaching with sulfuric acid to produce copper sulfate. In addition, some nickel was also leached and recovered as nickel sulfate in the *nickel sulfate plant*.

Solids generally passed through the nickel extraction process three times. The final residue was passed to the *concentrates plant* for acid leaching to recover base metals as sulfates and to yield a precious metal concentrate.

The basis of the exposure estimates is described in detail in the report of the international study (IWG, 1990). No reliable measurements of atmospheric dust are available before the 1970s, and levels of exposure in earlier years can be estimated only indirectly. The quantities and composition of the materials processed give quite good estimates of the proportions in which the various nickel species were likely to have been present. The absolute concentrations, however, could be estimated only by extrapolation from more recent conditions combined with written and verbal evidence provided by men employed at the time. These estimates are summarized in Table 46.1. It should be noted that sulfidic nickel was primarily nickel subsulfide (Ni_3S_2) but included some nickel sulfide (NiS).

3. COHORT AND METHODS

The study cohort was defined as all men with at least five years service at the refinery up to December 31, 1969, other than office staff. This enlarges

Table 46.1 Estimated Atmospheric Concentrations of Nickel in Different Working Areas of Refinery by Period of Operation

Location	Period	Estimated Ni concentration (mg/m³)	Estimate phase (%)				
			Ni	NiO	Ni/CuO	Ni_xS_y	Sol
Nickel Plants							
General	1920–1946	1–10	0–5	65–70	20–25	0–5	0–5
	1947–1965	1–5	0–5	75–80	5–10	0–5	0–5
	1966–1977	<5	0–5	—	85–90	0–5	0–5
Charge/discharge	1902–1959	10–25[a]	0–5	90–95	—	0–5	0–5
	1960–1977	5–15[a]	35–40	20–25	20–25	15–20	0–5
Plant cleaning	1902–1977	10–100[a]	70–75	—	15–20	10–15	0–5
Kiln Plant							
General	1969–1979	<1	35–40	15–20	15–20	15–20	0–5
Discharge	1969–1979	1–10[a]	30–45	5–10	30–35	15–20	0–5
Pellet Plant	1958–1978	<0.5	—	100			
	1979–1984	<1	80	20			
Powder Plant							
General	1941–1944	1–10	70–75	10–15	0–5	5–10	0–5
	1975–1977	1–5	100	—			
Sieving/mixing	1941–1978	5–25	80	20			
	1979–1984	1–5	100	—			

Process	Years						
Charge/discharge	1941–1974	10–25[a]	0–5	60–65	25–30	0–5	0–5
Calciners							
Linear	1902–1929	10–100	15–20	—	60–65	20–25	0–5
	1930–1936	10–100	10–15	—	45–50	35–40	0–5
Rotary	1931–1949	5–20	5–10	—	60–65	20–25	0–5
	1950–1954	5–20	10–15	20–25	40–45	10–15	0–5
	1955–1959	5–20	—	85–95	0–5	0–5	5–10
	1960–1968	10–15	5–10	30–35	40–45	5–10	10–15
	1969–1975	<5	5–10	30–35	40–45	5–10	10–15
Shed	1976–1979	<5	5–10	35–40	35–40	5–10	10–15
Centralized Grinding Plant	1936–1948	10–25	5–10	35–40	35–40	20–25	
	1949–1959	10–25	0–5	90–95	—	0–5	
	1960–1964	1–10	50–60	—	—	40–50	
Copper Plant	1902–1936	10–25	—	—	85–90	0–5	5–10
Nickel Sulfate							
General	1905–1946	<5	20–30	15–20	25–30	0–5	30–40
	1947–1979	<5	20–30	25–30	15–20	0–5	30–40
Packing	1905–1979	<5	—	—	—	—	100
Furnaces	1907–1963	10–25	35–40	10–25	10–25	15–20	0–5

[a] Higher levels of exposure experienced only occasionally.

Table 46.2 Follow-up Status, December 31, 1985

Year of First Employment	Alive	Dead	Lost to Follow-up	Total
Before 1920	3	523	253	779
1920–1929	49	438	75	562
1930+	728	399	56	1183
Total	780	1360	384	2524

somewhat the cohort of men employed prior to 1930, in comparison with that previously defined (Doll et al., 1970; Peto et al., 1984) in which men were identified from a limited number of pay sheets. A more important difference from previous studies, however, is the much larger number of men employed only since 1930.

Follow-up commenced in 1931. The cohort was followed up to the end of 1985 through company records and the National Health Service Central Register. A summary of follow-up details is given in Table 46.2. Due to inadequacies in identifying information, 24% of workers first employed before 1930 could not be traced. It is unclear whether tracing was more complete for men dying of lung or nasal sinus cancer. However, even if the cohort included all the lung and nasal sinus cancer deaths for men who would otherwise have been untraced, the standardized mortality ratios (SMRs) would only be overestimated by 32% for men first employed before 1920 and 13% for men first employed in the period 1920–1929. Five percent of individuals employed since 1930 have been lost to follow-up. All deaths were coded according to the seventh revision of the International Classification of Diseases. Expected numbers of deaths and SMRs were computed in the usual way from mortality rates for England and Wales. In addition, SMRs based on local mortality were also computed. These were obtained from national rates by multiplying by the SMRs for Glamorgan over the period 1968–1978 (Gardner et al., 1983, 1984).

4. OVERALL MORTALITY

Table 46.3 shows the lung and nasal cancer mortality by year of first employment. As in previous analyses, the excess for both cancers is largely confined to workers employed before 1930. In addition to the 67 nasal cancer deaths shown in this analysis for pre-1930 employees, 16 further cases of nasal sinus cancer have been identified: 3 cases were classified as cancer of the nasopharynx on the death certificate, 4 cases had nasal cancer mentioned only as a contributory cause of death in part II of the death certificate, and 9 deaths occurred before 1931 when follow-up began.

Table 46.3 Observed and Expected Deaths from Lung and Nasal Sinus Cancer (1931–1985) by Year of First Employment

Year First Employed	Lung Cancer			Nasal Cancer			Other Identified Nasal Cancers[a]
	Observed	Expected	SMR	Observed	Expected	SMR	
Before 1920	83	13.46	617[b]	55[b]	0.15	37647	13
1920–1929	88	28.04	314[b]	12[b]	0.17	7255	3
1930–1939	20	14.45	138	1	0.07	1434	0
1940–1949	14	11.88	118	0	0.06	—	0
1950 and later	9	10.74	84	0	0.06	—	0
Total	214	78.57	272[b]	68[b]	0.45	15111	16

[a]See text.
[b]$p < 0.001$.

There is some suggestion of a lung cancer excess in workers first employed between 1930 and 1939 (SMR 138). When the comparison is based on local rates, the excess is just significant (SMR 156, $p = 0.04$). In addition, one nasal cancer death has occurred in a tradesman first employed in 1937. The implication of these findings is discussed in Section 7.

Mortality from causes of death other than lung or nasal cancer is summarized in Table 46.4. There is little evidence of any excess for other cancer deaths or for nonmalignant causes. In workers employed before 1930, in whom the excess of respiratory cancers was greatest, there is an excess of prostate cancer (SMR 159), which just reaches statistical significance ($p < 0.05$) when adjusted for local rates, and nonsignificant excesses of bladder and colon cancer are also evident. Excesses at some sites are, of course, likely to occur by chance when many types of cancer are examined, and these excesses can be properly judged only after combining these data with that from other refinery cohorts, as has been done in the international study referred to previously (IWG, 1990)

In men employed before 1930, deaths from circulatory disease, respiratory disease, and other cancers were close to or below the numbers expected, whether the expected numbers were derived from local rates or corrected for locality. This was also true for men employed after 1930 with the exception of circulatory disease, the mortality for which was significantly elevated in comparison with national rates but not compared with local rates.

5. RISKS BY WORK AREA

Tables 46.5 and 46.6 show the lung and nasal cancer risks, respectively, by duration of employment in five *high-risk* work areas. The first four are those indicated as being associated with high risk from previous case control analyses (Kaldor et al., 1986). An additional group of reducer cleaners is included here because men in this small group are likely to have had a high exposure to metallic nickel (though in addition to other exposures).

For each of these five areas there is evidence of an increasing relative risk with duration of employment after allowing for time spent in other high-risk areas.

6. MODEL FITTING

It is apparent from Table 46.1 that employment in any of the high-risk areas would have entailed substantial exposure to more than one type of nickel compound. No category of workers exposed only to one nickel species could be defined, so that any attempt to quantify the effects of individual species on cancer mortality is faced with formidable problems. Inevitably some form

Table 46.4 Observed and Expected Deaths from Specific Causes of Death Other than Lung and Nasal Sinus Cancer by Year of First Employment[a]

Cause of Death	Workers First Employed Before 1930				Workers First Employed 1930 and After			
	Observed	Expected	SMR-1	SMR-2	Observed	Expected	SMR-1	SMR-2
Cancer of								
Buccal cavity and pharynx	4[b]	2.77	144	144	1	1.37	73	73
Larynx	2	2.06	97	98	1	0.92	151	110
Esophagus	2	4.11	49	44	6	2.60	231[c]	208
Stomach	15	20.94	72	60	7	9.89	71	58
Colon	16	10.47	153	149	8	5.50	145	141
Rectum	11	8.72	126	116	14	14.81	95	87
Prostate	13	8.18	159[c]	181[c]	2	3.83	51	59
Bladder	9	5.43	166	183	2	3.27	61	67
Kidney	1	1.81	55	55	1	1.63	61	61
Other	25	25.50	98	101	7	9.61	73	74
Cerebrovascular disease	81	66.46	122	93	29	28.30	102	78
Other circulatory disease	307	286.75	107	91	178	135.63	131[d]	111
Respiratory disease	99	96.39	103	96	42	40.79	103	95
Other causes	23	51.67	45	72	42	48.63	86	79
Total excluding lung and nasal sinus cancer	608	591.26	103	91	340	306.78[c]	111[c]	98

[a]Expected deaths are numbers at national mortality rates. SMR-1 is the SMR when compared with national mortality rates; SMR-2 is SMR adjusted for Glamorgan rates (see text).
[b]Includes three probable nasal sinus cancer deaths classified as nasopharyngeal cancer on the death certificate.
[c]$p < 0.05$.
[d]$p < 0.001$.

Table 46.5 Observed Lung Cancer Deaths and Relative Risks (RRs) Compared with National Mortality Rates by Duration of Employment in Different Jobs and Predicted RRs[a]

Area	Duration in Area	<1 yr[b]			1–5 yr[b]			>5 yr[b]		
		RR	Observed	Fitted RR[c]	RR	Observed	Fitted RR[c]	RR	Observed	Fitted RR[c]
Calcining	<1 yr	1.7	33	2.0	4.1	32	3.2	7.6	54	7.3
	1–5 yr	3.9	6	4.7	4.8	7	5.3	12.0	9	14.9
	>5 yr	13.3	14	8.1	16.0	9	13.8	8.8	3	14.7
Copper plant	<1 yr	1.7	33	2.1	5.7	23	4.1	8.4	36	8.6
	1–5 yr	3.4	17	2.9	4.3	7	4.9	15.1	11	12.3
	>5 yr	7.7	18	7.3	7.8	14	8.0	14.3	8	13.1
Furnaces	<1 yr	1.7	33	2.0	4.4	36	3.5	8.6	67	8.0
	1–5 yr	5.4	6	3.5	3.3	5	4.9	12.1	11	11.4
	>5 yr	3.3	2	9.8	21.9	5	14.8	18.1	2	14.3
Nickel sulfate	<1 yr	1.7	33	2.0	4.1	39	3.0	8.1	56	8.4
	1–5 yr	3.9	3	3.8	6.7	4	5.3	11.9	17	9.3
	>5 yr	5.2	3	5.7	11.3	6	8.3	14.0	7	13.6
Reducer cleaners	None	1.7	32	2.0	3.8	35	3.4	8.6	86	8.5
	Some	7.3	4	10.3	18.7	5	17.7	22.8	5	36.5

[a] See Section 6.
[b] Duration of employment in other high-risk areas.
[c] Do not show significant evidence of lack of fit.

Table 46.6 Observed Nasal Cancer Deaths and Relative Risks (RRs) Compared with National Mortality Rates by Duration of Employment in Different Jobs and Predicted RRs[a]

Area	Duration in Area	<1 yr[b]			1–5 yr[b]			>5 yr[b]		
		RR	Observed	Fitted RR[c]	RR	Observed	Fitted RR[c]	RR	Observed	Fitted RR[c]
Calcining	<1 yr	39.3	5	69.9	211.9	11	151.4	364.1	25	381.2
	1–5 yr	95.9	1	68.9	270.1	3	275.5	822.6	9	649.9
	>5 yr	389.6	3	424.7	390.9	2	602.0	1916.7	8	697.2
Copper plant	<1 yr	39.1	5	70.7	218.2	6	159.7	477.6	18	437.0
	1–5 yr	181.0	6	134.5	308.4	4	232.1	706.6	6	644.1
	>5 yr	322.5	7	380.6	413.4	8	374.1	850.2	7	697.2
Furnaces	<1 yr	39.4	5	69.7	173.6	10	153.7	460.9	36	442.9
	1–5 yr	—	0	155.5	190.5	2	221.9	730.9	7	430.5
	>5 yr	61.8	3	343.8	—	0	665.1	3714.0	4	7388.2
Nickel sulfate	<1 yr	39.7	5	69.1	142.9	9	155.3	452.3	7	394.0
	1–5 yr	353.5	2	183.8	359.6	2	267.9	482.0	8	509.1
	>5 yr	264.7	2	396.8	831.6	5	474.0	961.5	7	737.6
Reducer cleaners	None	39.9	5	68.5	174.6	11	155.0	484.2	48	440.7
	Some	332.5	1	181.2	386.7	2	206.9	—	0	289.8

[a] See Section 6.
[b] Duration of employment in other high-risk areas.
[c] Do not show significant evidence of lack of fit.

613

of statistical model must be employed, and all such models have their limitations.

In the model used here, the excess cancer risk is assumed to be linearly related to the average level of exposure to each species. From previous analyses it is clearly important that age at first exposure, duration of exposure and time since first employment are also taken into account (Peto et al., 1984; Kaldor et al., 1986). The full model is therefore of the form

$$\text{Excess risk} = A \times L \times D \times (a_0 + a_1 x_1 + a_2 x_2 + a_3 x_3 + a_4 x_4)$$

where A, L, and D are factors that depend on the age at first employment, time since first employment, and duration of employment. (For pre-1930 employees only employment up to 1935 is considered. Most of their exposure will have occurred by that date, after which exposure levels were generally lower.) The terms x_1, x_2, x_3, and x_4 are the average levels of exposure to metallic nickel, nickel oxide, nickel sulfides, and soluble nickel, respectively, and a_1, a_2, a_3, and a_4 are parameters to be estimated representing the relative potencies of the different species. The term a_0 is included to allow for inaccuracies in the exposure data and job classification. This term is necessary since a clear excess is observed in individuals employed only in areas of no known exposure. Since exposure estimates were available only in terms of ranges, the values used in this analysis are the geometric means of the highest and lowest values. The cases where nasal cancer had occurred but was not the underlying cause of death are not included as nasal cancer cases in this analysis. The model has been fitted by maximum-likelihood estimation using the packages GEMINI (Lalouel, 1979) and GLIM (Royal Statistical Society, 1985).

Our assumption of linear dose–response cannot be critically tested. From a regulatory point of view an advantage of a linear model is that the risk at lower exposure levels is likely to be over- rather than underestimated; but several carcinogens exhibit upward curvature in the dose–response curve in animal experiments, lung cancer incidence rates in smokers suggest a similar pattern (Doll and Peto, 1978), and a quadratic or higher power effect might be expected on theoretical grounds for any carcinogen that acts at more than one stage in multistage carcinogenesis.

The model was initially fitted to the data on men first employed before 1930, and its predictions were then tested on later employees. The parameter estimates for the four types of nickel compounds are shown in Table 46.7, and the remaining parameters are given in Table 46.8. For both types of cancer the best fitting model suggests risks for metallic and soluble nickel, with much less (if any) risk for nickel oxide or sulfides. The only significant effect, in the sense that assuming the particular species has no effect on risk gives a significantly poorer fit to the data, is the soluble nickel effect on nasal cancer risk. The effects of metallic nickel and of soluble nickel on lung cancer risk could be explained nearly as well by a model only involving the other

Table 46.7   Estimates of Risk by Species[a]

Model	Lung Cancer	Nasal Cancer
Best fitting		
Metallic	0.40 ($P = 0.10$)	0.23
Oxidic	0.028	0.0
Sulfidic	0.0	0.0
Soluble	0.64 ($P = 0.10$)	0.94 ($P = 0.006$)
Without metallic nickel[b]		
Metallic	0.0	0.0
Oxidic	0.057	0.0
Sulfidic	0.031	0.23
Soluble	0.59	0.90

[a] Per mg/m^3 dose per 1000 man-years (parameters a_1, a_2, a_3, a_4 in the model) for an individual aged under 25 at first employment, employed 5–10 years, 15–24 years since first employment.
[b] Not a significantly worse fit than the best fitting model for either cancer (see text).

species. This uncertainty in relation to the relative contributions to risk of different forms of nickel is illustrated by the analyses shown in the lower part of Table 46.7, in which metallic nickel has been excluded. The goodness of fit is only marginally worse (X_1^2 is 3.4 for lung cancer and 1.4 for nasal cancer), yet the conclusions are profoundly altered. The effect for nickel subsulfide, which was zero in the initial analysis, becomes large, while the previously marked effect for metallic nickel is of course eliminated.

Table 46.8   Excess Mortality Ratios for Lung and Nasal Cancer[a]

	Lung Cancer	Nasal Cancer
Duration of employment up to 1935		
5–9	1.0	1.0
10–14	4.28***	0.76
15–19	4.12***	1.71
>20	6.35***	2.94**
Age at first employment		
Under 25	1.0	1.0
25–34	1.11	1.80**
>35	0.79	6.23***
Time since first employment		
15–24	1.0	1.0
25–34	1.85**	1.77*
35–44	2.63***	1.61
>45	0.94	4.01***

[a] Compared with baseline category (5–9 years employment, age under 25 at first employment, 15–24 years since first employment): (*) $p < 0.05$; (**) $p < 0.01$; (***) $p < 0.001$.

Tables 46.5 and 46.6 show the relative risks predicted by the model by duration of employment in different high-risk areas. The risks by work area are generally predicted well by the model; for neither table is there statistically significant lack of fit.

Table 46.8 shows that for both cancers, risk is strongly related to duration of employment (though for nasal cancer this increase is only apparent in workers with at least 15 years of employment, the risk in men employed for 10–14 years being nonsignificantly lower than that for those employed 5–9 years). The risk of nasal cancer, but not lung cancer, increases markedly with age at first employment. The nasal cancer risk also increases with time since first employment for 45 years or more, with no risk appearing during the first 15 years. In contrast, the lung cancer risk shows no very clear pattern with time since first employment.

7. RISKS AFTER 1930

The uncertainties in our initial analysis, which suggested that the risk could be almost entirely due to metallic and soluble nickel, are further illustrated by the data on more recent employees. The initial analysis would predict that any excess would be strongest in individuals with long employment and high average levels of soluble nickel or metallic nickel exposure. The results for men first employed between 1930 and 1939 are summarized in Table 46.9. For lung cancer, the total number of observed cases (20) is very close to that predicted (21.0). For nasal cancer, however, more than four cases would have been predicted compared with just one observed. Moreover, combining the data for lung and nasal cancer among men with long ($>$10 years) service and high metallic or soluble nickel exposures, there were 4 deaths, compared with 11.74 predicted by the models. In contrast, the risk for other long-term workers was if anything underestimated (14 observed, 9.86 predicted). The most plausible interpretation, therefore, is that we have overestimated the risks for metallic (and possibly soluble) nickel and underestimated those for sulfides and/or oxides.

8. DISCUSSION

This study provides further confirmation that the cancer risks at this refinery are largely confined to workers employed before 1930, with some risk to individuals employed during the 1930s. Continuing follow-up of more recent employees is of course essential to monitor the effects of more recent working conditions.

We have emphasized the preliminary nature of our dose–response analysis. None of the exposure levels are known with any certainty, and neither are

Table 46.9 Observed, Expected, and Predicted Numbers of Lung and Nasal Cancer Deaths in Men First Employed 1930–1939[a]

	Lung Cancer			Nasal Cancer		
	Observed	Expected	Predicted	Observed	Expected	Predicted
<10 yr Service	3	2.28	3.68	0	0.012	0.11
10 or More yr Service Average level:						
Metallic nickel >1 or Soluble nickel >0.1	4	2.94	8.23	0	0.013	3.51
Other employees	13	6.93	9.09	1	0.033	0.77
Total	20	12.15	21.00	1	0.058	4.39

[a]Fifteen years or more since first employment. See model in Section 6 for predicted numbers.

617

the job histories without error, since changes of job and hence exposure were not always recorded. With these caveats, the results suggest that some of the risk, particularly of nasal cancer, is due to soluble nickel exposure. They also suggest that at least one insoluble species contributes to the risk, but the relative contributions of metallic nickel, nickel oxides, and nickel sulfides are very difficult to assess.

After reviewing evidence from this and other plants, the international working group reached a broadly similar conclusion (IWG, 1989). Although they found suggestive evidence implicating nickel oxide and to a lesser extent nickel subsulfide (but not metallic nickel), neither of these forms of nickel could be conclusively implicated or exonerated. In the absence of clearer evidence, it seems prudent to assume that exposure to any nickel compound (if of a particle size able to reach the nasal or bronchial tissue) may constitute a respiratory cancer hazard. In the light of our findings it might be prudent to consider that metallic nickel may also be a hazard.

The difficulty in drawing conclusions about specific insoluble forms of nickel is neatly illustrated in our study by a comparison of the copper plant and the nickel sulfate plant. Workers in these plants suffered substantial and similar cancer risks, and both had soluble nickel levels of about 1 mg/m^3, with little subsulfide or metallic nickel. The only marked difference was thus the nickel oxide level, which was over 10 mg/m^3 in the copper plant and only $1-2 \text{ mg/m}^3$ in the nickel sulfate plant. Superficially these observations suggest that the risk in both plants was due to soluble nickel exposure and that exposure to nickel oxide in the copper plant did not constitute a hazard. However, relatively minor differences in the exposure estimates (e.g., if the soluble nickel exposure in the nickel sulfate plant were underestimated by a factor of 2) could imply a substantial effect of nickel oxide.

A clearer picture might emerge if similar models were fitted to other refinery cohorts with somewhat different exposures. Such combined analysis should however include an assessment of sensitivity to inaccuracies in the exposure estimates. Equally important is the need to test the predictions of such models with data from other parts of the nickel industry where exposures are very much lower and, to date, there appears to be little evidence of any cancer risk.

ACKNOWLEDGMENTS

The authors thank INCO (Europe) for their cooperation in all aspects of this study. We thank Janis Davidson for her computing assistance and Sandra McVeigh for typing assistance. The Institute of Cancer Research receives support from the Cancer Research Campaign and the Medical Research Council.

REFERENCES

Bridge, J. C. (1933). *Annual Report of the Chief Inspector of Factories and Workshops for the Year 1932*. HMSO, London, pp. 103–104.

Doll, R., and Peto, R. (1978). Cigarette smoking and bronchial carcinoma: dose and time relationships among regular smokers and lifelong non-smokers. *J. Epid. Comm. Health* **32,** 303–313

Doll, R., Morgan, L. G., and Speizer, F. E. (1970). Cancer of the lung and nasal sinuses in nickel workers. *Br. J. Cancer* **24,** 623–632

Gardner, M. G., Winter, P. D., Taylor, C. P., and Acheson, E. D. (1983). *Atlas of Cancer Mortality in England and Wales 1968–1978*. Wiley, Chichester.

Gardner, M. G., Winter, P. D., and Barker, D.J.P. (1984). *Atlas of Mortality from Selected Diseases 1968–1978*. Wiley, Chichester.

IWG (1990). Report of the international committee on nickel carcinogenesis in men (Chairman: Sir Richard Doll). *Scand. J. Work Envir. Health*, **16,** 1–82.

Kaldor, J., Peto, J., Easton D. F., Doll, R., Hermon, C., and Morgan, L. G. (1986). Models for respiratory cancer in nickel refinery workers. *J. Nat. Can. Inst.* **77,** 841–848.

Lalouel, G. M. (1979). GEMINI—a computer program for optimisation of general non-linear functions. Department of Human Genetics Technical Report, No. 14. University of Utah, Salt Lake City.

Magnus, K., Andersen A., and Hogetveit A. C. (1982). Cancer of the respiratory organs among workers at a nickel refinery in Norway: second report. *Int. J. Cancer* **30,** 681–685.

Morgan, J. G. (1958). Some observations on the incidence of respiratory cancer in nickel workers. *Br. J. Ind. Med.* **15,** 224–234.

Peto, J., Cuckle, H., Doll, R., Hermon, C., and Morgan, L. G. (1984). Respiratory cancer mortality of Welsh nickel refiners. In F. W. Sunderman Jr. (Ed.), *Nickel in the Human Environment*. IARC Scientific Publication No 53, IARC, Lyon, pp. 37–46.

Roberts, R. S., Julian J. A., Muir D.C.F., and Shannon H. S. (1984). Cancer mortality associated with the high-temperature oxidation of nickel subsulphide. In F. W. Sunderman, Jr. (Ed.), *Nickel in the Human Environment*. IARC Scientific Publication No. 53, IARC, Lyon, pp. 23–35.

Royal Statistical Society (1985). *The GLIM System 3.77 Manual*. Numerical Algorithms Group, Oxford.

47

RECENT FOLLOW-UP OF NICKEL REFINERY WORKERS IN NORWAY AND RESPIRATORY CANCER

Aage Andersen

The Cancer Registry of Norway, Montebello, Oslo 3, Norway

1.   INTRODUCTION

In most occupational cancer studies in the field of epidemiology, it is of importance to determine dose–response relationships and time trends and if possible identify the carcinogenic agents. Our first two papers based on employees at the Norwegian nickel refinery were published in 1973 (Pedersen et al.) and 1982 (Magnus et al.) and showed an excess risk of respiratory cancer.

Nickel and Human Health: Current Perspectives, Edited by Evert Nieboer and Jerome O. Nriagu.
ISBN 0-471-50076-3   © 1992 John Wiley & Sons, Inc.

In studies where an excess risk of cancer is identified, continuous monitoring of the group until the risk is reduced to a minimum level seems feasible. Continuous observation of the Norwegian nickel refinery workers is an example of how a high-risk group can be followed up and how well collaboration between the health department at a refinery and a cancer registry can operate. This study was our first experience with such a cooperative effort, and it has thus far been successful; in fact, the collaboration has served as a model for other studies in Norway.

2. RECENT FINDINGS

Recent follow-up of the original cohort shows 370 new cases of all types of cancer, compared to the expected number based on incidence rates in the total Norwegian male population of 290 by the end of 1987. During the follow-up period 1953–1987, 163 new cases of respiratory cancer have been observed versus 42 expected (Table 47.1). A decreasing trend in ratio was seen between observed and expected cases, from 6.9 in 1926–1935 to 2.4 among those first employed in 1956–1965. In spite of this decrease, there is still an excess risk of respiratory cancer. To date there is no excess risk for other types of cancer. The observed number of cases is lower than the expected in all periods; when all periods are combined (1913–1965), a significant lower risk was shown, with 207 observed and 248 expected.

In Norway, cancer of the nose and nasal cavities is rare, with only 25 new cases per year in the entire country. Among refinery workers there are 26 new cases among all employees during the follow-up period 1953–1987 and the expected number is 1.07 (Table 47.2). A consistent decrease in the observed–expected ratio is found in all periods of follow-up, except the first 14 years, for which no cases have been observed. Among employees with first entry before 1956 new cases are still observed. So far no cases have been registered among employees with first entry after 1955, but the expected number is only 0.20.

During the 35 years of follow-up (1953–1987), 126 new cases of lung cancer have been observed, and the expected number was 36 (Table 47.3). An excess risk of lung cancer is still observed in all periods of employment and for all periods of follow-up. Table 47.3 shows a consistent decrease in the observed–expected ratio in all follow-up periods. However, it is difficult to evaluate such a trend in lung cancer because of a possible interaction between smoking and exposure to nickel. In the paper by Magnus et al. (1982) no interaction was demonstrated. If this is still true, is it necessary to take into account any change in smoking habits before a trend can be discussed. Of importance is that an excess risk of lung cancer is still observed.

In 1968, technological changes took place at the refinery, and the environmental conditions were improved. Another cohort of 1237 employees

Table 47.1 Observed (O) and Expected (E) Number of New Cases of Respiratory Cancer and All Other Types of Cancer Combined among 2247 Refinery Workers (1953–1987)

Year of First Employment	Number of Employees	Respiratory Cancer			All Other Types of Cancer		
		OBS	EXP	O/E	OBS	EXP	O/E
1913–1925	14	3	0.16	18.8	1	2.39	0.4
1926–1935	214	30	4.35	6.9	22	32.51	0.7
1936–1945	189	25	4.79	5.2	21	30.02	0.7
1946–1955	1197	84	23.92	3.5	117	134.47	0.9
1956+	623	21	8.80	2.4	46	48.98	0.9
1913–1965	2247	163	42.02	3.9	207	248.37	0.8

Table 47.2 Observed (O) and Expected (E) Number of Nasal Cancer among 2247 Male Refinery Workers by Number of Years Since First Exposure (1953–1987)

Year of First Employment	3–14 yr			15–24 yr			25–34 yr			35+ yr			Total		
	O	E	O/E	O	E	O/E	O	E	O/E	O	E	O/E	O	E	O/E
1913–1925	—	—	—	—	—	—	0	0	—	2	0.01	200	2	0.01	200
1926–1935	—	—	—	0	0.01	—	6	0.05	120	5	0.08	63	11	0.14	79
1936–1945	—	—	—	2	0.02	100	1	0.04	25	1	0.06	17	4	0.12	33
1946–1955	0	0.14	—	2	0.19	11	5	0.22	23	2	0.05	40	9	0.60	15
1956+	0	0.08	—	0	0.09	—	0	0.03	—	—	—	—	0	0.20	—
1913–1965	0	0.22	—	4	0.31	13	12	0.34	35	10	0.20	50	26	1.07	24

Table 47.3 Observed (O) and Expected (E) Number of Lung Cancer among 2247 Male Refinery Workers by Number of Years Since First Exposure (1953–1987)

Year of First Exposure	3–14 yr			15–24 yr			25–34 yr			35+ yr			Total		
	O	E	O/E	O	E	O/E	O	E	O/E	O	E	O/E	O	E	O/E
1913–1925	—	—	—	—	—	—	0	0	—	1	0.14	7.0	1	0.14	7.0
1926–1935	—	—	—	0	0.14	—	7	0.92	7.6	12	2.80	4.3	19	3.86	4.9
1936–1945	0	0.02	—	1	0.40	2.5	6	1.08	5.6	10	2.74	3.6	17	4.24	4.0
1946–1955	7	2.71	2.6	25	5.43	4.6	32	9.55	3.4	6	2.85	2.1	70	20.54	3.4
1956+	3	2.31	1.3	11	3.60	3.1	5	1.66	3.0	—	—	—	19	7.57	2.5
1913–1965	10	5.4	2.0	37	9.57	3.9	50	13.21	3.8	29	8.53	3.4	126	36.35	3.5

Table 47.4 Observed (O) and Expected (E) Number of New Cases of Lung Cancer among 1237 Male Refinery Workers by Year Since First Employment (1969–1987)

Years of Employment	1–14 yr			15+ yr			Total		
	O	E	O/E	O	E	O/E	O	E	O/E
1968–1972	4	2.76	1.4	3	0.97	3.1	7	3.73	1.9
1973–1977	1	1.22	0.8	—	—	—	1	1.22	0.8
1978–1983	0	0.35	—	—	—	—	0	0.35	—
1968–1983	5	4.33	1.2	3	0.97	3.1	8	5.30	1.5

with one year or more in total employment with first entry in 1968 or later has been followed up for total mortality and incidence of cancer to the end of 1987. Table 47.4 shows the results for the follow-up period 1969–1987. For lung cancer, 8 new cases were observed versus 5.3 expected. Among those with first entry in the period 1968–1972, there were 7 new cases versus the 3.7 expected. For a latency period of 15 years or more, 3 observed cases occurred *versus* 1 expected. This result indicates an excess risk of lung cancer also among those with first entry after 1968. Further follow-up of this group of employees for another 10–15 years is obviously needed in order to observe the exact excess risk.

3. CORRELATION OF EXPOSURE AND CANCER RISK

3.2. Classification of Workers and Exposure to Nickel Compounds

The classification of employees into broad departments depending on where the employees had their longest duration of employment presented in our first two reports has been criticized. A third cohort of employees with first entry in the period 1946–1968 has therefore been established. The aim was to classify the workers according to exposure levels of different nickel compounds. Employees included in our first study with employment before 1946 have been excluded, as it was impossible to classify these workers by all the departments in which they were employed.

The nickel refinery provided data on 3343 men with a total time of employment of 12 months or more. Excluded from the study were 93 men: 16 had died or emigrated before the start of follow-up, and another 77 employees were lost to complete follow-up. The study population therefore comprises 3250 employees, all alive January 1, 1953, and followed up to the end of 1984.

During the last 20 years, the environmental conditions at the refinery have changed as new processes have been introduced. Before 1968, the working conditions were gradually improved, but only infrequent measurements of the environment were recorded, especially before 1960. Estimates have been prepared for four different species of nickel compounds: all types of nickel sulfides, metallic nickel, all types of nickel oxides, and all soluble forms of nickel. For each of the 82 departments, specific exposures and quantitative ranges were calculated for the four different compounds during three different time periods.

The start and termination dates for each department in which an employee worked were assigned (coded). However, it was impossible to code individual jobs within the departments for each employee. Environmental data have been allocated to each employee's record as a cumulative amount of the given nickel species as well as the amount of total nickel.

3.2. Preliminary Results

Among those with first entry after 1945, seven cases of nasal cancer have so far been registered. Five cases were in the highest cumulative exposure group—to an insoluble nickel compound with an exposure percentile over 90%. For the remaining two workers, the cumulative amount of soluble nickel exceeded the 90% percentile (minor exposure was to nickel oxide).

A statistically significant excess risk of lung cancer is again confirmed with 96 new cases and an expected number of 34 (Table 47.5). Those with first entry before 1956 have the highest risk of this type of cancer; for later entries a decrease in the observed–expected ratio occurs (Table 47.5). It is important to emphasize that for a large proportion of the employees with first entry after 1960, the follow-up period is still short. It is known from the first study (Pedersen et al., 1973) that the average latency period for lung cancer is 25 years, and we must therefore be careful with any conclusion based on the decreasing observed–expected trend at this juncture in time. The risk was also tabulated according to latency: less than 15 years since first exposure and 15 years or more. No excess risk can be discerned in the first category, but in the 15 years or more group an excess risk is demonstrated with 82 observed *versus* 22 expected cases.

In the earlier studies, the employees were allocated to the department where they had been employed the longest. This classification represented a problem, as already mentioned. One of the important aims for the new cohort was to estimate dose–response relationships for different compounds of nickel. The results so far show a statistically significant excess risk among those who worked in the electrolysis department only (those with highest level of soluble nickel exposure) and those employed in roasting, smelting, and calcining only. The highest observed–expected ratio of 6.2 is observed among workers with the longest duration of employment in the electrolytical department.

Table 47.5   Observed (O) and Expected (E) Number of New Cases of Lung Cancer among 3250 Nickel Refinery Workers for Whom Exposure Estimates Have Been Made (1953–1984)

Year of First Employment	Number of Employees	O	E	O/E	95 % Confidence Interval
1946–1949	634	28	9.62	2.9	1.9–4.2
1950–1955	1257	51	15.61	3.3	2.4–4.3
1956–1959	436	9	4.21	2.1	1.0–4.0
1960–1968	923	8	4.71	1.7	0.7–3.4
1946–1968	3250	96	34.15	2.8	2.3–3.4

4. CONCLUSION

The present study reaffirms our results of previous reports that there is a risk of lung cancer for all nickel refinery workers, even among those with first entry after 1968. No cases of nasal cancer have been observed among employees after 1955, and the results show a decrease in risk for this disease. Most workers have had mixed exposure to more than one compound of nickel. It is still not possible to state with certainty which specific nickel compounds are human carcinogens. However, workers who have had their exposure mainly to nickel in the electrolysis department show a large excess of lung cancer. This evidence indicates that soluble nickel cannot be discounted as a human carcinogen.

REFERENCES

Pedersen, E., Høgetveit, A. C., and Andersen, A. (1973). Cancer of respiratory organs among workers at a nickel refinery in Norway. *Int. J. Cancer* **12,** 32–41.

Magnus, K., Andersen, A., and Høgetveit, A. C. (1982). Cancer of respiratory organs among workers at a nickel refinery in Norway. Second report. *Int. J. Cancer* **12,** 32–41.

48

CANCER MORTALITY IN ONTARIO NICKEL WORKERS: 1950–1984

Robin S. Roberts, James A. Julian, Ni Jadon, and David C. F. Muir

Occupational Health Program, McMaster University, Hamilton, Canada

Nickel and Human Health: Current Perspectives, Edited by Evert Nieboer and Jerome O. Nriagu.
ISBN 0-471-50076-3 © 1992 John Wiley & Sons, Inc.

1. INTRODUCTION

This chapter describes observed and expected mortality from cancer in a cohort of 54,509 nickel workers followed for 35 years. For analysis purposes, the cohort was subdivided into men with and without service in one of the three high-nickel-dust areas of the operation: the sinter plants at Copper Cliff and Coniston and the leaching, calcining, and sintering (LC&S) department at Port Colborne. At Copper Cliff, sinter plant workers experienced three times the expected number of lung cancer deaths; the standardized mortality ratio (SMR) rose steeply with increasing duration of service peaking at 943 with 10–15 years. A similar overall excess risk of lung cancer was seen in the smaller Coniston Sinter Plant, again with an indication of an exposure risk gradient. Men in the LC&S department at Port Colborne also experienced a dose-related excess risk of lung cancer death, which rose to an SMR of 806 with 20–25 years of service. Nasal cancer deaths were increased at both Copper Cliff Sinter Plant (6 deaths) and the LC&S department at Port Colborne (19 deaths), representing SMRs of 3704 and 7755, respectively, for this rare cancer. Laryngeal cancer and kidney cancer, both previously associated with nickel, were not in excess in these high-risk groups. A further exploration of death from these causes in the "lower exposure" remainder of the cohort revealed an epidemiologically modest elevation in lung cancer death in miners and in the copper refinery (both probably not nickel related). No evidence of laryngeal cancer excess was found. Kidney cancer death was slightly increased in miners, but as a whole was reasonably in line with expected. The results are consistent with either nickel subsulfide or nickel oxide, or both, being carcinogens.

Concern about the carcinogenic potential of nickel continues to abound in the scientific and regulatory communities around the world. The initial epidemiologic evidence (Doll, 1958) derived from workers exposed to various forms of nickel in INCO's Clydach refinery in South Wales has been complemented by a variety of later studies of occupationally exposed cohorts (Cox et al., 1981; Cragle et al., 1984; Enterline and Marsh, 1982; Lessard et al., 1978; Pedersen et al., 1973; Redmond, 1983; Shannon et al., 1984) and a host of animal experiments. Despite formal reviews of available data by NIOSH (1977) and IARC (1976), a firm scientific consensus as to which forms of nickel are carcinogenic remains elusive.

In an attempt to provide an authoritative opinion on this matter, various interested parties from government and industry recently initiated a collaborative project (Doll, 1990) designed to strengthen and pool all available data concerning the risk of cancer following exposure to "nickel." The specific aim was to use the pooled data to assess the carcinogenicity of individual species of nickel including insoluble (e.g., oxide, subsulfide), soluble (e.g., sulfate), and metallic forms. This initiative resulted in the formation of an international working group headed by Sir Richard Doll and comprised of scientists who had conducted relevant studies. As a key part of this collaborative venture, investigators were asked, wherever possible, to extend the follow-up of occupationally exposed cohorts and to provide the best available information on exposure in terms of detailed work histories and estimates of environmental contamination by nickel species. We report here the results of an updated mortality study of workers engaged in the mining, smelting, and refining of nickel in INCO's Ontario operation.

2. METHODS

We have conducted a historical prospective mortality study of 54,509 men employed by INCO in the Sudbury area of Ontario, Canada, and at the Port Colborne nickel refinery (PCNR). The cohort was determined from company records as all male workers or retirees with six months or more nonoffice service, known to be alive on or after January 1, 1950. Mortality follow-up was via computerized record linkage to the Canadian National Mortality Data Base of death certificates, a process we estimate successfully located about 95% of deaths in the group and yielded the official International Classification of Diseases Adapted (ICDA) coded cause of death in each case. Analysis was via a man-years at risk calculation that produced age and calendar time standardized mortality ratios (SMRs) with respect to mortality in the province of Ontario as a whole. Routine cause-specific analyses subdivided years at risk by accumulating duration of exposure and time since first exposure. In general, for deaths due to cancer a 15-year latency has been utilized so that only deaths and years at risk beyond 15 years from first exposure are reported in summary form. More details concerning the design of this study and the nature of the process at INCO have been published elsewhere (Roberts, 1989).

The study cohort was originally assembled in 1979, and the follow-up period was 1950–1976. Mortality during this interval has been reported earlier (Roberts, 1984). As participants in the Nickel Speciation Research Project described above, we have now extended follow-up to the end of 1984. In addition, we have attempted to provide the best available environmental estimates together with some refinements in the definition of some important occupational subgroups.

The focus of this analysis is mortality from cancer and its relationship to various occupational subgroups exposed to different mixes of nickel species. Following our previous approach, we first examine death from respiratory cancer and separate men with known high exposure to insoluble nickel through having worked in the sintering plants at Copper Cliff or Coniston (both in the Sudbury area) or in the leaching, calcining, and sintering (LC&S) department at PCNR. Respiratory cancer is then reviewed in the remaining occupational subgroups with special attention being paid to men exposed to soluble nickel in the electrolytic tank house at PCNR. Although preexisting evidence of a kidney cancer risk being associated with nickel is much less substantial, we present a fairly detailed analysis of mortality from this cause. Finally, we review mortality from cancer at other sites.

3. RESULTS

Mortality follow-up determined that 8387 men in the study cohort of 54,509 had died during the period 1950–1984. The additional eight years of follow-up produced an increase of almost 60% in the available mortality information. Overall, the *all-causes* SMR was 103 (observed 7382, expected 7138.3) for Sudbury area workers and 96 (observed 1005, expected 1044.7) for men who worked at the PCNR. The slight excess in Sudbury was due to increased mortality from accidental/violent causes.

3.1. Lung Cancer

Observed and expected mortality from lung cancer is displayed in Table 48.1 subdivided by time since first exposure and the primary occupational subgroups. In the Sudbury area we separated workers with any documented experience in the Copper Cliff sinter plant or Coniston sinter plant, both known to have operated with high airborne dust levels containing primarily a mixture of nickel oxide and sulfides. Similarly, at PCNR we separated men who have worked in the LC&S department during years when either calcining or sintering was being conducted, processes known to have caused high concentrations of subsulfide/oxide dust.

From Table 48.1 it is clear that an unusually high risk of mortality from lung cancer has occurred in workers at the Copper Cliff and Coniston sinter plants and in the LC&S department at PCNR. The observed excess risks are statistically unequivocal, even for the relatively small Coniston plant, and tend to be of similar magnitudes in the three high-risk subgroups. With the possible exception of Copper Cliff, excess mortality appears only after about 15 years has elapsed since first exposure to the relevant subprocess. The excess risks are thus temporally consistent with our understanding of the natural history of occupationally induced cancer. By contrast, no excess risk in death from lung cancer was apparent in men with out LC&S experience at PCNR, and only an epidemiologically modest, though statistically important, excess was found in the large nonsinter group at Sudbury.

In Table 48.2 we examine the risk of death from lung cancer by duration of service in the relevant subprocess(es) as a surrogate for "dose." Gradients of risk increasing with duration of service are seen for the Copper Cliff sinter plant and PCNR LC&S department groups, with perhaps the former having a steeper slope. No real evidence of an increasing pattern of risk is seen in non-LC&S department men at PCNR, but the large nonsinter group at Sudbury displays a shallow upward curving relationship with duration of exposure. The Coniston sinter plant group was too small to subdivide in this manner.

Table 48.1 Lung Cancer Mortality[a]

Time Since First Exposure	Sudbury Area			Port Colborne Nickel Refinery	
	Nonsinter Plant ($n = 47890$)	Copper Cliff Sinter Plant ($n = 1744$)	Coniston Sinter Plant ($n = 336$)	Non-LC&S Department ($n = 2616$)	LC&S Department ($n = 1671$)
<15 years					
Observed	19	4	0	1	1
Expected	21.6	1.5	0.5	1.7	1.6
SMR	88 (0.67)	265 (0.07)	—	57 (0.82)	65 (0.79)
95% CI	57–138	66–676	—	6–322	6–361
≥15 years					
Observed	485	63	8	30	72
Expected	433.3	20.3	2.7	32.3	29.9
SMR	112 (0.006)	311 ($<10^{-7}$)	298 (0.002)	93 (0.62)	241 ($<10^{-7}$)
95% CI	103–123	243–398	127–590	63–133	191–303

Source: ICDA9 162.
[a]Numbers in parentheses are probabilities. CI, confidence interval.

Table 48.2 Lung Cancer Mortality by Duration of Exposure[a]

Duration of Exposure (years)	Sudbury						Port Colborne					
	Nonsinter			Copper Cliff SP			Non-LC&S			LC&S		
	Observed	Expected	SMR (95% CI)	Observed	Expected	SMR (95% CI)	Observed	Expected	SMR (95% CI)	Observed	Expected	SMR (95% CI)
<1	23	21.0	109	15	12.0	125	1	2.4	42	22	12.0	183[b]
1–1.9	26	22.3	117	6	2.4	251	1	2.4	42	3	2.8	107
2–4.9	29	33.2	87	9	1.8	502[c]	1	2.4	42	9	4.7	191[b]
<5	78	76.6	102 (82,127)	30	16.2	185[c] (125,264)	1	2.4	42 (4,237)	34	20.0	174[c] (120,243)
5–9.9	21	25.2	83	12	1.5	797[c]	2	2.5	81	11	4.2	260[c]
10–14.9	16	16.8	95	16	1.7	943[c]	2	2.5	81	6	2.0	305[d]
15–19.9	37	34.3	108	4	0.9	465	10	8.4	119	5	1.1	442[c]
20–24.9	64	59.0	108	—	—	—	10	8.4	119	9	1.1	806[c]
25–29.9	89	77.2	115	—	—	—	17	19.1	89	7	1.9	363[c]
30–34.9	98	80.8	121[b]	—	—	—	17	19.1	89	7	1.9	363[c]
35–39.9	57	45.6	125	—	—	—	17	19.1	89	7	1.9	363[c]
≥40	25	16.5	152[b]	—	—	—	17	19.1	89	7	1.9	363[c]
≥5	407	355.4	115[c] (104,126)	32	4.1	790[c] (538,114)	29	29.9	97 (65,139)	38	10.4	366[c] (258,503)

[a]≥15 years since first exposure. CI, confidence interval.
[b]p < 0.05.
[c]p < 0.001 (one tailed).
[d]p < 0.01.

634

The Copper Cliff sinter plant operated from 1948 to 1964 and during the early years of operation experienced technical difficulties that increased ambient dust levels (see later environmental estimates). In Table 48.3 we examine the trend in excess risk in men first hired in the dusty period prior to 1952 compared to those first hired during 1952 or later. Although the point estimates of SMR tend to be lower in the late hired group, there remains evidence of excess risk in the longer term workers hired during 1952–1964. The relatively small size of this group results in quite wide confidence intervals for the SMR. It is thus not possible to make strong statistical statements about evidence of reduced lung cancer risk post-1951.

The oxidation of nickel subsulfide was conducted in a single building at PCNR from 1921 to 1973 initially via calcining and later supplemented with traveling grate sintering machines. In Table 48.4 we present an analysis of lung cancer SMRs by period of first hire and duration of service. The choice of decade for the period of first hire is somewhat arbitrary, dictated primarily by the number of men hired rather than known changes in the dustiness of the plant. The analysis shows a fairly constant pattern of excess risk over time that is consistent with recollections that conditions in the plant were reasonably unchanged throughout its life. The suggestion of a diminishing SMR with decade of hire in the $\geq$5-year service group is not strong statistically, as seen by the overlapping confidence intervals. If real, the trend probably represents longer average service in the earlier hired groups rather than a difference in risk at any service duration level.

The utilization of employment in the LC&S department at PCNR as a marker of exposure to high levels of insoluble nickel is not ideal in an epidemiological sense. We would have preferred to consider separately the sintering, calcining, and leaching components of this composite department but have found it difficult to define appropriate sets of process-specific job codes that would achieve the necessary split. In particular, the prevalent use of the noninformative job classification "process labor," especially in

Table 48.3   Lung Cancer Mortality by Date First Hired in Copper Cliff Sinter Plant[a]

Duration of Exposure (years)	First Exposure Before 1952			First Exposure 1952 or Later		
	Observed	Expected	SMR (95% CI)	Observed	Expected	SMR (95% CI)
<5	28	12.7	221[b] (146,318)	2	3.5	57 (6,206)
$\geq$5	30	3.4	884[b] (594,1259)	3	0.7	437[c] (86,1257)

[a] $\geq$15 years since first exposure. CI, confidence interval.
[b] $p < 0.001$ (one tailed).
[c] $p < 0.05$.

Table 48.4 Lung Cancer Mortality by Date First Hired in PCNR LC&S Department[a]

Duration of Exposure (years)	First Exposure Before 1930			First Exposure 1930–1939			First Exposure 1940 or Later		
	Observed	Expected	SMR (95% CI)	Observed	Expected	SMR (95% CI)	Observed	Expected	SMR (95% CI)
<5	6	4.5	133 (49,291)	10	2.9	346[b] (162,634)	18	12.2	148 (88,233)
≥5	20	4.7	422[b] (260,655)	11	3.0	373[c] (180,657)	7	2.7	260[d] (104,535)

[a] ≥15 years since first exposure. CI, confidence interval.
[b] $p < 0.001$ (one tailed).
[c] $p < 0.01$.
[d] $p < 0.05$.

Table 48.5 Lung Cancer Mortality in PCNR LC&S Department by Type of Activity[a]

Duration of Exposure (years)	Mainly Sinter			Mainly Calciner			Other Activity		
	Observed	Expected	SMR (95% CI)	Observed	Expected	SMR (95% CI)	Observed	Expected	SMR (95% CI)
<5	3	3.2	94 (19,275)	4	2.6	155 (38,392)	27	13.3	196[b] (129,284)
≥5	26	5.9	422[b] (288,644)	6	1.3	449[c] (169,1008)	6	3.2	190 (69,409)

[a] ≥15 years since first exposure. CI, confidence interval.
[b] $p < 0.001$ (one tailed).
[c] $p < 0.01$.

earlier years, frustrated this activity. In order to provide a partial solution to the problem, we have resorted to a subjective classification of men in the LC&S department provided by a retired supervisor. Using his personal recollection of the workers and plant, he reviewed job histories and subdivided the LC&S group into (a) those who had worked at some time on the sinter machines, (b) those who had worked at some time on the calciners but probably not on the sintering machines, and (c) all others. The analysis resulting from this subdivision is contained in Table 48.5 and shows similar patterns of risk for the sinter and calciner areas, both of which increase steeply with duration of service. The "other" group shows a smaller excess risk that does not appear to be related to duration of service. Although a little difficult to interpret, these findings are consistent with sintering and calcining contributing the primary risks, with the risk spilling over into the "other" category through imprecision in the classification process. Alternatively, leaching could contribute an independent risk or an apparent risk resulting from an environment contaminated with insoluble nickel dust from the other proximal processes.

In Table 48.6 we examined lung cancer mortality in the nonsinter group at Sudbury and non-LC&S group at PCNR. The analysis subdivides the cohorts by a series of secondary occupational groups but on an "ever worked" basis so that a man who held a variety of jobs will have contributed to more than one subgroup.

The modest excess in lung cancer mortality seen in the large Sudbury nonsinter group can be attributed to miners and a pocket of excess mortality in the copper refinery at Copper Cliff. Further examination of the miners suggests a falling pattern of risk over time with the bulk of the excess risk occurring in men with 25 years or more of mining experience first hired prior to 1930 (observed 24, expected 14.7, SMR 163). Similarly, more detailed analysis within the copper refinery highlights lead welders, tankhouse cranemen, and furnance workers as being at elevated risk.

Men without any documented LC&S department experience in PCNR experienced lung cancer mortality typical of the province of Ontario as a whole. Of particular interest are men who have worked in the electrolytic department at PCNR. The exposure to soluble nickel salts in this group has not produced evidence of excess lung cancer risk even in men with long histories of exposure.

3.2. Nasal Cancer

The analysis of mortality from nasal cancer presents a number of epidemiological problems. First, earlier versions of the ICDA were not very specific in locating the cancer site and in fact lump together cancer of the nose, nasal sinuses, and middle ear. As a result, we have used the somewhat broader category of ICDA9 code group 160 (or its equivalent in earlier versions)

Table 48.6 Lung Cancer Mortality by Secondary Subgroups[a]

Occupational Subgroup	All Exposures			≥25 years Exposure		
	Observed	Expected	SMR	Observed	Expected	SMR
A. Sudbury						
Miners	298	268.9	111[b]	111	85.6	130[b]
Other U/G	83	77.8	107	13	13.8	94
Milling	49	41.5	118	9	7.0	129
Furnaces C.C.	87	77.9	112	9	9.0	100
Furnaces coniston	14	13.1	107	2	1.5	130
Converters C.C.	49	49.2	100	9	7.6	118
Converters coniston	7	7.0	100	2	0.9	215
Smelter maintenance C.C.	68	65.9	103	16	13.0	123
Smelter maintenance coniston	5	7.2	69	3	2.0	154
Matte processing	37	42.5	87	4	6.0	67
Iron ore plant	13	12.3	106	1	0.4	279
Copper refinery	50	36.3	138[b]	26	13.3	196[c]
B. Port Colborne						
Nickel anode	7	7.7	91	2	1.4	144
Electrolytic	23	23.2	99	5	7.9	64
Yard/transportation	21	24.1	87	10	9.6	104

[a] ≥15 years since first exposure.
[b] $p < 0.05$.
[c] $p < 0.001$ (one tailed).

638

when investigating "nasal" cancer mortality. Second, primary nasal cancers can sometimes be attributed to other closely related sites if they originate on the border of the nasopharynx (then coded to nasopharynx ICDA9 147) or have invaded the ethmoid bone of the nose (then coded to ICDA9 170). In subsequent analyses we will note situations in which nasal cancer excesses have apparently spilled over into other cancer site classifications.

The mortality from nasal cancer has been summarized in Table 48.7 for the primary subgroups. Little evidence of excess is apparent in the nonsinter group at Sudbury, and no deaths from this cause were found in men who worked at the Coniston sinter plant or in men without LC&S experience at PCNR. By contrast, clear increased risks of death from nasal cancer were found in men who worked in the Copper Cliff sinter plant (observed 6, expected 0.2, SMR 3704) and the LC&S department at PCNR (observed 19, expected 0.25, SMR 7755). In both situations no excess in deaths occurred prior to 15 years since first exposure in the relevant plant. Trends in excess risk with increasing duration of service are shown in Table 48.8. In the LC&S department, risk of nasal cancer death can be seen to be elevated in men with less than 1 year of service. The risk rises steeply up to 20 years of service, when the observed number of deaths is more than 30 times that expected, and then falls off a little. The more sparse data from men working in the Copper Cliff sinter plant show a very similar picture but perhaps a somewhat less steeply rising risk.

3.3. Laryngeal Cancer

Observed mortality from cancer of the larynx was quite similar to that expected. For the three high-exposure subgroups 2 deaths occurred compared to 2.49 expected (SMR 80) in the period beyond 15 years since first exposure. In the remaining "lower exposure" groups the corresponding figures were 18 observed and 21.6 expected for an SMR of 83.

3.4. Kidney Cancer

Observed and expected mortality from kidney cancer are summarized in Table 48.9. No deaths from this cause were found in men with experience in the sinter plants at Copper Cliff or Coniston, and 3 deaths occurred in the PCNR LC&S department versus 2.33 expected. The large nonsinter subgroup at Sudbury registered 43 deaths from kidney cancer (beyond the latent period), which is in excess of the 34.37 expected but within the range that could be attributed to statistical sampling variation. Having said that, there is a suggestion of a pattern of increasing risk with duration of service and a somewhat more pronounced elevation in miners (observed 30, expected 21.2, SMR 141).

Table 48.7 Nasal Cancer Mortality[a]

Time Since First Exposure	Sudbury Area			Port Colborne Nickel Refinery	
	Nonsinter Plant ($n = 47890$)	Copper Cliff Sinter Plant ($n = 1744$)	Coniston Sinter Plant ($n = 336$)	Non-LC&S Department ($n = 2616$)	LC&S Department ($n = 1671$)
<15 years					
Observed	2	0	0	0	0
Expected	0.46	0.02	<0.01	0.03	0.03
SMR	431 (0.08)	—	—	—	—
95% CI	43–1570	—	—	—	—
≥15 years					
Observed	4	6	0	0	19
Expected	3.42	0.16	0.01	0.25	0.25
SMR	117 (0.44)	3704 ($<10^{-7}$)	—	—	7755 ($<10^{-7}$)
95% CI	29–299	1380–8190	—	—	4600–11800

Source: ICDA9 160.

[a]Numbers in parentheses are probabilities. CI, confidence interval.

Table 48.8 Nasal Cancer Mortality by Duration of Exposure[a]

Duration of Exposure (years)	Copper Cliff Sinter Plant			PCNR LC&S Department		
	Observed	Expected	SMR (95% CI)	Observed	Expected	SMR (95% CI)
<1	2	0.13	1493[b]	2	0.10	2062[c]
1–1.9	2	0.13	1493[b]	1	0.02	4348[b]
2–4.9	2	0.13	1493[b]	1	0.04	2703[b]
≤5	2	0.13	1493[c] (154,5538)	4	0.16	2548[d] (625,6375)
5–9.9	2	0.02	10000[d]	3	0.03	9091[d]
10–14.9	2	0.02	10000[d]	3	0.02	18750[d]
15–19.9	2	0.01	20000[d]	4	0.01	36364[d]
20–24.9				2	0.01	16667[d]
≥25				3	0.02	18750[d]
≥5	4	0.03	14286[d] (3333,34000)	15	0.09	17045[d] (9330,27600)

Source: ICDA9 160.0–160.9.

[a] ≥15 years since first exposure. CI, confidence interval.
[b] $p < 0.05$.
[c] $p < 0.01$.
[d] $p < 0.001$ (one tailed).

Table 48.9 Mortality from Cancer of the Kidney[a]

Time Since First Exposure (years)	Duration of Exposure (years)	Sudbury												Port Colborne							
		Nonsinter			Copper Cliff SP			Coniston SP				Non-LC&S			LC&S		
		O	E	SMR (95% CI)	O	E	SMR (95% CI)	O	E	SMR (95% CI)	O	E	SMR (95% CI)	O	E	SMR (95% CI)
<15	All	3	2.90	103 (21,303)	0	0.19	—	0	0.05	—	0	0.21	—	0	0.19	—
≥15	<5	7	6.60	106	0	1.32	—	0	0.22	—	0	0.22	—	3	1.54	195
	5–14.9	5	3.49	143	0	1.32	—	0	0.22	—	0	0.21	—	0	0.84	—
	15–24.9	16	8.13	197[b]	0	1.32	—	0	0.22	—	1	0.71	141	0	0.84	—
	≥25	15	16.16	93	—	—	—	0	0.22	—	2	1.41	142	0	0.84	—
	All exposures	43	34.37	125 (90,168)	0	1.63	—	0	0.22	—	3	2.33	129 (25,378)	3	2.38	126 (25,370)

[a] Abbreviations; O, observed; E, expected; CI, confidence interval.
[b] $p < 0.01$.

3.5. Cancer at Other Sites

Finally, in reviewing mortality from cancer in this cohort, we have surveyed a variety of other tumor sites in the primary subgroups as displayed in Table 48.10. Given the extensive multiple comparisons implicit in this table, it is reassuring that the observed occurrence of mortality from these other cancers are almost all quite consistent with expected numbers. Two possible anomalies are apparent in men with LC&S experience at PCNR for the lip, oral cavity, and pharynx classification and for mortality from bone cancer. In the former group of sites the excess is largely due to five nasopharyngeal cancer deaths and in the latter to two deaths from cancer of the bones of the face. We believe that these seven deaths are really cases of nasal cancer, or at least should be viewed as such, and thus do not constitute evidence of additional risks of cancer beyond those already discussed. It should be noted that two deaths from cancer in these closely related sites also occurred in men without LC&S experience at PCNR. While not statistically unusual, their appearance must be viewed as suspicious.

3.6. Environmental Assessments

The characterization of the environment in which this group of men worked has been very difficult both because of a lack of basic hygiene data and because of the complexity of the process.

As mentioned earlier, the Copper Cliff sinter plant experienced great operational difficulties in the early years requiring the repeated recycling of fine material. The consequently high levels of dust contained a mixture of insoluble oxide and subsulfide, and soluble nickel. Based on analysis of dust leaving via roof ventilators and one formal hygiene survey in 1960, the plant experienced a fall in dust levels from an average of perhaps 175 mg/m^3 of nickel prior to 1952 to about 50 mg/m^3 in the period between 1952 to the plant's closure in 1964. Limited data describe the composition of the nickel in the airborne dust as 60% oxide, 35% subsulfide, and 5% soluble forms. Process knowledge suggests that subsulfide should have been in relatively more abundance in the early years of operation.

The Copper Cliff and Port Colborne sinter plants were similar in design and operation and used the same subsulfide feedstock. We thus believe that the environments of the two plants were similar (at least after Copper Cliff overcame its early difficulties), and available hygiene surveys and roof monitoring support this. Prior to 1949, a coarser subsulfide feedstock was used at PCNR and the switch to the new finer materal resulted in more dust. No data exist for these early years, so post-1949 estimates, reduced by 40%, have been used to reflect the better early conditions.

The Coniston sinter plant was generally considered to be less dusty than the other two primarily because its role was to preheat and agglomerate concentrated nickel ore prior to smelting in the blast furnace, as opposed to

Table 48.10 Other Cancer Mortality[a]

Cancer Site	Sudbury Nonsinter			Copper Cliff Sinter		
	O	E	SMR	O	E	SMR
1. Lip, oral cavity, pharynx	30	41.58	72	2	2.06	97
2. Digestive	433	430.32	101	9	17.83	50
Esophagus	39	35.43	110	1	1.63	61
Stomach	103	96.80	106	3	3.59	84
Colon	117	112.63	104	0	4.67	—
Rectum	60	55.73	108	1	2.21	45
Liver	20	29.73	67	1	1.27	79
Pancreas	71	74.75	95	1	3.28	30
3. Bone	25	31.31	80	2	1.55	129
4. Male genitourinary	161	159.78	101	3	5.54	54
Prostate	83	79.86	104	2	2.27	88
Bladder	29	39.46	73	1	1.33	75
5. Other sites	89	95.66	93	3	4.74	63
6. Lymphatic/hemopoietic	113	115.17	98	4	5.20	77
Leukemia	48	46.33	104	2	2.04	98
All sites except lung, larynx, nose, and kidney	862	888.41	97	23	35.93	64

[a] ≥15 years since first exposure.
[b] Abbreviations: O, observed; E, expected.
[c] p < 0.001.

Coniston Sinter			PCNR Non-LC&S			PCNR LC&S		
O	E	SMR	O	E	SMR	O	E	SMR
0	0.25	—	1	3.08	33	10	2.85	351[c]
3	2.93	102	27	35.32	76	29	33.11	88
0	0.22	—	5	2.72	184	2	2.52	80
1	0.71	141	5	8.37	60	9	8.09	111
1	0.77	130	9	9.29	97	3	8.56	35
1	0.39	257	2	4.70	43	4	4.45	90
0	0.20	—	1	2.40	42	4	2.25	178
0	0.49	—	4	5.87	68	5	5.49	91
0	0.20	—	4	2.30	174	5	2.05	244
3	1.21	249	8	15.05	53	10	13.78	73
2	0.66	304	4	8.36	48	6	7.55	80
1	0.30	333	0	3.71	—	1	3.48	29
0	0.57	—	6	6.85	88	7	6.20	113
2	0.73	273	9	8.86	102	4	8.03	50
0	0.30	—	5	3.68	136	1	3.32	30
8	5.80	138	54	70	77	63	64.69	97

oxidizing fine nickel subsulfide. As such, the nickel composition of the dust was probably a mixture of the sulfitic form of nickel present in the ore (pentlandite) and oxide, although no formal chemical analysis of nickel constituents is available. Surveys at Coniston in the 1960s and 1970s reveal generally low levels averaging less than 3 mg/m^3 nickel, but isolated and intermittent high readings appeared in particular job locations, usually associated with the wearing of dust masks by workers. Together we believe that an average total nickel of 5 mg/m^3 in 1963 and later years, and perhaps twice this level prior to 1963, is a reasonable characterization of exposure at Coniston sinter plants.

4. DISCUSSION

From the mortality results presented in the previous section it is clear that the dustier parts of the INCO process have produced unusually high numbers of lung and nasal cancer deaths but not apparently any increase in mortality from laryngeal cancer. The evidence for excess risks of lung cancer at Copper Cliff and Coniston sinter plants and in the LC&S department at PCNR is statistically strong and epidemiologically persuasive because of the temporal sequence with initial exposure and the gradient of risk with increasing duration of service. Similar strong evidence exists for increased nasal cancer mortality, although as yet no deaths from this cause have occurred at Coniston.

Whether correspondingly elevated respiratory cancer risks exist beyond these dusty processes is a little less clear. No evidence of excess lung cancer deaths have appeared in the relatively large group of men working in other departments at Port Colborne. There is evidence of pockets of elevated lung cancer risk in miners and in the copper refinery. Both are epidemiologically modest and in a range where study bias is difficult to rule out, but both show evidence of increasing risk with service duration. We do not believe that, if real in a statistical sense, these risks are nickel related. The nickel content of the ore has averaged no more than 3%, and the prolonged use of dust suppression underground produced an airborne nickel level of <0.5 mg/m^3 during much of the time mining has occurred. In addition, recent studies (Muller et al., 1983) of Ontario gold miners have revealed a similarly elevated lung cancer risk, suggesting that this may be a general problem of hard rock mining rather than nickel per se. The copper refinery excess seems to have its roots in some specific occupations where nickel is again unlikely to be the culprit.

The analysis of mortality from cancer of the kidney and the survey of cancer mortality at other sites were unremarkable other than to uncover an additional number of deaths that probably should be viewed as due to nasal cancer.

The determination of which species of nickel have caused these cancer excesses is a difficult problem and unlikely to be resolved from this data set. None of the processes causing excess deaths represent relatively pure exposures to particular species. In fact, all three occurred in primarily mixed sulfide–oxide environments with a small proportion of soluble nickel "thrown in." An obvious interpretation is that either or both of nickel oxide and subsulfide are carcinogens with perhaps nickel oxide being the front runner since it could explain all three excesses, whereas the subsulfide (as opposed to the sulfide form of the ore) is probably relevant only to the sinter plants at Copper Cliff and PCNR. Alternatively, one could create a case for the excesses being due to soluble nickel. Although present as only 5% of the nickel in the airborne dust, the sinter plants had sufficient nickel in the air to produce a soluble nickel exposure many times greater than the relatively pure soluble nickel environment in the electrolytic department of Port Colborne. The lack of an increased lung cancer death rate in this department mitigates against the soluble nickel hypothesis, but the exposures may have been low enough to miss seeing an effect in this group.

Ultimately, the consensus regarding the carninogenic potential of these various nickel compounds will not hinge on this data set but on the pooling of all available data currently underway as the Nickel Speciation Research Project. We have welcomed the opportunity of contributing data and discussion to this important initiative.

ACKNOWLEDGMENTS

The authors acknowledge the financial support for this work from the Joint Occupational Health Committee of INCO and the Ontario Ministry of Labour.

REFERENCES

Cox, J. E., Doll, R., Scott, W. A., and Smith, S. (1981). Mortality of nickel workers: experience of men working with metallic nickel. *Br. J. Ind. Med.* **38,** 235–239.

Cragle, D. L., Hollis, D. R., Newport, T. H., and Shy, C. M. (1984). A retrospective cohort mortality study among workers occupationally exposed to metallic nickel dust at the Oak Ridge gaseous diffusion plant. IARC Scientific Publication No. 53, Lyon, France, 57.

Doll, R. (1958). Cancer of the lung and nose in nickel workers. *Br. J. Ind. Med.* **15,** 217–223.

Doll, R. (Ed-in-Chief) (1990). Report of the International Committee on nickel carcinogenesis in man. *Scand. J. work Environ. Health* **16,** 1–82.

Enterline, P. E., and Marsh, G. M. (1982). Mortality among wokers in a nickel refinery and alloy manufacturing plant in West Virginia. *J. Nat. Cancer Inst.* **68,** 925–933.

International Agency for Research on Cancer (1976). Cadmium, nickel, some epoxides, miscellaneous industrial chemicals, and general considerations on volatile anaesthetics. *IARC monograph 11*, International Agency for Research on Cancer, Lyon, France.

Lessard, R., Reed, D., Maheux, B., and Lambert, J.(1978). Lung Cancer in New Caledonia: a nickel smelting island. *J. Occ. Med.* **20,** 815–817.

Muller, J., et al. (1983). *Study of the Mortality of Ontario Miners*. Ontario Ministry of Labour, Toronto.

(NIOSH) National Institute for Occupational Safety and Health (1977). Criteria for a recommended standard: occupational exposure to inorganic nickel. DHEW (NIOSH) Publication No. 77-164.

Pedersen, E., Høgetveit, A. C., and Andersen, A. (1973). Cancer of the respiratory organs among workers at a nickel refinery in Norway. *Int. J. Cancer* **12,** 32–41.

Redmond, C. (1984). Site specific cancer mortality among workers involved in the production of high nickel alloys. In F. W. Sunderman (Ed.-in-Chief), Nickel in the Human Environment, IARC Scientific Publication No. 53, Lyon, France, 73.

Roberts, R. S., Julian, J. A., Muir, D.C.F., and Shannon, H. S. (1984). Cancer mortality associated with high temperature oxidation of nickel subsulphide. In F. W. Sunderman (Ed.-in-Chief), Nickel in the Human Environment, IARC Scientific Publication No. 53, Lyon, France, 23.

Roberts, R. S., Julian, J. A., Sweezey, D., Muir, D. C. F., Shannon, H. S., and Mastronmatteo, E. (1989). A study of workers engaged in the mining, smelting, and refining of nickel I: methodology and mortality by major cause groups. *Toxicol. Indust. Health*, **5,** 957–974.

Shannon, H. S., Julian, J. A., and Roberts, R. S. (1984). A mortality study of 11,500 nickel workers. *J. Nat. Can Inst.* **73**(6), 1251.

49

A SEVEN-YEAR SURVEY OF RESPIRATORY CANCERS AMONG NICKEL WORKERS IN NEW CALEDONIA (1978–1984)

M. Goldberg, P. Goldberg, A. Leclerc, J. F. Chastang, and M. J. Marne

INSERM U 88, Paris, France

J. Gueziec and F. Lavigne

Hopital Gaston Bourret, Noumea, New Caledonia

D. Dubourdieu and M. Huerre

Institut Pasteur, Noumea, New Caledonia

Nickel and Human Health: Current Perspectives, Edited by Evert Nieboer and Jerome O. Nriagu.
ISBN 0-471-50076-3   © 1992 John Wiley & Sons, Inc.

1. INTRODUCTION

Epidemiological studies have indicated that the lung cancer incidence rate is higher in New Caledonia than in most of the other South Pacific islands and is comparable to the rates observed in industrialized countries (Taylor et al., 1985). Various hypotheses have been proposed to explain this high incidence rate. Working in the nickel industry has been implicated by Lessard et al. (1978) and Langer et al. (1980). Indeed, a number of studies demonstrating a high incidence of cancers of nasal cavities and lung among nickel refinery workers in various countries are relevant in this respect (Morgan, 1958; Doll et al., 1977; Pedersen et al., 1973; Chovil et al., 1981). However, the findings concerning New Caldonia were based on an inaccurate assessment of lung cancer incidence, mainly due to poor health data collection. In addition, only data from the medical records of the Noumea hospital were available for the period examined.

A new epidemiological study was thus undertaken on the incidence of respiratory cancer in New Caledonia. The main objectives were (i) to compare the incidence rate of respiratory cancer among the population of nickel workers and the rest of the population and (ii) to look for a possible link between the occurrence of respiratory cancer and exposure to certain hazards specific to the nickel-refining industry.

The smelting/refining operations in New Caledonia are successively ore preparation (homogenization), drying, calcining, smelting, and refining into final products that are saleable ferronickels (20–25% nickel content) and high-grade nickel mattes (75% nickel) for further refining overseas (e.g., in France and Japan).

There are two production lines in the plant, one for matte and the other for ferronickel production (see Chapter 3). The matte line in the earlier decades (1931–1972) consisted of a calcining operation on a sintering grid, where coke and gypsum (a source of sulfur) were admixed with the ore. The resulting mixture was further smelted in waterjacket furnaces (1931–1962), then in blast furnaces (1962–1972) giving a crude ferrous nickel matte, and finally upgraded in converters into high-grade nickel matte.

More recently (1971 to present), the introduction of sulfur was modified, first, by direct introduction of gypsum into the electric furnace of the ferronickel line and, second, by direct introduction of molten sulfur to molten ferronickel. The ferronickel line includes a calcining operation in rotary kilns followed by a smelting step in electric furnaces. After the converting/refining operations, the ferronickel is cast into saleable ingots (see Fig. 5.2).

2. MATERIALS AND METHODS

An incidence survey was carried out on respiratory system cancers: lung and pleura [International Classification of Diseases (ICD): 162, 163, 165.0, 165.9] and certain sites of the upper respiratory tract (URT) including pharynx

(ICD: 146–148, 149.0) nasal cavities (ICD: 160) and larynx (ICD: 161). It covered the period from 1978 to 1984 and was restricted to men in view of the extremely small number of women working in the nickel industry.

Annual cumulative incidence rates were established for the population of nickel workers and the rest of the male population of New Caledonia. The numerator of the rates consisted of respiratory cancer cases occurring during the observation period. For each case of respiratory cancer in New Caledonia identified, a standardized medical record was established. These were subsequently sent to Paris where a "blind" review was carried out by a committee of experts on pulmonary and upper respiratory system pathology. Only those cases presenting a high degree of certainty as to the primary nature of the respiratory cancer were retained. On the basis of the personnel file of the only metallurgical nickel-producing company in New Caledonia, the cases were then separated into two categories: nickel workers and non–nickel workers. Only those persons who had been employed by the company for at least 10 years prior to the occurrence of their cancer were included in the nickel worker group.

Person-years were calculated as follows: the whole of the male population between the ages of 25 and 79 residing in New Caledonia between 1978 and 1984 was first estimated from data of the two general censuses carried out in 1976 and 1983. The person-years for the population of workers in the nickel company were established from the personnel file of the company in which the year of birth and year of hire of each employee has been recorded since 1930. A demographic model using mortality and emigration data (estimated from a sample of workers from the nickel company with the same age structure as the respiratory cancer cases) enabled the person-years of this population to be calculated for the study period (1978–1984) (Goldberg et al., 1987). The person-years for the non–nickel workers were calculated from the difference between the person-years for the total population and the person-years for the nickel workers.

The cumulative incidence rates were calculated for each population (Waterhouse et al., 1982). They were compared using a test based on the normal distribution. The relative risks were calculated as the ratios of cumulative incidence rates.

A case–control study within the group of nickel workers was also performed. This consisted of comparing the main occupational exposures of cases and controls accumulated throughout the career in the nickel company.

Five hazards have been studied, representing different groups of nickel-containing materials [*not specific compounds* like nickel oxide (NiO), nickel sulfate ($NiSO_4 \cdot 6H_2O$) or anhydrous $NiSO_4$].

The mineral groups considered were as follows (groups 1 and 2 being the most abundant, in most exposed workplaces):

1. Silicate oxides, general formula $SiO_2 \cdot x\text{-}MgO \cdot yFeO \cdot z\text{-}NiO$. They originate essentially from the ores.

2. Mixed or complex oxides, general formula x-$Fe_2O_3 \cdot y$-NiO, including NiO, which itself was present only in trace levels. They originate from the metallurgical operations involving molten phases.
3. Sulfides (presumed to be Ni_3S_2 but not excluding other sulfides).
4. Metallic forms, essentially nickel–iron alloys, the presence of pure nickel being highly unlikely.
5. Water-soluble forms *presumed to be mainly nickel sulfate*, but the presence of low levels of chloride is possible (proximity of the sea).

The workstations in the refinery were grouped into 105 workstation-periods, where the exposure to the hazards was uniform during specific periods of time. Measurements taken in the work environment provided estimates of the average levels encountered of the five hazards for each workstation-period. These estimates are rough, especially for periods involving industrial steps now obsolete. Because atmospheric monitoring was carried out relatively recently, especially speciation studies, all exposure data assessments involved consultation between metallurgists and industrial hygienists and were dependent on both recent speciation analysis of dusts sampled in the present New Caledonia smelter and semiquantitative past assessments determined by a study group of employees. The exposure estimates were finally merged with the occupational history of the cases and controls to quantify the exposures experienced by individual workers.

The cases and the controls were compared in three ways: Student or Fisher test for average exposures as well as a nonparametric test (Kruskal–Wallis); the chi-square test was used for the odds ratios.

3.   RESULTS

3.1.   Cumulative Incidence of Respiratory Cancers

Table 49.1 gives the cumulative incidence rates and the relative risks for both populations for the same period. For lung and pleura and URT cancers (including the nasal cavities, no case of which was observed in the nickel population), the rates appear similar in both populations. Furthermore, the relative risks are close to 1 and are not significant at the 5% level. Working in the nickel industry in New Caledonia, therefore, does not appear to increase the risk of respiratory cancer.

3.2.   Case–Control Study

We carried out a case–control study within the population of nickel workers. A group of 91 cases of respiratory cancer was compared with a group of 278 controls. The cases were all diagnosed during the 1978–1984 period and had worked in the nickel company for at least 10 years before the onset of their

Table 49.1 Cumulative[a] Incidence Rates of Respiratory Cancers and Relative Risk of Nickel Workers versus Non–Nickel Workers in New Caledonia (1978–1984)

	Non-Nickel Workers	Nickel Workers	Relative Risk[b]	Sig. Test[c]
Person-years	167,000	57,000	—	—
Lung, pleura				
Number of cases	131	65		
Incidence rates[d]	0.098	0.089	0.91	NS
Standard deviation	0.010	0.017		
URT				
Number of cases	55	26		
Incidence rates[d]	0.032	0.041	1.28	NS
Standard deviation	0.005	0.017		

[a] Age group 25–79 years.
[b] Ratio between nickel worker rate and non–nickel worker rate (confidence interval of this ratio cannot be established, its probability distribution being unknown).
[c] Normal law ($\alpha = 0.05$; NS, nonsignificant).
[d] Cumulative incidence rate.

cancer (see Table 49.1). The controls were selected from a file of all men working for the nickel company since January 1, 1930. Each case was matched with 2 subjects for age and 2 subjects for year of hiring by the company. Among the 364 potential controls selected in this way, 86 were excluded because they had died or had left New Caledonia before the beginning of the observation period or were lost to follow-up, yielding the total of 278 controls included in the study. The methodology of this study has been reported elsewhere (Goldberg et al., 1985).

Table 49.2 shows the principal characteristics of the cases and controls. Age and duration of employment in the company are not different for the two groups. Thus these factors could be excluded for the following analyses.

Cumulative exposures to the five species were calculated and are reported in Table 49.3. One can see that the most important exposures concern the oxides, considering both the number of exposed subjects as well as the exposure level.

For each case and control subject, the level of exposure for each hazard was calculated in two ways:

(a) Cumulative exposure with a 15-year period of latency and a 5-year lag period (exposures experienced during the 5 years before the occurrence of the cancer are not taken into account; the same procedure is used for the controls, the lag calculated as for the cases with which they are matched).
(b) Duration of exposure with a 15-year period of latency and a 5-year lag period.

Table 49.2   Means and Standard Deviations of Characteristics of Cases and Controls[a]

	Lung Pleura[b]	URT[b]	Controls[b]	Significance Test[c] Lung Cases	Significance Test[c] URT Cases
n	65	26	278	—	—
Age	54.8	52.8	53.1	NS	NS
	(11.2)	(8.6)	(9.8)		
Years worked at SLN	24.5	24	24.4	NS	NS
	(11.6)	(7.1)	(9.4)		
Latency	25.8	25.5	—	—	—
	(11.7)	(7.4)			

[a] Including office workers.
[b] Standard deviations are given in parentheses.
[c] NS, nonsignificant relative to controls.

Odds ratios were calculated for three exposure levels (weak, medium, strong), which also permitted equal distribution of the controls among these categories. None of the relative risks are significantly different from 1 at a 5% level for the lung and pleura cases, nor for the URT cases, when compared to the controls (Table 49.4). All confidence intervals include the value of 1 (except the odds ratio between medium and weak exposure groups for metallic forms and URT cancers; but the level of exposure is small; see Table 49.3).

These results confirm the negative findings of the incidence study.

Table 49.3   Cumulative Exposures[a]

	Silicate Oxide	Complex Oxide	Sulfide	Metallic	Soluble
Lung cases ($n = 58$)					
Number exposed	44	33	19	13	19
Mean exposure[b]	420	179	46	88	1.6
Standard deviation	432	409	105	144	2.9
URT cases ($n = 21$)					
Number exposed	20	15	11	11	11
Mean exposure[b]	766	222	19	56	9.3
Standard deviation	882	402	44	134	16.5
Controls ($n = 223$)					
Number exposed	182	151	94	76	94
Mean exposure[b]	715	211	9	91	6.3
Standard deviation	909	383	18	134	14.8

[a] Office workers excluded.
[b] Cumulative exposure among exposed workers, in milligrams of Ni, 15-year latency and 5-year lag period.

Table 49.4 Odds Ratios[a]

Nickel Species	Lung Pleura				URT			
	CE		DE		CE		DE	
	m/w	s/w	m/w	s/w	m/w	s/w	m/w	s/w
Silicate oxide	<1 (0.51–1.97)	<1	<1	<1	2.25 (0.66–7.63)	1.97 (0.57–6.84)	1.80 (0.58–5.63)	1.38 (0.42–4.55)
Complex oxide	<1	V	<1	<1	1.11 (0.35–3.45)	1.28 (0.42–3.87)	1.26 (0.42–3.82)	1.2 (0.36–3.49)
Sulfide	<1	<1	<1	<1	1.37 (0.45–4.22)	1.65 (0.57–4.78)	1.37 (0.45–4.22)	1.65 (0.57–4.78)
Metallic	V	V	V	V	3.18 (1.17–8.62)	1.13 (0.30–4.31)	1.99 (0.64–6.16)	2.26 (0.77–6.61)
Soluble	<1	<1	<1	<1	1.37 (0.45–4.22)	1.65 (0.57–4.78)	1.37 (0.45–4.22)	1.65 (0.57–4.78)

[a] Abbreviations: CE, cumulative exposure (mg of Ni), 15 year-latency and 5-year lag period; DE, cumulative duration (year), 15-year latency and 5-year lag period; m/w, medium against weak exposure subjects; s/w, strong against weak exposure subjects. Numbers in parentheses are confidence intervals.

4. DISCUSSION

This work is part of a broader survey aimed at establishing the incidence of respiratory cancers in New Caledonia over a 10-year period (1978–1987). The findings reported here, although incomplete, allow some conclusions to be drawn about the impact of the nickel industry on the epidemiology of respiratory cancers in New Caledonia. With regard to the role of a cancer hazard related to nickel exposure, our results point to the absence of a high risk in the New Caledonia nickel refinery.

Additional information on exposure and cancer incidence rates for this group of nickel workers is provided by Doll (1990).

ACKNOWLEDGMENTS

We thank all those who helped in this study, especially the Société-Le-Nickel, the Syndicat des Ouvriers et Employés de Nouvelle-Calédonie (SOENC), the director and physicians of the Hôpital Gaston Bourret, the Cancer Registry of New South Wales (Australia), the specialists of the Certification Committee (Professors J. Bignon, P. Brochard, B. Dautzenberg, A. Hirsch, and C. Sors and Doctors J. Brugère, J. Chomette, A. Chleq, M. Nébut, and J. Rodriguez).

REFERENCES

Chovil, A., Sutherland, R. B., and Halliday, M. (1981). Respiratory cancer in a cohort of nickel sinter plant workers. *Br. J. Ind. Med.* **38,** 327–333.

Doll, R., Mattews, J. D., and Morgan, L. G. (1977). Cancers of the lung and nasal-sinuses in nickel workers: a reassessment of the period at risk. *Br. J. Ind. Med.* **34,** 102–105.

Doll, R. (1990). Report of the International Committee on Nickel Carcinogenesis in Man. *Scand. J. Work Environ. Health* **16,** 1–82.

Goldberg, M., Fuhrer, R., Brodeur, J. M., Goldberg, P., Segnan, N., Leclerc, A., Chastang, J. F., Bourbonnais, R., and Francois, D. (1985). Respiratory tract cancer among workers in a nickel mining and refining company: a case-control study within a cohort. In S. S. Brown and F. W. Sunderman (Eds.), *Progress in Nickel Metabolism and Toxicology.* Blackwell, London, pp. 215–218.

Goldberg, M., Goldberg, P., Leclerc, A., Chastang, J. F., Fuhrer, R., Brodeur, J. M., Segnan, N., Floch, J. J., and Michel, G. (1987). Epidemiology of respiratory cancers related to nickel mining and refining in New Caledonia (1978–1984). *Int. J. Cancer* **40,** 300–304.

Langer, A. M., Rohl, A. N., Selikoff, I. J., Harlow, G. E., and Prinz, M. (1980). Asbestos as a cofactor in carcinogenesis among nickel-processing workers. *Science* **209,** 420–422.

Lessard, R., Reed, D., Maheux, B., and Lambert, J. (1978). Lung cancer in New Caledonia, a nickel smelting island. *J. Occup. Med.* **20,** 815–817.

Morgan, J. G. (1958). Some observations on the incidence of respiratory cancers in nickel workers. *Br. J. Ind. Med.* **15,** 224–234.

Pedersen, E., Høgetveit, A. C., and Andersen, A. (1973). Cancer of respiratory organs among workers in a nickel refinery in Norway. *Int. J. Cancer* **12,** 32–41.

Taylor, R., Levy, S., Henderson, B., Kolonel, L., and Lewis, N. (1985). Cancer in Pacific Island Countries. Doc. 554-85, South Pacific Commission, Nouméa.

Waterhouse, J., Calum, M., Shanmugaratam, K., and Powell, J. (1982). *Cancer Incidence in Five Continents*. IARC Scientific Publication M42, IARC, Lyon, pp. 668–670.

50

NASAL AND LUNG CANCER DEATH IN NICKEL REFINERS

D.C.F. Muir, R. S. Roberts, J. Julian, and N. Jadon

Occupational Health Program, McMaster University, Hamilton, Ontario, Canada

1. INTRODUCTION

An increased risk of death as a result of carcinoma of the lung and nasal sinuses has been documented by Roberts et al. (1989) in nickel sintering workers in Ontario. Although the raw material used in the refining process was similar in two locations (Sudbury and Port Colborne), the ratio of increased risk of lung cancer death compared to increased nasal cancer deaths was quite different. In Sudbury, the increased risk of lung cancer (compared to nasal cancer) was considerably greater than in Port Colborne.

Since both facilities are within a few hundred miles of each other in a single province of Canada, it is unlikely that this difference in the relative risks could be explained by varying diagnostic or therapeutic modalities or

Nickel and Human Health: Current Perspectives, Edited by Evert Nieboer and Jerome O. Nriagu.
ISBN 0-471-50076-3 © 1992 John Wiley & Sons, Inc.

by consistent differences in exposure to nonoccupational carcinogens in the ambient air or in cigarette smoke. It is possible that the difference could be due to the size of airborne particles in the two plants because the manufacturing processes were not identical even though both used the same ore. The purpose of this chapter is to present theoretical calculations of aerosol deposition in the nose and lung in order to examine the range of particle sizes that might account for the observed epidemiological data.

2. CALCULATION OF REGIONAL NICKEL DOSIMETRY IN THE RESPIRATORY TRACT

Regional deposition of airborne particles in the respiratory tract is determined by a number of factors of which the most important are the size and density of the particles. In the present problem, we want to know the fractional deposition of particles in two target areas, that is, the nose and the central areas of the bronchial tree. In the calculations that follow, the target area of the airway is taken to be the third bronchial generation since 60% of bronchial carcinomas in the general population are thought to occur close to the hilum (Garland et al., 1962). Specific information about the site of origin of such tumors in nickel workers is not available, but it is assumed that the pattern is similar to that observed in the general population. Data published by Yeh and Schum (1980) was used to calculate the fractional deposition of inhaled particles during nasal breathing in the nasopharynx (tidal volume 750 cm^3, breathing frequency 15 breaths/min) and those of Gerrity et al. (1979) to estimate deposition in the third generation of the bronchial tree (tidal volume 700 cm^3, breathing frequency 15.8 breaths/min). The calculations were carried out in a stepwise fashion assuming that the nose and airways represent filters arranged in series. These deposition rates were transformed to deposition per unit surface area by dividing by the estimated surface of the nose (Proctor, 1964) and of the third generation of the bronchi (Weibel, 1963). The results are shown in Table 50.1 and illustrated in Figure 50.1.

The calculations show that particle deposition per unit surface area of the nose and bronchial tree increases with particle size up to an aerodynamic diameter of about 7 μm. Above this size, the deposition in the nasal area continues to rise with increasing particle size, but the deposition in the airway falls. The explanation for this effect is that it results from the increasing deposition of larger particles in the nose as a result of impaction or settlement so that fewer particles are available for deposition in the airways. If the differential risks observed by Roberts et al. (1989) are a function of aerosol particle size, then the effect could be explained by a relatively greater concentration of particles having aerodynamic diameters greater than 7 or 8 μm in Port Colborne.

Table 50.1   Calculated Surface Concentration[a] in Nose and Third Bronchial Generation for each 1000 Particles Inhaled

Aerodynamic Size (μ)	Nose	Third-Generation Airway	Ratio
1	0.266	0.269	0.99
2	2.95	0.715	4.13
3	3.94	1.07	3.70
5	4.94	1.44	3.44
7	5.62	1.34	4.20
10	6.35	0.331	19.18

[a] In particles per square centimeter.

3. DISCUSSION

This analysis is, of necessity, speculative since it is no longer possible to obtain measurements of the environmental aerosol in sinter plants that closed some years ago. The calculations were made for nasal breathing because mouth breathing is associated with zero nasal deposition. It is for this reason that active cigarette smoking is unlikely to be associated with an increased risk of nasal carcinoma, although the same cannot be said for passive exposure to cigarette smoke.

For dosimetry to the respiratory tract, it is customary to combine deposition calculations with information about particle solubility and clearance. In the case of water-soluble substances, it is assumed that uptake is rapid and that it takes place at the site of particle deposition. In the case of relatively insoluble substances such as nickel oxide or subsulfide, the process is more complex. Individual particles may enter the mucosa itself. More probably, the particles remain on the moving layer of mucus and are removed by the cilia to the pharynx where they are swallowed. During this transport process, a certain amount of the compound is dissolved and is able to penetrate the

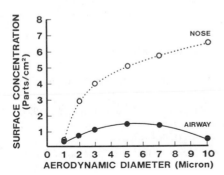

Figure 50.1. Calulated surface concentration (particles/cm^2) on the mucosa of the nose and third-generation airway for each 1000 particles inhaled.

mucosa. The rate of dissolution is a function of the solubility of the compound and particle size. Calculation of the amount that leaches into the mucosa is not straighforward because of the serial nature of clearance in the bronchial tree. As one region (such as the third generation of airways) is cleared, particles are transported into it from more distant sites. Although the calculations are complex, they have been carried out for radon progeny (NCRP, 1984). Such analyses are unlikely to be appropriate for nickel compounds because of an important effect of nickel salts on ciliary function. Adalis et al. (1978) have shown a strong ciliostatic effect of nickel solution on the cilia of the trachea in the hamster. If reduced ciliary function and mucosal clearance in the nose and airways of nickel workers results from long-term chronic exposure, then this might be an important determinant of cancer both from nickel compounds themselves and from other inhaled carcinogens such as tobacco smoke. At present, there is no information on the rate of ciliary clearance in nickel workers, and the regional deposition calculations presented above have not been extended to take account of clearance.

Because the critical aerodynamic particle size appears to be above 7 or 8 μm, conventional environmental sampling systems would not be appropriate for survey purposes. Some form of aerosol spectrometer would be required so that the concentration of nickel particles of 8, 9, or 10 μm aerodynamic diameter could be measured. However, as noted above, the sinter plants have now closed, and it is no longer possible to pursue the issue. Until it can be clearly demonstrated that no risk is present in other parts of the industry, it would appear to be prudent to obtain data on the size distribution of nickel in the workplace under present operating conditions.

Alternative explanations other than that presented here might include bimodal aerosols with different relative concentrations in the two plants. Whatever the final explanation, there seems good reason to characterize the aerosol in nickel industries in as much detail as possible. Measurements of the nickel concentration in the mucosa of the respiratory tract (Torjussen et al., 1978) cannot help since they provide no information of the size of the original airborne particles.

REFERENCES

Adalis, D., Gardner, D. E., and Miller, F. J. (1978). Cytotoxic affects of nickel on ciliated epithelium. *Am. Rev. Resp. Dis.* **118,** 347–354.

Garland, L. H., Beier, R. L., Coulson, W., Heald, J. H., and Stein, R. L. (1962). The apparent sites of origin of carcinomas of the lung. *Radiology* **78,** 1–11.

Gerrity, T. R., Lee, P. S., Hass, F. J., Marinelli, A., Werner, P., and Lourenco, R. V. (1979). Calculated deposition of inhaled particles in the airway generations of normal subjects. *J. Appl. Physiol. Respirat. Environ. Exercise Physiol.* **47**(4), 867–873.

(NCRP). National Council on Radiation Protection and Measurements. (1984). Report No. 78, Bethesda, Maryland.

Proctor, D. F. (1964). Physiology of the upper airway. In *Handbook of Physiology, Respiration*, Vol. I, American Physiological Society, Washington, DC., pp. 309–345.

Roberts, R. S. Julian, J. A., Muir, D. C. F., and Shannon, H. S. (1989). A study of mortality in workers engaged in the mining, smelting, and refining of nickel. II: Mortality from cancer of the respiratory tract and kidney. *Toxicol. Ind. Health*, **5**, 975–993.

Torjussen, W., Finn-Mogens, S. H., and Anderson, I. (1978). Concentration and distribution of heavy metals in nasal mucosa of nickel-exposed workers and of controls, studied with atomic absorbtion spectrophometric analysis and with Timm's sulphide silver method. *Acta Otolaryngol.* **86**, 449–463.

Weibel, E. R. (1963). *Morphometry of the Human Lung*. Academic Press, New York.

Yeh, H. C., and Schum, G. M. (1980). Models of human lung airways and their application to inhaled particle deposition. *Bull. Math. Biol.* **42**, 462–480.

INDEX